Juraj Hromkovič Ondrej Sýkora (Eds.)

Graph-Theoretic Concepts in Computer Science

24th International Workshop, WG'98
Smolenice Castle, Slovak Republic
June 18-20, 1998
Proceedings

Springer

Series Editors

Gerhard Goos, Karlsruhe University, Germany
Juris Hartmanis, Cornell University, NY, USA
Jan van Leeuwen, Utrecht University, The Netherlands

Volume Editors

Juraj Hromkovič
RWTH Aachen, Lehrstuhl für Informatik I
Ahornstr. 55, D-52074 Aachen, Germany
E-mail: jh@il.informatik.rwth-aachen.de

Ondrej Sýkora
Institute of Informatics, Slovak Academy of Sciences
P.O. Box 56, Dúbravská 9, 84000 Bratislava, Slovakia
E-mail: sykora@savba.sk

Cataloging-in-Publication data applied for

Die Deutsche Bibliothek - CIP-Einheitsaufnahme

Graph theoretic concepts in computer science : 24th international
workshop ; proceedings / WG '98, Smolenice Castle, Slovak
Republic, June 18 - 20, 1998 / Juraj Hromkovič ; Ondrej Sýkora
(ed.). - Berlin ; Heidelberg ; New York ; Barcelona ; Budapest ; Hong
Kong ; London ; Milan ; Paris ; Singapore ; Tokyo : Springer, 1998
  (Lecture notes in computer science ; Vol. 1517)
  ISBN 3-540-65195-0

CR Subject Classification (1998): G.2.2, F.2, F.1.2-3, F.3-4, E.1, I.3.5

ISSN 0302-9743
ISBN 3-540-65195-0 Springer-Verlag Berlin Heidelberg New York

Typesetting: Camera-ready by author
SPIN 10692760   06/3142 – 5 4 3 2 1 0   Printed on acid-free paper

Lecture Notes in Computer Science 1517

Edited by G. Goos, J. Hartmanis and J. van Leeuwen

Springer

Berlin
Heidelberg
New York
Barcelona
Hong Kong
London
Milan
Paris
Singapore
Tokyo

Preface

The International Workshop on Graph-Theoretic Concepts in Computer Science is one of the most traditional and high quality conferences in Computer Science. Previous conferences were organized at various places in Austria, Germany, Italy, and The Netherlands. The workshop aims at uniting theory and practice by demonstrating how graph-theoretic concepts can be applied to various areas in computer science, or by extracting new problems from applications. The goal is to present recent research results and to identify and explore directions of future research. The workshop is well-balanced with respect to established researchers and young scientists.

The 24th International Workshop on Graph-Theoretic Concepts in Computer Science (WG '98) was held at Smolenice Castle, near Bratislava, Slovak Republic, June 18–20, 1998. It was organized by the Slovak Academy of Computer Science in cooperation with the Department of Computer Science I at RWTH Aachen (Germany) and with Slovak Society for Computer Science. For the first time in its history, WG took place in a country of the former eastern block, in the Slovak Republic.

The program committee of WG'98 consisted of:

> H. Bodlaender, Utrecht (NL)
> A. Brandstädt, Rostock (D)
> M. Habib, Montpellier (F)
> J. Hromkovič, Aachen (D)
> L. Kirousis, Patras (GR)
> L. Kučera, Praha (CR)
> A. Marchetti-Spaccamela, Roma (I)
> E. Mayr, München (D)
> R. Moehring, Berlin (D)
> M. Nagl, Aachen (D)
> H. Noltemeier, Würzburg (D)
> F. Parisi Presicce, Roma (I)
> O. Sýkora, Bratislava (SK)
> G. Tinhofer, München (D)
> D. Wagner, Konstanz (D)
> P. Widmayer, Zürich (CH)

The call for papers for WG'98 was mailed and posted on several electronic bulletin boards. As a consequence, 61 papers were submitted and reviewed by the program committee. All submissions were carefully refereed by four or more members of the program committee. We would like to thank very much all members of the program committee and all subreferees for their cooperation and for exhaustive and detailed reports and comments.

After collecting, combining, and weighting all reports the program committee selected 30 papers for presentation at the workshop. The average level of submissions was high, with 38 papers scoring above the average, and we were sorry to have to reject some of them. We were impressed by the scientific quality of the graph-theoretic applications they covered. Thanks to all authors for their submissions.

The workshop took place June 18–20, 1998, and was attended by 66 scientists from the following countries: Brazil, Canada, Czech Republic, France, Germany, Greece, Hungary, Israel, Italy, Japan, The Netherlands, Russia, Slovakia, United Kingdom, and United States of America.

This volume contains all contributed papers from the workshop. They have all undergone careful revision after the meeting, based on the discussions and comments from the audience and the referees.

Special thanks have to go to Martin Bečka, Sylvia Gavorová, Tomáš Hrúz, Robert Szelepcsényi, and Imrich Vrťo for the excellent organization of the workshop in Smolenice castle as well as for the nice social program in the forest around the castle. We would like to thank Walter Unger for the organization of the electronic program committee meeting and for the help with LaTeX during the whole time, starting with submissions and finishing with the preparation of this volume.

Aachen, August, 1998   Juraj Hromkovič, Ondrej Sýkora

Referees

Agostino, Sergio De
Alimonti, P.
d'Amore, Fabrizio
Böckenhauer, H.-J.
Babel, Luitpold
Becka, Martin
Bodlaender, Hans L.
Brandenburg, Franz J.
Brandes, Ulrik
Brandstädt, Andreas
Calamoneri, Tiziana
Capelle, Christian
Dahlhaus, Elias
Di Ianni, Miriam
Dragan, Feodor
Eidenbenz, Stephan
Flammini, Michele
Franciosa, Paolo Giulio
Gustedt, Jens
Guttmann, Sven
Habib, Michel
Handke, Dagmar
Hromkovič, Juraj
Hruz, Tomas
Kirousis, Lefteris M.
Klasing, Ralf
Kloks, Ton
Koehler, Ekkehard
Korner, Janos
Kratsch, Dieter
Kreuter, B.

Kučera, Ludek
Labahn, Roger
Lanlignel, J.M.
Le, Van Bang
van Leeuwen, Jan
Liebers, Annegret
Linnhoff-Popien, Claudia
Liotta, Giuseppe
Möhring, Rolf
Müller, Haiko
Malvestuto, F.
Marchetti-Spaccamela, Alberto
Mayr, Ernst W.
Meyer, B.
Monti, Angelo
Moscarini, Marina
Nagl, Manfred
Neyer, Gabriele
Noltemeier,
Oberschelp, Walter
Olariu, S.
Paredaens, Jan
Paul, C.
Piperno, Adolfo
Presicce, F. Parisi
Remmele, Peter
Roos, Thomas
Schied, G.
Schiermeyer, I.
Schlude, Konrad

Schmidt, Bertil
Schulz, Andreas S.
Schïr, A.
Seese, Detlef
Seibert, Sebastian
Silvestri, Riccardo
Skutella, Martin
Stacho, Ladislav
Stamatiou, Yannis C.
Sterbini, Andrea
Stork, Frederik
Sýkora, Ondrej
Szelepcsenyi, Robert
Szymczak, Thomas
Tel, Gerard
Thiele, Lothar
Thilikos, Dimitrios M.
Tinhofer, Gottfried
Triesch, E.
Ulber, Roland
Veldhorst, Marinus
Verriet, Jacques
Vrťo, Imrich
Wagner, Annika
Wagner, Dorothea
Wattenhofer, Roger
Weihe, Karsten
Widmayer, Peter

Table of Contents

Linear Time Solvable Optimization Problems on Graphs of Bounded Clique Width
(Extended Abstract)

B. Courcelle[1], J.A. Makowsky[2*], and U. Rotics[2]

[1] Laboratoire d'Informatique
Université Bordeaux-I
33405 Talence, France
Bruno.Courcelle@@labri.u-bordeaux.fr

[2] Department of Computer Science
Technion–Israel Institute of Technology
32000 Haifa, Israel
{janos,rotics}@@cs.technion.ac.il

Abstract. Graphs of clique-width at most k were introduced by Courcelle, Engelfriet and Rozenberg (1993) as graphs which can be defined by k-expressions based on graph operations which use k vertex labels. In this paper we show that the $(q, q - 4)$ graphs are of clique width at most q and P_4-tidy graphs are of clique-width at most 4. Furthermore, the k-expression (for $k = 4$ or $k = q$) associated with such a graph can be found in linear time.

$(q, q - 4)$ graphs were introduced by Babel and Olariu (1995) and extends the class of P_4-sparse graphs. P_4-sparse graphs were introduced by Hoàng (1985) and are widely studied because of their applications in areas such as scheduling, clustering and computational semantics. Another extension of P_4-sparse graphs are the P_4-tidy graphs which were introduced by Rusu (1995).

Furthermore, we show that the class of LinEMSOL($\tau_{1,L}$) optimization problems is solvable in $O(f(|V|, |E|))$ time on a class of graphs of clique-width at most k in which for every graph G an expression defining it can be constructed in $O(f(|V|, |E|))$ time. By the above this applies in particular to $(q, q - 4)$ graphs, P_4-tidy graphs and P_4-sparse graphs with f linear.

Finally, we show that the above results cannot be extended to MSOL(τ_2) decision and optimization problems on the vocabulary τ_2 which allow edges to be considered as elements of the domains of the graphs in question, and by that, allow quantifying over edges in addition to quantifying over vertices.

* Partially supported by a Grant of the Israeli Ministry of Science for French-Israeli Cooperation (1994), a Grant of the German-Israeli Binational Foundation (1995-1996), and by the Fund for Promotion of Research of the Technion–Israeli Institute of Technology

1   Introduction

The class of P_4-sparse graphs was introduced by Hoàng in his doctoral dissertation [Hoà85], as the class of graphs for which every set of five vertices induces at most one P_4 (i.e., a path of length 4). This class contains the class of P_4-reducible graphs introduced by Jamison and Olariu in [JO89], as the class of graphs for which no vertex belongs to more than one induced P_4. These two classes contain the class of cographs, and has been studied intensively in the recent years, motivated by the practical applications of these classes in areas such as scheduling, clustering and computational semantics. In [JO89] and in [JO92b] a unique tree representation is proposed for the classes of P_4-reducible and P_4-sparse graphs respectively. These tree representations are used later in [JO95a] and in [JO92a] to develop linear $O(|V|+|E|)$ recognition algorithms for these classes. In [JO95b] linear $O(|V| + |E|)$ time algorithms are proposed for solving five optimization problems on the class of P_4-sparse graphs: maximum size clique, maximum size stable set, minimum coloring, minimum covering by cliques, and minimum fill-in. If the tree representation of the P_4-sparse graph is also given as input, then the running time of these algorithms is just $O(|V|)$ independently of the number of edges in the graph. They conclude their paper with

Problem 1 ([JO95b]). Find other optimization problems which can be solved in linear time on the class of P_4-sparse graphs.

Giakoumakis and Vanherpe in [GV97] took up this line of research. They used the modular decomposition tree representation of a graph, to obtain linear $O(|V| + |E|)$ time algorithms for the maximum weight clique and for the maximum weight stable set problems in the case P_4-sparse graphs, for the optimal weighted coloring and for the minimum weight clique cover problems in the case of P_4-reducible graphs. If the modular decomposition of the graph is given an input, then the running time of these algorithms is just $O(|V|)$.

Giakoumakis and Vanherpe also introduced in [GV97] the classes of extended P_4-sparse and extended P_4-reducible graphs, and showed how to extend their results to these two classes of graphs, with a minimal additional work. P_4-sparse graphs (P_4-reducible graphs) can be characterized by a set of seven (nine) forbidden induced subgraphs $Z_1, \ldots, Z_7$ ($Z_1, \ldots, Z_9$), [GV97]. The class of extended P_4-sparse (P_4-reducible) graphs is defined by the set $Z_2, \ldots, Z_7$ ($Z_2, \ldots, Z_9$) of forbidden induced subgraphs.

Babel and Olariu introduced in [BO95] the class of (q,t) graphs which, for $t = q - 4$, extends the class of P_4-sparse graphs. In such a graph no set of at most q vertices is allowed to induce more than t distinct P_4's. Clearly, we assume that $q \geq 4$. $(4,0)$ graphs are exactly the cographs.

Rusu, cf. [GRT97], introduced the class of P_4-tidy graphs which extends the class of extended P_4-sparse graphs. Let G be a graph and X be an induced P_4. A vertex v outside X is a *partner of X* if X and v together induce at least two P_4's. A graph is P_4-*tidy* if any induced P_4 has at most one partner.

In this paper we show that a wide class of decision and optimization problems on the classes of $(q, q - 4)$ graphs and P_4-tidy graphs is solvable in time $O(|V| + |E|)$ or in time $O(|V|)$ assuming that the modular decomposition of the graph is given as input. These problems are characterized by their expressibility in certain variations of Monadic Second Order Logic $MSOL(\tau_{1,L})$ (for decision problems) or $LinEMSOL(\tau_{1,L})$ (for optimization problems), the study of which was initiated by B. Courcelle and others in a sequence of papers [Cou90,Cou91,Cou94,Cou95,Cou96,CM93,ALS91]. Roughly speaking, $MSOL(\tau_1)$ is Monadic Second Order Logic with quantification over subsets of vertices, but not of edges; $MSOL(\tau_{1,L})$ is the extension of $MSOL(\tau_1)$ with the addition of labels added to the vertices. $LinEMSOL(\tau_{1,L})$ is the extension of $MSOL(\tau_{1,L})$ which allows to search for sets of vertices which are optimal with respect to some linear evaluation function. The precise definitions will be given in section 2. A typical $MSOL(\tau_{1,L})$ decision problem is k-colorability for fixed k. The maximum weight clique and the maximum weight stable set problems are $LinEMSOL(\tau_{1,L})$ definable. The optimal (weighted) coloring problem is not $LinEMSOL(\tau_{1,L})$ definable, cf. [Lau93]. We show that:

Theorem 1. *Every $LinEMSOL(\tau_{1,L})$ problem on the classes of $(q, q-4)$ graphs and P_4-tidy graphs can be solved in time $O(|V| + |E|)$ and the corresponding algorithm can be effectively constructed from its $LinEMSOL(\tau_{1,L})$ definition. If the modular decomposition of the graph is given as input then the running time of the algorithm is $O(|V|)$.*
In particular this also holds for the class of (extended) P_4-sparse graphs, for the class of (extended) P_4-reducible graphs and for the class of cographs.

For example, in the terminology and numbering of [GJ79], all the following problems are $LinEMSOL(\tau_{1,L})$ definable. So we have:

Corollary 1. *The following problems can be solved in linear time on the classes of $(q, q - 4)$ graphs, P_4-tidy graphs (and any of their subclasses): vertex cover [GT1], dominating set [GT2] domatic number for fixed k [GT3], k-colorability for fixed k [GT4], partition into cliques for fixed k [GT15], clique [GT19], independent set [GT20], induced path [GT23] and unweighted Steiner trees [ND12].*

Remark 1. This includes some of the results of [BO98] which were obtained independently and without analyzing the logical form of the problems.

In section 3 we extend Theorem 1 above to the class of graphs of bounded clique-width, introduced by Courcelle et al. [CER93]. We recall the notions of graph operations and clique-width presented in [CO].

A k-graph is a labeled graph with vertex labels in $\{1, 2 \ldots, k\}$. We shall use 3 types of graph operations on k-graphs denoted $\oplus$, $\eta_{i,j}$, and $\rho_{i \rightarrow j}$. Informally, $G_1 \oplus G_2$ is the disjoint union of the k-graphs G_1 and G_2, $\eta_{i,j}(G)$ is the k-graph obtained by adding to G undirected edges connecting all vertices labeled i to all the vertices labeled j in G, and $\rho_{i \rightarrow j}(G)$ is the k-graph obtained by changing all the i labels to j labels in G. A formal definition of these graph operations is given in section 4.

With every graph G one can associate an algebraic expression built using the 3 type of operations mentioned above which defines G. We call such an expression a k-expression defining G, if all the labels in the expression are in $\{1, \ldots, k\}$. Clearly, for every graph G, there is an n-expression which defines G, where n is the number of vertices of G. Let $C(k)$ be the class of graphs which can be defined by k-expressions. The clique-width of a graph G, denoted $cwd(G)$, is defined by: $cwd(G) = Min\{k : G \in C(k)\}$. With these definitions we show:

Theorem 2. *Let C be a class of graphs of clique-width at most k (i.e., $C \subseteq C(k)$) such that there is a (known) $O(f(|E|, |V|))$ algorithm, which for each graph G in C, constructs a k-expression defining it. Then every $LinEMSOL(\tau_{1,L})$ problem on C can be solved (constructively) in time $O(f(|E|, |V|))$.*

Theorem 2 applies to any class of graphs of bounded clique-width. There are many such classes. For example, the cliques, the cographs, and any class of graph of tree width at most k. We show that:

Proposition 1. *$(q, q-4)$ graphs and P_4-tidy graphs have clique-width at most q and 4 respectively, and for each $(q, q-4)$ (P_4-tidy) graph G, a q-expression (4-expression) defining it can be constructed in Linear $O(|V| + |E|)$ time.*

Theorem 1 above follows from Theorem 2 and Proposition 1.

Courcelle and Mosbah [CM93] also considered the logics $MSOL(\tau_2)$ which is Monadic Second Order Logic with quantification over subsets of vertices or edges, $MSOL(\tau_{2,L})$ which is the extension of $MSOL(\tau_2)$ by the addition of labels added to the vertices and edges, and $LinEMSOL(\tau_{2,L})$ which is the extension of $MSOL(\tau_{2,L})$ to optimization problems. They showed that Theorem 1 also holds for all the $LinEMSOL(\tau_{2,L})$ optimization problems on each class of graphs of tree width at most k. However, our next result shows that for the classes of $(q, q-4)$, P_4-tidy, (extended) P_4-sparse, (extended) P_4-reducible, cographs and all graph classes which contains the cliques, Theorem 1 above is best possible, provided that $\mathbf{P} \neq \mathbf{NP}$.

For labeled graphs this is easy to see as every graph can be represented as a labeled clique with exactly the original edges labeled with one unary predicate. But theorem 1 is also best possible for unlabeled graphs, provided that $\mathbf{P} \neq \mathbf{NP}$ on unary languages. More precisely, denote by $\mathbf{P_1}$ ($\mathbf{NP_1}$) the class of languages over one letter (also called tally languages), which are in $\mathbf{P}$ ($\mathbf{NP}$). In section 4 we show that:

Theorem 3. *If $\mathbf{P_1} \neq \mathbf{NP_1}$ then there is an $MSOL(\tau_2)$ definable decision problem over the class of cliques which is not solvable in polynomial time.*

Corollary 2. *If $\mathbf{P_1} \neq \mathbf{NP_1}$ then over all the classes which contain the cliques, in particular, the classes $C(k)$ for each k, $(q, q-4)$ graphs, P_4-tidy graphs, (extended) P_4-sparse, (extended) P_4-reducible, and cographs, there are $MSOL(\tau_2)$ definable decision problems which are not solvable in polynomial time.*

In this extended abstract we just sketch the proofs of the theorems mentioned above. The detailed proofs will be presented in the full paper.

2 Background

2.1 Monadic Second Order Logic decision and optimization problems

In what follows, we will use the term *graph* for finite undirected graphs without self loops or multiple edges. We will use the term *labeled graph* for graphs having labels which are associated with their vertices or edges.

The following are the two most common (labeled) graph presentations, for logically oriented work:

Definition 1 ((The vocabularies τ_1 and $\tau_{1,L}$)). *We denote by $\tau_{1,L}$ the vocabulary of the form: $E, U_1, \ldots, U_k$ for some fixed integer k. For a labeled graph G, we denote by $G(\tau_{1,L})$ the presentation of G as a logical structure $\langle V, E, U_1, \ldots, U_k \rangle$, where V is the domain of the logical structure which consists of the set of vertices of G, E is the binary relation corresponding to the incidence matrix of G, and $U_1, \ldots, U_k$ are the unary predicates corresponding to the labels of the vertices of G. In the case when $k = 0$ we denote this vocabulary as τ_1.*

Definition 2 ((The vocabularies τ_2 and $\tau_{2,L}$)). *We denote by $\tau_{2,L}$ the vocabulary: $\{R, P_E, P_V, U_1, \ldots, U_k\}$ for some fixed integer k. For a labeled graph G, we denote by $G(\tau_{2,L})$ the presentation of G as a logical structure $\langle V \cup E, R, P_E, P_V, U_1, \ldots, U_k \rangle$, where the domain of the logical structure consists of the set of vertices and edges of G, R is a binary relation, such that (e, v) is in R if v is a vertex of G which is incident with the edge e of G, P_V (P_E) is a unary predicate such that v (e) is in P_V (P_E) if v (e) is a vertex (an edge) of G, and $U_1, \ldots, U_k$ are the unary predicates corresponding to the labels of the vertices and edges of G. In the case when $k = 0$ we denote this vocabulary as τ_2.*

We recall that Second Order Logic (SOL) is like first order logic, but allows also variables and quantification over relation variables of various but fixed arities. Monadic Second Order Logic ($MSOL$) is the sub-logic of SOL where relation symbols are restricted to be unary. More details on the definition of $MSOL$ can be found in [Cou97,EF95,Pap94]. For a set variable X and a first order variable u, we denote by $X(u)$ the atomic formula indicating that $u \in X$.

Graphs are a special case of finite structures. Therefore, before concentrating on graphs, we start with the following definitions and facts concerning finite structures. In what follows we will be concerned only with finite structures, therefore whenever we use the term *structure* we mean *finite structure*. Let τ denote any vocabulary consisting of a finite set of relation symbols, and let K be any class of $\tau-structures$. We denote by $Str(\tau)$ the class of all $\tau-structures$, and for every closed MSOL formula φ over τ we denote by $Mod(\varphi)$ the class of all τ-structures in which φ holds.

Definition 3 ((MSOL(τ) decision problem over K)). *We say that a decision problem, is an MSOL(τ) decision problem over K, if it can be expressed in the following form: Given a τ-structure $A \in K$ does $A \models \varphi$? where φ is any* closed MSOL formula over τ. Note that φ and K are not part of the problem instance, which consists just of $\mathcal{A}$. In the case that the class K consists of all $\tau - structures$, $K = Str(\tau)$, we will say that a problem which can be stated as above is an $MSOL(\tau)$ decision problem.

Let $f_1, f_2, \ldots, f_m$, be m function symbols for some fixed integer m. For a set variable X_i and an assignment z we use $|z(X_i)|_j$ as a short notation for: $\sum_{a \in z(X_i)} f_j(a)$. We denote by $|A|$ the cardinality of a finite set A.

Definition 4 ((LinEMSOL(τ) optimization problems over K)). *We say that an optimization problem P, is a* LinEMSOL(τ) *optimization problem over K, if it can be expressed in the following form: Given a τ-structure $\mathcal{A} \in K$, and m evaluation functions $f_1, \ldots, f_m$ associating integer values to the elements of $\mathcal{A}$, find an assignment z to the free variables in θ such that:*

$$\sum_{\substack{1 \leq i \leq l \\ 1 \leq j \leq m}} a_{ij}|z(X_i)|_j = opt\{ \sum_{\substack{1 \leq i \leq l \\ 1 \leq j \leq m}} a_{ij}|z'(X_i)|_j : \langle \mathcal{A}, z' \rangle \models \theta(X_1, \ldots, X_l)\}$$

where θ is an MSOL(τ) formula having free set variables $X_1, \ldots, X_l$, opt is either Min or Max, and $\{a_{ij} : 1 \leq i \leq l, 1 \leq j \leq m\}$ are any integers. Since the coefficients a_{ij} can be negative we shall only deal with Max. A minimization is obtained from a maximization with negated coefficients. Note that $\theta(X_1, \ldots, X_l), K$ and the constants $\{a_{ij}\}$ are not part of the problem instance, which consists just of $\mathcal{A}$ and the evaluation functions $f_1, \ldots, f_m$.

For any assignment z to the free variables of θ which satisfies the above condition, we say that z realizes a solution to the problem P on $\mathcal{A}$ with evaluation functions $f_1, \ldots, f_m$.

In the case that the class K consists of all the τ-structures, $K = Str(\tau)$, we denote a LinEMSOL(τ) optimization problem over K shortly as a LinEMSOL(τ) problem. Note that the syntax of every LinEMSOL(τ) problem is completely defined by τ, $\theta(X_1, \ldots, X_l)$, the constants $\{a_{ij}\}$ and m.

Example 1. The maximum weight clique problem is to find for a given graph G, with weights assigned to its vertices, a clique C of G such that the total weights of the vertices of C is maximum. This problem is a $LinEMSOL(\tau_1)$ problem since it can be expressed as follows: Given a graph G represented as a τ_1-structure, $G(\tau_1)$ and one evaluation functions f_1 associating integer weight values to the vertices of $G(\tau_1)$, find an assignment z to the free set variable X_1 in θ such that:

$$|z(X_1)|_1 = Max\{|z'(X_1)|_1 : \langle G(\tau_1), z' \rangle \models \theta(X_1)\}$$

where $\theta(X_1)$ is defined by:

$$\theta(X_1) = \forall u, v((X_1(v) \wedge X_1(u) \wedge u \neq v) \longrightarrow E(u, v))$$

Remark 2. Every $MSOL(\tau)$ decision problem can be expressed also as a $LinEMSOL(\tau))$ optimization problem. Thus, in the sequel we will be concerned only with $LinEMSOL(\tau)$ optimization problems.

2.2   The modular decomposition of $(q, q-4)$ and P_4-tidy graphs

A subset M of vertices of a graph G is called a *module* of G if every vertex outside M is either adjacent to all vertices in M or to none of them. A module M is called *strong*, if for any module M_1 either $M \cap M_1 = \emptyset$, or one module contains the other.

The *modular decomposition* of a graph G, is a tree denoted as $T(G)$. The leaves of $T(G)$ are the vertices of G, and the set of leaves associated with the subtree rooted at an internal node, induce a strong module of G. An internal node is labeled by either P, S or N standing for Parallel, Series and Neighborhood, respectively, and it can be shown that for every graph G the tree $T(G)$ is unique up to isomorphism. More details on how the tree $T(G)$ is constructed can be found in [GV97,BM83,CH94].

Let h be an internal node of $T(G)$, we denote by $M(h)$ the module corresponding to h which consists of the set of vertices of G of the subtree of $T(G)$ rooted at h. Let $\{h_1, \ldots, h_r\}$ be the set of sons of h in $T(G)$, we denote by $G(h) = \langle V(h), E(h)\rangle$ the *representative graph* of the module $M(h)$ defined by: $V(h) = \{h_1, \ldots, h_r\}$ and

$$E(h) = \{(h_i, h_j) \mid \exists u, v(u \in M(h_i) \wedge v \in M(h_j) \wedge (u, v) \in E)\}$$

Note that by the definition of a module, if a vertex of $M(h_i)$ is adjacent to a vertex of $M(h_j)$ then every vertex of $M(h_i)$ will be adjacent to every vertex of $M(h_j)$. From the construction of $T(G)$ it follows that:

Proposition 2. *Let G be any graph and let h be an internal node of $T(G)$:*

(i) if h is an S-node then $G(h)$ is a complete graph.
(ii) if h is a P-node then $G(h)$ is edge-less.

Recall that the neighborhood $N(v)$ of a vertex v of G is defined as the set of vertices of G adjacent to v, i.e.: $N(v) = \{u | (u, v) \in E\}$.

Definition 5 ((Prime spider)). *A graph G is a* prime spider *if the vertex set of G can be partitioned into sets S, K and R such that:*

(i) S is a stable set (i.e. no vertex in S is adjacent to the other), K is a clique and $|S| = |K| \geq 2$.
(ii) R contains at most one vertex, i.e. $|R| \leq 1$, and if R contains one vertex say r, then r is adjacent to all the vertices in K and is not adjacent to any of the vertices in S.
(iii) There exist a bijection f between S and K such that either $N(x) = \{f(x)\}$ for all vertices x in S or else $N(x) = K - \{f(x)\}$ for all vertices x in S.

The triple (S, K, R) is called the spider partition *of G.*

The following proposition follows from [GRT97]:

Proposition 3 ((Giakoumakis, Roussel and Thuillier)). *Let G be a P_4-tidy graph and let h be an internal N-node of $T(G)$, then $G(h)$ is either isomorphic to a prime spider, to a cycle of five vertices C_5, to a path of five vertices P_5, or to the complement of a path of five vertices $\overline{P_5}$.*

The following proposition follows from [BO95]:

Proposition 4 ((Babel and Olariu)). *Let G be a $(q, q-4)$ graph and let h be an internal N-node of $T(G)$, then $G(h)$ is either isomorphic to a prime spider, or to a graph with at most q vertices.*

3 Graphs of bounded clique-width

3.1 Graph operations and clique-width

A k-*graph* is a labeled graph with (vertex) labels in $\{1, 2, \ldots, k\}$. A k-graph G, is represented as a structure $\langle V, E, V_1, \ldots V_k \rangle$, where V and E are the sets of vertices and edges respectively, and $V_1, \ldots, V_k$ form a partition of V, such that V_i is the set of vertices labeled i in G. Note that some V_i's may be empty. A non-labeled graph $G = \langle V, E \rangle$, will be considered as a 1-graph such that all the vertices of G are labeled by 1.

For k-graphs G, H such that $G = \langle V, E, V_1, \ldots, V_k \rangle$ and $H = \langle V', E', V_1', \ldots, V_k' \rangle$ and $V \cap V' = \emptyset$ (if this is not the case then replace H with a disjoint copy of H), we denote by $G \oplus H$, the disjoint union of G and H such that:

$$G \oplus H = \langle V \cup V', E \cup E', V_1 \cup V_1', \ldots, V_k \cup V_k' \rangle$$

Note that $G \oplus G \neq G$.

For a k-graph G as above we denote by $\eta_{i,j}(G)$, where $i \neq j$, the k-graph obtained by connecting all the vertices labeled i to all the vertices labeled j in G. Formally:

$$\eta_{i,j}(G) = \langle V, E', V_1, \ldots V_k \rangle \,, \ where$$

$$E' = E \cup \{(u, v) : u \in V_i, \ v \in V_j, \ u \neq v\}$$

For a k-graph G as above we denote by $\rho_{i \to j}(G)$ the renaming of i into j in G such that:

$$\rho_{i \to j}(G) = \langle V, E, V_1', \ldots V_k' \rangle, \ where$$

$V_i' = \emptyset$, $V_j' = V_j \cup V_i$, and $V_p' = V_p$ for $p \neq i, j$.

These graph operations have been introduced in [CER93] for characterizing graph grammars. For every vertex v of a graph G and $i \in \{1, \ldots k\}$, we denote by $i(v)$ the k-graph consisting of one vertex v labeled by i.

Example 2. A clique with four vertices u, v, w, x can be expressed as:

$$\rho_{2 \to 1}(\eta_{1,2}(2(u) \oplus \rho_{2 \to 1}(\eta_{1,2}(2(v) \oplus \rho_{2 \to 1}(\eta_{1,2}(1(w) \oplus 2(x)))))))$$

With every graph G one can associate an algebraic expression built using the 3 type of operations mentioned above which defines G. We call such an expression a k-expression defining G, if all the labels in the expression are in $\{1, \ldots, k\}$. Clearly, for every graph G, there is an n-expression which defines G, where n is the number of vertices of G. Let $C(k)$ be the class of graphs which can be defined by k-expressions. The clique-width of a graph G, denoted $cwd(G)$, is defined by: $cwd(G) = Min\{k : G \in C(k)\}$. This value is a complexity measure on graphs somewhat similar to tree width, which yields efficient graph algorithms provides the graph is given with its k-expression (for fixed k). A related notion has been introduced by Wanke [Wan94] in connection with graph grammars. $C(1)$ is the class of edge-less graphs.

Cographs are exactly the graphs of clique width at most 2, cf. [CO]. Trees have clique width at most 3.

Problem 2. Find characterization of graphs of clique width at most $k, k \geq 3$. Does there exist a polynomial time algorithms for recognizing the classes $C(k)$, $k \geq 4$?

A polynomial time algorithm for recognizing the class $C(3)$ is presented in [Rot98].

3.2 P_4-tidy graphs are of $cwd \leq 4$ and $(q, q-4)$ graphs are of $cwd \leq q$

Let G and H be two disjoint graphs and let v be a vertex of G. We denote by $G[H/v]$ the graph K obtained by the substitution in G of H for v. Formally, $V(K) = V(G) \cup V(H) - \{v\}$, and

$$E(K) = E(H) \cup \{e : e \in E(G) \text{ and } e \text{ is not incident with } v\} \cup \{(u, w) : u \in V(H), w \in V(G) \text{ and } w \text{ is adjacent to } v \text{ in } G\}$$

Proposition 5. *For every disjoint graphs G, H, and for every vertex v of G, $cwd(G[H/v]) = max\{cwd(G), cwd(H)\}$.*

Recall that for any graph G, we denote by $T(G)$ the modular decomposition of G (which is a tree), and for each internal node h of $T(G)$ we denote by $G(h)$ the representative graph of h defined in section 2.2.

Proposition 6. *For every graph G, $cwd(G) = max\{cwd(H) : H \text{ is a representative graph of an internal node } h \text{ in the modular decomposition of } G\}$.*

Proposition 7. *For every prime spider G, $cwd(G) \leq 4$.*

Proposition 8. *The P_4-tidy graphs have a clique width at most 4, and for each P_4-tidy graph G, a 4-expression defining it can be constructed in time $O(|V| + |E|)$.*

Proof. Let G be a P_4-tidy graph and let $T(G)$ be the modular decomposition of G. By proposition 6 above in order to show that $cwd(G) \leq 4$, it is enough to show that for each internal node h of $T(G)$, $cwd(G(h)) \leq 4$, where $G(h)$ is the representative graph of h in $T(G)$. If h is a P-node (S-node) then $G(h)$ is an edge-less graph (a clique), and has a clique width equals to 1 (2). If h is an N-node then by proposition 3 above $G(h)$ is either a prime spider, a cycle of five vertices C_5, a path of five vertices P_5 or its complement $\overline{P_5}$. Since C_5, P_5, and $\overline{P_5}$ have $cwd \leq 4$, and prime spiders have $cwd \leq 4$ by proposition 7 above we have shown that $cwd(G) \leq 4$. A 4-expression defining G can be constructed in linear time as follows:

(i) Construct the modular decomposition of G, $T(G)$ in time $O(|V| + |E|)$ by classical methods, as shown in [GV97].
(ii) From the modular decomposition $T(G)$ construct an expression consisting of a sequence of vertex substitutions which defines G, as follows from proposition 6 above. Since the number of vertices in $T(G)$ is $O(|V|)$ (as proved in [Spi92]), this step can be done in time $O(|V| + |E|)$.
(iii) Convert the expression of vertex substitutions obtained at the previous step, to a 4-expression for G as follows from proposition 5. This step can be done in time $O(|V|+|E|)$, since each graph H used in the substitutions is either an edge-less graph, a clique, a C_5 cycle, a P_5 path, its complement $\overline{P_5}$, or a prime spider for which a 4-expression can be constructed in $O(|V(H)| + E(|H|))$ as can be shown easily for the first 5 cases and as shown in proposition 7 for the case of prime spiders.

Remark 3. The graph Z_8 in [GV97] is P_4-sparse (and hence P_4-tidy) and of clique-width exactly 4. So the bound on the clique-width in propositions 8 and 9 are best possible. Likewise C_5 is an extended P_4-reducible graph and has clique width exactly 4. So the bound on the clique-width in proposition 8 is also best possible for the class of extended P_4-reducible graphs. This is in contrast to P_4-reducible graphs which can be shown, using [GV97, theorem 4.2], to be of clique-width ≤ 3.

Proposition 9. *The $(q, q-4)$ graphs have a clique width at most q, and for each $(q, q-4)$ graph G, a q-expression defining it can be constructed in time $O(|V| + |E|)$.*

Proof. Along the same lines of the proof of Proposition 8 above using Proposition 4 instead of Proposition 3.

Remark 4. Proposition 6 can be used to define a new class of graphs of bounded clique width C_1 from an other class of bounded clique width C. be setting C_1 to be the class of graphs whose prime graphs in the modular decomposition are in C. This might be useful in an attempt to solve problem 2.

3.3   The Feferman-Vaught Theorem

In the proof of Theorem 2 we shall use a version of the Feferman-Vaught Theorem, [FV59] adapted to $MSOL$. It is not clear who observed first that this adaptation to $MSOL$ is true, but it is already in [Läu68,She75] and follows from [Fef,Ehr61]. For a good exposition, cf. [Gur79,Gur85].

We review some notation from [CM93].

Definition 6. *Let $\mathcal{A}$ be a τ-structure, let A be the domain of $\mathcal{A}$ and let φ be a $MSOL(\tau)$-formula with free set variables $X_1, \ldots, X_n$. We denote by $sat(\mathcal{A}, \varphi)$ the set of subsets of A^n for which φ holds in $\mathcal{A}$. Formally:*
$$sat(\mathcal{A}, \varphi) = \{(D_1, \ldots, D_n) : D_i \subseteq A, (A, D_1, \ldots, D_n) \models \varphi(X_1, \ldots, X_n)\};$$
$sat(\mathcal{A}, \varphi)$ is the family of tuples of sets (corresponding to free set variables in φ) which make φ true in $\mathcal{A}$.

Lemma 1. *Let k, h and n be fixed non-negative integers, then there are finitely many $MSOL(\tau_{1,L})$-formulas with free variables in $\{X_1, \ldots X_n\}$ of quantifier depth $\leq h$ in the language expressing properties of k-graphs, up to tautological equivalence.*

Proof. Follows from lemma 4.2 of [Cou90], adapted to k-graphs.

Lemma 2. *For each k, for each operation $f \in \{\rho_{i \rightarrow j}, \eta_{i,j} : i, j \in \{1 \ldots k\}, i \neq j\}$ over k-graphs, for every $MSOL(\tau_{1,L})$ formula θ, there is an $MSOL(\tau_{1,L})$ formula $\theta^\natural$ such that for every k-graph G represented over $\tau_{1,L}$, $sat(f(G), \theta) = sat(G, \theta^\natural))$.*

For any set D we denote by $\mathcal{P}(D)$ the power set of D, i.e. the set of all subsets of D. Let E, F be two subsets of D such that $E \cap F = \emptyset$, let $A \subseteq \mathcal{P}(E)^n$, and let $B \subseteq \mathcal{P}(F)^n$, (we call such A and B separated), we define $A \boxtimes B$ by:

$$A \boxtimes B = \{(D_1 \cup D_1', \ldots, D_n \cup D_n') : (D_1, \ldots, D_n) \in A, (D_1', \ldots, D_n') \in B)\}$$

Theorem 4 ((Feferman-Vaught for $MSOL$)). *For each k and for every $MSOL(\tau_{1,L})$ formula θ with free variables $X_1, \ldots, X_n$, two lists of $MSOL(\tau_{1,L})$ formulas $\varphi_1, \ldots, \varphi_m$ and $\psi_1, \ldots, \psi_m$ can be constructed such that all the formulas have the same free variables as θ and have quantifier depth no larger than the quantifier depth of θ, and for every two k-graphs G and H represented over $\tau_{1,L}$ such that $V(G) \cap V(H) = \emptyset$,*

$$sat(G \oplus H, \theta) = \bigcup_{1 \leq i \leq m} sat(G, \varphi_i) \boxtimes sat(H, \psi_i).$$

Proof. Immediate reformulation of the result by Feferman-Vaught as discussed in [Gur85]. The result can also be proved directly using pebble games for $MSOL$.

3.4 Linear algorithms for optimization problems on bounded clique-width graphs

Let G be a graph, let $f_1, \ldots, f_m$ be m evaluation functions associating integer values to the vertices of G, let $D_1, \ldots, D_l \subseteq V(G)$ and let

$$h(D_1, \ldots, D_l) = \sum_{\substack{1 \leq i \leq l \\ 1 \leq j \leq m}} a_{ij} |D_i|_j$$

where $\{a_{ij} : 1 \le i \le l, 1 \le j \le m\}$ are any integers, and $|D_i|_j$ is a short notation for: $\sum_{a \in D_i} f_j(a)$. For $A \subseteq \mathcal{P}(V(G))^l$, let

$$Max\_h(A) = Max\{h(D_1, \ldots, D_l) : (D_1, \ldots, D_l) \in A\}$$

It is clear that for separated A and B:

$$Max\_h(A \boxtimes B) = Max\_h(A) + Max\_h(B)$$

and for general A and B:

$$Max\_h(A \cup B) = Max\{Max\_h(A), \ Max\_h(B)\}$$

From definition 4 above it follows that a LinEMSOL$(\tau_{1,L})$ optimization problem over a class of graphs K can be formulated as the computation of $Max\_h(sat(G, \theta))$ for a given graph $G \in K$ represented over $\tau_{1,L}$, where θ is a fixed MSOL$(\tau_{1,L})$ formula.

For each k-expression g we denote by $Tree(g)$ the labeled tree corresponding to g. The leaves of $Tree(g)$ are the singletons in g (the basic graphs) labeled by their initial label from $\{1, \ldots, k\}$, and the internal nodes of $Tree(g)$ corresponds to the operations appearing in g. For each internal node x of $Tree(g)$, we denote by $Graph(x)$ the k-graph defined by the k-expression corresponding to the subtree of $Tree(g)$ rooted at x.

Theorem 2. *Let $\mathcal{C}$ be a class of graphs of clique-width at most k, $\mathcal{C} \subseteq \mathcal{C}(k)$, such that there is a (known) $O(f(|E|, |V|))$ algorithm, which for each graph G in $\mathcal{C}$, constructs a k-expression defining it. Then every LinEMSOL$(\tau_{1,L})$ problem on $\mathcal{C}$ can be solved (constructively) in time $O(f(|E|, |V|))$.*

Proof ((Sketch)). Let P be a LinEMSOL$(\tau_{1,L})$ optimization problem over a class of graphs $\mathcal{C} \subseteq \mathcal{C}(k)$. As mentioned above P can be formulated as the computation of $Max\_h(sat(G, \theta))$ for a given graph $G \in \mathcal{C}$ represented over $\tau_{1,L}$. Since $G \in \mathcal{C}$ there is a k-expression g which defines G. The computation of $Max\_h(sat(G, \theta))$ can be done as follows:

(i) Traverse $Tree(g)$ from top to bottom starting at the root assigning formulas to the internal nodes of the tree using lemma 2 and Theorem 4, such that θ is assigned to the root and for each internal node x and a formula φ assigned to x, $sat(Graph(x), \varphi)$ can be calculated from the sons of x and the formulas assigned to them. In other words, $sat(Graph(x), \varphi)$ can be calculated from $sat(Graph(y_i), \psi_{i,j})$ where y_i are the sons of x and $\psi_{i,j}$ are the formulas assigned to y_i.

(ii) Traverse $Tree(g)$ from bottom to top and, at each node x and for each formula φ assigned to x by the previous step, compute $Max\_h(sat(Graph(x), \varphi))$ using the values obtained so far for the sons of x and the formulas assigned to them.

For the complexity, the total time for handling the input graph G is $O(f(|V|, |E|))$ for constructing the k-expression g plus the total time for applying the above procedure. First note that the size of the tree $Tree(g)$ is $O(|V(G)|)$. In step (i) of the above process the number of formulas assigned to each node is bounded by a constant (which does not depend on the size of the input graph G) since by lemma 2 and Theorem 4 all these formulas are of quantifier depth no larger than the quantifier depth of θ, and by lemma 1 the number of such formulas is bounded (up to tautological equivalence) by a constant which depends just on the size of θ and k. Hence, the computation done at each node by the above procedure is bounded by a constant (with the uniform cost measure), and the total time of the above procedure is bounded by $O(|V|)$. Therefore the total complexity of handling the input graph G is $O(f(|V|, |E|)) + O(|V|) = O(f(|V|, |E|))$.

4   Conclusion and open problems

We have shown that every LINEMSOL($\tau_{1,L}$) problem on the class of graphs of clique width at most k has a linear time solution. Can we extend this result for the vocabulary τ_2?

For labeled graphs over the vocabulary $\tau_{2,L}$ (with unary predicates on vertices and edges) it is easy to see that the answer is no, unless $\mathbf{P} = \mathbf{NP}$.

Proposition 10. *If* $\mathbf{P} \neq \mathbf{NP}$ *then there is a* $MSOL(\tau_{2,L})$ *definable decision problem over the class of cliques which is not solvable in polynomial time.*

Proof ((Sketch)). Assume that the vocabulary $\tau_{2,L}$ has one unary predicate U_1 which is used to label edges of the graph in question. Let $G(\tau_{2,L})$ be a graph represented over $\tau_{2,L}$, One can construct an MSOL formula φ_{3col-2} over $\tau_{2,L}$ which states that there is a partition of the vertices of G into three sets X_1, X_2, X_3 such such that these sets form a 3-coloring of the subgraph of G constructed from G by omitting all the edges which are not labeled by U_1. If we could decide in polynomial time for a given clique K, whether $K(\tau_{2,L}) \models \varphi_{3col-2}$, we could solve the 3-colorability problem on general graphs in polynomial time, a contradiction.

However, coding arbitrary graphs as definable subgraphs of cliques seems to be cheating. The genuine question is, whether we can get the same without this trick. If $\mathbf{P} \neq \mathbf{NP}$ already on unary languages the answer is yes. More precisely, let τ_0 denote the empty vocabulary. A *spectrum* is a set of structures over τ_0 definable by a formula φ of the form $\exists X_1, X_2, ..., X_l \sigma$, such that σ is first order. BIN denotes the set of all spectra definable by formulas using only one binary predicate symbol which represents a graph relation, i.e a relation which is irreflexive and symmetric. Recall that $\mathbf{P}_1$ ($\mathbf{NP}_1$) denotes the class of languages over one letter (also called tally languages), which are in $\mathbf{P}$ ($\mathbf{NP}$). In [Fag74] it was proved that $\mathbf{P}_1 = \mathbf{NP}_1$ if and only if $BIN \subseteq \mathbf{P}_1$.

Theorem 3. *If $P_1 \neq NP_1$ then there is a $MSOL(\tau_2)$ definable decision problem over the class of cliques which is not solvable in polynomial time.*

Proof (Sketch). Let $\mathcal{A}$ be a structure over $\tau_\emptyset$, we denote by $K_{\mathcal{A}}$, the clique corresponding to $\mathcal{A}$, such that the number of elements in the universe of $\mathcal{A}$ equals to the number of vertices of the clique $K_{\mathcal{A}}$.

Let φ be an $SOL(\tau_\emptyset)$ sentence, we denote by $\varphi^\sharp$ the $MSOL(\tau_2)$ sentence corresponding to φ, which is constructed from φ, by replacing every binary set predicate $U(x,y)$ by the formula: $\exists t (U(t) \wedge R(t,x) \wedge R(t,y))$. Since in a clique all the edges between all pair of vertices exist, each pair of vertices (x,y) can be identified by the unique edge t, incident to both x and y. Therefore, quantification over pair of vertices in cliques can be replaced by quantification over edges, as indicated by the above formula which replaces a binary set predicated $U(x,y)$. Therefore, for every structure $\mathcal{A}$ over $\tau_\emptyset$ and every spectrum S defined by a sentence φ one can decide whether $\mathcal{A} \in S$ by deciding whether $K_{\mathcal{A}} \models \varphi^\sharp$. Therefore, assuming that, over the class of cliques every $MSOL(\tau_2)$-definable decision problem can be solved in polynomial time, we get that $BIN \subseteq P_1$. By [Fag74] this implies that $P_1 = NP_1$, a contradiction.

Problem 3. Can we prove Theorem 3 above by replacing the condition $P_1 \neq NP_1$ by the condition $P \neq NP$?

We might ask the following:

Question 1. Is there any class of graphs C, such that C is not of bounded clique width and every $MSOL(\tau_1)$ problem on C has a polynomial time solution?

As an anonymous referee pointed out, the answer is yes. By combining small (of size $\log n$) graphs of clique width $O(\log n)$ with graphs of size n and bounded clique width we can construct a C where we still can apply our proof technique and show that every $MSOL(\tau_1)$ problem on C has a polynomial time solution.

However, the following problem remains open:

Problem 4. Is there any class of graphs C, such that C is not of clique width bounded by $O(\log n)$ and every $MSOL(\tau_1)$ problem on C has a polynomial time solution?

We have discussed in this paper optimization problems. In a forthcoming paper we shall show how our techniques can be extended to $MSOL(\tau_1)$ counting (or enumeration) problems, such as counting the number of maximal cliques.

Acknowledgments

We are indebted to Luitpold Babel who made us aware of the $(q, q-4)$ graphs and the P_4-tidy graphs and the prime graphs associated with their corresponding modular decompositions.

We are also indebted to an anonymous referee who suggested the improved formulation of problem 4.

References

[ALS91] S. Arnborg, J. Lagergren, and D. Seese. Easy problems for tree decomposable graphs. *Journal of Algorithms*, 12:308–340, 1991.

[BM83] H. Buer and R.H. Möhring. A fast algorithm for the decomposition of graphs and posets. *Math. Oper. Res.*, 8:170–184, 1983.

[BO95] L. Babel and S. Olariu. On the isomorphism of graphs with few P_4s. In M. Nagl, editor, *Graph Theoretic Concepts in Computer Science, 21th International Workshop, WG'95*, volume 1017 of *Lecture Notes in Computer Science*, pages 24–36. Springer Verlag, 1995.

[BO98] L. Babel and S. Olariu. Domination and steiner tree problems on graphs with few P_4's. These proceedings, 1998.

[CER93] B. Courcelle, J. Engelfriet, and G. Rozenberg. Handle-rewriting hypergraph grammars. *J. Comput. System Sci.*, 46:218–270, 1993.

[CH94] A. Cournier and M. Habib. A new linear algorithm for modular decomposition. *Lecture Notes in Computer Science*, 787:68–84, 1994.

[CM93] B. Courcelle and M. Mosbah. Monadic second–order evaluations on tree-decomposable graphs. *Theoretical Computer Science*, 109:49–82, 1993.

[CO] B. Courcelle and S. Olariu. Upper bounds to the clique-width of graphs. submitted for publication (http://dept-info.labri.u-bordeaux.fr/~courcell/ActSci.html).

[Cou90] B. Courcelle. The monadic second-order logic of graphs i: Recognizable sets of finite graphs. *Information and Computation*, 85:12–75, 1990.

[Cou91] B. Courcelle. The monadic second-order logic of graphs V: On closing the gap between definability and recognizability. *Theoret. Comput. Sci.*, 80:153–202, 1991.

[Cou94] B. Courcelle. The monadic second-order logic of graphs VI: On several representations of graphs by relational structures. *Disc. Appl. Math.*, 54:117–149, 1994.

[Cou95] B. Courcelle. The monadic second-order logic of graphs VIII: Orientations. *Annals Pure Applied Logic*, 72:103–143, 1995.

[Cou96] B. Courcelle. The monadic second-order logic of graphs X: Linear orders. *Theoret. Comput. Sci.*, 160:87–143, 1996.

[Cou97] B. Courcelle. The expression of graph properties and graph transformations in monadic second-order logic. In G. Rozenberg, editor, *Handbook of graph grammars and computing by graph transformations, Vol. 1 : Foundations*, chapter 5, pages 313–400. World Scientific, 1997.

[EF95] H.D. Ebbinghaus and J. Flum. *Finite Model Theory*. Perspectives in Mathematical Logic. Springer, 1995.

[Ehr61] A. Ehrenfeucht. An application of games to the completeness problem for formalized theories. *Fundamenta Mathematicae*, 49:129–141, 1961.

[Fag74] R. Fagin. Generalized first-order spectra and polynomial time recognizable sets. *American Math. Society Proc.*, 7:27–41, 1974.

[Fef] S. Feferman. Some recent work of Ehrenfeucht and Fraïssé. Proceedings of the Summer Institute of Symbolic Logic, Ithaca 1957, pp. 201-209.

[FV59] S. Feferman and R. Vaught. The first order properties of algebraic systems. *Fundamenta Mathematicae*, 47:57–103, 1959.

[GJ79] M.G. Garey and D.S. Johnson. *Computers and Intractability*. Mathematical Series. W.H. Freeman and Company, 1979.

[GRT97] V. Giakoumakis, , F. Roussel, and H. Thuillier. On P_4-tidy graphs. *Discrete Mathematics and Theoretical Computer Science*, 1:17–41, 1997.

[Gur79] Y. Gurevich. Modest theory of short chains, I. *Journal of Symbolic Logic*, 44:481–490, 1979.

[Gur85] Y. Gurevich. Monadic second order theories. In *Model-Theoretic Logics*, Perspectives in Mathematical Logic, chapter 14. Springer Verlag, 1985.

[GV97] V. Giakoumakis and J. Vanherpe. On extended P_4-reducible and extended P_4-sparse graphs. *Theoret. Comput. Sci.*, 180:269–286, 1997.

[Hoà85] C. Hoàng. Doctoral thesis. McGill University, Montreal, 1985.

[JO89] B. Jamison and S. Olariu. P_4-reducible graphs a class of tree representable graphs. *Studies Appl. Math.*, 81:79–87, 1989.

[JO92a] B. Jamison and S. Olariu. A linear-time recognition algorithm for P_4-sparse graphs. *SIAM J. Comput.*, 21:381–406, 1992.

[JO92b] B. Jamison and S. Olariu. A unique tree representation for P_4-sparse graphs. *Discrete Appl. Math.*, 35:115–129, 1992.

[JO95a] B. Jamison and S. Olariu. A linear-time algorithm to recognize P_4-reducible graphs. *Theoret. Comput. Sci.*, 145:329–344, 1995.

[JO95b] B. Jamison and S. Olariu. Linear-time optimization algorithms for P_4-sparse graphs. *Discrete Appl. Math.*, 61:155–175, 1995.

[Läu68] H. Läuchli. A decision procedure for the weak second order theory of linear order. In *Logic Colloquium '66*, pages 189–197. North Holland, 1968.

[Lau93] C. Lautemann. Logical definability of NP-optimization problems with monadic auxiliary predicates. *Lecture Notes in Computer Science*, 702:327–339, 1993.

[Pap94] C. Papadimitriou. *Computational Complexity*. Addison Wesley, 1994.

[Rot98] U. Rotics. *Efficient Algorithms for Generally Intractable Graph Problems Restricted to Specific Classes of Graphs*. PhD thesis, Technion- Israel Institute of Technology, 1998.

[She75] S. Shelah. The monadic theory of order. *Annals of Mathematics*, 102:379–419, 1975.

[Spi92] J. Spinrad. P_4-trees and substitution decomposition. *Discrete Appl. Math.*, 39:263–291, 1992.

[Wan94] E. Wanke. k-NLC graphs and polynomial algorithms. *Discrete Appl. Math.*, 54:251–266, 1994.

Minus Domination in Small-Degree Graphs
(extended abstract)

Peter Damaschke

FernUniversität, Theoretische Informatik II
58084 Hagen, Germany
Peter.Damaschke@fernuni-hagen.de

Abstract. Minus domination in graphs is a variant of domination where the vertices must be labeled $-1, 0, +1$ such that the sum of labels in each $N[v]$ is positive. (As usual, $N[v]$ means the set containing v together with its neighbors.) The minus domination number γ^- is the minimum total sum of labels that can be achieved. In this paper we prove linear lower bounds for γ^- in graphs either with $\Delta \leq 3$, or with $\Delta \leq 4$ but without vertices of degree 2. The central section is concerned with complexity results for $\Delta \leq 4$: We show that computing γ^- is NP-hard and MAX SNP-hard there, but that γ^- can be approximated in linear time within some constant factor. Finally, our approach also applies to signed domination (where the labels are $-1, +1$ only) in small-degree graphs.

1 The Notion of Minus Domination

Let $G = (V, E)$ be a simple undirected graph with n vertices. $N[v]$ denotes the closed neighborhood of vertex v, that is, v together with all adjacent vertices. A minus dominating function is a labeling of the vertices by values $-1, 0, +1$ such that the sum of labels in each $N[v]$ is positive. This generalizes the classical and extensively studied concept of domination in graphs, where only labels 0 and $+1$ are permitted. Similar to the domination number $\gamma(G)$, the minus domination number $\gamma^-(G)$ is defined to be the minimum total sum of labels for minus dominating functions. We simply write γ^- for $\gamma^-(G)$. In contrast to γ, the minus domination number γ^- can be negative. Note that there always exist minus dominating functions with any total sum from γ^- to n.

A "sociological" motivation of minus domination has been suggested by [5]. Let our graph be the model of a network of people. An edge means that the joined vertices are somehow related (acquaintances, neighbors, colleagues, or the like). Assume that every "vertex" gives his opinion about some controverse question, which may be negative, undecided, or positive, indicated by label $-1, 0, +1$, respectively. Then it can happen that *every* vertex observes a majority of positive opinions in his neighborhood (in case of a minus dominating function), nevertheless the total excess of positive votes is very small compared to the graph size, or even the negative votes abound! One may brood about possible consequences of this effect in social networks. Other motivations may come from facility location problems.

A purely graph-theoretic motivation is that the minus domination problem can be seen, in a clear sense, as a proper generalization of the classical domination problem; cf. the remark in Section 2.

A similar concept is signed domination where only labels $+1$ and -1 are allowed [6] [7]. The signed domination number γ_s is defined to be the minimum total sum of labels for signed dominating functions (i.e. the sum of labels in each $N[v]$ must be positive again). Obviously we have $\gamma^- \leq \gamma_s$, but there is no such relation between γ_s and γ.

Here we study minus domination in graphs of small maximum vertex degree Δ, which is a natural restriction. Our work is inspired by some observations in [5] which we shall extend and refine.

It can be simply inferred from [5] that $\gamma^- \geq n/3$ for $\Delta \leq 2$ (unions of paths and cycles). Further, it has been shown there that $\gamma^- \geq 0$ for $\Delta \leq 5$, and $\gamma^- \geq 1$ for $\Delta \leq 3$. For $\Delta = 6$ or larger, γ^- can be negative. Moreover, for Δ-regular graphs, the sharper result $\gamma^- \geq n/(\Delta + 1)$ has been proven in [4]. Only a few results are known about the complexity of minus domination [3]: The problem is NP-complete for bipartite graphs and chordal graphs, and efficiently solvable for trees.

In the present paper we prove some lower bounds for γ^-, but mainly we study the complexity aspects in case $\Delta \leq 4$.

2   Preliminaries

Considering some fixed minus dominating function, let P, Z, and M be the sets of vertices with labels $+1$, 0, and -1, also called positive, neutral, and negative vertices, respectively. Let P_i denote the set of positive vertices having exactly i negative neighbors. Similarly, M_i is defined to be the set of negative vertices having exactly i positive neighbors. Clearly, $M_0 = M_1 = \emptyset$. Let D_i be the set of vertices of degree i. Further, let be $M_2^0 = M_2 \cap D_2$ and $M_2' = M_2 \setminus M_2^0$. The lower case symbols $p, z, m, p_i, m_i, d_i, m_2^0, m_2'$ denote the cardinalities of the so defined sets. Note that $\gamma^- = \min(p - m)$ where the minimum is taken over all minus dominating functions.

If X, Y are disjoint sets of vertices of G, symbol $[X, Y]$ denotes the bipartite subgraph of G consisting of the parts X, Y and of all edges between X, Y inherited from G; edges within X or Y are ignored. Our main tool throughout the paper is to derive useful equations and inequalities by counting the edges of carefully chosen bipartite subgraphs in two ways, namely as the sum of degrees (i.e. with respect to the subgraph) in both X and Y.

The first central lemma describes $p - m$ in $\Delta \leq 4$ graphs in nonnegative terms.

Lemma 1. *Any minus dominating function in a $\Delta \leq 4$ graph satisfies*
$p - m = p_0 + p_1/2 + m_3/2 + m_4.$

Proof. $\Delta \leq 4$ implies $p = p_0 + p_1 + p_2$ and $m = m_2 + m_3 + m_4$. In $[P, M]$ we see $p_1 + 2p_2 = 2m_2 + 3m_3 + 4m_4$. Now eliminate $-m_2$ in $p - m$. $\Diamond$

Lemma 2. *In a $\Delta \leq 4$ graph we have $m'_2 \leq z \leq 4p_0 + 2p_1$ and, moreover, $m'_2 \leq 2p_0 + p_1$.*

Proof. Consider $[M'_2, Z]$. Clearly, every M'_2 vertex has a neutral neighbor. Since $\Delta \leq 4$, a neutral vertex has at most one negative neighbor, and the first inequality follows.

Consider $[Z, P_0 \cup P_1]$. Every neutral vertex has a positive neighbor that must be in P_0 or P_1, due to $\Delta \leq 4$. Every vertex from P_0 and P_1 has at most 4 and 2 neutral neighbors, respectively. This proves the second inequality.

We can reduce our estimate for m'_2. For this let Z' (of cardinality z') be the set of vertices from Z having a negative neighbor. Similarly as above, consider $[M'_2, Z']$ and $[Z', P_0 \cup P_1]$. Since every Z' vertex has at least two positive neighbors, we even have $m'_2 \leq z' \leq 2p_0 + p_1$. $\Diamond$

The next lemma holds in any graph and is useful for reductions.

Lemma 3. *If $x \in D_1$ and w is the unique neighbor of x then we may assume, in an optimal minus dominating function, that $x \in Z$ and $w \in P$.*

Proof. Since x must be dominated, at least one of x, w is positive. Assume $x \in P$. Then either $w \in Z$ or $w \in P$. In the former case we may exchange the labels of x and w, deteriorating neither minus domination nor $p - m$. If $w \in P$ then w must have some neighbor $w' \in M$, otherwise we could reset the label of x to 0, contradicting optimality. But now we can make both x and w' neutral. $\Diamond$

As a consequence, the classical domination problem can be reduced to special instances of minus domination: If H is any graph of n vertices then append a path of length 3 to every vertex u of H, say (u, v, w, x). Let G be the augmented graph. We may assume $x \in Z$, $w \in P$, and, by similar arguments, $v \in Z$ and $u \notin M$. This yields $\gamma^-(G) = n + \gamma(H)$. Thus many hardness results for domination immediately translate to minus domination.

3   Linear Lower Bounds

First we strengthen the result of [5] for $\Delta \leq 3$.

Theorem 1. *If $\Delta \leq 3$ then $\gamma^- \geq n/5$, and this bound is tight.*

Proof. Consider an optimal minus dominating function. Since $\Delta \leq 3$, Lemma 1 simplifies to $\gamma^- = p_0 + p_1/2 + m_3/2$, and we also have $p_2 = 0$. Note that $n = p_0 + p_1 + m_2 + m_3 + z$.

From $[M_2, P_1]$ we immediately see $2m_2 \leq p_1$. Next consider $[Z, P]$. Every neutral vertex has a positive neighbor. Every P_0 vertex has at most 3 neighbors in Z, and every P_1 vertex has at most one neighbor in Z. So $z \leq 3p_0 + p_1$. Replacing m_2 and z with these estimates we obtain $n \leq 4p_0 + 5p_1/2 + m_3 \leq 5\gamma^-$.

On the other hand, we present arbitrarily large (connected) graphs with $\Delta = 3$ and $\gamma^- = n/5$: Take a cycle of length divisible by 3 and append a further vertex of degree 1 to every vertex of this cycle, except every third vertex. Sets P and M are obvious. $\Diamond$

In contrast, there exist arbitrarily large connected graphs with $\Delta = 4$ but $\gamma^- = 0$: Take a 2-regular graph (union of cycles) whose $n/2$ vertices are positive. Then add an independent set of $n/2$ negative vertices in such a way that $P = P_2$, $M = M_2$. This can be easily done in many different ways, even if the graph is required to be connected.

Remember that linear lower bounds for γ^- hold in regular graphs [4]. So it is interesting to relax this condition and to consider graphs with given minimum degree δ and maximum degree Δ. The next result refers to the case $\delta = 3$, $\Delta = 4$. However it turns out that we only need the assumption $d_2 = 0$.

Proposition 1. *If $\Delta \leq 4$ and $d_2 = 0$ then $\gamma^- \geq n/12$.*

Proof. Consider an optimal minus dominating function. Since $d_2 = 0$ by assumption, we have $m_2 = m_2'$, and hence $n = p_0 + p_1 + p_2 + m_2' + m_3 + m_4 + z$.

From $[P_2, M]$ and $\Delta \leq 4$ we see $p_2 \leq 2m \leq 2m_2' + 2m_3 + 2m_4$. Lemma 2 yields $p_2 \leq 4p_0 + 2p_1 + 2m_3 + 2m_4$, $m_2' \leq 2p_0 = p_1$, and $z \leq 4p_0 + 2p_1$. Eliminating these summands in n gives $n \leq 11p_0 + 6p_1 + 3m_3 + 3m_4$ so $n \leq 12\gamma^-$ by Lemma 1. $\Diamond$

We conjecture that the constant can be slightly raised, but a limit is given by $1/9$, since it is not hard to construct arbitrarily large connected graphs with $\gamma^- = n/9$. It remains open whether a linear lower bound holds for $\Delta = 5$ and $d_2 = 0$.

We may further relax our assumptions and allow a few vertices of degree 2. Modifying the above proof we see that a (smaller) positive linear lower bound remains. This applies, for example, to the complete grid graphs which answers a question from [5].

4   Complexity Results for Degree 4

In the following, $\alpha(H)$ denotes the size of a maximum independent vertex set in graph H.

Theorem 2. *Minus domination is NP-hard for planar $\Delta \leq 4$ graphs.*

Proof. We give a reduction from the maximum independent set problem for planar $\Delta \leq 3$ graphs which is NP-hard [8]. Let H be such a graph with n vertices and e edges. W.l.o.g. we may assume that no vertex has degree 1. Replace every edge uv of H with a component consisting of new vertices x, w, y, z and edges xw, wy, yz, and link w to both u and v. The so obtained planar graph G satisfies $\Delta \leq 4$.

By Lemma 3 we may assume $x, z \in Z$ and $w, y \in P$. Regardless of the labeling of the original vertices from H, all vertices in G fulfill the minus domination condition, except perhaps the vertices w subdividing the edges of H. Note that at most one of u, v can be negative, and that vertices from G that should be positive can be made neutral. Since M forms an independent set in H, we have $\gamma^-(G) = 2e - \alpha(H)$. $\Diamond$

Moreover, we can use our reduction (but without planarity) to show:

Theorem 3. *For some $\epsilon > 0$, the minus domination number in $\Delta \leq 4$ graphs cannot be approximated in polynomial time within a factor $1+\epsilon$, unless $P = NP$.*

Proof. By [1], the maximum independent set problem in $\Delta \leq 3$ graphs is MAX SNP-complete, so there exists $\epsilon > 0$ such that $\alpha(H)$ cannot be approximated for such H in polynomial time within, say, $1 + 12\epsilon$ unless $P = NP$. Assume that γ^- in $\Delta \leq 4$ graphs can be approximated within factor $1 + \epsilon$. Since $2e \leq 3n$ and $\alpha(H) \geq n/4$, this could be used to approximate $\alpha(H) = 2e - \gamma^-(G)$ within $1 + 12\epsilon$, a contradiction. $\Diamond$

Despite of NP-hardness in general, we can simply characterize the graphs with $\Delta \leq 4$ and $\gamma^- = 0$. Namely, these are exactly the graphs addressed in the remark before Theorem 1.

Theorem 4. *A graph with $\Delta \leq 4$ satisfies $\gamma^- = 0$ if and only if the following holds: $d_4 = d_2 = n/2$, every vertex of D_2 has its two neighbors in D_4, and every vertex of D_4 has two neighbors in D_4 and D_2, respectively.*

Proof. If the conditions are fulfilled then $P = D_4$ and $M = D_2$ gives a minus dominating function.

Conversely, consider a graph with $\Delta \leq 4$ and a minus dominating function with $p - m = 0$ on it. Lemma 1 gives $p = p_2 = m = m_2$. Moreover we have $m_2' = z = 0$, otherwise Lemma 2 would imply the existence of positive vertices outside P_2, which is a contradiction here. So $P = P_2 = D_4$ and $M = M_2^0 = D_2$ have the asserted properties. $\Diamond$

Consequently, we can simply decide in linear time whether a $\Delta \leq 4$ graph satisfies $\gamma^- = 0$. More elaboration of the idea of Theorem 4 leads to the constant approximability of γ^- in $\Delta \leq 4$ graphs. The existence of such a constant for $\Delta \leq 4$ is not evident, since γ^-/n can be arbitrarily small.

Theorem 5. *In $\Delta \leq 4$ graphs, γ^- can be approximated in linear time within some constant factor (here 20).*

Proof. We may consider graphs with $\gamma^- > 0$, since the case $\gamma^- = 0$ can be easily recognized, by Theorem 4.

Before going into the details, we roughly describe the idea leading to our approximation. The graph must have a structure similar to that in Theorem 4, with only $O(\gamma^-)$ exceptions. To be more precise, consider an optimal minus dominating function. By Lemma 1 we have $\gamma^- = p_0 + p_1/2 + m_3/2 + m_4$, so $n - p_2 - m_2^0 = p_0 + p_1 + m_2' + m_3 + m_4 + z \leq 7p_0 + 4p_1 + m_3 + m_4 \leq 8\gamma^-$ where $m_2' + z$ is estimated due to Lemma 2. That means, all but $8\gamma^-$ vertices belong to $P_2 \cup M_2^0$. Moreover, these two sets must have "nearly" equal size. Note that $P_2 \subseteq D_4$ and $M_2^0 \subseteq D_2$. Thus, in our approximation we will make "almost" all vertices of D_4 and D_2 positive and negative, respectively. However, in order to guarantee both the minus domination property and a small excess of positive vertices, we have to consider the vertex neighborhoods, too.

In the following we collect some useful inequalities.

First let us bound $p_2 - m_2^0$. It is clear that at most $2m_2' + 3m_3 + 4m_4$ vertices of P_2 may have negative neighbors outside M_2^0. From $[P_2, M_2^0]$ we see that $2(p_2 - 2m_2' - 3m_3 - 4m_4) \leq 2m_2^0$, hence

$$p_2 - m_2^0 \leq 2m_2' + 3m_3 + 4m_4.$$

Similarly, at most $2p_0 + p_1$ vertices of P_2 may have neighbors in $P \cap D_2$. We define Q to be the set of D_4 vertices having at most two neighbors in D_2. For the cardinality of Q we get $q \geq p_2 - 2p_0 - p_1$.

Next, at most p_1 vertices of M_2^0 may have neighbors outside P_2. Let U be the set of D_2 vertices having both neighbors in D_4. So the cardinality of U satisfies

$$u \geq m_2^0 - p_1.$$

In other words, all but p_1 vertices of M_2^0 have both neighbors in P_2, and we have seen above that at most $2p_0 + p_1$ of them are not in Q. Hence at most $4p_0 + 2p_1$ vertices of M_2^0 may have neighbors in $P_2 \setminus Q$. That means, at least $m_2^0 - 4p_0 - 3p_1$ vertices of M_2^0 have both neighbors in Q.

Now let T be the set of D_2 vertices having both neighbors in Q. (Clearly $T \subseteq U$.) For the cardinality of T we have

$$t \geq m_2^0 - 4p_0 - 3p_1.$$

The nice point is that our sets Q, U, T can be constructed in linear time, since their definitions rely on local degree conditions only. Moreover, they are close enough to the original (and hard-to-compute) sets P and M.

Now define a new labeling by $\tilde{M} = T$, $\tilde{Z} = U \setminus T$, and $\tilde{P} = V \setminus U$. First we make sure that this is a minus dominating function. Since $\tilde{Z} \cup \tilde{M} \subseteq U \subseteq D_2$, both neighbors of any negative or neutral vertex are in $\tilde{P}$. Next consider a vertex $v \in \tilde{P}$. If v has a neighbor $w \in \tilde{M}$ at all, then $w \in T$. This implies that v itself belongs to Q, and v has at least two neighbors in $\tilde{P}$. Therefore the sum of labels in each vertex neighborhood is positive.

Using Lemma 1 and 2, and the highlighted inequalities we get the following estimation:

$$\tilde{p} - \tilde{m} = n - u - t$$
$$\leq p_0 + p_1 + p_2 + m_2^0 + m_2' + m_3 + m_4 + z - u - t$$
$$\leq 5p_0 + 5p_1 + p_2 - m_2^0 + m_2' + m_3 + m_4 + z$$
$$\leq 5p_0 + 5p_1 + 3m_2' + 4m_3 + 5m_4 + z$$
$$\leq 15p_0 + 10p_1 + 4m_3 + 5m_4$$
$$\leq 20\gamma^- \ \Diamond$$

We have already spent some effort to get a ratio being not too large, but further improvements are desirable. However, large approximation ratios are not untypical in the field. For a recent example, an $8k$-approximation algorithm for the minimum fill-in problem where k is the optimal solution, see [10]. The ratio remains large even in the bounded-degree case. (This result is of similar spirit: First it is shown that only an easily detectable subset of $O(k^2)$ vertices is "interesting".) We hope that a better ratio can be achieved by utilizing graph factors [9], since $[P_2, M_2^0]$ forms a large subgraph of degree at most 2.

Our approach yields another generalization of Theorem 4:

Proposition 2. *For every fixed k there is a polynomial algorithm deciding whether a given $\Delta \leq 4$ graph satisfies $\gamma^- \leq k$.*

Proof. Assume $\gamma^- \leq k$. We observed above that at most $8\gamma^- \leq 8k$ vertices are outside $P_2 \cup M_2^0$. A naive algorithm might try all 3^{8k} labelings of all $\binom{n}{8k}$ candidate subsets, and label all other vertices by $+1$ and -1. Since $P_2 \subseteq D_4$ and $M_2^0 \subseteq D_2$. the labeling of $n - 8k$ vertices is uniquely determined. $\Diamond$

Admittedly, the last result is of rather academic interest in the present form, because of the unreasonable time bound. We conjecture that easily recognizable large subsets of D_4 and D_2 have labels $+1$ and -1, respectively, in any optimal solution. This would drastically reduce the number of candidate subsets. Furthermore, it would be nice to characterize the $\Delta \leq 4$ graphs with $\gamma^- = 1, 2, 3, \ldots$ by degree conditions, too. More interestingly, it is not clear whether analogues of the previous results remain true for $\Delta = 5$.

5   Signed Domination

We have already defined the signed domination number γ_s in Section 1. Our approach works also for signed domination, moreover, since label 0 disappears, some things become easier and some results can be extended to $\Delta \leq 5$. In this informal section we report the γ^s counterparts of our γ^- results. The very similar proofs are omitted, subject to some hints.

Lemma 1 remains unchanged. For $\Delta \leq 5$ we get an additional $3m_5/2$ term.
Lemma 2 becomes trivial, since $m_2' = z = 0$ in arbitrary graphs.
Lemma 3: Both any degree-1 vertex and its neighbor must be positive.
Theorem 1: Note that now $n = p_0 + p_1 + m_2 + m_3 \leq p_0 + 3p_1/2 + m_3 \leq 3\gamma_s$, thus $\gamma_s \geq n/3$, and this bound is tight: Consider a cycle of even length, color the edges alternatingly red and blue, and finally, for every red edge add a vertex being adjacent to both endpoints of this red edge. Moreover, if a graph contains only vertices of degree 2 and 3 then we have $\gamma_s = n - 2\pi$, where π means the neighborhood packing number.
Proposition 1 can be extended to $\Delta \leq 5$. First note that $p_2 \leq 2m_3 + 2m_4 + 2m_5$, which yields $n = p_0 + p_1 + p_2 + m_3 + m_4 + m_5 \leq p_0 + p_1 + 3m_3 + 3m_4 + 3m_5 \leq 6\gamma_s$.
Theorem 2: NP-completeness is obvious for planar $\Delta \leq 3$ graphs, using the modified Lemma 3.
Theorem 3 therefore holds for $\Delta \leq 3$.
Theorem 4: We can simply decide whether $\gamma_s = 0$ in $\Delta \leq 5$ graphs. The only difference is that $P_2 = D_4 \cup D_5$.

Theorem 5 : Constant approximability holds even for $\Delta \leq 5$, however since there is no label 0, we must choose a more "bulky" function which gives a poor ratio.

Proposition 2 : For $\Delta \leq 5$ and any fixed k, one can decide whether $\gamma_s \leq k$ in $O(n^{2k})$ time.

6   Concluding Remarks

Our final remarks are not restricted to small-degree graphs.

We mention a nice identity that holds in many cases. A subset S of vertices is called an independent perfect dominating set (IPDS) if every vertex of the graph belongs to exactly one set $N[v]$, $v \in S$. For a bibliography on perfect domination cf. e.g. [2].

Theorem 6. *If a graph admits an IPDS then all IPDS have the same cardinality, namely $\gamma^- = \gamma$.*

Proof. Let S be an IPDS of size s. In a minus dominating function, every $N[v]$ must contain more positive than negative vertices, hence $\gamma^- \geq s$. On the other hand, there exists a minus dominating function with $p - m = s$: Choose $P = S$ and $M = \emptyset$. Note that S is also a dominating set, hence $\gamma \leq \gamma^-$, and the reverse inequality holds, trivially, in any graph [5]. So all three invariants are equal. Since this consideration is true for any IPDS, the assertion follows completely. ◊

Note that domination is the hitting set problem in the hypergraph whose hyperedges are the vertex neighboroods in a graph, or equivalently, the covering problem in the dual hypergraph. So it should be attractive to investigate the minus counterpart of the covering problem in general hypergraphs and in hypergraphs of bounded rank.

References

1. P.Berman, T.Fujito: On approximation properties of the independent set problem for degree 3 graphs, *4th WADS'95, LNCS* 955 (Springer), 449-460
2. G.J.Chang, C.Pandu Rangan, S.R.Coorg: Weighted independent perfect domination on cocomparability graphs, *4th ISAAC'93*, LNCS 762 (Springer), 506-515
3. J.Dunbar, W.Goddard, S.T.Hedetniemi, M.A.Henning, A.McRae: The algorithmic complexity of minus domination in graphs, *Discrete Applied Math.* 68 (1996), 73-84
4. J.Dunbar, S.T.Hedetniemi, M.A.Henning, A.McRae: Minus domination in regular graphs, *Discrete Math.* 149 (1996), 311-312
5. J.Dunbar, S.T.Hedetniemi, M.A.Henning, A.McRae: Minus domination in graphs, submitted to *Discrete Math.*
6. J.Dunbar, S.T.Hedetniemi, M.A.Henning, P.J.Slater: Signed domination in graphs, in: *Graph Theory, Combinatorics and Applications*, Wiley 1995, 311-322
7. O.Favaron: Signed domination in regular graphs, *Discrete Math.* 158 (1996), 287-293

8. M.R.Garey, D.S.Johnson: *Computers and Intractability – a Guide to the Theory of NP-Completeness*, Freeman 1979

9. P.Hell, D.G.Kirkpatrick: Algorithms for degree constrained graph factors of minimum deficiency, J. Algorithms 14 (1993), 115-138

10. A.Natanzon, R.Shamir, R.Sharon: A polynomial approximation algorithm for the minimum fill-in problem, *30th ACM STOC'98*, 41-47

11. C.H.Papadimitriou, M.Yannakakis: Optimization, approximation, and complexity classes, *J. Comp. System Sc.* 43 (1991), 425-440

The Vertex-Disjoint Triangles Problem

Venkatesan Guruswami[1] *, C. Pandu Rangan[1] **, M. S. Chang[2] ***,
G. J. Chang[3], and C. K. Wong[4]

[1] Dept. of Computer Science & Engg., Indian Institute of Technology, Madras-600
036, INDIA.
[2] Dept. of Comp. Sci. and Information Engg., National Chung Cheng University,
Ming-Hsiun, Chiayi 621, Taiwan, Republic of China.
[3] Dept. of Computer Sci., National Chio-Tung University, Taiwan, Republic of China.
[4] Dept. of Computer Sci. and Engg, Chinese University of Hong Kong, Hong Kong.

Abstract. The vertex-disjoint triangles (VDT) problem asks for a set of
maximum number of pairwise vertex-disjoint triangles in a given graph
G. The triangle cover problem asks for the existence of a *perfect* triangle
packing in a graph G. It is known that the triangle cover problem is NP-
complete on general graphs with clique number 3 [6]. The VDT problem
is MAX SNP-hard on graphs with maximum degree four, while it can be
approximated within $3/2 + \epsilon$, for any $\epsilon > 0$, in polynomial time [11].
We prove that the VDT problem is NP-complete even when the input
graphs are chordal, planar, line or total graphs. We present an $O(m\sqrt{n})$
algorithm for the VDT problem on split graphs and an $O(n^3)$ algorithm
for the VDT problem on cographs. A linear algorithm for the triangle
cover problem on strongly chordal graphs is also presented. Finally, the
notion of packing-hardness, which may be crucial to the understanding
of the *difficulty* of generalized matching problems, is defined.

1 Introduction

The **triangle cover** problem is defined as follows: Given a graph $G = (V, E)$ with
$|V| = 3n$, are there n vertex-disjoint triangles in G ? This problem is known to be
NP-complete on general graphs [6]: another related problem called the Vertex-
disjoint Triangles **(VDT)** problem asks for a set of maximum number of vertex-
disjoint triangles in a given graph G. This problem is also NP-hard (since the
more restricted triangle cover problem is NP-complete), but can be approximated
within $3/2+\epsilon$ for any $\epsilon > 0$ [11]. The problem is MAX SNP-hard even on degree-
bounded graphs [12], but admits a polynomial time approximation scheme on
planar graphs [1] and λ-precision unit disk graphs [10]. There are, however,
apparently no hardness results or exact algorithms known for this problem on
natural restricted families of graphs, and it is the purpose of this paper to initiate
an investigation of such issues.

---

* Current Email: `venkat@lcs.mit.edu`
** Email: `rangan@iitm.ernet.in`
*** Email: `mschang@cs.ccu.edu.tw`

We are attempting to solve a specific instance of a generalized matching problem, namely tripartite matching. The classical matching problem asks for a set of maximum number of independent edges in a given graph. The notion of a matching in a graph not only has a beautiful mathematical theory associated with it, but also has numerous applications in such diverse fields as transversal theory, assignment problems, network flows, multiprocessor scheduling, and the Chinese postman and traveling salesman problems.

Generalization of the *classical* matching problem is motivated by both theoretical and practical constraints and has also motivated a lot of research though most of them have only negative NP-completeness results [9, 13]. As mentioned in [13], the problem seems to be especially relevant in the context of examination scheduling, where in addition to assignment of courses to examinations without any first-order conflicts (essentially a graph coloring problem), there is the additional objective of minimizing second-order conflicts (or inconveniences) like a student writing two examinations on the same day.

It is well known that a maximum matching of a general graph can be found in polynomial time [5, 14], however the generalized matching problem seems to be very difficult [13]. Thus, research in this direction also helps in narrowing down the perceived gap between the classes P and NP with respect to graph packing problems.

One other possible application for the K_m-packing problem (for a fixed m), mentioned in [3] as the *orgy problem*, is that: Given a group of people and the affinities and dislikes between them, is it possible to divide them into groups of m members each such that the persons in each group are all mutually compatible (if such a partition is not possible, find one that leaves the least number of isolated persons). Note that a special case of the K_3-packing (or VDT) problem which we are dealing with in this paper, is the classical *3-dimensional matching problem* (3DM), which given a set of n boys, n girls and n houses and their mutual affinities, asks for the existence of a perfect *marriage* in which each boy lives with a distinct girl in a house acceptable to both of them.

We consider the VDT problem on some interesting classes of graphs. We prove that the vertex-disjoint triangles problem is NP-complete on **chordal** graphs, **planar** graphs, **line** graphs and **total** graphs. A linear algorithm for the triangle cover problem is presented for the class of strongly chordal graphs. We also provide a polynomial time algorithm for the VDT problem when the input graph is restricted to **split** graphs and **cographs**. The algorithm for cographs is based on dynamic programming and runs in $O(n^3)$ time.

We now review some definitions relating to special classes of graphs. A graph is *chordal* if it contains no induced cycle of length greater than three. A graph is *strongly chordal* if it is chordal and every even cycle of length greater than 4 has an odd chord. A strongly chordal graph $G = (V, E)$ has an associated *strong elimination ordering* of vertices as $v_1, v_2, \ldots, v_n$ that satisfies, for each i, $(1 \leq i \leq n)$, $N_i[v_j] \subseteq N_i[v_k]$ whenever $v_j, v_k \in N_i[v_i]$ and $j < k$. ($N_i[v]$ stands for the closed neighborhood of v in the subgraph G_i of G induced by $\{v_i, v_{i+1}, \ldots, v_n\}$). A *cograph* is a graph that has no induced P_4. A graph is said to be *split* if its vertex set can be partitioned into a clique and an independent set. For a comprehensive treatment of these classes of graphs, see [7].

The line graph of an undirected simple graph G, denoted $L(G)$, has vertex set equal to the edge set of G and two vertices in $L(G)$ are adjacent if their corresponding edges in G are incident upon a common vertex of G. The total graph of a graph $G = (V, E)$, denoted $T(G)$, has vertex set $V \cup E$ and two vertices of $T(G)$ are adjacent if the corresponding vertices and edges of G are adjacent. Note that $T(G)$ will have both G and $L(G)$ as induced subgraphs. For all graph-theoretic terms not defined explicitly here, refer [8].

Let us fix some notation that will be used throughout this paper. For a graph G, we use $\alpha(G)$, $\omega(G)$ and $\gamma(G)$ to stand for the size of the largest independent set in G, clique number of G and the size of a minimum vertex-cover of G respectively. Also the parameter that is our main concern in this paper, namely the maximum number of pairwise vertex-disjoint triangles in G, will be denoted by $t(G)$.

2   NP-completeness results

2.1   Chordal Graphs

In the following, we will prove that the VDT problem is NP-hard on chordal graphs[1]. We prove this result by reducing the 3-satisfiability problem to the VDT problem on chordal graphs.

Problem. The 3-satisfiability problem (3SAT).

Instance: Collection $C = \{c_1, c_2, \cdots, c_m\}$ of clauses on a finite set U of variables such that $|c_i| = 3$ for $1 \le i \le m$.

Question: Is there a truth assignment for U that satisfies all the clauses in C?

We will first show how to construct a chordal graph from a 3SAT instance. Given an instance of 3SAT where $C = \{c_1, c_2, \cdots, c_m\}$ and $U = \{u_1, u_2, \cdots, u_n\}$, we will construct a chordal graph.

- For each variable u_i, we construct
  $W_i = \{w[i, j], \bar{w}[i, j] : 1 \le j \le m\}$,
  $X_i = \{x[i, j], \bar{x}[i, j] : 1 \le j \le m\}$,
  $S_i = \{s[i, j] : 1 \le j \le m\}$, and
  $T_i = \{t[i, j] : 1 \le j \le m\}$.
  Let $W = \cup_{i=1}^n W_i$, $X = \cup_{i=1}^n X_i$, $S = \cup_{i=1}^n S_i$, and $T = \cup_{i=1}^n T_i$.
- For each clause c_j, we construct a vertex $y[j]$. Let $Y = \{y[j] : 1 \le j \le m\}$.

We will construct a chordal graph $G_C = (X \cup W \cup K, KK \cup WK \cup XW \cup XY)$ where $K = S \cup T \cup Y$ is a clique, both W and X are independent sets, KK is the set of all edges connecting two vertices of K, XW is the set of edges connecting a vertex in X and a vertex in W and so on. We have shown how to construct S, T, and K. Totally, vertex set K has $2mn + m$ vertices and both vertex sets W

---

[1] After presentation of this paper, we found out about reference [4] where the VDT problem on chordal graphs is mentioned as an open question.

and X have $2mn$ vertices each. Next, we show how to construct XW, XY, and WK.

- Each vertex $w[i,j]$ (resp. $\bar{w}[i,j]$) is connected to a vertex $x[i,j]$ (resp. $\bar{x}[i,j]$). Let
$$XW = \{(x[i,j], w[i,j]), (\bar{x}[i,j], \bar{w}[i,j]) : 1 \le i \le n, 1 \le j \le m\}.$$
- Each vertex $w[i,j]$ is connected to vertex $s[i,j+1]$ if $1 \le j < m$ and each vertex $w[i,m]$ is connected to vertex $s[i,1]$. Each vertex $\bar{w}[i,j]$ is connected to vertex $s[i,j]$. Each vertex $t[i,j]$ is connected to vertices $w[i,j]$ and $\bar{w}[i,j]$. That is, for each variable u_i, we construct the following two sets of edges:
$$WS_i = \{(w[i,j], s[i,j+1]), (\bar{w}[i,j], s[i,j]) : 1 \le j < m\} \cup$$
$$\{(w[i,m], s[i,1]), (\bar{w}[i,m], s[i,m])\} \text{ and,}$$
$$WT_i = \{(t[i,j], w[i,j]), (t[i,j], \bar{w}[i,j]) : 1 \le j \le m\}.$$
  Let $WS = \cup_{i=1}^n WS_i$ and $WT = \cup_{i=1}^n WT_i$.
- For each $u_i \in c_j$ (resp. $\bar{u}_i \in c_j$), $y[j]$ is connected to $w[i,j]$ and $x[i,j]$ (resp. $\bar{w}[i,j]$ and $\bar{x}[i,j]$).
$$WY = \{(w[i,j], y[j]) : u_i \in c_j\} \cup \{(\bar{w}[i,j], y[j]) : \bar{u}_i \in c_j\}, \text{ and}$$
$$XY = \{(x[i,j], y[j]) : u_i \in c_j\} \cup \{(\bar{x}[i,j], y[j]) : \bar{u}_i \in c_j\}.$$
- Let $WK = WS \cup WT \cup WY$.

It is straightforward to verify that $G_C = (X \cup W \cup K, KK \cup WK \cup XY \cup XW)$ is a chordal graph and can be constructed in polynomial time.

Totally, there are $4mn + m(2n+1)$ vertices in G_C. We now state the following lemma and sketch its proof (the complete details of the proof may be found in the full version of the paper).

Lemma 1. *The 3SAT instance has a truth assignment for U that satisfies all clauses in C if and only if G_C has $m(n+1)$ vertex-disjoint triangles.*

Proof: (*Sketch :*) We outline the proof of the "if" part, the other part follows similarly. Suppose G_C has a set T of $m(n+1)$ vertex-disjoint triangles. It can be shown that each triangle in T contains a vertex $y \in Y$ or two vertices in K with one in S and the other in T.

Consider the subgraph $G_C[i]$ induced by $W_i \cup S_i \cup T_i$. The set of triangles T_i in $G_C[i]$ can be partitioned into two sets:
$$T_i^t = \{\{\bar{w}[i,j], s[i,j], t[i,j]\} : 1 \le j \le m\} \text{ and,}$$
$$T_i^f = \{\{w[i,j], s[i,j+1], t[i,j]\} : 1 \le j < m\} \cup \{\{w[i,m], s[i,1], t[i,m]\}\}.$$
It can also be shown that T will have to include exactly m triangles from T_i, either all triangles in T_i^t or all triangles in T_i^f. Thus, in general T specifies a truth assignment for U, with the variable u_i being set true if and only if $T \cap T_i = T_i^t$. It is easy to see that owing to the existence of m triangles in T with one vertex each from Y, W and X, this truth assignment will indeed satisfy all clauses in C. ∎

Using lemma 1 and the fact that G_C can be constructed in polynomial time from the 3SAT instance, we immediately have:

Theorem 1. *The vertex-disjoint triangle problem is NP-complete on chordal graphs.*

2.2   Planar Graphs

Theorem 2. *The vertex-disjoint triangle problem is NP-complete on planar graphs.*

Proof : The reduction is from the the independent set problem on planar cubic graphs [6]. Let $H = (V, E)$ be an arbitrary planar cubic graph. We reduce the problem of determining $\alpha(H)$ to the problem of finding $t(G)$, for a suitably constructed planar graph G.

Let $|V| = n$ and $|E| = m$. First form $H' = (V', E')$ (H' will not be simple) from H as follows: For $1 \leq i \leq m$, insert two new vertices u_i and w_i in the edge e_i of H and add one more edge between u_i and w_i. More precisely, if $V = \{1, 2, \cdots, n\}$ and $E = \{e_1, e_2, \cdots, e_m\}$ where $e_i = (f(i), g(i))$ for $1 \leq i \leq m$ Define $V' = V \cup \{u_i, w_i | 1 \leq i \leq m\}$ and $E_i = \{(f(i), u_i), (w_i, g(i)), p_i, q_i\}$ where p_i, q_i are the two edges between u_i and w_i in H'. Now set $E' = \cup_{i=1}^{m} E_i$.

Clearly H' has no triangle, is planar and is 3-regular. Though, line graph are defined in the strict sense only for simple graphs, we can speak of the line graph of $H' - L(H')$ whose vertices correspond to edges of H' and two vertices in $L(H')$ are adjacent iff the corresponding edges share <u>at least</u> one vertex in common. We set $G = L(H')$. It is not difficult to see that G will also be planar and $\Delta(G) = 4$. Since H' is triangle-free and 3-regular, it is easy to see that $t(G) = \alpha(H')$. From our construction of H', it is possible to verify that $\alpha(H') = \alpha(H) + m$. Hence we have:
$$t(G) = \alpha(H) + m.$$
The conclusion now follows since G is planar with $\Delta = 4$ and can be constructed from H in polynomial time. ∎

2.3   Line Graphs

We now prove that the VDT problem is NP-complete on line graphs (Note that line graphs are defined *only* for simple graphs).

Lemma 2. *If G is a triangle-free 3-regular simple graph and $L(G)$ is its line graph, then $t(L(G)) = \alpha(G)$.*

Theorem 3. *The independent set problem is NP-complete on triangle-free 3-regular graphs.*

Proof (*Sketch :*) The reduction is once again from the independent set problem on planar cubic graphs [6]. Let $H = (V', E')$ be an arbitrary planar cubic graph. We reduce the problem of determining $\alpha(H)$ to the problem of finding $\alpha(G)$ for a suitably constructed triangle-free cubic graph G. Let $V' = \{1, 2, \cdots, n\}$ and $E' = \{e_1, e_2, \cdots, e_m\}$ and $e_i = (f(i), g(i))$ for $1 \leq i \leq m$. Define $G = (V, E)$ as follows: For $1 \leq i \leq m$, set
$U_i = \{u_{ij} | 1 \leq j \leq 8\}$ and
$E_i = \{(f(i), u_{i1}), (g(i), u_{i8})\} \cup \{(u_{ij}, u_{i,j+1}) | 1 \leq j < 8\} \cup \{(u_{i8}, u_{i1})\}$
$\qquad \cup \{(u_{ij}, u_{i,j+3}) | j = 2, 3, 4\}$

Now define: $E = \cup_{i=1}^m E_i$ and $V = V' \cup (\cup_{i=1}^m U_i)$. Note that $|V| = 8m + n$ and $|E| = 13m$.

Informally, in G we attach a 8-cycle with certain specific diagonals instead of each edge of H. From the construction it is clear that G is 3-regular and has no triangle. By our construction of G, it can be shown that $\alpha(G) = \alpha(H) + 4m$ and the result follows since G can clearly be obtained in polynomial time from H. $\blacksquare$

Theorem 4. *The vertex-disjoint triangle problem is NP-complete on line graphs.*

Proof : Follows from lemma 2 and theorem 3. $\blacksquare$

2.4   Total graphs

Recall that the total graph of $G = (V, E)$ is given by $T(G) = (V \cup E, E'')$ where:

$$E'' = E \cup \{(e_1, e_2) | e_1, e_2 \in E \text{ and } e_1, e_2 \text{ are adjacent in G } \}$$
$$\cup \{(v, e) | v \in V, e \in E \text{ and } v \text{ is one of the ends of } e \text{ in } G\}$$

Theorem 5. *The vertex-disjoint triangles problem is NP-complete even when restricted to total graphs.*

Proof: The problem is clearly in NP. To prove NP-completeness, we employ a reduction from the independent set problem on a 3-regular triangle-free graph $H = (V, E)$ which is NP-complete due to theorem 3. Let $V = \{v_1, v_2, \ldots, v_n\}$.

From H obtain a graph H' by attaching a new pendant vertex to each vertex of H. More formally, $H' = (V', E')$ where:
$$V' = V \cup \{u_i | 1 \leq i \leq n\} \text{ and,}$$
$$E' = E \cup \{q_i = (u_i, v_i) | 1 \leq i \leq n\}.$$
Now, let $G = T(H')$ be the total graph of H'. Let P be a set of maximum number of pairwise vertex-disjoint triangles of G. By the nature of the graph H', it is easy to see that P may be modified without reducing its cardinality to include all the triangles T_i of the form (u_i, q_i, v_i) of G for $1 \leq i \leq n$. Since $V' \subset \cup_{i=1}^n T_i$, clearly the optimum way to find the remaining triangles for P, is to find a set of maximum number of vertex-disjoint triangles of $L(H)$. Thus, we have:
$$t(T(H')) = n + t(L(H)) \text{ and using lemma 2, } t(T(H')) = n + \alpha(H).$$
The result now follows since determining $\alpha(H)$ is NP-complete and $G = T(H')$ can be constructed in polynomial time from H. $\blacksquare$

3   Triangle Cover on Strongly Chordal Graphs

In this section we present a linear algorithm for the triangle cover problem on strongly chordal graphs given the strong elimination ordering[2]. Let $v_1, v_2, \cdots, v_n$

---

[2] We found out, after presentation of this paper, that this result also appears in [4] –
in fact a more general algorithm that works for all chordal graphs is presented there.

be the strong elimination ordering of a strongly chordal graph $G = (V, E), |V| = n, |E| = m$. Let $N(v_1) = \{v_{i_1}, \cdots, v_{i_k}\}$ with $1 < i_1 < \cdots < i_k$. The following lemma forms the basis of our algorithm:

Lemma 3. *If G has a triangle cover, then G has a triangle cover comprising the triangle (v_1, v_{i_1}, v_{i_2}).*

Proof : Since G has a triangle cover T, we must have $k \geq 2$ ($k = |N(v_1)|$), let (v_1, v_{i_p}, v_{i_q}) belong to the triangle cover ($p < q$). Then by strong elimination ordering , $N(v_{i_1}) \subseteq N(v_{i_p}), N(v_{i_2}) \subseteq N(v_{i_q})$. So v_{i_p} and v_{i_q} can be substituted for v_{i_1} and v_{i_2} respectively in the triangles in which they occur in T to give another triangle cover T' such that $(v_1, v_{i_1}, v_{i_2}) \in T'$. ∎

Using the above lemma, we devise the following algorithm to find a triangle cover of a strongly chordal graph G if one exists:
Algorithm TC-SCG :
Input : A strongly chordal graph $G = (V, E)$ along with its strong elimination ordering $v_1, v_2, \cdots, v_n$.
Output : Whether or not G has a triangle cover.

1. Initially all $v_i \in V$ are unmarked.
2. While there exists an unmarked vertex in V
   (a) Choose the smallest such vertex, say $v_k \in V$
   (b) If v_k has more than one neighbor v_i with $i > k$ then
       − Let v_p, v_q be the smallest such neighbors
       − Mark v_k, v_p, v_q and continue
       else output that G has no triangle cover.
3. Output that G has a triangle cover.

The proof of correctness follows from lemma 3. It is straightforward to modify the algorithm so that it outputs a triangle cover if one exists. The complexity is clearly linear once the ordering of vertices is available. Hence, we get,

Theorem 6. *For a strongly chordal graph, a triangle cover, if one exists, can be obtained in linear time once the strong elimination ordering of vertices is available.*

4   Vertex-disjoint triangles of Split Graphs

We now prove that if the input graph is restricted to split graphs, the vertex-disjoint triangles problem can be solved in polynomial time. Suppose $G = (V, E)$ is a split graph where V is the *disjoint* union of an independent set S and a clique K. Clearly it is enough to find the maximum number of vertex-disjoint triangles in G such that each triangle has one vertex from S and two from K (the optimal way to find some more triangles all of whose vertices are in K to form the optimal triangle packing is obvious) and so we will consider this problem from now on. Henceforth, we shall denote by $tS(G)$ the maximum number of vertex-disjoint triangles in G each of which has one vertex from S.

Since K induces a clique, the following lemma is obvious (recall that $t(G)$ denotes the maximum number of vertex-disjoint triangles in G):

Lemma 4. *If $G = (V, E)$ is a split graph, then:*

$$t(G) = tS(G) + \left\lfloor \frac{|K| - 2 \times tS(G)}{3} \right\rfloor$$

We now proceed to map the vertex-disjoint problem on a split graph $G = (V, E)$ with $V = S \cup K$ as above, to the *maximum matching* problem on a suitably defined graph H. The construction of the graph H is described below:

For each $u \in S \subseteq V$, add a new vertex u' that is adjacent to all and only vertices to which u is adjacent in G. Also "join" u and u' by an edge. Also make K an independent set in H. Let H be the resulting graph. More formally, $H = (V', E')$ where:

$$V' = V \cup \{u' | u \in S \subseteq V\} \text{ and,}$$
$$E' = \{(u, u') | u \in S\} \cup \{(u, v), (u', v) | u \in S, v \in K \text{ and } (u, v) \in E\}.$$

Lemma 5. *If $m(H)$ denotes the matching number of H, then $tS(G) = m(H) - |S|$.*

Proof: Let M be a maximum matching of H, so that $|M| = m(H)$. A matching M is said to saturate a vertex v, if v is one of the ends of some edge in M. If for some $u \in S \subseteq V$, M saturates u without saturating u' or saturates u' without saturating u, then clearly the matching M may be modified (without changing its cardinality) so that it includes the edge (u, u'). Thus repeating this process we will obtain a matching M' in which for each pair of vertices $\{u, u'\}$ where $u \in S$, either the edge (u, u') is used by M' or both u, u' are saturated by *different* edges of M'.

By our construction, the neighborhoods of u, u' coincide in H, it therefore follows that $tS(G)$ equals the number of pairs (u, u') such that u and u' are saturated by *different* edges of M' and this can be easily seen to be

$$|M'| - |S| = |M| - |S| = m(H) - |S| \quad \blacksquare$$

Theorem 7. *The vertex-disjoint triangles problem can be solved in $O(m\sqrt{n})$ time on split graphs.*

Proof: Clearly it suffices to consider connected split graphs $G = (V, E)$. Based on the proof of lemma 5, we find that a maximum number of pairwise vertex-disjoint triangles in G may be obtained from <u>any</u> maximum matching M of $H = (V', E')$ as follows:

1. For each $u \in S$ such that u and u' are both saturated by *different* edges (u, v_1) and (u', v_2) of M where v_1, v_2 are *distinct* vertices in K, add the triangle (u, v_1, v_2) to the triangle-packing.
2. Pack as many triangles, all of whose three vertices will lie in K, using the vertices of K left unused by step 1.

Clearly the process can be done in linear time, if for each vertex $v \in V(H)$, its neighbor if any in M is stored and presented along with M. Since $|V'| = O(n)$ and $|E'| = O(m + n)$, the entire process may be implemented in $O(m\sqrt{n})$ time [14]. $\blacksquare$

5   Cographs

In this section, we devise a polynomial algorithm for the VDT cover problem on cographs based on dynamic programming. Cographs (or Complement reducible graphs) are graphs with no induced P_4. The class of cographs may also be defined recursively as follows:

- K_1 is a cograph.
- If G_1, G_2 are cographs, then so is $G_1 \cup G_2$ and $G_1 \times G_2$ $((G_1^c \cup G_2^c)^c)$.

Lemma 6. Let $G = G_1 \times G_2, G_1 = (V_1, E_1), G_2 = (V_2, E_2)$ and let T be a set of p vertex-disjoint $K_m s$ in G $(m \geq 2)$. Then there exists T', a set of p vertex-disjoint $K_m s$ such that T' covers the same vertices as does T and T' does not contain two $K_m s$ C_1, C_2 such that $C_1 \cap V_1 = \phi$ and $C_2 \cap V_2 = \phi$.

Lemma 7. If $m = 2$ in lemma 6 above and $|V_1| \leq |V_2|$, then T' can be chosen so that every edge in T' intersects V_2.

Let $G = (V, E)$, $|V| = n$.
Define $f(G, p)$ for $0 \leq p \leq n$ as follows:

$$f(G, p) = \begin{cases} -1 \text{ if } G \text{ does not have } p \text{ vertex-disjoint triangles} \\ q \quad \text{if } G \text{ has } p \text{ vertex-disjoint triangles and a matching of size } q \text{ disjoint} \\ \quad \text{with the } p \text{ triangles } (q \geq 0) \text{ and } q \text{ is the largest such number} \end{cases}$$

5.1   Recurrence equations for $f(G, p)$

We now give recurrence equations for $f(G, p)$ for the two cases $G = G_1 \cup G_2$ and $G = G_1 \times G_2$, $G_i = (V_i, E_i), |V_i| = n_i$ for $i = 1, 2$.
Case 1: $G = G_1 \cup G_2$
Since p triangles in G have to be chosen as i triangles from G_1 and $p - i$ triangles from G_2 for some $0 \leq i \leq p$, we clearly have

$$f(G, p) = \max_{0 \leq i \leq p} (f(G_1, i) + f(G_2, p - i))$$

Case 2: $G = G_1 \times G_2$
This case is more tricky as triangles can have $0, 1$ or 2 vertices from V_1 and V_2. However by lemma 6, we may assume that in choosing the p triangles either no triangle with all three vertices in V_1 is chosen or no triangle with all three vertices in V_2 is chosen, hence following sub-cases arise :
Case 2.1: No triangle with all three vertices in V_1 is chosen.
Let s, r, t denote the number of triangles with $2, 1$ and 0 vertices in V_1 respectively that are chosen with $s + r + t = p$, $s, r, t \geq 0$.
Denote by $\alpha(s, r, t)$ the maximum number of independent edges disjoint with p triangles with s, r, t number of them (the triangles) having $2, 1$ and 0 vertices from V_1 respectively. Let k_1 and k_2 be the number of vertices in V_1, V_2 respectively that are "free" for choosing edges disjoint with the p triangles. Then, $k_1 = n_1 - r - 2s$ and $k_2 = n_2 - s - 2r - 3t$.

By lemma 6 and lemma 7, we may assume that the chosen maximum number of edges comprise exactly $\min\{k_1, k_2\}$ edges from $V_1 \times V_2$. It follows that $\alpha(s, r, t)$ is given as follows:

- $k_1 < 0$ or $k_2 < 0$ or $f(G_2, t) < r$ or $f(G_1, 0) < s$: $\alpha(s, r, t) = -1$
- $f(G_2, t) \geq r, f(G_1, 0) \geq s, 0 \leq k_1 \leq k_2$

$$\alpha(s, r, t) = \begin{cases} k_1 + f(G_2, t) - r & \text{if } 2(f(G_2, t) - r) \leq k_2 - k_1 \\ k_1 + \lfloor \frac{k_2 - k_1}{2} \rfloor & \text{otherwise} \end{cases}$$

- $f(G_2, t) \geq r, f(G_1, 0) \geq s, 0 \leq k_2 < k_1$

$$\alpha(s, r, t) = \begin{cases} k_2 + f(G_1, 0) - s & \text{if } 2(f(G_1, 0) - s) \leq k_1 - k_2 \\ k_2 + \lfloor \frac{k_1 - k_2}{2} \rfloor & \text{otherwise} \end{cases}$$

In order that $\alpha(s, r, t) \geq 0$, the value of r should satisfy the following constraints: $k_1 \geq 0$, $k_2 \geq 0$, $f(G_2, t) \geq r$, and $f(G_1, 0) \geq s = p - t - r$. In other words, since $k_1 = n_1 - r - 2s = n_1 - r - 2(p - t - r) = (n_1 - 2p + 2t) + r$ and $k_2 = n_2 - s - 2r - 3t = n_2 - (p - t - r) - 2r - 3t = (n_2 - p - 2t) - r$, we must have: $\min\{n_2 - p - 2t, f(G_2, t)\} \geq r \geq \max\{-(n_1 - 2p + 2t), p - t - f(G_1, 0)\}$.

Now it is a straight-forward computation to show that $(k_2 - k_1) - 2(f(G_2, t) - r)$, $(k_1 - k_2) - 2(f(G_1, 0) - s)$, $k_1 + f(G_2, t) - r$, $k_2 + f(G_1, 0) - s$, $k_1 + \lfloor \frac{k_2 - k_1}{2} \rfloor$ and $k_2 + \lfloor \frac{k_1 - k_2}{2} \rfloor$ are *all* independent of the value of r (as well as s) and depend only on n_1, n_2, p, t, $f(G_1, 0)$ and $f(G_2, t)$.

Define α'_p to be the maximum number of independent edges that can be chosen such that they are vertex-disjoint with p triangles none of which has all three vertices in V_1. Clearly, we have: $\alpha'_p = \max\{\alpha(s, r, t) | 0 \leq s, r, t \leq n, s + r + t = p\}$.

For a fixed triple (s, r, t), it is clear that $\alpha(s, r, t)$ can be computed in $O(1)$ time, and hence from the above discussions, it is easy to see that α'_p can be computed in $O(n)$ time.

Case 2.2: No triangle with all three vertices in V_2 is chosen.
In this one can define $\beta(s, r, t)$ and β'_p in a manner similar to case 2.1 and set up similar equations. It follows also that β'_p can be computed in $O(n)$ time as well. Clearly we have, for the case of join operation, $f(G, p) = \max\{\alpha'_p, \beta'_p\}$.

5.2   Complexity Analysis

Using the results of section 5.1, the fact that the *parse tree* associated with a cograph G can be obtained in linear time [2] and the fact $t(G) = \max\{p : f(G, p) \geq 0\}$, the following results may be shown:

Lemma 8. *If $G = G_1 \cup G_2$ or $G = G_1 \times G_2$, then given $f(G_1, p)$ and $f(G_2, p)$ for $0 \leq p \leq n$, computation of $f(G, p)$ for $0 \leq p \leq n$ can be done in $O(n^2)$ time.*

Theorem 8. *The maximum number of vertex-disjoint triangles of a cograph can be computed in $O(n^3)$ time.*

6   Conclusions and Open Questions

We have established that though matching is polynomial-time solvable even on general graphs, *tripartite matching* or the vertex-disjoint triangle problem is NP-complete even when restricted to chordal, planar, line and total graphs. Thus, for most natural classes of graphs, packing a graph H (i.e finding vertex-disjoint induced copies of H) in a graph G of that class becomes *difficult* when H has three vertices.

This motivates the definition of the notion of **packing-hardness number** of a class C of graphs as follows: The packing-hardness number of a class C of graphs, denoted $PH(C)$, is the smallest integer n such that there exists a n–vertex graph H such that packing H in graphs of the class C becomes NP-complete. Since matching is polynomial-time solvable, it follows that $PH(C) \geq 3$ for all classes C of graphs. The packing-hardness number is thus an excellent measure of the difficulty of generalized matching problems on any given class of graphs. It is proved in [13] that packing any connected graph with more than two vertices is NP-complete on general graphs and consequently the packing-hardness number of general graphs equals three. Also, it follows from our definition that, if the VDT problem is NP-complete on a class C of graphs, then $PH(C) = 3$. Thus, this paper proves that the packing-hardness number of chordal, planar, line and total graphs is each equal to three. This paper is thus the stepping stone for further research on the complexity of generalized matching problems on special classes of graphs. The determination of the packing-hardness number for several other classes of graphs will be presented as part of a separate document.

The VDT problem itself remains open on several other interesting classes of graphs like permutation, distance-hereditary and comparability graphs. It is easy to prove that the VDT problem is solvable in linear time on complements of bipartite graphs. It however remains open on the super-class of cocomparability graphs.

References

1. B.S.Baker. Approximation algorithms for NP-complete problems on planar graphs. *Journal of the ACM*, 41(1994), 153-180.
2. D.G.Corneil, Y.Perl and L.K.Stewart. A Linear recognition algorithm for cographs. *SIAM Jl. on Computing*, 14(1985), pp 926-934.
3. G.Cornuejols, D.Hartvigsen and W.Pulleyblank. Packing Subgraphs in a Graph. *Operations Research Letters*, 1(1982), pp 139-143.
4. E.Dahlhaus and M.Karpinski. Matching and Multidimensional Matching in Chordal and Strongly Chordal Graphs, *Discerete Applied Math.*, (84)1998, pp 79-91.
5. J.Edmonds. Paths, trees and flowers. *Canadian J. Math.*, 17(1965), pp 449-469.
6. M.R.Garey and D.S.Johnson. Computers and Intractability: A guide to the theory of NP-completeness. *Freeman, San Francisco*, 1979.
7. M.C.Golumbic. Algorithmic graph theory and Perfect graphs. *Academic Press, New York*, 1980.

8. F.Harary. Graph Theory. *Addison-Wesley, Reading, MA*, 1969.

9. P.Hell and D.G.Kirkpatrick. On generalized matching problems. *Info. Proc. Letters*, 12(1981), pp 33-35.

10. H.B.Hunt III, M.V.Marathe, V.Radhakrishnan, S.S.Ravi, D.J.Rosenkrantz and R.E.Stearns. A unified approach to approximation schemes for NP- and PSPACE-hard problems for geometric graphs. *Proc. 2nd Ann. European Symp. on Algorithms (1994)*, Lecture Notes in Comput. Sci. 855, Springer-Verlag, 424-435.

11. C.A.J.Hurkens and A.Schrijver. On the size of systems of sets every t of which have an SDR, with an application to the worst-case ratio of heuristics for packing problems. *SIAM J. Discrete Mathematics*, 2 (1989), 68-72.

12. V.Kann. Maximum bounded 3-dimensional matching is MAX SNP-complete. *Information Processing Letters*, 37(1991), pp 27-35.

13. D.G.Kirkpatrick and P.Hell. On the complexity of general graph factor problems. *SIAM Jl. on Computing*, 12(1983), pp 601-609.

14. S.Micali and V.V.Vazirani. An $O(\sqrt{|V|}|E|)$ algorithm for finding maximum matching in general graphs. *Proc. 21st Annual Symposium on the foundation of Comp. Sci.*, (1980), pp 17-27.

Communication in the Two-Way Listen-in Vertex-disjoint Paths Mode
(Extended Abstract)

Hans-Joachim Böckenhauer

Lehrstuhl für Informatik I, RWTH Aachen, Ahornstr. 55, 52056 Aachen, Germany,
hjb@i1.informatik.rwth-aachen.de

Abstract. A new communication mode for the dissemination of information among processors of interconnection networks via vertex-disjoint paths is introduced and investigated. In this communication mode, in one communication step two processors communicating via a path P send their pieces of information to all other processors on this path, too. The complexity of a communication algorithm is measured by the number of communication steps (rounds).

The main results are optimal (or nearly optimal, up to one round) broadcast, accumulation, and gossip algorithms for paths, cycles, complete graphs, two-dimensional grids, hypercubes and hypercube-like networks.

1 Introduction and Definitions

The study and the comparison of the computational power of distinct interconnection networks as candidates for the use as parallel architectures for existing parallel computers is an intensively investigated research branch of current theory of parallel computing. One of the fundamental approaches helping to search for the best (most effective) structures of interconnection networks is the study of the communication facilities of networks (i.e., of the complexity (effectivity) of solving fundamental communication tasks of information dissemination).

Some of the basic communication tasks are broadcast, accumulation and gossip (an overview of the study of their complexity according to one-way and two-way communication modes can be found in [6, 8, 9]).

Broadcast, accumulation, and *gossip* can be described as follows. Assume that each vertex (processor) x in a graph (network) G has some piece of information $I(x)$. The *cumulative message* $I(G)$ of G is the set of all pieces of information originally distributed in all vertices of G. To solve the *broadcast [accumulation]* problem for a given graph G and a vertex u of G, we have to find a communication strategy (using the edges of G as communication links) such that all vertices in G learn the piece of information residing in u [that u learns the cumulative message of G]. To solve the *gossip* problem for a given graph G, a communication strategy such that all vertices in G learn the cumulative message of G must be found. Since the above stated communication problems are solvable only in connected graphs, we note that from now on we use the notion "graph" for connected undirected graphs.

The meaning of a "communication strategy" depends on the communication mode. A communication strategy is realized by a *communication algorithm* consisting of a number of *communication steps (rounds)*. The rules describing what can happen in one communication step (round) are defined exactly by the communication mode. In this paper we consider the *two-way listen-in vertex-disjoint paths mode (2LVDP mode)*. In this mode one round can be described as a set $P = \{P_1, \ldots, P_k\}$ of vertex-disjoint (simple) paths, where $P_i = x_{i,1}, \ldots, x_{i,\ell_i}$. In this round the two endpoints $x_{i,1}$ and x_{i,ℓ_i} communicate with each other via the path P_i and send there complete information to all other nodes on this path, i. e. after this round all nodes on the path P_i know the complete information of $x_{i,1}$ and x_{i,ℓ_i}.

This communication mode was not considered in the literature before (as far as we know), but several other disjoint paths modes were introduced and investigated in [4, 5, 10–13, 15, 14].

The complexity of a communication algorithm $A = A_1, \ldots, A_r$, denoted by $\text{com}(A)$, is the number r of rounds of A. A communication algorithm that solves the broadcast [accumulation] problem for a graph G and a vertex $v \in V(G)$ is called a *broadcast [accumulation] algorithm* for G and v. A communication algorithm that solves the gossip problem for a graph G is called a *gossip algorithm* for G.

For a graph G and a vertex $v \in V(G)$ the *broadcast complexity for v and G in the 2LVDP mode* is defined as

$$B^{2lv}(v, G) := \min\{\text{com}(A) \mid A \text{ is a broadcast algorithm in the 2LVDP mode}$$
$$\text{for } v \text{ and } G\}.$$

Furthermore the *broadcast complexity for G in the 2LVDP mode* is defined as

$$B^{2lv}(G) := \max\{B^{2lv}(v, G) \mid v \in V(G)\},$$

and the *minimum broadcast complexity for G in the 2LVDP mode* is defined as

$$B^{2lv}_{\min}(G) := \min\{B^{2lv}(v, G) \mid v \in V(G)\}.$$

The *accumulation complexity for v and G in the 2LVDP mode* is defined as

$$A^{2lv}(v, G) := \min\{\text{com}(A) \mid A \text{ is an accumulation algorithm in the 2LVDP}$$
$$\text{mode for } v \text{ and } G\}.$$

Furthermore the *accumulation complexity for G in the 2LVDP mode* is defined as

$$A^{2lv}(G) := \max\{A^{2lv}(v, G) \mid v \in V(G)\},$$

and the *minimum accumulation complexity for G in the 2LVDP mode* is defined as

$$A^{2lv}_{\min}(G) := \min\{A^{2lv}(v, G) \mid v \in V(G)\}.$$

The *gossip complexity for G in the 2LVDP mode* is defined as

$$R^{2lv}(G) := \min\{\text{com}(A) \mid A \text{ is a gossip algorithm in the 2LVDP mode for } G\}.$$

For a detailed analysis of the communication algorithms we need the following notations:

Let G be a graph, let $A = A_1, \ldots, A_r$ be a communication algorithm for G with r rounds. For any $x \in V(G)$, $0 \leq i \leq r$, we define $I_i(x)$ as the set of pieces of information that the vertex x knows after i rounds of A. Particularly, $I_0(x) = I(x)$ holds. Furthermore we define $I_i(M) := \bigcup_{x \in M} I_i(x)$ for any $M \subseteq V(G)$, $0 \leq i \leq r$. A node that knows the cumulative message $I(G)$ after k rounds of A is called an *accumulation point* or *accumulation node* of G after k rounds of A. A set of nodes M with $\bigcup_{x \in M} I_k(x) = I(G)$ is called a *cumulative set* of G after k rounds of A.

This extended abstract is organized as follows: In Section 2 we will prove strict (or nearly strict) lower and upper general bounds on the broadcast, accumulation, and gossip complexity. In Section 3 we will present without proofs optimal (or nearly optimal) bounds on the broadcast, accumulation, and gossip complexity for several classes of interconnection networks.

2   General Bounds

In this section we will prove some general bounds on the communication in the 2LVDP mode which hold for any graph. First, we show the following technical lemma:

Lemma 1. *Let $G = (V, E)$ be a graph, and let A be a communication algorithm for G in the 2LVDP mode. For any $x \in V$ and $k \geq 0$ the following holds:*

a) $|I_{k+1}(x)| \leq |I_k(x)| + 2 \cdot \max\{|I_k(v)| \mid v \in V\}$,
b) $|I_k(x)| \leq 3^k$.

Proof. **a):** Since the active paths in one round of A are vertex-disjoint, every node can receive at most two messages in one round, namely from the two endpoints of the active path. Now we consider an active path $x_1, \ldots, x_\ell$ in round $k+1$ with $x = x_i$ for some $i \in \{1, \ldots, \ell\}$. Then $I_{k+1}(x) = I_k(x) \cup I_k(x_1) \cup I_k(x_\ell)$ holds. This implies

$$|I_{k+1}(x)| \leq |I_k(x)| + |I_k(x_1)| + |I_k(x_\ell)|$$
$$\leq |I_k(x)| + 2 \cdot \max\{|I_k(v)| \mid v \in V\}.$$

b): This follows from a) with induction over k, since $|I_0(v)| = 1$ holds for all $v \in V$. $\square$

Observation 1. *For any $n \geq 3$ and any graph $G = (V, E)$ with n vertices:*

a) $1 \leq B^{2lv}(G) \leq n - 1$,

b) $1 \leq B_{\min}^{2lv}(G) \leq n - 2$.

Proof. The lower bounds in a) and b) are obvious.
There exists an algorithm for broadcast from a vertex x that needs at most $n - 1$ rounds: In every round x sends its information to another still uninformed vertex. Thus, $B^{2lv}(G) \leq n - 1$.
It remains to show that $B_{\min}^{2lv}(G) \leq n - 2$. There exists at least one path $P = x, y, z$ of length 2 with different x, y, z in G. Thus, an optimal algorithm for broadcasting from the node x needs at most $n - 2$ rounds: In the first round x and z communicate, after this round y also knows the information. In the other $n - 3$ rounds the vertex x sends its information to the other $n - 3$ nodes of G. □

Observation 2. *The bounds of Observation 1 are strict.*

Proof. For the lower bounds we consider the cycle C_n with n nodes. C_n is defined by $V(C_n) := \{0, \ldots, n-1\}$ and $E(C_n) := \{\{i, i+1\} \mid 0 \leq i < n-1\} \cup \{\{n-1, 0\}\}$. $B^{2lv}(C_n) = B_{\min}^{2lv}(C_n)$ holds, since C_n is vertex-symmetric. Thus, it suffices to consider broadcasting from the node 0: One communication on the path $0, 1, \ldots, n - 1$ obviously solves the broadcast problem for 0 and C_n.
For the upper bounds we consider the star S_n with n nodes. S_n is defined by $V(S_n) := \{0, 1, \ldots, n - 1\}$ and $E(S_n) := \{\{0, i\} \mid 1 \leq i \leq n - 1\}$. There are no vertex-disjoint paths in S_n. Thus, there exists only one active path in each round, and in each round of a broadcast algorithm only one leaf of S_n can learn the message. Thus, for broadcasting from a leaf v at least $n - 2$ rounds are needed to inform the other $n - 2$ leaves, and for broadcasting from the center node at least $n - 1$ rounds are necessary to inform the $n - 1$ leaves. Thus, $B^{2lv}(S_n) \geq n - 1$ and $B_{\min}^{2lv}(S_n) \geq n - 2$. □

Theorem 1. *For $n \geq 2$ and for any graph $G = (V, E)$ with n nodes:*

a) $\lceil \log_3 n \rceil \leq A_{\min}^{2lv}(G) \leq \lfloor \frac{n}{2} \rfloor$,
b) $\lceil \log_3 n \rceil \leq A^{2lv}(G) \leq \lfloor \frac{n}{2} \rfloor + 1$.

Proof. The lower bounds follow directly from Lemma 1.
a): For the proof of $A_{\min}^{2lv}(G) \leq \lfloor \frac{n}{2} \rfloor$ we consider a spanning tree T of G. It suffices to show that $A_{\min}^{2lv}(T) \leq \lfloor \frac{n}{2} \rfloor$ holds. To prove this, we consider the following communication algorithm A for T:

1. $\tilde{T} := T$
2. If $|V(\tilde{T})| \leq 1$, then stop.
3. Choose two different leaves u and v of $\tilde{T}$
4. Communicate between u and v.
5. Remove u and v from $\tilde{T}$ and continue with 2.

It remains to show that A is an accumulation algorithm for T with $\lfloor \frac{n}{2} \rfloor$ rounds: In Step 5 of A the tree $\tilde{T}$ is reduced by exactly two nodes in every loop. The algorithm stops, if $|V(\tilde{T})| \leq 1$ holds. Thus, Step 4 is executed $\lfloor \frac{n}{2} \rfloor$ times, and therefore A is a communication algorithm with $\lfloor \frac{n}{2} \rfloor$ rounds.

Furthermore, the nodes of $\tilde{T}$ form a cumulative set after every loop iteration. If $|V(\tilde{T})| = 2$ in Step 4, then u and v are accumulation nodes after Step 4. If $|V(\tilde{T})| = 3$ holds with $V(\tilde{T}) = \{u, v, x\}$ in Step 4, then u and v communicate with each other via x. Thus, x is an accumulation node after Step 4. One of these two cases will occur since $|V(\tilde{T})|$ is reduced by exactly 2 in each loop. Thus, A is an accumulation algorithm for T and also for G.

b): $A^{2lv}(G) \leq A_{min}^{2lv}(G) + 1$ is obvious: To accumulate in any vertex v accumulate in a vertex x with $A^{2lv}(x, G) = A_{min}^{2lv}(G)$ and communicate in one additional round between v and x.   □

Observation 3. *The bounds from Theorem 1 are strict.*

Proof. The strictness of the lower bounds follows directly from the results for the complete graph, as presented in Subsection 3.1.

For the strictness of the upper bound from Theorem 1 a) we consider the star S_n. The size of a minimal cumulative set can be reduced at most by 2 in each round. Thus, at least $\lceil \frac{n-1}{2} \rceil = \lfloor \frac{n}{2} \rfloor$ rounds are necessary to obtain an accumulation node.

For the strictness of the upper bound from Theorem 1 b) we consider the accumulation in a leaf v of S_n for any odd $n \geq 3$. After $\frac{n-3}{2}$ rounds there exist at least two vertices $x, y \neq v$ that have not sent their information. It obviously takes at least two rounds to send $I(x)$ and $I(y)$ to v. Thus, $A^{2lv}(v, S_n) \geq \frac{n-3}{2} + 2 = \lfloor \frac{n}{2} \rfloor + 1$.   □

Theorem 2. *For any $n \geq 2$ and for any graph $G = (V, E)$ with n nodes:*

$$\lceil \log_3 n \rceil \leq R^{2lv}(G) \leq n + \left\lfloor \frac{n}{2} \right\rfloor - 1.$$

Proof. $R^{2lv}(G) \geq \lceil \log_3 n \rceil$ is obvious, since gossip is at least as hard as accumulation.

The upper bound follows from the gossip algorithm consisting of accumulation in a vertex v and broadcasting from v. Thus, $R^{2lv}(G) \leq A_{min}^{2lv}(G) + B^{2lv}(G) \leq \lfloor \frac{n}{2} \rfloor + n - 1$.   □

Observation 4. *The lower bound from Theorem 2 is strict and the upper bound from Theorem 2 is almost (i. e. up to one round) strict.*

Proof. The strictness of the lower bound follows directly from the results for the complete graph, as presented in Subsection 3.1.

For the strictness of the upper bound we consider the star S_n for any odd $n \geq 3$. Let A be a gossip algorithm for S_n. In each round of A at most two nodes send their information. Thus, there exists a vertex $x \in V(S_n)$ that has not sent in the first $\frac{n-1}{2} = \lfloor \frac{n}{2} \rfloor$ rounds. Thus, $I(x)$ has still to be sent to all other nodes after $\lfloor \frac{n}{2} \rfloor$ rounds. According to Observation 1, this needs at least $n - 2$ additional rounds. Thus, $R^{2lv}(S_n) \geq n + \lfloor \frac{n}{2} \rfloor - 2$.   □

We conjecture that it could be possible to improve the general upper bound on the gossip complexity by 1, but we are not able to prove this. Thus, we leave it as an open problem here.

3   Communication in Several Classes of Networks

In this chapter we will present our results for several classes of networks. Due to space limitations most of the proofs are omitted in this extended abstract.

3.1   Communication in Complete Graphs

In this section we will determine the exact broadcast, accumulation, and gossip complexity in the 2LVDP mode for the complete graph.

The complete graph with n nodes K_n is defined by $V(K_n) := \{1, \ldots, n\}$ and $E(K_n) := \{\{i, j\} \mid i, j \in \{1, \ldots, n\}, i \neq j\}$.

Theorem 3. *For any $n \geq 2$:*

a) $B^{2lv}(K_n) = B^{2lv}_{min}(K_n) = 1$,

b) $A^{2lv}(K_n) = A^{2lv}_{min}(K_n) = \lceil \log_3 n \rceil$,

c) $R^{2lv}(K_n) = \begin{cases} \lceil \log_3 n \rceil & \text{if some } m \in \mathbb{N} \text{ exists with } 3^m < n \leq 2 \cdot 3^m \\ \lceil \log_3 n \rceil + 1 & \text{if some } m \in \mathbb{N} \text{ exists with } 2 \cdot 3^m < n \leq 3^{m+1} \end{cases}$

Sketch of the Proof.
a): Broadcast in K_n can be done by one communication on a Hamiltonian cycle of K_n.
b): A cumulative set S of $3 \cdot m$ nodes after round i can be reduced to a cumulative set of m nodes after round $i + 1$ by communicating on m vertex-disjoint paths, each consisting of three nodes of S.
c): The lower bounds follow from b) and Lemma 1. The upper bound in the first case can be shown using the following gossip algorithm:

1. Divide $V(K_n)$ into two disjoint subsets V_1, V_2 with $|V_1|, |V_2| \leq 3^m$.
2. Accumulate in the subgraph induced by V_i in m rounds in a vertex v_i for $i \in \{1, 2\}$.
3. Communicate in one round on a Hamiltonian path of K_n with the endpoints v_1 and v_2.

$\square$

3.2   Communication in the Path

In this section we will determine the exact broadcast, accumulation, and gossip complexity of the path P_n. We will see that even in this simple network the communication tasks can be solved nearly optimally.

The path with n nodes P_n is defined by $V(P_n) := \{1, \ldots, n\}$ and $E(P_n) := \{\{i, i+1\} \mid 1 \leq i < n\}$.

Theorem 4. *For any $n \geq 3$:*

a) $B^{2lv}(P_n) = 2$,

b) $B^{2lv}_{\min}(P_n) = 1$,

c) $A^{2lv}_{\min}(P_n) = \lceil \log_3 n \rceil$,

d) $A^{2lv}(P_n) = \begin{cases} \lceil \log_3 n \rceil & \text{if some } m \in \mathbb{N} \text{ exists with } 3^m < n \leq \frac{3^{m+1}+1}{2} \\ \lceil \log_3 n \rceil + 1 & \text{if some } m \in \mathbb{N} \text{ exists with } \frac{3^{m+1}+1}{2} < n \leq 3^{m+1} \end{cases}$

e) $R^{2lv}(P_n) = \begin{cases} \lceil \log_3 n \rceil & \text{if some } m \in \mathbb{N} \text{ exists with } n = 3^m + 1 \\ \lceil \log_3 n \rceil + 1 & \text{if some } m \in \mathbb{N} \text{ exists with } 3^m + 1 < n \leq 3^{m+1} \end{cases}$

Proof.

a), b): Proof omitted.

c): $A^{2lv}_{\min}(P_n) \geq \lceil \log_3 n \rceil$ follows directly from Theorem 1 a).
We prove $A^{2lv}_{\min}(P_n) \leq \lceil \log_3 n \rceil$ by induction over $m := \lceil \log_3 n \rceil$.
The claim obviously holds for $m = 1$, since $m = 1$ implies $n = 2$ or $n = 3$.
As induction hypothesis, let $A^{2lv}_{\min}(P_n) \leq \lceil \log_3 n \rceil$ for all $n \leq 3^m$.
For the induction step let $3^m < n \leq 3^{m+1}$. This implies $\lceil \log_3 n \rceil = m + 1$.
Let $\text{path}(i,j) := (\{i, \ldots, j\}, \{\{k, k+1\} \mid i \leq k < j\})$ denote the subpath of P_n
containing exactly the vertices $i, \ldots, j$ for any $i, j \in \{1, \ldots, m\}$ with $i \leq j$.
If $3^m < n \leq 2 \cdot 3^m$, then divide P_n into the two subpaths $P^L := \text{path}(1, 3^m)$
and $P^R := \text{path}(3^m + 1, n)$. By induction hypothesis there exist vertices $x \in$
$V(P^L)$ and $y \in V(P^R)$ with $A^{2lv}(x, P^L) = A^{2lv}_{\min}(P^L) \leq m$ and $A^{2lv}(y, P^R) =$
$A^{2lv}_{\min}(P^R) \leq m$. Accumulate in m rounds in P^L in the vertex x and in P^R in
the vertex y and communicate in one additional round between x and y. Then
x, y, and all vertices between x and y are accumulation points of P_n after $m + 1$
rounds.
If $2 \cdot 3^m < n \leq 3^{m+1}$, then divide P_n into the three subpaths $P^L := \text{path}(1, 3^m)$,
$P^M := \text{path}(3^m + 1, 2 \cdot 3^m)$, and $P^R := \text{path}(2 \cdot 3^m + 1, n)$. By induction hypothesis
there exist vertices $x \in V(P^L)$, $y \in V(P^M)$, and $z \in V(P^R)$ with $A^{2lv}(x, P^L) =$
$A^{2lv}_{\min}(P^L) \leq m$, $A^{2lv}(y, P^M) = A^{2lv}_{\min}(P^M) \leq m$, and $A^{2lv}(z, P^R) = A^{2lv}_{\min}(P^R) \leq$
m. Accumulate in m rounds in P^L in the vertex x, in P^M in the vertex y, and
in P^R in the vertex z and communicate in one additional round between x and
z. Then y is an accumulation point of P_n after $m + 1$ rounds.

d): Proof omitted.

e): If $n = 3^m + 1$ for some $m \in \mathbb{N}$, then $R^{2lv}(P_n) \geq \lceil \log_3 n \rceil$ follows directly
from Theorem 2. For the proof of $R^{2lv}(P_n) \leq \lceil \log_3 n \rceil$ we consider the following
gossip algorithm: Divide the path P_n into the two subpaths $P := \text{path}(1, \frac{n}{2})$ and
$P' := \text{path}(\frac{n}{2} + 1, n)$, and accumulate in these subpaths in the endpoints 1 and
n. Since P and P' have $\frac{n}{2} = \frac{3^m+1}{2}$ vertices each, this accumulation is possible in
$\lceil \log_3 \frac{n}{2} \rceil = m$ rounds. One additional communication between the nodes 1 and
n completes the gossip. Thus, $R^{2lv}(P_n) \leq m + 1 = \lceil \log_3 n \rceil$.
If $3^m + 1 < n \leq 3^{m+1}$ for some $m \in \mathbb{N}$, first, we prove by contradiction that
$R^{2lv}(P_n) \geq m + 2 = \lceil \log_3 n \rceil + 1$ holds:
Suppose that $R^{2lv}(P_n) \leq m + 1 = \lceil \log_3 n \rceil$. Theorem 4 c) implies, that $A^{2lv}_{\min}(P_n)$
$= m + 1$. Thus, there exists no accumulation point after m rounds. This implies

that every node lies on an active path in round $m + 1$. We show the following claim:

There exists only one active path in round $m + 1$ with the nodes 1 and n as senders. (1)

Proof of the claim (1): We prove (1) by contradiction: Suppose that the two endpoints of P_n lie on two different active paths in round $m + 1$. Let this be path$(1, x)$ and path(y, n). Let $P_x := \text{path}(x, n)$ and $P_y := \text{path}(1, y)$. This situation is shown in Fig.1.

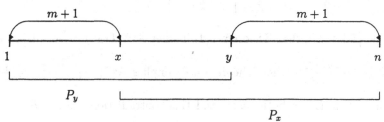

Figure 1. The paths in the proof of (*) in Theorem 4 e)

The vertices 1 and x, and also the vertices y and n, form a cumulative set of P_n after m rounds. Thus, x is an accumulation point of P_x after m rounds, since 1 and x are a cumulative set of P_n after m rounds, and every information from P_x that is known to 1 has been sent to 1 via x. With the same arguments, y is an accumulation point of P_y after m rounds. But y is also an accumulation point of $P_x \setminus P_y$ since every information from $P_x \setminus P_y$ has been sent to x via y. Thus, y is an accumulation point of P_n after m rounds in contradiction to Theorem 4 c). This completes the proof of (1).

From (1) we know that the two endpoints of P_n form a cumulative set after m rounds. This implies that one of these endpoints knows at least $\lceil \frac{n}{2} \rceil$ pieces of information. W. l. o. g. we assume that 1 is this endpoint.

Let A be an accumulation algorithm for P_n s. t. the vertex 1 knows $\ell \geq \lceil \frac{n}{2} \rceil$ pieces of information after m rounds of A, namely $I(i_1), I(i_2), \ldots, I(i_\ell)$ for some nodes $i_1 < i_2 < \ldots < i_\ell$.

For all $1 \leq j \leq \ell$ there is a vertex $v \in \{i_1, i_2, \ldots, i_\ell\}$ that sends the information $I(i_j)$ to the vertex 1. Note that if 1 would get the information $I(i_j)$ from a vertex $u \notin \{i_1, i_2, \ldots, i_\ell\}$ then it would get the information $I(u)$ at the same time. This contradicts $u \notin \{i_1, i_2, \ldots, i_\ell\}$.

Thus, for the accumulation in the vertex 1 it suffices to consider those active paths that contain an endpoint in $\{i_1, i_2, \ldots, i_\ell\}$. We can construct an accumulation algorithm B for $G = (\{i_1, i_2, \ldots, i_\ell\}, \{\{i_j, i_{j+1}\} \mid 1 \leq j < \ell\})$ and the vertex $1 = i_1$ from the algorithm A. Obviously, G is isomorphic to P_ℓ. Thus, there exists an accumulation algorithm for P_ℓ and 1 with m rounds. In the same way we can construct an accumulation algorithm for $P_{\lceil \frac{n}{2} \rceil}$ and 1 from B. Thus, we get

$$A^{2lv}(1, P_{\lceil \frac{n}{2} \rceil}) \leq m. \tag{2}$$

Note that a construction as above is necessary to obtain (1) from the fact $I_m(1) \geq \lceil \frac{n}{2} \rceil$, since in general $A^{2lv}(G_0) \leq A^{2lv}(G)$ does not hold for every induced subgraph G_0 of a graph G.

Together with the fact $A^{2lv}(1, P_n) = A^{2lv}(n, P_n) = A^{2lv}(P_n)$ this implies $A^{2lv}(P_{\lceil \frac{n}{2} \rceil}) \leq m$.

But $3^m + 2 \leq n \leq 3^{m+1}$ holds and thus:

$$\frac{3^m + 3}{2} = \left\lceil \frac{3^m + 2}{2} \right\rceil \leq \left\lceil \frac{n}{2} \right\rceil \leq \left\lceil \frac{3^{m+1}}{2} \right\rceil = \frac{3^{m+1} + 1}{2}.$$

If $\frac{3^m + 3}{2} \leq \lceil \frac{n}{2} \rceil \leq 3^m$, then Theorem 4 d) implies $A^{2lv}(P_{\lceil \frac{n}{2} \rceil}) = \lceil \log_3 \lceil \frac{n}{2} \rceil \rceil + 1 = m + 1$.

If $3^m < \lceil \frac{n}{2} \rceil \leq \frac{3^{m+1}+1}{2}$, then Theorem 4 d) implies $A^{2lv}(P_{\lceil \frac{n}{2} \rceil}) = \lceil \log_3 \lceil \frac{n}{2} \rceil \rceil = m + 1$.

This contradicts (2) in both cases, and this contradiction implies $R^{2lv}(P_n) \geq m + 2$.

It remains to show that $R^{2lv}(P_n) \leq m + 2$.

If $3^m + 1 < n \leq \frac{3^{m+1}+1}{2}$, we know from Theorem 4 d), that $A^{2lv}(P_n) = m+1$. This implies $R^{2lv}(P_n) \leq m + 2$, since there exists a gossip algorithm that accumulates in one endpoint and communicates between the two endpoints in one additional round.

If $\frac{3^{m+1}+1}{2} < n \leq 3^{m+1}$, we consider the following gossip algorithm: Divide P_n into the two subpaths $P := \text{path}(1, \lfloor \frac{n}{2} \rfloor)$ and $P' := \text{path}(\lfloor \frac{n}{2} \rfloor + 1, n)$, and accumulate in these subpaths in the nodes 1 and n. This can be done in $m + 1$ rounds, according to Theorem 4 d), since P and P' have at most $\lceil \frac{n}{2} \rceil \leq \frac{3^{m+1}+1}{2}$ nodes each. Then communicate in one additional round between 1 and n. This algorithm implies $R^{2lv}(P_n) \leq m + 2$. □

3.3   Communication in the Cycle

In this section we determine the exact broadcast, accumulation, and gossip complexity of the cycle C_n. The broadcast and accumulation complexity of C_n follow directly from the results for the path.

Theorem 5. *For any $n \geq 3$:*

a) $B_{min}^{2lv}(C_n) = B^{2lv}(C_n) = 1,$

b) $A_{min}^{2lv}(C_n) = A^{2lv}(C_n) = \lceil \log_3 n \rceil,$

c) $R^{2lv}(C_n) = \begin{cases} \lceil \log_3 n \rceil & \text{if some } m \in \mathbb{N} \text{ exists with } 3^m < n < 2 \cdot 3^m - 1 \\ \lceil \log_3 n \rceil + 1 & \text{if some } m \in \mathbb{N} \text{ ex. with } 2 \cdot 3^m - 1 \leq n \leq 3^{m+1} \end{cases}$

Proof. Proof omitted.

3.4   Communication in some Hypercube-like Networks

In this section we will use the results for the cycle to determine the broadcast, accumulation, and gossip complexity for the hypercube network H_k, the cube-connected-cycles network CCC_k, the butterfly network BF_k, the shuffle-exchange network SE_k, and the DeBruijn network DB_k.
The formal definitions of these networks and a discussion of their properties can be found in [9, 14, 16].

Theorem 6. *For any $k \geq 2$ and for $X_k \in \{H_k, CCC_k, BF_k, DB_k\}$ and for $n := |V(X_k)|$:*

a) $B^{2lv}(X_k) = B^{2lv}_{\min}(X_k) = 1$,

b) $A^{2lv}(X_k) = A^{2lv}_{\min}(X_k) = \lceil \log_3 n \rceil$,

c) $R^{2lv}(X_k) = \begin{cases} \lceil \log_3(n) \rceil & \text{if some } m \in \mathbb{N} \text{ exists with } 3^m < n < 2 \cdot 3^m - 1 \\ \lceil \log_3(n) \rceil + 1 & \text{if some } m \in \mathbb{N} \text{ ex. with } 2 \cdot 3^m - 1 \leq n \leq 3^{m+1} \end{cases}$

Sketch of the Proof. All of these networks contain a Hamiltonian cycle [16]. Furthermore, $n \neq 2 \cdot 3^m$ and $n \neq 2 \cdot 3^m - 1$ holds for all $m \in \mathbb{N}$ and for all of these networks. Thus, Theorem 6 follows from the results for the cycle.   □

Theorem 7. *For any $k \geq 2$:*

a) $B^{2lv}_{\min}(SE_k) = 1$,

b) $B^{2lv}(SE_k) = 2$,

c) $A^{2lv}_{\min}(SE_k) = \lceil \log_3(2^k) \rceil$,

d) $A^{2lv}(SE_k) = \begin{cases} \lceil \log_3(2^k) \rceil & \text{if some } m \in \mathbb{N} \text{ exists with } 3^m < 2^k \leq \frac{3^{m+1}+1}{2} \\ \lceil \log_3(2^k) \rceil + 1 & \text{if some } m \in \mathbb{N} \text{ ex. with } \frac{3^{m+1}+1}{2} < 2^k \leq 3^{m+1} \end{cases}$

Sketch of the Proof. SE_k contains a Hamiltonian path [7]. Therefore Theorem 7 follows from the results for the path and from the fact, that SE_k contains two nodes of degree 1.   □

Theorem 8. *For any $k \geq 2$:*

a) If $2^k = 3^m + 1$ for some $m \in \mathbb{N}$, then $R^{2lv}(SE_k) = \lceil \log_3(2^k) \rceil$.

b) If $\frac{3^{m+1}+1}{2} < 2^k \leq 3^{m+1}$ for some $m \in \mathbb{N}$, then $R^{2lv}(SE_k) = \lceil \log_3(2^k) \rceil + 1$.

c) If $3^m + 1 < 2^k \leq \frac{3^{m+1}+1}{2}$ for some $m \in \mathbb{N}$, then

$$\lceil \log_3(2^k) \rceil \leq R^{2lv}(SE_k) \leq \lceil \log_3(2^k) \rceil + 1.$$

Sketch of the Proof. Theorem 8 follows from the results for the path.   □

We were not able to determine the exact gossip complexity for the shuffle-exchange network SE_k, if $3^m + 1 < 2^k \leq \frac{3^{m+1}+1}{2}$ for some $m \in \mathbb{N}$, and we leave it as an open problem here.

3.5   Communication in the Two-dimensional Grid

In this section we will determine the accumulation and gossip complexity for the two-dimensional grid $G_{k,\ell}$, using the results for P_n and C_n.

The two-dimensional $k \times \ell$-grid $G_{k,\ell}$ is defined by

$$V(G_{k,\ell}) := \{(i,j) \mid 1 \leq i \leq k, 1 \leq j \leq \ell\} \text{ and}$$
$$E(G_{k,\ell}) := \{\{(i,j),(i+1,j)\} \mid 1 \leq i < k, 1 \leq j \leq \ell\}$$
$$\cup \{\{(i,j),(i,j+1)\} \mid 1 \leq i \leq k, 1 \leq j < \ell\}.$$

Since every two-dimensional grid contains a Hamiltonian path, we know $B_{\min}^{2lv}(G_{k,\ell}) = 1$ and $B^{2lv}(G_{k,\ell}) \leq 2$.

If k or ℓ is even, $G_{k,\ell}$ contains a Hamiltonian cycle, and thus, $B^{2lv}(G_{k,\ell}) = 1$ holds. If k and ℓ are odd, then there exists a vertex that cannot be the endpoint of a Hamiltonian path, e. g. the vertex (1,2). Thus, $B^{2lv}(G_{k,\ell}) = 2$ holds.

Theorem 9.

> a) $A_{\min}^{2lv}(G_{k,\ell}) = \lceil \log_3(k \cdot \ell) \rceil$ for any $k, \ell \geq 1$,
> b) $A^{2lv}(G_{k,\ell}) = \lceil \log_3(k \cdot \ell) \rceil$ for any $k, \ell \geq 2$.

Proof. Proof omitted.

Theorem 10. *For any $k, \ell \geq 2$:*

a) *If $k \cdot \ell = 2 \cdot 3^m - 1$ for some $m \in \mathbb{N}$, then*

$$\lceil \log_3 k \cdot \ell \rceil \leq R^{2lv}(G_{k,\ell}) \leq \lceil \log_3 k \cdot \ell \rceil + 1.$$

b) *If $k \cdot \ell \neq 2 \cdot 3^m - 1$ for all $m \in \mathbb{N}$, then*

$$R^{2lv}(G_{k,\ell}) = \begin{cases} \lceil \log_3(k \cdot \ell) \rceil & \text{if some } m \in \mathbb{N} \text{ exists with } 3^m \leq k \cdot \ell \leq 2 \cdot 3^m \\ \lceil \log_3(k \cdot \ell) \rceil + 1 & \text{if } \exists\, m \in \mathbb{N} \text{ with } 2 \cdot 3^m + 1 \leq k \cdot \ell \leq 3^{m+1} \end{cases}$$

Proof. Proof omitted.

We were not able to determine the exact gossip complexity of $G_{k,\ell}$ in the case $k \cdot \ell = 2 \cdot 3^m - 1$, and we leave it as an open problem here.

4   Conclusion

In this paper, we have constructed accumulation and gossip algorithms in the $2LVDP$ mode for many networks and we have proven their optimality.

We have seen that the accumulation and gossip problem can be solved optimally or nearly optimally for all networks containing a Hamiltonian path.

The main open problem we see left is the design of optimal accumulation and gossip algorithms for networks that do not contain a Hamiltonian path, e. g. for complete k-ary trees.

References

1. B. Bollobas: Graph Theory. Springer-Verlag, New York, 1979.
2. T. H. Cormen, C. E. Leiserson, R. L. Rivest: Introduction to Algorithms. MIT Press 1990.
3. S. Even: Graph Algorithms. Pitman, London, 1979.
4. A. M. Farley: Minimum-Time Line Broadcast Networks. *Networks* 10 (1980), 59–70.
5. R. Feldmann, J. Hromkovič, S. Madhavapeddy, B. Monien, P. Mysliwietz: Optimal Algorithms for Dissemination of Information in Generalized Communication Modes. *Discrete Applied Mathematics* 53 (1994), No. 1-3, 55–78.
6. P. Fraigniaud, E. Lazard: Methods and problems of communication in usual networks. *Discrete Applied Mathematics* 53 (1994), No. 1-3, 79–133.
7. R. Feldmann, P. Mysliwietz: The Shuffle Exchange Network has a Hamiltonian Path. *Mathematical Systems Theory* 29(5) (1996), 471–485.
8. S. M. Hedetniemi, S. T. Hedetniemi, A. L. Liestman: A Survey of Gossiping and Broadcasting in Communication Networks. *Networks* 18 (1988), 319–349.
9. J. Hromkovič, R. Klasing, B. Monien, R. Peine: Dissemination of Information in Interconnection Networks (Broadcasting and Gossiping). In: D.-Z. Du and F. Hsu, editors, *Combinatorial Network Theory*, Kluwer Academic Publishers (1995), 125–212.
10. J. Hromkovič, R. Klasing, E. A. Stöhr: Dissemination of Information in Vertex-Disjoint Paths Mode. *Computers and Artificial Intelligence* 15 (1996), No. 4, 295–318.
11. J. Hromkovič, R. Klasing, E. A. Stöhr, H. Wagener: Gossiping in Vertex-Disjoint Paths Mode in d-Dimensional Grids and Planar Graphs. *Information and Computation* 123 (1995), No. 1, 17–28.
12. J. Hromkovič, R. Klasing, W. Unger, H. Wagener: Optimal Algorithms for Broadcast and Gossip in the Edge-Disjoint Path Modes. *Information and Computation* 133 (1997), No. 1, 1–33.
13. J. Hromkovič, K. Loryś, P. Kanarek, R. Klasing, W. Unger, H. Wagener: On the Sizes of Permutation Networks and Consequences for Efficient Simulation of Hypercube Algorithms on Bounded-Degree Networks. *Proc. of the 12th Symposium on Theoretical Aspects of Computer Science (STACS'95)*, Springer LNCS 900, 255–266.
14. R. Klasing: On the Complexity of Broadcast and Gossip in Different Communication Modes (Dissertation). *Shaker Verlag*, Aachen, 1996.
15. R. Klasing: The Relationship Between Gossiping in Vertex-Disjoint Paths Mode and Bisection Width. *Discrete Applied Mathematics* 82 (1998), No. 1-3, 227–244.
16. F. T. Leighton: Introduction to Parallel Algorithms and Architectures: Arrays, Trees, Hypercubes. *Morgan Kaufmann Publishers*, 1992.

Broadcasting on Anonymous Unoriented Tori*

Stefan Dobrev and Peter Ružička

Institute of Informatics
Comenius University
Slovakia
E–mail:{dobrev,ruzicka}@dcs.fmph.uniba.sk

Abstract. We consider broadcasting on asynchronous anonymous totally unoriented $n \times m$ torus, where $N = n \cdot m$ is the number of nodes. We present a broadcasting algorithm with message complexity $1.43N + O(n+m)$ and prove the lower bound in the form $1.14N - O(1)$. This is an improvement over the previous $2N + O(\sqrt{N})$ upper bound and $1.04N - O(\sqrt{N})$ lower bound achieved by Diks, Kranakis and Pelc [DKP96]. Unlike the algorithm from [DKP96], our algorithm works also on non-square tori, does not require the knowledge of sizes n and m and uses only messages of size $O(1)$ bits. This is the first known broadcasting algorithm on unoriented tori that does not use all edges.

1 Introduction

Broadcasting is one of the most fundamental communication tasks in parallel and distributed computing. One node of the network, called *source*, has a message which has to be transmitted to all other nodes in the network.

The complexity of broadcasting strongly depends on the amount of topological information available at nodes. If links of a network are globally consistently labelled, forming *sense of direction* [FMS95,Tel95a], broadcasting is possible using only linear number of messages w.r.t. the number of nodes [FMS96]. But if a network is unoriented (no consistent global labels on links are available), then the lower bound for general networks is linear in the number of links [FMS96]. This lower bound is achievable by the naive broadcasting algorithm, in which a node immediately spreads the message to all neighbours except to the one from which it received it.

While this strategy cannot be improved on general networks, broadcasting algorithms might exploit the knowledge of special topologies to reduce the number of messages. For example, on the complete unoriented N–node network the broadcasting is trivially accomplished by sending only $N - 1$ messages from the source to all its neighbours. Another, not so trivial example, is the class of unoriented N–node chordal ring networks with chords leading to $2k$ closest neighbours in the ring, where the broadcasting can be performed using only $O(N)$ messages [Pel97].

---

* The research was partially supported by EU Grant No. INCO-COP 96-0195 "ALTEC-KIT" and by the Slovak VEGA project 1/4315/97.

Other results have been obtained for broadcasting on special topologies without orientation. It was stated as an open question (see [Tel95a], cf. also [Mans96]) whether there exists a broadcasting algorithm on unoriented N–node hypercubes using only $O(N)$ messages in the worst case. This question was positively answered only recently in [DR97,DKP96,DDKPR98], where two different independently obtained linear message algorithms for broadcasting on unoriented anonymous hypercubes were presented. Further improvements of these results in the time and bit complexity can be found in [DRT98].

We are interested in broadcasting on unoriented tori. Since tori have constant degree, the naive broadcasting algorithm using $3N+1$ messages is asymptotically optimal. However, it is possible to improve the constant factor, as documented by the algorithm from [DKP96], which uses $2N+O(\sqrt{N})$ messages on $\sqrt{N}\times\sqrt{N}$ tori. In [DKP96] it was also shown that the lower bound is non-trivial ($1.04N - O(\sqrt{N})$ messages). We further improve these results, both in upper and lower bounds. We present $1.43N + O(n+m)$ message broadcasting algorithm which, unlike the algorithm from [DKP96], works also on non-square torus, does not need to know its size and uses only messages of size $O(1)$. This is the first know broadcasting algorithm working on unlabeled tori that does not use every link. We also present an improved lower bound $1.14N - O(1)$, using a technique with potential for further improvement of this lower bound. Our lower bound is on the number of used links and it applies also for the case of synchronous communication, regardless whether vertices know the size of torus.

The paper is organized as follows. In Section 2 we present the computational model. In Section 3 we show $1.43N + O(n+m)$ broadcasting algorithm. In Section 4 we prove an $1.14N - O(1)$ lower bound for broadcasting on unlabeled N–node tori.

2   The model

The computational model is the standard model of asynchronous distributed computing [Tel94b]. Every message will be delivered in a finite but unbounded time. FIFO on links is not required. All processors are identical and run the same algorithm.

The underlying topology of the network is anonymous unoriented torus of size $n \times m$. Anonymity means that processors do not have given distinct identities. Unorientation means that each processor can distinguish its links only by uninterpreted labels 1, 2, 3 and 4. However, this labeling is arbitrary at each processor and labels are thus without any topological meaning. That means that if a processor sends a message on an unused link, the actual link (from the set of yet unused links) is chosen by the adversary, as all unused links look alike to the sender and we are interested in the worst case behaviour.

We are interested in *communication complexity*, expressed by the number of messages sent in the worst case. The worst case refers to the adversary decisions concerning choices of yet unused links, and to the worst possible message delays.

We are considering the problem of *broadcasting*. At the beginning there is a single active processor – the source of information. Other processors will became active only after receiving a message. Only active processors can send messages. At the end of computation we require each processor to be active (it has received an information).

3   Upper Bound

In this section we present a broadcasting algorithm on asynchronous anonymous totally unoriented $n \times m$ tori using $\frac{10}{7}n \cdot m + O(n + m)$ messages in the worst case.

3.1   Informal description of the algorithm

Our algorithm $\mathcal{A}$ starts from the source s by sending an initial message in one direction until it returns back to s. This can be done using *handrail* technique from [Pet85,Mans96b] with $O(n)$ messages of size $O(1)$, where n is the size of the torus in the direction of the initial message. Created path circling around tori is called *equator*. From the handrail technique follows that vertices on the equator can consistently distinguish between their *north* and *south* sides.

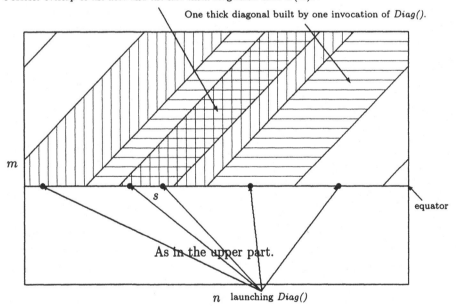

Figure 1.

In the second phase, another message is sent along the equator eastward and at each 7th vertex it launches northward the subroutine *Diag()* until it returns back to the source. The first launch of *Diag()*, denoted as $Diag^1()$, is done using marked messages, so it will not interfere with the last one, which may overlap with it. *Diag()* procedure broadcasts on thick (7 vertex) diagonal in north-east direction until it returns back to the equator.

The overall message complexity of the broadcasting algorithm is $\frac{10}{7}N + O(n+m)$ and it follows from these facts:

- The cost of the start-up (building the equator and launching *Diag()* procedures) is $O(n)$.
- *Diag()* uses $\frac{10}{7}$ messages of size $O(1)$ bits per each reached vertex.
- The whole torus can be covered by disjoint thick diagonals built by *Diag()*. Since *Diag()* stops on the equator, different invocations of *Diag()* do not overlap. The only exception is the first and the last invocation of *Diag()*, which can overlap for n non-multiple of 7, where n is the length of the equator. There are at most $6m$ vertices in the intersection, that means totally $\frac{60}{7}m$ additional messages. (See Figure 1)

3.2   Detailed description of the algorithm

StartUp

At the source on start up:

0. For each incident link h:
   Send(S_x^1), where x is label of h at the source vertex;

Figure 2.

At arbitrary vertex:

	Upon receiving	Send	See Figure
1.	S_x^1	S_x^2 on all remaining links	3
2.	S_1^2 on h and S_x^2	S_x^3 on h	4
3.	S_x^3 on h_1 and S_y^3 on h_2, let $x < y$	U_1 on h_1, B_1 on h_2 and M_1 on link on which S_1^2 was sent, but nothing received.	5

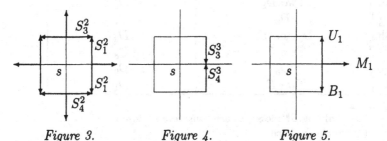

Figure 3.   *Figure 4.*   *Figure 5.*

Building the equator:

	After receiving message(s)	Send	See Figure
4a.	U_1	U on unused links	6
4b.	B_1	B on unused links	6
4c.	M_1, not at source	M on unused links [1]	5,6 and 9
4d.	M_1 at source	L on link with label 1 [2]	10
5a.	U on h and M	U on all links except h	7
5b.	B on h and M	B on all links except h	7
5c.	U and B	M_1 on links on which no M_1, U or B was received	8

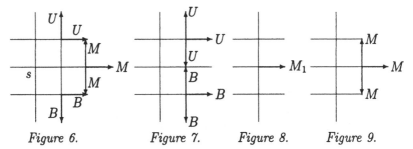

Figure 6.   Figure 7.   Figure 8.   Figure 9.

Launching phase

	After receiving message(s)	Send	See Figure
6.	L	D_0' where U_1 was sent	5 and 10
		L_1 where M_1 was sent	
7.	L_i, not at source	$L_{(i+1) \bmod 7}$ where M_1 was sent	8 and 10
		if $i = 0$, send D_0 from where	7 and 10
		U message came	

Diag() procedure:

	After receiving message(s)	Send to all unused links [3]
8a.	D_0	D_1 [4]
8b.	D_1 and not M received before	D_2
8c.	D_2 and (D_9 or S_x^2)	D_3
8d.	Two D_2	D_4
8e.	D_4	D_5
8f.	Two D_5	D_6
8g.	D_6	D_7
8h.	D_5 and D_7	D_3
8i.	Two D_7	D_8
8j.	D_8	D_9
8k.	Two D_9	D_1

---

[1] Links used only by S_x^y messages are considered unused.
[2] Starting launching phase.

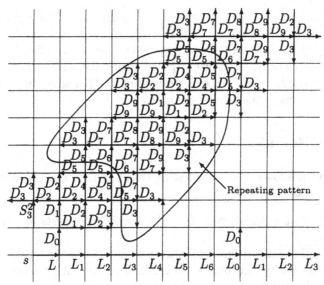

Figure 10.

If D'_0 message is received, D'_i messages will be used, to avoid possible interference between the first and the last invocation of $Diag()$.

It is easy to see that computation cyclically proceeds in cycle $8a \to 8b \to 8d \to$ $\to 8e \to 8f \to 8g \to 8i \to 8j \to 8k \to 8b$ with concurrent steps 8c and 8h. In one such cycle altogether 28 new vertices are added to the thick diagonal using 40 messages, resulting in overall $\frac{10}{7}$ messages per vertex.

$Diag()$ terminates when it returns back to the equator from the south – a vertex that has received M and B message will not send any $Diag()$ message.

3.3   Making termination explicit

The algorithm presented in subsection 3.2 terminates implicitly. One way to make the termination explicit within the same complexity bound is the following:

- Vertices reached by messages of $Diag()$ terminate when they finish their work in Diag().
- Vertices of the equator terminate when the launching token of the second phase has passed them.
- The only problem are vertices of the first – $Diag^1()$ and the last – $Diag^q()$ thick diagonal. The problem is how to terminate in order to nonblock broadcasting on the second thick diagonal. One possible solution is the following:
  - When $Diag^1()$ returns back to the equator, it returns k steps to the west and launches $Diag^q()$ to south-west. Vertices reached by $Diag^q()$

---

[3] Unused links – unused by D_i and M messages. No messages are sent from vertices that received a B and M message – termination of $Diag()$ when equator was reached from the south.

[4] Messages are sent along two links on which no U, D_0 or S^y_x message arrived. (see Figure 6)

but not by $Diag^1\,()$ will terminate after finishing their work in $Diag^q\,()$. When $Diag^q\,()$ reaches the equator, it goes k steps eastward and launches $Diag^1\,()$ to north-east. Now vertices can terminate after finishing their work in $Diag^1\,()$. k can be computed during the construction of the equator as the length of the equator modulo 7. This additional computation can be done using $O(m)$ messages.

4   Lower Bound

We will prove the lower bound by letting the algorithm and the adversary play the following game:

- At the beginning the domain D of the algorithm consists of a single vertex – the source.
- The goal of the algorithm is to extend its domain to the whole torus.
- The game proceeds in rounds. Each round begins with a move of the algorithm, which is followed by a move of the adversary.
- The algorithm knows the graph D representing its domain, but does not know how it is embedded into torus.
- The algorithm specifies (during its move) for each vertex of its domain the number of yet unexplored links it wants to explore.
- The adversary chooses an embedding of D into torus and decides which of yet unexplored links at given vertex will be explored. (According to the orders given by the algorithm.)
- At the end of each round all explored links are added to the domain of the algorithm.
- The game terminates when the domain spans the whole torus.
- The goal of the adversary is to maximize the number of explored links.

Since the game proceeds in synchronous rounds, the lower bound applies also for the case of synchronous communication. Moreover, this is the lower bound on the number of used links, not only on the number of used messages.

Our adversary tests all possible embeddings, and for each embedding it tests all possible ways of choosing explored links. The embedding (and the choice of explored links) which leads to the smallest new domain is chosen. (See Figure 11)

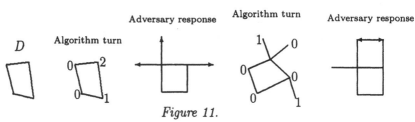

Figure 11.

Let D_i be the domain of the algorithm after the round i. D_i can be divided to the *core graph* C_i and *hanging trees* T_i. The *core graph* can be defined as the maximal subgraph C_i of D_i such that $\forall v \in C_i$, v has at least two neighbours in C_i. T_i is the rest of the domain. T_i can be viewed as a forest of trees with roots in C_i. (These roots are not in T_i). We denote these graphs after the termination of the game by D, C and T.

The following lemma is crucial:

Lemma 1. *If $C_i \neq \emptyset$, then there are no hanging trees of depth greater than 2.*

Proof. By contradiction. Consider the first round (say r) in which there is a hanging tree $T^{(3)}$ of depth 3. That means that in the round $r-1$ there was a hanging tree $T^{(2)}$ of depth 2 and from one of its leaves at depth 2 (say from a vertex v) a new link was explored. Let the root of $T^{(2)}$ be a vertex $t \in C_{r-1}$. It has two neighbours in C_{r-1}. Take the embedding $\mathcal{E}$ used by the adversary at round $r-1$ and modify it locally to embed $T^{(2)}$ such as in Figure 12. This modification is indeed possible, because the place for the vertex v is either free or is occupied by another branch of $T^{(2)}$ (which can be exchanged with the branch leading to v). If this place belongs to C_{r-1} or to another tree, then in the previous round the adversary should have directed the growth of $T^{(2)}$ to this place, thus constructing smaller domain.

The resulting embedding $\mathcal{E}'$ (together with the choice of explored link (v, w)) results in smaller domain compared with $\mathcal{E}$, which is the contradiction with respect to our choice of the adversary.

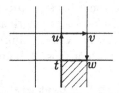

Figure 12.

It is easy to see that $C_i = \emptyset$ holds only at the very beginning of the computation. Using similar arguments as in the proof above we can show that once there is a tree of depth 4, the adversary will turn its branches to form a cycle.

So far we have shown that the adversary can limit the depth of hanging trees by looping their branches back to the core graph. Another way to reduce the depth of hanging trees is to connect two trees, thus eliminating trees that grow deep. We will refine our adversary by making it prefer the following option:

- If there are several possibilities (for embedment and choice of explored links) resulting to the domains of the same size, prefer the option in which a tree branches loop back to the core graph rather than the option in which trees come in touch. (See Figure 13)

Algorithm turn

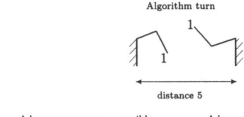

distance 5

Adversary response - possible   Adversary response - preferred

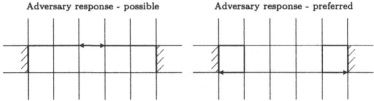

Figure 13.

Lemma 1 allows us to examine how the core graph grows:

Lemma 2. *If $C_i \neq \emptyset$, then there exists an ear decomposition of $C_{i+1} - C_i$ such that each ear has the length at most 4.*

More precisely, there exists a sequence of graphs $C_i = C_{i,0}, C_{i,1}, \ldots, C_{i,k} = C_{i+1}$ such that for each j, $0 \leq j < k$, $C_{i,j+1} - C_{i,j}$ is a path of length at most 4 which starts and ends in $C_{i,j}$.

Proof. First note that there is an ear decomposition with ears of length at most 6. See that new vertices are added to the core graph only when branches of some tree loop back to the core graph or when two trees meet. (The third possibility occurs when two branches of the same tree meet. But this case is not possible under our adversary, because the adversary preferes to loop branches back to the core graph.) In the first case an ear of length at most 3 is formed (since trees are of depth at most 2, plus newly explored link), in the second case each tree can contribute by a path of length at most 3, bounding the overall length of the new created ear by 6.

To form a *long ear* of length 5 or 6, at least one tree must contribute by a path of length 3. Our adversary prefers looping back branches of trees to the core graph. It may happen that adversary cannot loop back branch of length 3, because that will increase the size of the domain. One such situation is shown on Figure 14.

Algorithm turn   Adversary response

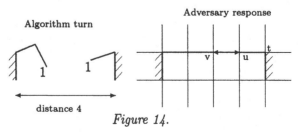

distance 4

Figure 14.

However, that is possible because the tree of the vertex u contributed by only 2 links $((t, u), (u, v))$ and there were overlapping links from different trees $((u, v)$

and (v, u)). If there are not overlapping links (there is no request for exploration from u), looping back to the core graph is possible, because it will not enlarge the domain. Similarly, looping back is possible if both trees contribute by paths of length 3, as shown in Figure 13.

That means that if there is still a long ear, then it must also have small loop(s) at its end(s). We can first add an ear formed by this (these) small loop(s), decreasing the overall length of the long ear to at most 4. (It is easy to see that an ear of length 6 must have small loops at both of its ends.)

Let E_C be the number of links in the core graph C. We will prove our lower bound by proving the lower bound on the expression

$$\frac{E_C + |T|}{|C| + |T|} = \frac{\text{total number of explored links}}{N} \quad (1)$$

Let C_0 be the initial core graph (formed when the first loop is closed) and let E_{C_0} be the number of links in it. Let C_i be the core graph after adding i ears according to Lemma 2 and let E_{C_i} be the number of links in it. Let T_i be the set of hanging trees at that moment. Following Lemma 1, we can bound $|T_i|$ by the number of vertices at the distance at most 2 from C_i.

We are interested in the ratio

$$\frac{E_{C_k} + |T_k|}{|C_k| + |T_k|} = \frac{E_{C_0} + |T_0| + \sum_{i=1}^{k}(E_{C_i} - E_{C_{i-1}} + |T_i| - |T_{i-1}|)}{|C_0| + |T_0| + \sum_{i=1}^{k}(|C_i| - |C_{i-1}| + |T_i| - |T_{i-1}|)} \quad (2)$$

Denote $E_{C_i} - E_{C_{i-1}}$ by e_i, it represents the number of links in the i-th ear. Similarly $|C_i| - |C_{i-1}| = v_i$, the number of inner vertices in the i-th ear. Clearly $e_i = v_i + 1$. Let t_i be the number of vertices that are at distance at most 2 from C_i, but they are at greater distance from C_{i-1}. $|T_i| - |T_{i-1}|$ can be estimated as $t_i - v_i$, since v_i inner vertices of the i-th ear are transferred from T_{i-1} to C_i. (Note that all inner vertices of the i-th ear are inside T_{i-1}, because the ear has length at most 4.)

We can rewrite (2):

$$\frac{E_{C_0} + |T_0| + \sum_{i=1}^{k}(e_i + (t_i - v_i))}{|C_0| + |T_0| + \sum_{i=1}^{k}(v_i + (t_i - v_i))} = \frac{E_{C_0} + |T_0| + \sum_{i=1}^{k}(t_i + 1)}{|C_0| + |T_0| + \sum_{i=1}^{k}t_i} \quad (3)$$

Let $t = max_{1 \leq i \leq k}t_i$. Because $\sum_{i=1}^{k}(t_i + 1)/\sum_{i=1}^{k}t_i \geq (t+1)/t$, we get

$$\frac{E_{C_k} + |T_k|}{|C_k| + |T_k|} \geq \frac{E_{C_0} + |T_0| + k(t+1)}{|C_0| + |T_0| + kt} \quad (4)$$

Since $|C_0|$, E_{C_0}, $|T_0|$ and t are in $O(1)$, we can write $|C_0| + |T_0| + kt = k't$ and $E_{C_0} + |T_0| + k(t+1) = k'(t+1) - O(1)$ for some $k' > k$. Applying to (4) we get

$$E_{C_k} + |T_k| \geq \frac{k'(t+1) - O(1)}{k't}(|C_k| + |T_k|) = \frac{t+1}{t} \cdot (|C_k| + |T_k|) - O(1) \quad (5)$$

since $k't \geq |C_k| + |T_k|$.

Note that this expression is in more general form than the simple lower bound for ceased broadcasting. It says that any algorithm that has reached r vertices must have used $r(t+1)/t - O(1)$ links.

All we need now is to bound t:

Lemma 3. t can be bounded by 7 for ears of length at most 4.

Proof. First note that there is only a finite (and not really high) number of possible cases which can be tested by computer. We perform the case analysis.

We will use the fact that each vertex of C has at least two neighbours in C.

Consider an ear of 4 links and 3 vertices u, v and w being added to C. These three vertices either lie on a line or not. In the first case, all possible situations (up to the symmetry) are shown in Figure 15.

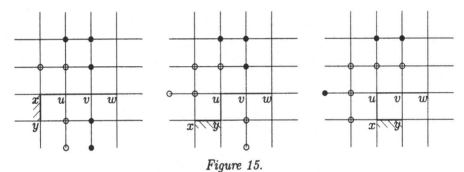

Figure 15.

x and y are vertices in C. Empty circles represent vertices in 2–neighbourhood of C. Full circles represent vertices potentially added to the 2–neighbourhood of C by adding the ear uvw.

Due to the symmetry only the left part from the vertical axis passing through v is shown. t is in these cases bounded by 6, 4 and 5, respectively. (If the right part is mirror image of the left part. Otherwise t is even smaller.)

If u, v and w don't lie on a line, then four possible cases are shown in Figure 16. Again it is sufficient to consider only one half (left bottom) of the situation.

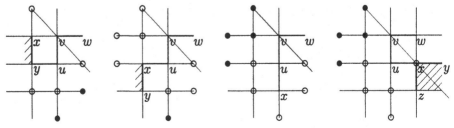

Figure 16.

t is in these cases bounded by 4, 2, 7 and 7, respectively.

Ears of smaller length are handled similarly. All possible cases can be drawn on previous figures, just with smaller number of t – vertices.

Now we can apply (5):

Proposition 1. *Any broadcasting algorithm on unoriented tori that reached r vertices must have used at least $8/7r - O(1)$ links in the worst case.*

Corollary 1. *Broadcasting on unoriented N–node tori requires the use of at least $8/7N - O(1)$ links, even in synchronous case.*

5 Conclusions

We have presented improved upper and lower bounds for broadcasting on an unoriented torus. The main question is how to narrow or close the gap between these bounds.

We believe that the lower bound can be improved. A possible way of improvement can be obtained by the analysis of the following situation. The highest t $(t = 7)$ is reached by adding ears that are not sustainable. Including such an ear prevents from further inclusion of an ear with $t = 7$ on involved vertices. Ears with smaller t must be added to prepare the ground for another ear with $t = 7$. Further improvement can be based on the following hypothesis: *There exists an adversary such that in each completed computation there are no hanging trees of depth 2 or more (although during the computation there could be some).*

We note that the algorithm from [DKP96] can be modified to compute relative addresses (w.r.t. the starting vertex) of all vertices in the torus. Our algorithm cannot be modified in such a way. It would be interesting to prove (or disprove) the lower bound of $2N - o(N)$ messages for this problem.

References

[DKP96]   Diks, K. – Kranakis, E. – Pelc, A.: *Broadcasting in Unlabeled Tori.* Départment d'Informatique, Université du Québec á Hull, Technical Report RR 96/12-5, 1996.

[DR97]   Dobrev, S. - Ružička, P.: *Linear Broadcasting and N log log N Election in Unoriented Hypercubes.* Proc. of the 4th International Colloquium on Structural Information and Communication Complexity (SIROCCO'97), Carleton Press, Ascona, Switzerland, July 1997, pp. 55–73.

[DRT98]   Dobrev, S. - Ružička, P. - Tel, G.: *Time and Bit Optimal Broadcasting on Anonymous Unoriented Hypercubes.* Proc. of the 5th International Colloquium on Structural Information and Communication Complexity (SIROCCO'98), Carleton Press, Amalfi, Italy, June 1998.

[DDKPR98] Diks, K. - Dobrev, S. - Kranakis, E. - Pelc, A. - Ružička, P.: *Broadcasting in Unlabeled Hypercubes with Linear Number of Messages.* To appear in Information Processing Letters, 1998.

[FMS95]   Flocchini, P. – Mans, B. – Santoro, N.: *Sense of Direction: Formal Definitions and Properties.* Proc. of the 1st International Colloquium on Structural Information and Communication Complexity (SIROCCO'94), Carleton Press, 1995, pp. 9–34.

[FMS96]   Flocchini, P. – Mans, B. – Santoro, N.: *On the Impact of Sense of Direction on Communication Complexity.* Information Processing Letters, Vol. 63(1), July 1997, pp. 23–31.

[Mans96]   Mans, B.: *Broadcast, Traversal and Election in Unlabelled Hypercube.* Proc. of the 3rd International Colloquium on Structural Information and Communication Complexity (SIROCCO'96), Carleton Press, Siena, Italy, June 1996, pp. 333–334.

[Mans96b] Mans, B.: *Optimal Distributed Algorithms in Unlabeled Tori and Chordal Rings.* Proc. of the 3rd International Colloquium on Structural Information and Communication Complexity (SIROCCO'96), Carleton Press, Siena, Italy, June 1996, pp. 17–31.

[Pel97]   Peleg, D.: Personal communication at the Sienna Research School'97 on "Compact Routing and Sense of Direction", Siena, Italy, June 1997.

[Pet85]   Peterson, G. L.: *Efficient algorithms for elections in meshes and complete networks.* Technical Report TR140, Dept. of Computer Science, Univ. of Rochester, Rochester, NY 14627, 1985.

[Tel94a]   Tel, G.: *Network Orientation.* International Journal of Foundations of Computer Science 5, 1994, pp. 23–57.

[Tel94b]   Tel, G.: *Introduction to Distributed Algorithms.* Cambridge University Press, Cambridge, 1994.

[Tel95a]   Tel, G.: *Sense of Direction in Processor Networks.* In: SOFSEM'95, Theory and Practise of Informatics, LNCS 1012, Springer–Verlag, 1995, pp. 50–82.

Families of Graphs Having Broadcasting and Gossiping Properties

Guillaume Fertin[1], André Raspaud[1]

LaBRI U.M.R. C.N.R.S. 5800, Université Bordeaux I
351 Cours de la Libération, F33405 Talence Cedex
{fertin,raspaud}@labri.u-bordeaux.fr

Abstract. Broadcasting and gossiping are two problems of information dissemination described in a group of individuals connected by a communication network. In broadcasting (resp. gossiping), one node (resp. every node) has a piece of information and needs to transmit it to everyone else in the network. These communications patterns find their main applications in the field of interconnection networks for parallel architectures. In this paper, we are interested in Minimum Broadcast (resp. Gossip, Linear Gossip) Graphs (resp. Digraphs), that is graphs (resp. digraphs) that can achieve broadcasting (resp. gossiping, linear gossiping) in minimum time, and with a minimum number of edges. Many papers have investigated these subjects, but only a few general results on the size of graphs of order n are known. In this paper, we take the census of all the known non-isomorphic families of graphs (resp. digraphs) which are Minimum Broadcast Graphs, Minimum Gossip Graphs, Minimum Linear Gossip Graphs and/or Minimum Broadcast Digraphs, and we show that in most cases, the proposed minimum graphs that can be found in the literature are Knödel graphs [10, 7].

Keywords : Broadcasting, gossiping, minimum broadcast graphs, minimum gossip graphs, Knödel graphs, circulant graphs, hypercubes.

1 Introduction

Broadcasting and gossiping are two problems of information dissemination described in a group of individuals connected by a communication network. In *broadcasting* (resp. *gossiping*), one node (resp. every node) knows a piece of information and needs to transmit it to everyone else. This is achieved by placing communication calls over the communication lines of the network. Throughout this paper (except in one case, which, for readability reasons, is treated separately below), we will consider a *1-port* and *unit cost* model, that is a node can communicate with at most one of its neighbours at any given time, and a communication between two nodes takes one unit of time. This model implies that we will deal with connected graphs without loops or multiple edges

---

[1] This research was supported by ALTEC-KIT, Project no INCO-COP 96-0195

to model the communication networks. Note also that, depending on the cases, we will either consider a *half-duplex* or a *full-duplex* model. In the latter, when a communication takes place along a communication line, the information flows in both directions, while in the former only one direction is allowed. Hence, in the *half-duplex* model, we will deal with directed graphs, while we will consider undirected graphs in the *full-duplex* model.

Let us first consider the *full-duplex* model. Let G be a graph modelling an interconnection network. We will denote by $b(v)$ the *broadcast time* of v, that is the time to achieve broadcasting from a vertex v of G in the network. Moreover, $b(G)$, the *broadcast time* of G, is defined as follows : $b(G) = max\{b(v) \mid v \in V(G)\}$. If we consider the complete graph of order n, K_n, it is not difficult to see that $b(K_n) = \lceil \log_2(n) \rceil$. Any graph G such that $b(G) = b(K_n) = \lceil \log_2(n) \rceil$ is called a *broadcast graph*. Note that it is not necessary to consider K_n to get a broadcast graph. Hence we call *Minimum Broadcast Graph* of order n, or MBG_n, any broadcast graph G having the minimum number of edges. This number is denoted by $B(n)$.

Similarly, in the *half-duplex* model, we have the following : $\vec{b}(v)$ is the *broadcast time of vertex* v and $\vec{b}(G) = max\{\vec{b}(v) \mid v \in V(G)\}$ is the *broadcast time of digraph* G. Liestman and Peters [13] have shown that $\vec{b}(K_n^*) = \lceil \log_2(n) \rceil$, where K_n^* is the complete directed graph of order n, that is the complete graph K_n where each undirected edge uv has been replaced by a pair of symmetric edges (u, v) and (v, u). Any digraph G such that $\vec{b}(G) = \vec{b}(K_n^*)$ is called a *broadcast digraph*, and, similarly to the undirected case, it is not necessary to consider K_n^* to get a broadcast digraph. Hence we will call *Minimum Broadcast Digraph* of order n, or MBD_n, any broadcast digraph with the minimum number of edges. This number is denoted by $\vec{B}(n)$.

The gossiping problem, back in the *full-duplex* model, relies on analogous definitions. Let $g(G)$ be the time to gossip in a graph G. Knödel [10] has shown that for the complete graph K_n, we have :

- $g(K_n) = \lceil \log_2(n) \rceil$ for any even n ;
- $g(K_n) = \lceil \log_2(n) \rceil + 1$ for any odd n ;

Any graph G such that $g(G) = g(K_n)$ is called a *gossip graph*. As previously, it is not necessary to consider K_n to get a gossip graph. Consequently, we call a *Minimum Gossip Graph* of order n, or MGG_n, any gossip graph with a minimum number of edges. This number is denoted by $G(n)$.

Remark 1.1. In this paper, we will deliberately not mention *Minimum Gossip Digraphs* (i.e. graphs achieving gossiping in the *half-duplex* model), since very little is known about these digraphs. In particular, no general result concerning its number of directed edges is known.

Now suppose we do not consider a *unit cost* model, but a *linear cost* one, that is each communication implies a fixed start-up time β, and a propagation time

τ proportional to the amount of information exchanged. We then define $g_{\beta,\tau}(G)$ to be the time to gossip in the graph G, and $g_{\beta,\tau}(n)$ to be the time to gossip in K_n. Fraigniaud and Peters [7] proved that $g_{\beta,\tau}(n) = \lceil \log_2(n) \rceil \beta + (n-1)\tau$ for any $\beta \geq 0$ and $\tau \geq 0$. A *Linear Gossip Graph* will denote any graph G able to gossip in $g_{\beta,\tau}(n)$, and a *Minimum Linear Gossip Graph*, or *MLGG*, is a Linear Gossip Graph with the minimum number of edges, noted $G_{\beta,\tau}(n)$.

In this paper, we intend to survey the general known results concerning the size - that is, the number of edges - of a *MGG* (resp. *MLGG*, *MBG*, *MBD*) with n nodes, and mainly to point out several non-isomorphic families of graphs which are *MGG* (resp. *MLGG*, *MBG*, *MBD*) of order n.

In Sect. 3, we will survey the general results concerning Minimum Broadcast Graphs of order n (e.g. $n = 2^k$ and $n = 2^k - 2$). We will first show three non-isomorphic families of graphs which are *MBGs* of order 2^k. Moreover, we will show that the examples of *MBGs* of order $2^k - 2$ given independently by Khachatrian et al. [9] and Dinneen et al. [2] are both isomorphic to the family of Knödel graphs of degree $k - 1$.

In Sect. 4, we will survey the known general results concerning Minimum Broadcast Digraphs of order n (e.g. $n = 2^k$, $n = 2^k - 1$ and $n = 2^k - 2$) and give different non-isomorphic families of digraphs which are *MBDs* of order n for these three cases. In particular, for $n = 2^k$, we give $k + 3$ non-isomorphic families of *MBGs*.

In Sect. 5, we will focus on Minimum Gossip Graphs. The size of such graphs is known in the general case for $n = 2^k$, $n = 2^k - 2$ and $n = 2^k - 4$ (see [11]). For each of the three cases, we will survey the results ; we will first show that, in the case $n = 2^k$, the three families of *MBGs* of order n given in Sect. 3 remain *MGGs*. Moreover, we will show that the family of graphs pointed out by Labahn [11] as *MGGs* of order $2^k - 2$ (resp. $2^k - 4$) is in fact the family of Knödel graphs of degree $k - 1$.

Section 6 will be devoted to *MLGGs*. We will survey the results of [7], and we will show that, for $n = 2^k$, the three non-isomorphic families which are *MGGs* remain *MLGGs*. Moreover, for even n such that $2^k - 6 \leq n \leq 2^k - 2$ with $k \geq 4$, the Knödel graphs of degree $k - 1$ are *MLGGs*.

Finally, Sect. 7 recalls the different results given in the paper and gives a general method to obtain recursively families of *MBGs* (resp. *MBDs*, *MGGs*, *MLGGs*) of order 2^k, families of *MBDs* of order $2^k - 2$ and families of *MGGs* (resp. *MLGGs*) of order $2^k - 4$.

2 Definitions

In the following, we will consider different families of graphs that have been defined in the literature, and we will show they are either *MGGs*, *MLGGs*, *MBGs* and/or *MBDs*. In this section, we give the definitions of each of the family we are going to use.

Definition 2.1 (Hypercube of dimension k). *The hypercube of dimension k, H_k, has $n = 2^k$ vertices. Each vertex of the vertex set V is of the form $(x_0, x_1, \ldots, x_{k-1})$ with $x_i \in \{0, 1\}$ for any $0 \le i \le k - 1$. The edge set E is of cardinality $k \cdot 2^{k-1}$ and is of the form $E = \{((x_0, x_1, \ldots, x_{i-1}, 0, x_{i+1}, \ldots, x_{k-1}), (x_0, x_1, \ldots, x_{i-1}, 1, x_{i+1}, \ldots, x_{k-1}))\}$ for every $i \in \{0, \ldots, k-1\}$.*

Definition 2.2 (Circulant graphs and digraphs).

- *A circulant graph on n vertices $C_n(a_1, a_2, \ldots a_p)$ $(a_i \in \mathbb{N}^*)$, $a_1 < a_2 < \ldots < a_p$, has vertex set $V = \{v_0, v_1, \ldots, v_{n-1}\}$ and edge set $E = \{(v_x, v_y) \mid \exists\, a_i, 1 \le i \le p$ such that $x + a_i \equiv y \pmod{n}\}$.*
- *A circulant digraph on n vertices $C'_n(a_1, a_2, \ldots, a_p)$ $(a_i \in \mathbb{N}^*)$, $a_1 < a_2 < \ldots < a_p$, has vertex set $V = \{v_0, v_1, \ldots, v_{n-1}\}$ and edge set $E = \{(v_x, v_y) \mid \exists\, a_i, 1 \le i \le p$ such that $x + a_i \equiv y \pmod{n}\}$.*

Remark 2.1. In this paper, we will often focus on circulant digraphs, and particularly on some specific families of such digraphs, for instance $C'_{2^k}(1, 3, \ldots, 2^k - 1)$ [16, 5]. We refer to Fig. 1 for some examples of circulant digraphs. Note also that we will be interested by a certain family of circulant (undirected) graphs, namely $G(n, d)$, which is defined below.

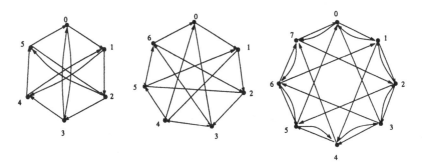

Fig. 1. $C'_6(1, 3)$, $C'_7(1, 3)$ and $C'_8(1, 3, 7)$ (from left to right)

Definition 2.3 (Circulant Graphs $G(n, d)$ [17]). *The circulant graph $G(n, d)$ with $d \ge 2$, is defined as follows. The vertex set is $V = \{0, 1, 2, \ldots n - 1\}$, and the edge set is $E = \{(u, v) \mid \exists i, 0 \le i \le \lceil \log_d(n) \rceil - 1$, such that $u + d^i \equiv v \pmod{n}\}$.*

Remark 2.2. Note that $G(n, d)$ is a circulant graph $C_n(d^0, d^1, \ldots d^{\lceil \log_d(n) \rceil - 1})$. In [17], Park and Chwa proved that if $n = cd^m$ with $1 \le c < d$, then $G(n, d)$ can be constructed recursively, using d copies of $G(cd^{m-1}, d)$. Note also that for our purpose, we will consider the family of circulant graphs $G(2^k, 4)$, such as the one displayed in Fig. 2. In that case, $G(2^k, 4)$ can be considered as a $G(cd^m, d)$, where $c = 1$ when k is even ($k = 2m$), and $c = 2$ when k is odd ($k = 2m + 1$).

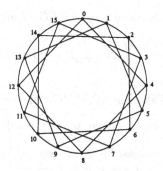

Fig. 2. The recursive circulant graph $G(16, 4)$

Definition 2.4 (Knödel graph $W_{\Delta,n}$ [10, 7]). *The Knödel graph on $n \geq 2$ vertices (n even) and of maximum degree $\Delta \geq 1$ is denoted $W_{\Delta,n}$. The vertices of $W_{\Delta,n}$ are the couples (i, j) with $i=1,2$ and $0 \leq j \leq \frac{n}{2} - 1$. For every j, $0 \leq j \leq \frac{n}{2} - 1$, there is an edge between vertex $(1, j)$ and every vertex $(2, j + 2^k - 1$ mod $\frac{n}{2})$, for $k = 0, \ldots, \Delta - 1$.*

Remark 2.3. For $0 \leq k \leq \Delta - 1$, an edge of $W_{\Delta,n}$ which connects a vertex $(1, j)$ to the vertex $(2, j + 2^k - 1$ mod $\frac{n}{2})$ is said to be *in dimension k*.

Note that when $\Delta \geq 2$, the Knödel graphs can also be defined as Cayley graphs on the semi-direct product $G = \mathbb{Z}_{\frac{n}{2}} \ltimes \mathbb{Z}_2$ for the multiplicative law : $(x, y)(x', y') = (x + x', y + (-1)^x y')$, with $x, x' \in \mathbb{Z}_2$ and $y, y' \in \mathbb{Z}_{\frac{n}{2}}$, and with the set of generators $S = \{(1, 2^i - 1), 0 \leq i \leq \Delta - 1\}$ [8].

In the following, we will be mostly interested by $W_{k-1,2^k-4}$, $W_{k-1,2^k-2}$ and $W_{k,2^k}$. For a better understanding, we give two examples of such Knödel graphs (cf. Fig. 3).

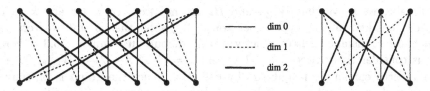

Fig. 3. $W_{3,14}$ and (right) $W_{3,8}$

3 Minimum Broadcast Graphs

3.1 General Results

Let us first recall the only known general results that exist concerning $MBGs$.

Theorem 3.1. *Let $B(n)$ be the size of a MBG of order n. Then :*

- $B(2^k) = k \cdot 2^{k-1}$ *for all* k *[4];*
- $B(2^k - 2) = (k - 1) \cdot (2^{k-1} - 1)$ *for all* $k \geq 3$ *[2, 9].*

3.2   Non-Isomorphic Families of $MBGs$ of Order $n = 2^k$

Thanks to the results above, we can prove the following Theorem.

Theorem 3.2. *There are at least three non-isomorphic families of graphs which are MBGs of order $n = 2^k$ for any $k \geq 4$. They are the following :*

- *The hypercube of dimension k, H_k ;*
- *The recursive circulant graph $G(2^k, 4)$;*
- *The Knödel graph $W_{k,n}$.*

Proof. We need to prove first that each family of graphs is MBG of order 2^k. As seen in the previous Theorem, the size $B(n)$ of a MBG of order $n = 2^k$ verifies $B(2^k) = k \cdot 2^{k-1}$ [4]. Since the broadcast time and the gossip time are equal when the number of nodes is even, and since $B(2^k) = G(2^k)$ for any k [11], we conclude that all the graphs which are MGG_{2^k} are MBG_{2^k} as well. Since it is well-known that those three families of graphs are $MGGs$ of order 2^k (cf. for this Proof of Theorem 5.2), we know that they are also $MBGs$.

Now let us prove that these three families of graphs are not isomorphic. It is easy to see that $G(2^k, 4)$ is not isomorphic to H_k for any $k \geq 2$ since, in that case, their diameter differ. Indeed, Park and Chwa [17] have proved that $Diam(G(2^k, 4)) = \lceil \frac{3k-1}{4} \rceil$, while it is well-known that $Diam(H_k) = k$.

Similarly, it is not difficult to see that $W_{k,2^k}$ and $G(2^k, 4)$ are not isomorphic for any $k \geq 3$. Indeed, Knödel graphs are bipartite by definition, and it is well known that there cannot exist cycles of odd length in bipartite graphs (see for instance [1], p.6). However, there always exists cycles of odd length in $G(2^k, 4)$ when $k \geq 3$. For instance, consider the cycle of length 5, $(0 - 1 - 2 - 3 - 4 - 0)$.

Finally, we prove that $W_{k,2^k}$ and H_k are not isomorphic for any $k \geq 4$. Indeed, it is known that hypercubes are $(0, 2)$-graphs, that is any two vertices which are at distance 2 have exactly two neighbours in common [15, 12]. Now let us consider the vertices $u = (1, 1)$ and $v = (1, 2^{k-1} - 2)$ in $W_{k,2^k}$. Those two vertices cannot be neighbours by definition, since the graph is bipartite. Actually, they are at distance 2, because there is an edge $(1, 1)(2, 1)$ (dimension 0) and an edge $(2, 1)(1, 2^{k-1} - 2)$ (dimension 2). Let N_u (resp. N_v) be the set of neighbours of u (resp. of v) in $W_{k,2^k}$. We have :

- $N_u = (2, 0) \cup \{(2, 2^p) \mid 0 \leq p \leq k - 2\}$;
- $N_v = (2, 2^{k-1} - 2) \cup (2, 2^{k-1} - 1) \cup \{(2, 2^p - 3) \mid 2 \leq p \leq k - 1\}$.

In that case, let us count the number of neighbours u and v have in common. Standard calculations show that, for any $k \geq 4$, u and v have only one neighbour in common, which is $(2, 1)$. Hence $W_{k,2^k}$ is not a $(0, 2)$-graph, which yields that it is not isomorphic to H_k for any $k \geq 4$. $\qquad\square$

Remark 3.1. In their paper, Fraigniaud and Peters [7] said that $W_{k,2^k}$ and H_k were not isomorphic for any $k \geq 4$, saying that there are no 4-cycles in $W_{k,2^k}$ with $k \geq 4$. Though the result is correct, the proof is not since there are always 4-cycles in $W_{k,2^k}$ for any $k \geq 2$. Take for instance the cycle $(1,0) - (2,1) - (1,1) - (2,0) - (1,0)$.

3.3 Families of $MBGs$ of Order $n = 2^k - 2$

Khachatrian et al. [9] and Dinneen et al. [2] have shown independently that the size of a MBG of order $n = 2^k - 2$ for any $k \geq 3$ is $B(2^k - 2) = (k-1) \cdot (2^{k-1} - 1)$. For this they used different proofs, but we can show that the family of graphs they gave as $MBGs$ is in both cases isomorphic to the Knödel graph $W_{k-1,2^k-2}$. Hence the following propositions.

Proposition 3.1. *The family of graphs given in [9] as $MBGs$ on $2^k - 2$ vertices is isomorphic to the Knödel graphs $W_{k-1,2^k-2}$.*

Proof. In [9], the authors gave a family of graphs which were $MBGs$ of order $n = 2^k - 2$. Let KH_n be this family of graphs. It is defined as follows. Let the vertex set be $V = \{v_0, v_1, \ldots v_{2^k-3}\}$, and the edge set $E = \{(v_i, v_j) \mid i+j \equiv 2^r - 1 \bmod (n)\}$ with $r \in \{1, 2, \ldots k-1\}$. Let us prove that this definition is equivalent to the one of $W_{k-1,2^k-2}$. First, note that KH_n is a bipartite graph : indeed, we can partition the set of vertices V in V_1 and V_2 as follows : let $V_1 = \{v_{2p} \mid 0 \leq p \leq \frac{n}{2} - 1\}$ and $V_2 = \{v_{2p+1} \mid 0 \leq p \leq \frac{n}{2} - 1\}$. Let us now identify V_i of KH_n ($i \in \{1, 2\}$) to the sets of vertices $V'_i = \{(i, p) \mid 0 \leq p \leq \frac{n}{2} - 1\}$ of $W_{k-1,n}$. For this, let us rename the vertices v_i ($i \in \{0, \ldots n-1\}$) of V in KH_n the following way :

- Let us rename v_{2p} as the couple $(1, \frac{n}{2} - p \bmod \frac{n}{2})$ for every $0 \leq p \leq \frac{n}{2} - 1$;
- Let us rename v_{2p+1} as the couple $(2, p)$ for every $0 \leq p \leq \frac{n}{2} - 1$;

In that case, it is not difficult to see that KH_n and $W_{k-1,n}$ are isomorphic. Indeed, let us come back to the definition of Khachatrian et al., i.e. there is an edge (v_i, v_j) iff $i + j \equiv 2^r - 1 \bmod n$, with $r \in \{1, 2, \ldots k-1\}$. W.l.o.g., let us consider i even, and consequently j odd. Let then $i = 2m$ and $j = 2l + 1$. By definition, v_i and v_j are neighbours iff $2m + 2l + 1 \equiv 2^r - 1 \bmod n$ for some $r \in \{1, \ldots k-1\}$. That is $m + l \equiv 2^{r'} - 1 \bmod \frac{n}{2}$ for some $r' \in \{0, \ldots, k-2\}$. Now let us replace v_i and v_j by their identifications above. We conclude that there is an edge $((1, m), (2, l))$ iff $l \equiv m + 2^{r'} - 1 \bmod \frac{n}{2}$ for some $r' \in \{0, \ldots, k - 2\}$. This is exactly the definition of the Knödel graph $W_{k-1,n}$. $\square$

Proposition 3.2. *The family of graphs given in [2] as $MBGs$ on $2^k - 2$ vertices is isomorphic to the Knödel graphs $W_{k-1,2^k-2}$.*

Proof. Dinneen et al. [2] gave $MBGs$ of order $2^k - 2$ by defining them as Cayley graphs on the dihedral group $D_{\frac{n}{2}}$ (with $n = 2^k - 2$), with generators $\alpha, \alpha\beta^1, \alpha\beta^3, \ldots, \alpha\beta^{2^{k-2}-1}$, and where $\alpha^2 = e$, $\alpha\beta\alpha^{-1} = \beta^{-1}$ and $\beta^{2^{k-1}-1} = e$.

To prove that this Cayley graph is isomorphic to $W_{k-1,2^k-2}$, it suffices to prove that this is a semi-direct product $\mathbb{Z}_{\frac{n}{2}} \ltimes \mathbb{Z}_2$ (with $n = 2^k - 2$) as defined in Remark 2.3. Identifiying $\alpha\beta^i$ to $(1,i)$ and β^i to $(0, \frac{n}{2} - i \bmod (\frac{n}{2}))$ for all $0 \leq i \leq \frac{n}{2} - 1$ gives us directly the result. $\qquad\square$

4   Minimum Broadcast Digraphs

4.1   General Results

Let us first recall the only known general results that exist concerning MBDs.

Theorem 4.1. *Let $\vec{B}(n)$ be the size of a MBD of order n. Then :*

- $\vec{B}(2^k) = k \cdot 2^k$ *for all k [13];*
- $\vec{B}(2^k - 1) = (k-1) \cdot (2^k - 1)$ *for all $k \geq 3$ [5];*
- $\vec{B}(2^k - 2) = (k-1) \cdot (2^k - 2)$ *for all $k \geq 3$ [5].*

4.2   Non-Isomorphic Families of MBDs of Order n for $2^k - 2 \leq n \leq 2^k$

Since we deal with directed graphs in this section, we need to introduce the following notion.

Definition 4.1. *Let us consider an undirected graph G. We will denote by G^* the directed graph obtained by replacing each edge of G by two arcs in opposite directions.*

Before proving the main Theorem (Theorem 4.3), let us focus on a special case of circulant digraphs of order $n = 2^k$, and let us show that they are MBDs. This is done in Theorem 4.2.

Theorem 4.2. *Let $n = 2^k$ and let $a_i = 2^i$ for all $0 \leq i \leq k - 1$. Let us choose $k-1$ distinct a_i among the k existing ones. We call them the b_j ($1 \leq j \leq k-1$), such that $b_1 < b_2 \ldots < b_{k-1}$. Then $C'_n(1, 1+b_1, 1+b_1+b_2, \ldots, 1+\sum_{j=1}^{k-1} b_j)$ is a MBD.*

Proof. First, note that each of those digraphs are of size $k \cdot 2^k$. Since $\vec{B}(2^k) = k \cdot 2^k$ by Theorem 4.1, it remains to show that any vertex can broadcast in minimum time in each of these digraphs. This result can be proved by induction on k. When $k = 2$, we have $a_0 = 1$ and $a_1 = 2$. Then we can build two circulant digraphs, namely $C'_4(1,2)$ (when $b_1 = a_0$) and $C'_4(1,3)$ (when $b_1 = a_1$). It is not difficult to see that those two digraphs are MBDs. Now suppose that the Theorem is true for k, and let us show it is then true for $k + 1$. The key idea here is to partition the vertex set in two distinct subsets, such that each of the digraphs induced by each of the subsets is one of the MBDs of order 2^k constructed as above. For this, let us distinguish two cases :

- $b_1 = a_1 = 2$. In that case, we know that $b_i = a_i$ for any $1 \leq i \leq k$, that is the circulant digraph is $C'_{2^{k+1}}(1, 3, 7 \ldots 2^{k+1} - 1)$. In [5], it has been shown that this family of circulant digraphs is MBD for any k.
- $b_1 = a_0 = 1$. In that case, we can see that the digraph constructed will be of the form : $C'_{2^{k+1}}(1, 2, \alpha_1, \ldots, \alpha_{k-1})$, where each of the α_i is even. Let us then partition the set V of vertices in two distinct subsets V_0 and V_1 as follows : $V_0 = \{v_{2i} \mid 0 \leq i \leq 2^{k-1} - 1\}$ and $V_1 = \{v_{2i+1} \mid 0 \leq i \leq 2^{k-1} - 1\}$. It is interesting to note that, in that case, the digraph induced by V_0 (resp. V_1) is isomorphic to one of the circulant digraphs of order 2^k constructed by our method. Indeed, let us distinguish two more cases.

  1. If $b_2 = a_2 = 4$, the circulant digraph constructed will be $C'_{2^{k+1}}(1, 2, 6, 14, \ldots, 2^{k+1} - 2)$, and the digraph induced by V_0 (resp. V_1) is $C'_{2^k}(1, 3, 7, \ldots 2^k - 1)$.
  2. Otherwise, the digraph induced by V_0 (resp. V_1) is the circulant digraph of order 2^k constructed with the parameters $b'_i = \frac{b_{i+1}}{2}$ for all $1 \leq i \leq k$.

In all cases, we see that either the circulant digraph constructed is $C'_{2^{k+1}}(1, 3, 7 \ldots 2^{k+1} - 1)$, which is known to be MBD, or that it can be recursively constructed with two copies of a circulant digraph of order 2^k (constructed by our method) joined by the following Hamiltonian circuit : $v_0 - v_1 - v_2 - \ldots - v_{2^{k+1}-1} - v_0$. Since we supposed that those two copies are $MBDs$ of order 2^k, we can prove now the Theorem : indeed, we deal with circulant digraphs, hence they are vertex-transitive. Consequently, we can focus on broadcasting from one particular vertex, say v_0. In that case, the broadcast scheme from v_0 is the following : first, broadcast to v_1 during the first time unit. This is possible since there is a directed edge (v_0, v_1) by definition. Now, since $v_0 \in V_0$ and $v_1 \in V_1$, let v_0 (resp. v_1) broadcast in the copy of the MBD of order 2^k induced by V_0 (resp. V_1) between time units 2 and $k+1$. This completes broadcasting from v_0 in $k+1$ time units. Hence, each of those digraphs of dimension $k+1$ is a MBD. □

Proposition 4.1. *For $n = 2^k$, there exist k circulant digraphs constructed as in Theorem 4.2. Moreover, those k digraphs are non-isomorphic one to another.*

Proof. Since we choose $k - 1$ distinct b_i among k a_i, it directly follows that we have k possibilities, hence we can construct k circulant digraphs of order 2^k. Let us call them the $C'^i_{2^k}$, with $1 \leq i \leq k$, such that $a_{i-1} \notin \{b_j \mid 1 \leq j \leq k - 1\}$. To prove that any two such digraphs are non-isomorphic, let us focus on those directed edges (u, v) such that there is a directed edge (v, u). Let us call them *symmetric edges*. Let us now consider a circulant digraph $C'^i_{2^k}$; let us delete all the non-symmetric directed edges, and replace all the pairs of symmetric edges by a single undirected edge. In that case, if we show that the (undirected) graph $G^i_{2^k}$ obtained this way is non-isomorphic to any of the other $G^j_{2^k}$ for $j \neq i$, it follows directly that each $C'^i_{2^k}$ is non-isomorphic to any $C'^j_{2^k}$ with $j \neq i$.

Let us show, by induction on k, that $G^{i+1}_{2^k}$ is composed of 2^i copies of cycles of length 2^{k-i} for all $0 \leq i \leq k-1$ (note that , when $i = k-1$, the cycle of length 2 will be considered here to be K_2). When $k = 2$, C'^1_4 is $C'_4(1, 3)$ and C'^2_4 is $C'_4(1, 2)$.

Then G_4^1 is the cycle of length 4, and G_4^2 is composed of two copies of the complete graph K_2. Now suppose the hypothesis is true for any k, and let us prove it for $k + 1$. As seen in the proof of Theorem 4.2, k of the $k + 1$ circulant digraphs of order 2^{k+1} can be built using two copies of a circulant digraph of order 2^k, and joining them by the hamiltonian circuit $v_0 - v_1 - v_2 - \ldots - v_{2^{k+1}-1} - v_0$; the $k + 1$-th digraph being $C'_{2^{k+1}}(1, 3, 7, \ldots 2^{k+1} - 1)$. For each $1 \leq i \leq k$, we can see that $C'^{i+1}_{2^{k+1}}$ is the digraph constructed using two copies of $C'^i_{2^k}$, and that $C'^1_{2^{k+1}}$ is $C'_{2^{k+1}}(1, 3, 7, \ldots 2^{k+1} - 1)$. In that case, we know by construction that $G^1_{2^{k+1}}$ is the cycle of length 2^{k+1}, and since each $C'^i_{2^{k+1}}$ is constructed from two copies of $C'^{i-1}_{2^k}$ ($2 \leq i \leq k + 1$) joined by a hamiltonian circuit, $G^i_{2^{k+1}}$ will be the union of two copies of $G^{i-1}_{2^k}$ for all $2 \leq i \leq k + 1$. Hence each $G^i_{2^{k+1}}$ is composed of 2^{i-1} copies of cycles of length $2^{k+1-(i-1)}$ for all $1 \leq i \leq k + 1$, and the result is proved by induction. □

Thanks to the Theorem and Proposition above, and to some results given in [13, 17, 5], we have the following Theorem.

Theorem 4.3 ([13, 17, 5]). *For the following values of n, the families of digraphs listed below are non-isomorphic families of MBDs :*

- *For $n = 2^k$, the directed hypercube H_k^*, the directed Knödel graph $W_{k,n}^*$, the directed recursive circulant graph $G^*(n, 4)$, and the k circulant digraphs C'^i_n with $1 \leq i \leq k$ are MBDs ;*
- *For $n = 2^k - 1$, the circulant digraph $C'_n(1, 3, \ldots 2^{k-1} - 1)$ is a MBD ;*
- *For $n = 2^k - 2$, the directed Knödel graph $W_{k-1,n}^*$ and the circulant digraph $C'_n(1, 3, \ldots 2^{k-1} - 1)$ are MBDs.*

Proof. Most of these properties have already been shown in [13], [5], Theorem 4.2 and Proposition 4.1. Some others remain to be proved, as done in the following. Let us prove first that $G^*(2^k, 4)$ is a MBD. This is straightforward since $\vec{B}(2^k) = k \cdot 2^k$, and since $\vec{b}(G^*(2^k, 4)) = b(G(2^k, 4))$ for any n (indeed, the broadcast scheme used in $G(2^k, 4)$ remains valid in $G^*(2^k, 4)$).

Now let us prove that any $C'^i_{2^k}$ ($1 \leq i \leq k$) is not isomorphic to H_k^* (resp. $G^*(2^k, 4)$, $W_{k,2^k}^*$). For this, note that for any directed edge (u, v) in H_k^* (resp. $G^*(2^k, 4)$, $W_{k,2^k}^*$), there exists a directed edge (v, u) by definition. However, in $C'^1_{2^k}$, there is a directed edge (v_0, v_3) and no directed edge (v_3, v_0) for any $k \geq 3$; moreover, in any $C'^i_{2^k}$ with $2 \leq i \leq k$, there is a directed edge (v_0, v_1) and no directed edge (v_1, v_0) for any $k \geq 2$. Hence there is no isomorphism between H_k^* (resp. $G^*(2^k, 4)$, $W_{k,2^k}^*$) and $C'^i_{2^k}$ for any $1 \leq i \leq k$. This completes the proof of the Theorem. □

5   Minimum Gossip Graphs

5.1   General Results

Let us first recall the only known general results that exist concerning MGGs. These results can be found in [11].

Theorem 5.1. *Let $G(n)$ be the size of a MGG of order n. Then :*

- $G(2^k) = k \cdot 2^{k-1}$ *for all k ;*
- $G(2^k - 2) = (k-1) \cdot (2^{k-1} - 1)$ *for all $k \geq 4$;*
- $G(2^k - 4) = (k-1) \cdot (2^{k-1} - 2)$ *for all $k \geq 6$;*

5.2   Non-Isomorphic Families of MGGs on 2^k vertices

Thanks to the results above, and thanks to Theorem 3.2, we can prove the following Theorem.

Theorem 5.2. *There are at least three non-isomorphic families of graphs which are MGGs of order $n = 2^k$ for any $k \geq 4$. They are the following :*

- *The hypercube of dimension k, H_k ;*
- *The recursive circulant graph $G(2^k, 4)$;*
- *The Knödel graph $W_{k,n}$.*

Proof. In Theorem 3.2, the non-isomorphism of those three families has been proved. It remains to prove here that those three families of graphs are MGGs of order 2^k. For the hypercube, it is a very well-known property (see for instance [11]). The recursive circulant $G(2^k, 4)$ is also a MGG. For this, we refer to [17, 14]. Moreover, $W_{k,2^k}$ is also a MGG on $n = 2^k$ nodes, since $W_{k,2^k}$ is underlying Knödel's proof [10] that it is possible to gossip in $\lceil \log_2(n) \rceil$ time units when n is even.   □

5.3   Families of MGGs of Order $n = 2^k - 2$ and $n = 2^k - 4$

In [11], Labahn proved the exact value of $G(n)$ for $n = 2^k - 2$ and $n = 2^k - 4$. For this, he displayed graphs which were gossip graphs and had the minimum number of edges. It appears that these graphs he gave as examples of MGGs are isomorphic to the Knödel graphs $W_{k-1,n}$, as proved in the following Proposition.

Proposition 5.1.
The family of graphs given in [11] as MGGs on n vertices :

- *Is isomorphic to the Knödel graphs $W_{k-1,2^k-2}$ when $n = 2^k - 2$;*
- *Is isomorphic to the Knödel graphs $W_{k-1,2^k-4}$ when $n = 2^k - 4$.*

Proof. In [11], Labahn gave a family of graphs that were MGGs of order $n = 2^k - 2$. He pointed out that these graphs are isomorphic to the Cayley graphs on the dihedral group $D_{\frac{n}{2}}$ that Dinneen et al. [2] gave as examples of MBGs of the same order. By Proposition 3.2, the result follows directly : this graph is isomorphic to the Knödel graph $W_{k-1,2^k-2}$.

In the case $n = 2^k - 4$, Labahn [11] showed that the graphs he gave as MGGs of order n are Cayley graphs defined on the dihedral group $D_{\frac{n}{2}}$, this time with $n = 2^k - 4$, and with the same set of generators as above. Since this remains a semi-direct product as defined in Remark 2.3 (the only difference here being that $n = 2^k - 4$), it follows that this graph is isomorphic to the Knödel graph $W_{k-1,2^k-4}$.   □

6   Minimum Linear Gossip Graphs ($MLGGS$)

These graphs have been studied by Fraigniaud and Peters [7]. They proved that $G_{\beta,\tau}(n) = G_{1,1}(n)$ for any $\beta > 0$ and $\tau > 0$, that is the structure of a $MLGG$ does not depend on β and τ. In this section, we consider $\beta \neq 0$ and $\tau \neq 0$. Fraigniaud and Peters proved the following.

Theorem 6.1 ([7]). *For the following values of even n, we have :*

- *For $n = 2^k$, $G_{\beta,\tau}(n) = k \cdot 2^{k-1}$, and H_k and $W_{k,2^k}$ are $MLGGs$;*
- *For all $2^k - 6 \leq n \leq 2^k - 2$ with $k \geq 4$, $G_{\beta,\tau}(n) = \frac{n(k-1)}{2}$ and $W_{k-1,n}$ is a $MLGG$.*

Moreover, we have the following proposition.

Proposition 6.1. *For all $n = 2^k$, $G(2^k, 4)$ is a Minimum Linear Gossip Graph.*

Proof. It appears that the scheme used in [14] to achieve gossiping in minimum time in the *unit cost* model in $G(2^k, 4)$ also allows to achieve gossiping in minimum time in the *linear cost* model. This can be proved by induction on n. Indeed, it is not difficult to see that it is possible to gossip in minimum time, that is in $g_{\beta,\tau}(2^k) = k\beta + (2^k - 1)\tau$, in $G(2^k, 4)$ when $k = 1$ and $k = 2$.

Now suppose it is possible to gossip in time $g_{\beta,\tau}(2^k)$ in $G(2^k, 4)$ for some $k \geq 1$. We use the fact that $G(2^{k+2}, 4)$ can be built from four copies of $G(2^k, 4)$, as stated in [17, 14]. Let us gossip independently in each of the four copies of $G(2^k, 4)$. This takes $g_{\beta,\tau}(2^k)$, and at the end of the process, each vertex of each copy G_i ($i \in [1; 4]$) of $G(2^k, 4)$ knows all the information contained in its own copy. During the next step, let each vertex of G_0 (resp. of G_2) communicate with the vertex of G_1 (resp. of G_3) to which it is adjacent. This takes $\beta + 2^k\tau$, and, after that round, each vertex knows 2^{k+1} different pieces of information. Now, during the last step, let each vertex of G_0 (resp. of G_1) communicate with the vertex of G_2 (resp. of G_3) to which it is adjacent. This takes $\beta + 2^{k+1}\tau$, and at the end of this step, gossiping is completed.

On the whole, the time used to gossip in $G(2^{k+2}, 4)$ is $g_{\beta,\tau}(2^k) + \beta + 2^k\tau + \beta + 2^{k+1}\tau$. Since $g_{\beta,\tau}(2^k) = k\beta + 2^k\tau$ by hypothesis, we know that gossiping can be achieved in minimum time in $G(2^{k+2}, 4)$. This proves the result by induction. □

7   Summary of the Results

The different results concerning non-isomorphic families of graphs (resp. digraphs) which are MBG, $MBDs$, $MGGs$ and/or $MLGGs$ are displayed in Figs. 4 and 5. In addition to these results, we can note the following : in the undirected case, suppose we have two graphs of order n which are $MBGs$, namely G_1 and G_2. If $B(2n) = 2 \cdot B(n) + n$, then any graph obtained by joining the vertices of G_1 to the vertices of G_2 by a perfect matching is still a MBD of order $2n$. Indeed, it has $B(2n)$ edges, and any vertex will be able to broadcast in minimum time (broadcast through the edge of the perfect matching first, then broadcast in each of the two MBGs of order n). This is similar to Farley's construction [3], except that the perfect matching here could correspond to any permutation of the indices between the n vertices u_i of G_1 and the n vertices v_i of G_2 ($1 \leq i \leq n$). In particular, when $n = 2^k$, this method can be applied. Moreover, it can be applied recursively, with the possibility to take different permutations from one step to another. This allows us to get many non-isomorphic MBGs of order 2^k. It is easy to see that this method also works for $n = 2^k$ in the case of MGGs and $MLGG$s (in that case, it is necessary to gossip first in each copy, before using the last time unit to exchange information using the edges of the perfect matching).

This method also works in the directed case for MBDs, still with $n = 2^k$. In that case, instead of considering a perfect matching between two copies, it is necessary to consider two perfect matchings : one with arcs going from G_1 to G_2, the other with arcs going from G_2 to G_1.

Finally, we also see that this method can be applied, this time only once, when $n = 2^k - 2$ to obtain MGGs and $MLGG$s of order $2^{k+1} - 4$, and when $n = 2^k - 1$ to obtain MBDs of order $2^{k+1} - 2$.

Hence we have the following Proposition.

Proposition 7.1.

- *In the case $n = 2^k$, the method described above gives us recursively families of MBGs (resp. MBDs, MGGs, $MLGG$s) of order 2^{k+1} ;*
- *In the case $n = 2^k - 1$, the method above, used once, gives us families of MBDs of order $2^{k+1} - 2$;*
- *In the case $n = 2^k - 2$, the method described above, used once, gives us families of MGGs (resp. $MLGG$s) of order $2^{k+1} - 4$.*

Note that the problem of determining how many of the families constructed as above are non-isomorphic is still open. However, since there exists $n!$ permutations on n vertices, and since it is possible to use different permutations when the graphs are constructed recursively, one may think that the number of non-isomorphic graphs obtained is relatively high.

8   Conclusion

This paper aims to gather information concerning the known general values of $G(n)$, $G_{\beta,\tau}(n)$, $B(n)$ and $\tilde{B}(n)$, and above all, to give in each possible case as many non-isomorphic families of graphs which are MGGs, $MLGG$s, MBGs and/or MBDs as possible. This has been done by gathering the results from various authors. Moreover, it appears that, in the undirected case, for $n = 2^k - 2$ and $n = 2^k - 4$, the families of graphs given by the authors (namely [11] for the MGGs, and [9] and [2] for the MBDs) are always isomorphic to $W_{k-1,n}$. Though we know very little about the size of MGGs, $MLGG$s, MBGs and MBDs in

	Gossiping			
	MGG		MLGG	
n	$G(n)$	Graphs	$G_{\beta,\tau}(n)$	Graphs
2^k	$\frac{nk}{2}$	H_k $G(2^k,4)$ $W_{k,2^k}$	$\frac{nk}{2}$	H_k $G(2^k,4)$ $W_{k,2^k}$
2^k-2	$\frac{n(k-1)}{2}$	$W_{k-1,2^k-2}$	$\frac{n(k-1)}{2}$	$W_{k-1,2^k-2}$
2^k-4	$\frac{n(k-1)}{2}$	$W_{k-1,2^k-4}$	$\frac{n(k-1)}{2}$	$W_{k-1,2^k-4}$
2^k-6			$\frac{n(k-1)}{2}$	$W_{k-1,2^k-6}$

Fig. 4. Sum-up of the results : Gossiping

	Broadcasting			
	MBG		MBD	
n	$B(n)$	Graphs	$\vec{B}(n)$	Graphs
2^k	$\frac{nk}{2}$	H_k $G(2^k,4)$ $W_{k,2^k}$	nk	H_k^* $G^*(2^k,4)$ $W_{k,2^k}^*$ $C_{2^k}'^{i}$ with $1\le i \le k$
2^k-1			$n(k-1)$	$C_{2^k-1}'(1,3,\ldots 2^{k-1}-1)$
2^k-2	$\frac{n(k-1)}{2}$	$W_{k-1,2^k-2}$	$n(k-1)$	$W_{k-1,2^k-2}^*$ $C_{2^k-2}'(1,3,\ldots 2^{k-1}-1)$

Fig. 5. Sum-up of the results : Broadcasting

the general case, we found it interesting to sum-up the results and give, as far as we know, all the non-isomorphic families of graphs that respect these properties.

It is interesting to note too that Knödel graphs seem to play an important role in these communications patterns, since they are $MGGs$, $MLGGs$, $MBGs$ and $MBDs$ in every known case of even order. Moreover, we believe that this family of graphs has many more interesting characteristics, such as the fact that $W_{k-2,n}$ is a gossip (hence broadcast) graph for any even n such that $2^{k-1}+2 \le n \le 3\cdot 2^{k-2}-4$ (cf. [6]).

In a word, we believe our study could be useful as a handbook of non-isomorphic graphs being either $MGGs$, $MLGGs$, $MBDs$ and/or $MBDs$.

References

1. J. de Rumeur. *Communications dans les réseaux d'interconnexion.* Masson, 1994.
2. M.J. Dinneen, M.R. Fellows, and V. Faber. Algebraic constructions of efficient broadcast networks. *Proc. of Applied Algebra, Algorithms and Error Correcting Codes (AAECC'91)*, 539:152–158, 1991. Lectures Notes in Computer Science, Springer-Verlag, Berlin.
3. A. Farley. Minimal broadcast networks. *Networks*, 9:313–332, 1979.

4. A. Farley, S. Hedetniemi, S. Mitchell, and A. Proskurowski. Minimum broadcast graphs. *Discrete Mathematics*, 25:189–193, 1979.
5. G. Fertin. On the structure of minimum broadcast digraphs. Technical Report RR-1173-97, Laboratoire Bordelais de Recherche en Informatique, 1997. Submitted.
6. G. Fertin. A study of minimum gossip graphs. Technical Report RR-1172-97, Laboratoire Bordelais de Recherche en Informatique, 1997. Submitted.
7. P. Fraigniaud and J.G. Peters. Minimum linear gossip graphs and maximal linear (δ, k)-gossip graphs. Technical Report CMPT TR 94-06, Simon Fraser University, Burnaby, B.C., 1994.
8. M-C. Heydemann. Private communication, 1998.
9. L.H Khachatrian and H.S. Haroutunian. Construction of new classes of minimal broadcast networks. *Proc. Third International Colloquium on Coding Theory*, pages 69–77, 1990.
10. W. Knodel. New gossips and telephones. *Discrete Mathematics*, 13:95, 1975.
11. R. Labahn. Some minimum gossip graphs. *Networks*, 23:333–341, 1993.
12. J.-M. Laborde and S.P. Rao Hebbare. Another characterization of hypercubes. *Discrete Math.*, 39:161–166, 1982.
13. A.L. Liestman and J.G. Peters. Minimum broadcast digraphs. *Discrete Applied Mathematics*, 37/38:401–419, 1992.
14. C. Micheneau. *Graphes Récursifs Circulants : Structure et Communications. Communications Vagabondes et Simulation*. PhD thesis, Université Bordeaux I, 1996.
15. M. Mulder. $(0, \lambda)$-graphs and n-cubes. *Discrete Math.*, 28:179–188, 1979.
16. J-H. Park and K-Y. Chwa. On the construction of regular minimal broadcast digraphs. *Theoretical Computer Science*, 124:329–342, 1994.
17. J.-H. Park and K.-Y. Chwa. Recursive circulant : a new topology for multicomputers networks (extended abstract). In *Proc. Int. Symp. Parallel Architectures, Algorithms and Networks ISPAN'94, Kanazawa, Japan*, pages 73–80, 1994.

Optical All-to-All Communication in Inflated Networks

Olivier Togni*

LaBRI, UMR 5800, Université Bordeaux I
351 cours de la libération
33405 TALENCE (FRANCE)
e-mail: togni@labri.u-bordeaux.fr

Abstract. The problem of all-to-all communication in all-optical networks consist of designing directed paths between all ordered pairs of nodes and assigning them a wavelength, such that every two dipaths sharing a link have distinct wavelengths. The parameter to be minimised is the number of colors. We determine bounds on the number of wavelengths needed for optical networks that are obtained by replacing each node of a network by a complete subgraph.

1 Introduction

In all-optical networks, the information, once transmitted as light, reaches its final destination directly without being converted to electronic form in between, thus allowing a high speed transmission. Multiple messages can be transmitted across the same channel simultaneously as long as they use distinct wavelengths. This technique is known as Wavelength-Division Multiplexing (WDM). In practice, each link can support a limited number of wavelengths (actually between 4 and 100). Therefore, a typical problem is to assign wavelengths (or colors) to each path of a routing in the network with the constraint that no two paths sharing a link get the same color. Thus for a given graph, one must find a routing and a coloring function that minimize the number of wavelengths needed. A good survey on results and problems concerning communications in all-optical networks has been written by B. Beauquier et al. [3].

An interesting graph-transformation is to replace each vertex of a graph by a complete subgraph, then obtaining what is called an *inflated* graph.

In this paper, we determine bounds on the number of wavelengths needed for an inflated network, according to the number of wavelengths needed for the network itself.

2 Definitions and Notation

The network is represented by a directed symmetric graph $G = (V, A)$, where the set of vertices V represents the nodes of the network and the set of arcs A represents optical links.

* This research was supported by the ALTEC-KIT, Project $n°$ INCO-COP 96-0195

Sometimes, to simplify, we will represent the network by an undirected graph (for the figures, in particular), but we have to keep in mind that an edge represent two opposite arcs.

Let $G = (V, A)$ be a directed symmetric graph of order n.

A routing R is a set of $n(n-1)$ dipaths between all pair of vertices: $R = \{R(x, y),\ x, y \in V(G),\ x \neq y\}$.

For a routing R, the load $\vec{\pi}(a, R)$ of an arc a is the number of paths of R going through a. The *arc-forwarding index* $\vec{\pi}(R)$ of a routing R is the maximum number of paths going through any arc of G: $\vec{\pi}(R) = \max_{a \in A} \vec{\pi}(a, R)$. The arc-forwarding index $\vec{\pi}(G)$ of G is defined as $\vec{\pi}(G) = \min_R \vec{\pi}(R)$.

Let $\vec{w}(R)$ denote the minimum number of colors necessary to color all the paths of the routing R, with the condition that no two dipaths sharing an arc have the same color. We denote by $\vec{w}(G) = \min_R \vec{w}(R)$ the minimum number of colors needed over all routings R of G.

Given a graph G and a routing R, the *conflict* graph $G_c(R)$ is an undirected graph with vertex set R, and such that two dipaths are adjacent in $G_c(R)$ if and only if they share an arc of G. Therefore, the chromatic number of the conflict graph gives us the number of wavelengths needed to color all the paths of the routing R: $\vec{w}(R) = \chi(G_c(R))$.

Trivially, $\vec{w}(G) \geq \vec{\pi}(G)$, and in [6], the authors asked the following question: Does the equality $\vec{w}(G) = \vec{\pi}(G)$ holds for all symmetric graph G?

This equality has been proved for paths, cycles, hypercube [6], toroidal mesh of even side [2], trees [9], clique-compound graphs [1], cartesian product of cycles and paths [12].

In [7], Dunbar and Haynes called *inflated* graph, the graph obtained by the transformation shown in Fig. 1:

Definition 1 *The inflated graph $I(G)$ of a graph G is the graph obtained by replacing each vertex x of G of degree $d(x)$ by a complete graph K_x of order $d(x)$ and each edge $xy \in E(G)$ is replaced by an edge between K_x and K_y such that for all $x \in V(G)$, each vertex in K_x has degree $d(x)$.*

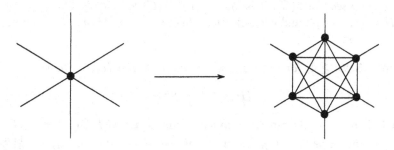

Fig. 1. Vertex in G and corresponding clique in $I(G)$

This is an interesting transformation since it preserves the degree of the graph and increases the order.

Note that if $S(G)$ denote the subdivided graph of G (the graph obtained by replacing each edge by a path of length 2), the graph $I(G)$ can be seen as the line-graph of $S(G)$: $I(G) = L(S(G))$.

Definition 2 *A graph G is $[k, w]$-colorable if there exists a routing R and a coloring function c of the paths of R, c: $R \rightarrow \{1, 2, \ldots, w\}$, such that:*

i) c is a correct coloring, i-e two paths sharing a same arc have different colors.
ii) $\forall v \in \{1, \ldots, w\}$, $\forall x \in V(G)$, there is at most k paths of color v that begin at x and at most k paths of color v that end at x.

Our definition of $[k, w]$-colorability slightly differ from the one given in [13] which had one more condition.

Note that if G is $[k, w]$-colorable then $k \leq r$ and $\vec{w}(G) \leq w$.

3 Number of wavelengths for Inflated graphs

Proposition 3 *Let G be an r-regular $[k, w]$-colorable graph, $r \geq 3$.*

- *If $r \geq k + \sqrt{2k - 1}$ then $I(G)$ is $[k, r^2 w]$-colorable*
- *If $r < k + \sqrt{2k - 1}$ then $\vec{w}(I(G)) \leq (r^2 + 2r - 1)\vec{w}(G)$.*

Let K_p be the complete graph of order p and let K_p^n denote the n-dimensional generalized hypercube, defined as $K_p^n = K_p^{n-1} \times K_p$ and $K_p^1 = K_p$.

As shown in the following proposition, there exists an infinite family of graphs for which Proposition 3 determines the exact value of the parameter $\vec{w}$.

Proposition 4 $\forall n \geq \frac{p+\sqrt{2p-1}}{p-1}$, *for the n-dimensional generalized hypercube K_p^n, we have*

$$\vec{w}(I(K_p^n)) = (n(p-1))^2 p^{n-1}$$

Proof. K_p^n is $[p, p^{n-1}]$-colorable (see [13]), so by Proposition 3, if $n(p-1) \geq p + \sqrt{2p - 1}$, we have the inequality $\vec{w}(I(K_p^n)) \leq (n(p-1))^2 p^{n-1}$. If G has order N and edge-bisection $\beta(G)$, it is known that $\vec{w}(G) \times \beta(G) \geq N^2/4$. $I(K_p^n)$ has order $N = n(p-1)p^n$ and edge-bisection $\beta(I(K_p^n)) = p^{n+1}/4$, so we obtain the desired relation. $\square$

Corollary 5 $\forall n \geq 4$, *for the hypercube of dimension n H_n, we have*

$$\vec{w}(I(H_n)) = n^2 2^{n-1}$$

Remark 1 *there exists r-regular graphs G for which $\vec{w}(I(G)) > r^2 \vec{w}(G)$. An example is the cycle $G = C_{4p+1}$. In [6] it was shown that $\vec{w}(C_n) = \lceil \frac{1}{2} \lfloor \frac{n^2}{4} \rfloor \rceil$. We have $r = 2$ and $\vec{w}(G) = p(2p + 1)$. $I(G)$ is isomorphic to C_{8p+2} therefore, $\vec{w}(I(G)) = 8p^2 + 4p + 1 > 4(p(2p + 1)) = r^2 \vec{w}(G)$.*

4 Proof of Proposition 3

Let G be a $[k, w]$-colorable, r-regular graph of order n. Let R and c be respectively a routing in G and a coloring function which realize the $[k, w]$-colorability.

Let $G' = I(G)$. The idea of the proof is to exhibit a routing R' and a coloring function c' of G'.

Let $N(x)$ denote the set of neighbors of x.

Set $V(G') = \{(x, i), x \in V(G), 1 \le i \le r\}$ and
$$A(G') = \{((x, i), (x, j)), \ 1 \le i \ne j \le r\} \cup \{((x, i_{xy}), (y, i_{yx})) \, / (x, y) \in A(G)\},$$
with $\{i_{xy}, y \in N(x)\} = \{1, \dots, r\}$.

Routing R' of G':

A natural way of obtaining a routing R' of G' is to extend the routing R as follow:

- $R'((x, i), (x, j)) = (x, i), (x, j)$,
- $R'((x, i), (y, j)) = (x, i), (x, i_{xx_1}), (x_1, i_{x_1x}), (x_1, i_{x_1x_2}), \dots, (y, i_{yx_{p-1}}), (y, j)$
  where $R(x, y) = x, x_1, x_2, \dots, x_p = y$.

To each path in G, correspond r^2 paths in G', as illustrated in Fig. 2.

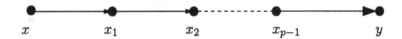

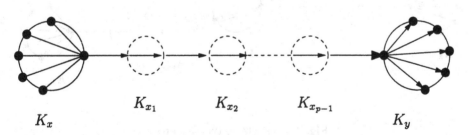

Fig. 2. Path in G and the r^2 corresponding paths in G' (circles represent r-cliques)

Coloring c' of R':

The idea is to color a the set of paths $\{R'((x, i), (y, j)); \ 1 \le i, j \le r\}$ with the set of colors $\{(v, \alpha); \ 1 \le \alpha \le r^2\}$, if the path $R(x, y)$ has color v in G.

Let $G'_c = G'_c(R')$ be the conflict graph associated with R'.

Let $P_v = \{(x,y), x, y \in V(G), c(R(x,y)) = v\}$ and let $\Gamma_v = \{R'((x,i),(y,j))\,/(x,y) \in P_v,\ 1 \le i,j \le r\}$. We denote by G'_v the subgraph of G'_c induced by Γ_v.

Since the paths of P_v have the same color in G, it is easily seen that if two paths $\alpha = R'((x_1,i_1),(y_1,j_1))$ and $\beta = R'((x_2,i_2),(y_2,j_2))$ share a common arc in G' then $(x_1 = y_1$ and $x_2 = y_2)$ or $(y_1 = x_2)$ or $(y_2 = x_1)$.

For all $x,y \in V(G)$, since G is $[k,w]$-colorable, there are $k' \le k$ paths of color v that end at x and $k'' \le k$ paths of color v that start at y.

If $(t_l,x) \in P_v$ for $1 \le l \le k'$ and $(y,z_l) \in P_v$ for $1 \le l \le k''$, suppose $R(t_l,x) = (t_l, \dots, \theta_l, x)$, $R(y,z_l) = (y, \omega_l, \dots, z_l)$ and $R(x,y) = (x, x_1, \dots, x_p = y)$.

So, given $(x,y) \in P_v$, $\{R'((x,i),(y,j)),\ 1 \le i,j \le r\}$ form a complete graph of order r^2 in G'_v that we denote by K_{xy} and $\{R'((t_l,i),(x,i_{xx_1})), R'((x,i_{x\theta_l}),(y,j)),\ 1 \le i,j \le r\}$ form a complete graph K_{2r}. $\{R'((x,i),(y,i_{y\omega_l})), R'((y,i_{yx_{p-1}}),(z_l,j)),\ 1 \le i,j \le r\}$ also form a complete graph K_{2r}. The conflict arcs can be seen in Fig. 3.

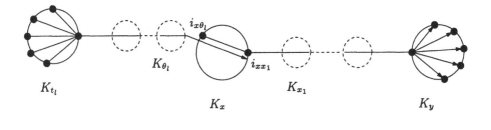

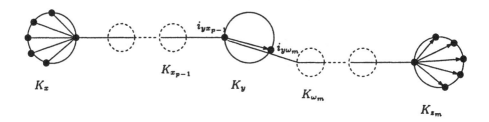

Fig. 3. Conflicts on type K_r arcs

Let $I'(x) = \{i_{x\theta_l}, 1 \le l \le k'\}$ and let $I(x)$ be any subset of $\{1, \dots, r\}$ verifying $I(x) \supset I'(x)$, $|I(x)| = k$. Let $J'(y) = \{i_{y\omega_l}, 1 \le l \le k''\}$ and $J(y) \subset \{1, \dots, r\}$, $J(y) \supset J'(y)$, $|J(y)| = k$.

We are going to partition the set of r^2 paths corresponding to a path $R(x,y)$ of G in three types of paths A, B and C, that we will color separately.

For $i \in I(x)$, $A_i(x,y) = \{R'((x,i),(y,j)), j \notin J(y)\}$

For $j \in J(y)$, $A'_j(x,y) = \{R'((x,i),(y,j)), i \notin I(x)\}$

$A(x, y) = (\cup_{i \in I(x)} A_i(x, y)) \bigcup (\cup_{j \in J(y)} A'_j(x, y))$.

$B(x, y) = \{R'((x, i), (y, j)), \ i \in I(x), j \in J(y)\}$.

$C(x, y) = \{R'((x, i), (y, j)), \ i \notin I(x), j \notin J(y)\}$.

For $(x, y) \in P_v$, $A(x, y) \cup B(x, y) \cup C(x, y)$ is a partition of K_{xy}.

Let $A = \cup_{(x,y) \in P_v} A(x, y)$, $B = \cup_{(x,y) \in P_v} B(x, y)$, and $C = \cup_{(x,y) \in P_v} C(x, y)$. $A \cup B \cup C$ is then a partition of G'_v.

Coloring type A paths: Let G'_A be the graph obtained by contracting all the sets $A_i(x, y)$ and $A'_i(x, y)$ in G'_v, i-e $V(G'_A) = \{A_i(x, y), i \in I(x), (x, y) \in P_v\} \cup \{A'_j(x, y), j \in J(y), (x, y) \in P_v\}$, two vertices of G'_A are adjacent if at least one element of a set is adjacent in G'_v with at least one element of the other set.

$\forall i \in I(x)$, the vertex $A_i(x, y)$ is adjacent with $A_j(x, y), \forall j \neq i, j \in I(x)$, with $A'_j(x, y), \forall j \in J(y)$, and with $A'_j(t_l, x)$, if $i = i_{x\theta_l}$, $j = i_{xx_1} \in J(x)$. So $A_i(x, y)$ has at most $|I(x)| - 1 + |J(y)| = 2k - 1$ neighbors in K_{xy} and one neighbor not in K_{xy}.

$\forall j \in J(y)$, the vertex $A'_j(x, y)$ is adjacent with $A_i(x, y), \forall i \in I(x)$, with $A'_i(x, y), \forall i \neq j, i \in J(y)$, and with $A_k(y, z_l)$, if $j = i_{y\omega_l}$, $k = i_{yx_{p-1}} \in I(y)$. So $A_j(x, y)$ has at most $|I(x)| + |J(y)| - 1 = 2k - 1$ neighbors in K_{xy} and one neighbor not in K_{xy}. See Fig. 4 for an example.

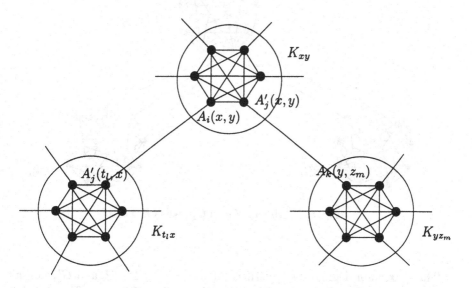

Fig. 4. The graph G'_A, when $k = 3$

Therefore, the maximum degree of G'_A verifies: $\Delta(G'_A) = 2k - 1 + 1 = 2k$ and G'_A doesn't contain a $(2k+1)$-clique. So, by a Theorem of Brooks ([4] Th. 6 page 326), the chromatic number of G'_A verifies $\chi(G'_A) \leq \Delta(G'_A) \leq 2k$.

Let p be such a coloring with values in $\{0, 1, \ldots 2k-1\}$ and $\forall (x, y) \in P_v$, index the paths of $A_i(x, y)$ and $A'_j(x, y)$ in this way: $A_i(x, y) = \{\alpha_1, \alpha_2, \ldots, \alpha_{r-k}\}$, $A'_j(x, y) = \{\beta_1, \beta_2, \ldots, \beta_{r-k}\}$.

The coloring of the paths of A is :

$$c'(\alpha_l) = (v, p(A_i(x, y))(r-k) + l), \quad 1 \leq l \leq r - k$$

$$c'(\beta_l) = (v, p(A'_j(x, y))(r-k) + l), \quad 1 \leq l \leq r - k.$$

Coloring type B paths: Let G'_B be the subgraph of G'_v induced by B. In G'_B, the vertex $\alpha = R''((x, i), (y, j))$ is adjacent with the $k^2 - 1$ other vertices of $B(x, y)$, with k vertices $R'((t_l, i'), (x, i_{xx_1})) \in B(t_l, x)$, $i' \in I(t_l)$, if $i = i_{x\theta_l}$ and with k vertices $R'((y, i_{yx_{p-1}}), (z_m, j')) \in B(y, z_m)$, $j' \in J(z_m)$, if $j = i_{y\omega_l}$. An illustration is given in Fig. 5.

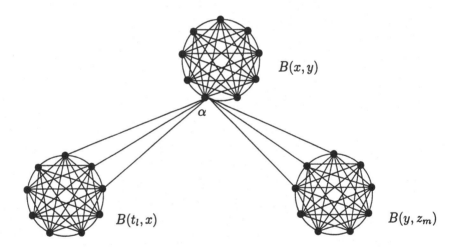

Fig. 5. Neighbors of α in G'_B, when $k = 3$

The maximum degree of G'_B verifies $\Delta(G'_B) = k^2 - 1 + 2k$ and G'_B doesn't contain a $(k^2 + 2k)$-clique. So, the chromatic number of G'_B verifies $\chi(G'_B) \leq \Delta(G'_B) \leq k^2 + 2k - 1$.

Let $q : V(G'_B) \rightarrow \{2k(r-k) + 1, 2k(r-k) + 2, \ldots, 2k(r-k) + k^2 + 2k - 1\}$ be a coloring of the vertices of G'_B and set $\forall \alpha \in B$:

$$c'(\alpha) = (v, q(\alpha))$$

Coloring type C paths: Let $(x, y) \in P_v$. We have colored $|A(x, y)| + |B(x, y)| = 2k(r - k) + k^2$ of the r^2 vertices of K_{xy} using a total of $2k(r - k) + k^2 + 2k - 1$ colors. We have still to color $m = |C(x, y)| = (r - k)^2$ vertices in K_{xy}. These m vertices are not adjacent with other r^2-cliques therefore they can be colored using the $2k - 1$ colors unused in K_{xy}. If $m > 2k - 1$, the remaining uncolored vertices are given new colors.

Coloring the paths $R'((x, i), (x, j))$

Lemma 6 *If G is $[k, w]$-colorable, there exists a routing R_0 for which G is still $[k, w]$-colorable and for each arc (x, x_1) of G, $R_0(x, x_1) = (x, x_1)$.*

Proof. Let R be a routing and c be a coloring, for which G is $[k, w]$-colorable.

For each arc (x, x_1) such that $R(x, x_1) \neq (x, x_1)$, we modify the routing in this way:

Let $v = c(R(x, x_1))$.

If there exists a path $R(a, b) = (a, \ldots, a_p, x, x_1, b_1, \ldots, b_k, b)$ through the arc (x, x_1) with color $c(R(a, b)) = v$, then we modify the routing in this way:
$$\begin{cases} R_0(a, b) = (a, \ldots, a_p, R(x, x_1), b_1, \ldots, b_k, b) \\ R_0(x, x_1) = (x, x_1) \end{cases}$$
and we assign them the color v.

If no path $R(a, b)$ through the arc (x, x_1) has color v, then we can set:
$R_0(x, x_1) = (x, x_1)$ and $c(R_0(x, x_1)) = v$.

In both cases, the number of paths with color v that begin at x still unchanged. □

This lemma will help us to prove that we can choose a color for the arc $((x, i), (x, j))$, in order the following condition is satisfied :

$\forall x \in V(G')$, for all color (v, α), there are at most k paths of color (v, α) beginning at x and at most k paths of color (v, α) ending at x.

Let x_1 be the vertex such that $j = i_{xx_1}$. According to Lemma 6, we can suppose $R(x, x_1) = (x, x_1)$. Let $v = c(R(x, x_1))$ be the color of the arc (x, x_1) in G. There are r paths beginning at (x, i) through the arc $((x, i), (x, j))$ with color (v, α) for some $\alpha \in D_i \subset \{1, \ldots, r^2\}$.

There are at most kr paths $R'((t_l, i'), (x, j))$ ending at (x, j) with color (v, α) for some $\alpha \in F \subset \{1, \ldots, r^2\}$(at most r of them through the arc $((x, i), (x, j))$).

If $r \geq k + \lceil \sqrt{2k - 1} \rceil$, $r^2 - kr - r \geq r$ and we give different colors (v, β_i) to the $(r - 1)$ arcs $((x, i), (x, j))$, $1 \leq i \leq r$, $i \neq j$ ending at (x, j), $\beta_i \notin D_i \cup F$.

Else, $r < k + \sqrt{2k - 1}$ and there are at most $2r$ paths that go through the arc $((x, i), (x, j))$, with color (v, α) for some α, $1 \leq \alpha \leq r^2$. So we give to $R'((x, i), (x, j))$ one of the $r^2 - 2r$ remaining colors (v, β). $r^2 - 2r \geq 1$ since $r \geq 3$.

$[k, r^2w]$-colorability of G': Suppose $r \geq k + \sqrt{2k - 1}$.

Number of paths colored (v, α_0) *beginning at* (x, i): If a path $R'((x, i), (y, j))$ is colored (v, α_0) for $y \neq x$, the path $R(x, y)$ is colored v in G.

There are at most k paths $R(x, y_l)$, $1 \leq l \leq k' \leq k$ in G with color v and then at most k paths $R'((x, i), (y_l, j))$ in G' colored (v, α_0): at most one in each K_{xy_l}.

If $R'((x, i), (x, j))$ is colored (v, α_0), one of the y_l is a neighbor x_1 of x, where $R(x, x_1)$ is colored v in G. In this case, no path $R'((x, i), (x_1, j'))$ has color (v, α_0).

Conclusion: Given (v, α_0) and (x, i) there exist at most k paths beginning at (x, i) colored (v, α_0).

Number of paths colored (v, α_0) *ending at* (x, j): If a path $R'((t, i), (x, j))$ is colored (v, α_0) with $t \neq x$, the path $R(t, x)$ has color v in G. There are at most k paths $R(t_l, x)$, $1 \leq l \leq k'' \leq k$ in G with color v and then at most k paths $R'((t_l, i'), (x, j))$ in G' colored (v, α_0): at most one in each $K_{t_l x}$.

If $R'((x, i), (x, j))$ is colored (v, α_0), by definition of the coloring no path in $K_{t_l x}$ ending at (x, j) is colored (v, α_0). And no other path $R'((x, i'), (x, j))$, $i' \neq i$ is colored (v, α_0). There exist only one path ending at (x, j) colored (v, α_0).

Conclusion: Given (v, α_0) and (x, j) there exist at most k paths ending at (x, j) colored (v, α_0).

Number of colors used: We have used $w' = \max\{r^2, 2k(r - k) + k^2 + 2k - 1\}$ colors. So if $r^2 \geq 2k(r - k) + k^2 + 2k - 1$, i-e. $r \geq k + \sqrt{2k - 1}$, we have used $w' = r^2 w$ colors. This coloring is clearly minimal (according to the routing R' considered) since G'_c contain r^2-cliques.

As $2k(r - k) + k^2 \leq r^2$, if $r^2 < 2k(r - k) + k^2 + 2k - 1$, we have used at most $r^2 + 2k - 1 \leq r^2 + 2r - 1$ colors.

Moreover, we have assigned different colors to vertices adjacent in G'_v, and the colors of a vertex of G'_{v_1} and a vertex of G'_{v_2} differ by the first coordinate.

This ends the proof of Proposition 3. $\square$

5 Conclusion

We have shown an upper bound for the number of wavelengths for inflated graphs.

We can define, in an analogous way, the *cycle-inflation*, to be the transformation consisting in replacing each node of an r-regular network by a cycle of length r.

The cube-connected-cycle CCC_n is an example of a cycle-inflation of the hypercube H_n.

Is-it possible to derive similar results for the number of wavelengths for cycle-inflated graphs?

References

1. D. Amar, A. Raspaud, and O. Togni. All-to-all wavelength routing in all-optical compound networks. submited.
2. B. Beauquier. All-to-all communication in some wavelength-routed all-optical networks. submited, 1996.
3. B. Beauquier, J-C. Bermond, L. Gargano, P. Hell, S. Perennes, and U. Vaccaro. Graphs problems arising from wavelength-routing in all-optical networks. In *In Proc. WOCS'97*, 1997.
4. C. Berge. *Graphes et hypergraphes*. DUNOD, 1973.
5. J.-C. Bermond, F. Comellas, and D. F. Hsu. Distributed loop computer networks: A survey. *Journal of Parallel and Distrubuted Computing*, 24:2–10, 1995.
6. J.-C. Bermond, L. Gargano, S. Perennes, A. A. Rescigno, and U. Vaccaro. Efficient collective communication in optical networks. *Lecture Notes in Computer Science*, 1099:574–585, 1996.
7. J. E. Dunbar and T. W. Haines. Domination in inflated graphs. *Congr. Numerantium*, 118:143–154, 1996.
8. Erlebach and Jansen. Call scheduling in trees, rings and meshes. In *HICSS: Hawaii International Conference on System Sciences*. IEEE Computer Society Press, 1997.
9. L. Gargano, P. Hell, and S. Perenes. Colouring paths in directed symetric trees with applications to WDM routing. In LNCS, editor, *in Proc. of ICALP'97*, volume 1256, pages 505–515. Springer-Verlag, 1997.
10. M. C. Heydemmann, J. C. Meyer, and D. Sotteau. On forwarding indices of networks. *Discrete Applied Math.*, 23:103–123, 1989.
11. C. Kaklamanis, P. Persiano, T. Erlebach, and K. Jansen. Constrained bipartite edge coloring with applications to wavelength routing. *Lecture Notes in Computer Science*, 1256:493–504, 1997.
12. H. Schroeder, O. Sykora, and I. Vrt'o. Optical all-to-all communication for some product graphs. *Lecture Notes in Computer Science*, 1338:555–??, 1997. in Proc. of SOFSEM'97.
13. O. Togni. *Force des graphes, indice optique des réseaux*. PhD thesis, Université de Bordeaux I, 1998. In french.

A Generalization of AT-free Graphs and a Generic Algorithm for Solving Treewidth, Minimum Fill-In and Vertex Ranking

Hajo Broersma[1], Ton Kloks[1], Dieter Kratsch[2], and Haiko Müller[2]

[1] Faculty of Applied Mathematics
University of Twente, P.O. Box 217
7500 AE Enschede
the Netherlands
{H.J.Broersma,A.J.J.Kloks}@math.utwente.nl
[2] Fakultät für Mathematik und Informatik
Friedrich-Schiller-Universität
07740 Jena
Germany
{kratsch,hm}@minet.uni-jena.de

Abstract. A subset A of the vertices of a graph G is an *asteroidal set* if for each vertex $a \in A$, the set $A \setminus \{a\}$ is contained in one component of $G - N[a]$. An asteroidal set of cardinality three is called *asteroidal triple* and graphs without an asteroidal triple are called *AT-free*. The maximum cardinality of an asteroidal set of G, denoted by $\mathsf{an}(G)$, is said to be the *asteroidal number* of G. We present a scheme for designing algorithms for triangulation problems on graphs. As a consequence, we obtain algorithms to compute graph parameters such as treewidth, minimum fill-in and vertex ranking number. The running time of these algorithms is a polynomial (of degree asteroidal number plus a small constant) in the number of vertices and the number of minimal separators of the input graph.

1 Introduction

Graphs without an asteroidal triple are called asteroidal triple-free graphs (short AT-free graphs) and attained much attention recently. Möhring has shown that every minimal triangulation of an AT-free graph is an interval graph which implies that for every AT-free graph the treewidth and the pathwidth of the graph are equal [25]. Furthermore a collection of interesting structural and algorithmic properties of AT-free graphs has been obtained by Corneil, Olariu and Stewart, among them an existence theorem for so-called dominating pairs in connected AT-free graphs and a linear time algorithm to compute a dominating pair for connected AT-free graphs (see [10, 11]).

The class of graphs with bounded asteroidal number extends the class of AT-free graphs, based on a natural way of generalizing the concept of asteroidal triples to so-called asteroidal sets, first given by Walter [29]. A set of vertices

A of a graph G is called an asteroidal set if for every vertex $a \in A$ all vertices of $A \setminus \{a\}$ are contained in the same component of $G - N[a]$. Walter, Prisner and Lin et al. used asteroidal sets to characterize certain subclasses of the class of chordal graphs [23, 27, 29]. We introduce the asteroidal number of a graph G, denoted by $\mathsf{an}(G)$, as the maximum cardinality of an asteroidal set in G. Thus AT-free graphs are exactly those graphs G with $\mathsf{an}(G) \leq 2$.

In this paper we consider the NP-complete graph problems TREEWIDTH, MINIMUM FILL-IN and VERTEX RANKING that all remain NP-complete when restricted to AT-free graphs. In fact, each of the three problems remains NP-complete on cobipartite graphs [2, 6, 30], that form a small subclass of the class of AT-free graphs. TREEWIDTH has been studied in numerous recent papers, mainly since many NP-complete graph problems become solvable in polynomial time or even linear time when restricted to the class of graphs with bounded treewidth [1, 4, 16]. In this respect it is interesting, that for each constant k, there is a linear time algorithm that determines whether a given graph has treewidth at most k [5, 16]. The MINIMUM FILL-IN problem stems from the optimal performance of Gaussian elimination on sparse matrices and has important applications in this area. Both TREEWIDTH and MINIMUM FILL-IN ask for a certain chordal embedding of the given graph. This often allows the design of similar algorithms for both problems, when graphs of some special class are considered. The VERTEX RANKING problem received much attention lately because of the growing number of applications. The problem of finding an optimal vertex ranking is equivalent to the problem of finding a minimum-height elimination tree of a graph [12]. This measure is of importance for the parallel Cholesky factorization of matrices [7, 24]. Other applications lie in the field of VLSI-layout design [21].

Using an algorithm of [17] to list all minimal separators of a given graph G in time $O(n^5 r)$, where n is the number of vertices of G and r is the number of minimal separators of G, one has designed algorithms with a running time bounded by a polynomial in the number of vertices and the number of minimal separators of the input graph, that compute the treewidth and the minimum fill-in [19] as well as the vertex ranking number [20] on AT-free graphs. It is worth mentioning that the running time of these algorithms is not bounded by a polynomial in the input length, since AT-free graphs may have 'exponentially' many minimal separators.

We generalize the method used in [19, 20]. To be more precise, we focus on certain sets of minimal separators called blocking sets. We show that these blocking sets have at most $\mathsf{an}(G)$ elements, and that they decompose the graph into a number of so-called blocks, which is bounded by a polynomial of order $\mathsf{an}(G)$ in the number of minimal separators of G. We consider graphs H obtained from a block of G by making the separators of the blocking set complete, and establish a relation between the blocks of H and the blocks of G. Together with some known recurrence relations for the three aforementioned problems in terms of the minimal separators S of G and the components of $G - S$, this enables us to give a scheme for recursive algorithms. In this way, for each of the three problems, we obtain an algorithm that solves the corresponding problem for all graphs G in time $O(n^5 r + m + kr^{k+1}(n+m)n\log n)$, where $k = \text{an}(G)$ and r is the number of minimal separators of G. Moreover, the algorithms can be implemented without knowing the asteroidal number or the number of minimal separators of the input graphs in advance. In that case, the algorithms will generate the correct answers, within the stated timebound. This is of importance, since computing the asteroidal number in general is NP-complete [18].

2   Preliminaries

Throughout the paper, let G denote a graph with vertex set V and edge set E. We denote the number of vertices of G by n, the number of edges of G by m, and the size of a maximum clique in G by $\omega(G)$. For a proper subset $W \subset V$, $G - W$ denotes the subgraph of G obtained by removing the vertices of W. For a vertex $x \in V$, we write $G - x$ instead of $G - \{x\}$. For $\emptyset \neq W \subseteq V$, $G[W]$ denotes the subgraph of G induced by the vertices of W. For any set S, we denote by $S^{[2]}$ the set of all subsets of S of cardinality 2. For any set $\mathfrak{S}$ whose elements are sets itself, we use $\bigcup \mathfrak{S}$ to denote $\bigcup_{S \in \mathfrak{S}} S$. For a vertex $x \in V$, $N(x)$ is the neighborhood of x and $N[x] = \{x\} \cup N(x)$ is the closed neighborhood of x. We say that a sequence $P = (u_0, u_1, \ldots, u_l)$ of pairwise distinct vertices of G is a u,v-path of G if $u = u_0$, $v = u_l$, and for $i = 1, \ldots, l$ there is an edge $\{u_{i-1}, u_i\} \in E$.

Definition 1. *A subset* $A \subseteq V$ *is called an* asteroidal set *of G if for each* $a \in A$ *the vertices of* $A \setminus \{a\}$ *are contained in one component of* $G - N[a]$. *The maximum cardinality of an asteroidal set of G is denoted by* $\text{an}(G)$, *and is called the* asteroidal number *of G.*

By definition the vertices of an asteroidal set are pairwise nonadjacent. Hence $\text{an}(G) \leq \alpha(G)$, where $\alpha(G)$ denotes the maximum cardinality of an independent set in G. Furthermore for every k there exist graphs of asteroidal number k, e.g., $\text{an}(C_{2k}) = k$ for $k \geq 2$, where C_n is the chordless cycle on n vertices. Notice that every subset of an asteroidal set is itself asteroidal.

An asteroidal set of cardinality three was called an asteroidal triple (short AT) in [22], where it was shown that chordal graphs without AT are exactly those that are interval graphs.

There are polynomial time algorithms to compute the asteroidal number for graphs in some special classes like HHD-free graphs (including all chordal graphs), claw-free graphs, circular-arc graphs and circular permutation graphs. However the corresponding decision problem remains NP-complete on triangle-free 3-connected 3-regular planar graphs [18].

Definition 2. *A graph H is* chordal *(or* triangulated*) if it does not contain a chordless cycle of length at least four as an induced subgraph.*

Definition 3. *A* triangulation *of G is a graph H with the same vertex set as G such that H is chordal and G is a subgraph of H. A triangulation H of G is called* minimal *if there is no proper subgraph H' of H which is also a triangulation of G.*

Definition 4. *The* treewidth *of G, denoted by $\mathrm{tw}(G)$, is the minimum of $\omega(H) - 1$ taken over all triangulations H of G.*

Definition 5. *The* minimum fill-in *of G, denoted by $\mathrm{mfi}(G)$, is the minimum of $|E(H) \setminus E|$ taken over all triangulations H of G.*

Definition 6. *Let t be an integer. A* (vertex) t-ranking *of G is a coloring $c : V \to \{1, \ldots, t\}$ such that for every pair of vertices x and y with $c(x) = c(y)$ and for every path between x and y there is a vertex z on this path with $c(z) > c(x)$. The* vertex ranking number *of G, denoted by $\chi_r(G)$, is the smallest value t for which the graph G admits a t-ranking.*

A proper subset $S \subseteq V$ is a *separator* of G if $G - S$ is disconnected.

Definition 7. *A vertex set $S \subset V$ is an a, b-separator of G if the removal of S separates a and b in distinct components of $G - S$. If no proper subset of an a, b-separator S is an a, b-separator then S is a minimal a, b-separator. A vertex set $S \subset V$ is a minimal separator of G if there exist nonadjacent vertices a and b of G such that S is a minimal a, b-separator of G.*

We define $\mathrm{Comp}(G) = \{X : \varnothing \neq X \subseteq V \text{ and } G[X] \text{ is a component of } G\}$. By $\mathrm{Sep}(G)$ we denote the set of all minimal separators of G. The following lemma is well-known and was rediscovered many times (see, e.g., [14]).

Definition 8. *Let S be a separator of G. A component H of $G - S$ is* full *(w.r.t. S) if every vertex of S has at least one neighbor in H.*

Lemma 1. *A set S of vertices of G is a minimal separator of G if and only if $G - S$ has at least two full components.*

Notice that Lemma 1 enables the design of a linear time algorithm that decides whether a given vertex set S is a minimal separator of a given graph G.

Dirac established the following characterization of chordal graphs [13].

Theorem 1. *G is a chordal graph if and only if every minimal separator of G is a clique.*

Definition 9. *Let $\mathfrak{S}$ be any set of vertex subsets of G. Then $G_{\mathfrak{S}} = (V, E \cup \bigcup_{S \in \mathfrak{S}} S^{[2]})$ is the graph obtained from G by adding exactly those edges, which are not present in G and which are edges of a complete graph on some $S \in \mathfrak{S}$.*

Now we can state a characterization of minimal triangulations.

Theorem 2. *A graph H is a minimal triangulation of G if and only if $H = G_{\mathrm{Sep}(H)}$.*

In the following lemma we mention two useful characteristics of minimal triangulations (see e.g. [19]).

Lemma 2. *If H is a minimal triangulation of a graph G then*

1. *If a and b are nonadjacent in H, then every minimal a, b-separator in H is also a minimal a, b-separator in G.*
2. *If S is a minimal separator in H and if C is the vertex set of a component of $H - S$, then C induces also a component in $G - S$.*

3   Recurrence relations and minimal separators

Some well-known graph parameters can be computed by applying recurrence relations involving the set of all minimal separators of the graph under consideration. The most prominent examples concern the treewidth, minimum fill-in and vertex ranking, and appeared in [19], [19], and [12] respectively. In the next theorem, $G(\{S\}, C) = G_{\{S\}}[S \cup C]$ and $\text{fill}(S) = \binom{|S|}{2} - |E(G[S])|$.

Theorem 3. *Let G be a graph which is not complete. Then*

$$\text{tw}(G) = \min_{S \in \text{Sep}(G)} \max_{C \in \text{Comp}(G-S)} \text{tw}(G(\{S\}, C)).$$

$$\text{mfi}(G) = \min_{S \in \text{Sep}(G)} \left(\text{fill}(S) + \sum_{C \in \text{Comp}(G-S)} \big(\text{mfi}(G(\{S\}, C)) - \text{fill}(S) \big) \right).$$

$$\chi_{\text{r}}(G) = \min_{S \in \text{Sep}(G)} \left(|S| + \max_{C \in \text{Comp}(G-S)} \chi_{\text{r}}(G[C]) \right).$$

Besides many efficient algorithms on special graph classes for the three problems, one has obtained algorithms for AT-free graphs, that are based on the abovementioned recurrence relations, in [19] and [20].

Our major goal in the remainder of this paper is to generalize the approach for AT-free graphs to obtain a general scheme for designing recursive algorithms on graphs which is applicable as soon as there is a recurrence relation for computing the graph parameter under consideration similar to those in Theorem 3. Because of space limitation, we omit all proofs in this extended abstract.

4   Blocks

Blocking sets and blocks are central concepts for the recursive algorithms and the corresponding decompositions.

Definition 10. *A set $\mathfrak{S}$ of minimal separators of G is a* blocking set *if the elements of $\mathfrak{S}$ (except for possibly the empty set) are incomparable with respect to set inclusion and for all $S \in \mathfrak{S}$ the vertex set $\bigcup \mathfrak{S} \setminus S$ is contained in one component of $G - S$.*

Note that in particular any minimal separator of G is a blocking set.

Definition 11. *Let $\mathfrak{S}$ be a blocking set of G with $|\mathfrak{S}| \geq 2$. Then a vertex $v \in V \setminus \bigcup \mathfrak{S}$ is said to be* in the interior *of $\mathfrak{S}$ if, for every $S \in \mathfrak{S}$, the vertex v and the vertex set $\bigcup \mathfrak{S} \setminus S$ are contained in one component of the subgraph $G - S$.*

Lemma 3. *For every blocking set $\mathfrak{S}$ of G, $|\mathfrak{S}| \leq \text{an}(G)$.*

Definition 12. *A pair $(\mathfrak{S}, C)$ is a* block *of G if $\mathfrak{S}$ is a blocking set of G, $C \subseteq V$ and one of the following conditions is fulfilled.*

- *$|\mathfrak{S}| \geq 2$ and the set C is the set of all vertices in the interior of $\mathfrak{S}$.*

- If $\mathfrak{S}$ contains exactly one element S, then C is the vertex set of a component of $G - S$ or $C = \varnothing$.
- $\mathfrak{S} = \varnothing$ and C is the vertex set of a component of G.

The definition and Lemma 3 immediately imply

Observation 1 *The number of different blocks of G is at most*

$$(|\mathsf{Sep}(G)| + 1) \cdot |V| + \sum_{k=2}^{\mathsf{an}(G)} \binom{|\mathsf{Sep}(G)|}{k}.$$

The following definition is motivated by the recurrence relations in Section 3 and Theorems 1 and 2.

Definition 13. *The realization $G(\mathfrak{S}, C)$ of a block $(\mathfrak{S}, C)$ of G is the graph $G_{\mathfrak{S}}[C \cup \bigcup \mathfrak{S}]$.*

The definition implies that the realization of any block is a connected graph.

5   Decomposing blocks

We consider a block $(\mathfrak{S}, C)$ of G, its realization $H = G(\mathfrak{S}, C)$ and a minimal separator T of H. Then for an arbitrary component $H[D]$ of $H - T$, the pair $(\{T\}, D)$ is a block of H. Our major goal in this section is to prove a claim stating that any block $(\{T\}, D)$ of H can be described as a block of G in the following sense: For any block $(\{T\}, D)$ of $H = G(\mathfrak{S}, C)$, there is a block $(\mathfrak{T}, D')$ of G such that the corresponding realizations are exactly the same graphs, i.e., $G(\mathfrak{T}, D') = H(\{T\}, D)$.
   The consequence is that any algorithm, which recursively computes a minimal separator T for the current graph H and then calls itself on the realization of the block $(\{T\}, D)$ for each component D of $H - T$ until the current graph is complete, will only work on realizations of blocks of the input graph G. Together with Lemma 3 and Observation 1 this implies, that each recursive algorithm of this type checks at most $O(|\mathsf{Sep}(G)|^{\mathsf{an}(G)})$ realizations of blocks of the input graph G.
   We start with two lemmas that are essential for this section. First we consider minimal separators of realizations.

Lemma 4. *Let $(\mathfrak{S}, C)$ be a block of G and let a and b be nonadjacent vertices in $G(\mathfrak{S}, C)$. Then every minimal a, b-separator in $G(\mathfrak{S}, C)$ is a minimal a, b-separator in G.*

The next lemma classifies the minimal separators of realizations into three types.

Lemma 5. *Let $(\mathfrak{S}, C)$ be a block of G and let T be a minimal separator of $H = G(\mathfrak{S}, C)$. Then exactly one of the following three conditions holds:*

Type 1: *there are distinct minimal separators $S_1, S_2 \in \mathfrak{S}$ with $T \subset S_1$ and $T \subset S_2$,*

Type 2: *there is exactly one separator $S_0 \in \mathfrak{S}$ such that $T \subset S_0$,*

Type 3: $T \setminus S \neq \emptyset$ *for all $S \in \mathfrak{S}$.*

Furthermore, in Types 1 and 2 the graph $H - T$ has exactly two components.

Proposition 1 (Type 1). *Let $(\mathfrak{S}, C)$ be a block of G and let T be a minimal separator of $H = G(\mathfrak{S}, C)$ such that there exist at least two different minimal separators in $\mathfrak{S}$ containing T. Then $C = \emptyset$, $|\mathfrak{S}| = 2$ and for each $S \in \mathfrak{S}$ we have $H(\{T\}, S \setminus T) = G(\{S\}, \emptyset)$.*

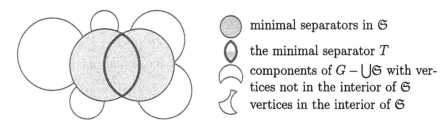

minimal separators in $\mathfrak{S}$

the minimal separator T

components of $G - \bigcup \mathfrak{S}$ with vertices not in the interior of $\mathfrak{S}$

vertices in the interior of $\mathfrak{S}$

Fig. 1. Type 1

Let $(\{S_1, S_2\}, \emptyset)$ be a block of G. By Proposition 1 the unique minimal separator $T = S_1 \cap S_2$ of $G(\{S_1, S_2\}, \emptyset)$ decomposes $(\{S_1, S_2\}, \emptyset)$ into two other blocks of G. We define the decomposition of $(\{S_1, S_2\}, \emptyset)$ by

$$\mathsf{Dec}(\{S_1, S_2\}, \emptyset, T) = \{(\{S_1\}, \emptyset), (\{S_2\}, \emptyset)\}.$$

Proposition 2 (Type 2). *Let $(\mathfrak{S}, C)$ be a block of G and let T be a minimal separator of $H = G(\mathfrak{S}, C)$ such that there is a unique separator $S_0 \in \mathfrak{S}$ with $T \subset S_0$. Let $\mathfrak{T} = \mathfrak{S} \setminus \{S_0\}$ and $D = C \cup \bigcup \mathfrak{T}$. Then $H[D]$ and $H[S_0 \setminus T]$ are the components of $H - T$. Furthermore $(\{T\} \cup \mathfrak{T}, C)$ is a block of G with $G(\{T\} \cup \mathfrak{T}, C) = H(\{T\}, D)$, and $(\{S_0\}, \emptyset)$ is a block of G with $G(\{S_0\}, \emptyset) = H(\{T\}, S_0 \setminus T)$.*

Let $(\mathfrak{S}, C)$ be a block of G and let T be a minimal separator of $H = G(\mathfrak{S}, C)$ such that there is a unique separator $S_0 \in \mathfrak{S}$ with $T \subset S_0$. Based on Proposition 2 we define

$$\mathsf{Dec}(\mathfrak{S}, C, T) = \{(\{S_0\}, \emptyset), (\{T\} \cup \mathfrak{S} \setminus \{S_0\}, C)\}.$$

Proposition 3 (Type 3). *Let $(\mathfrak{S}, C)$ be a block of G and let T be a minimal separator of $H = G(\mathfrak{S}, C)$ such that $T \setminus S \neq \emptyset$ for all $S \in \mathfrak{S}$. Let $H[D]$ be a component of $H - T$. Let $\mathfrak{T} = \{S : S \in \mathfrak{S} \text{ and } S \setminus T \subseteq D\}$ and $D' = D \setminus \bigcup \mathfrak{T}$. Then $(\{T\} \cup \mathfrak{T}, D')$ is a block of G and $G(\{T\} \cup \mathfrak{T}, D') = H(\{T\}, D)$.*

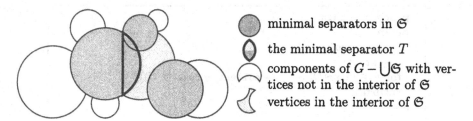

minimal separators in $\mathfrak{S}$

the minimal separator T

components of $G - \bigcup\mathfrak{S}$ with vertices not in the interior of $\mathfrak{S}$

vertices in the interior of $\mathfrak{S}$

Fig. 2. Type 2

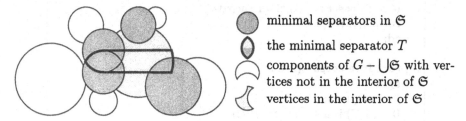

minimal separators in $\mathfrak{S}$

the minimal separator T

components of $G - \bigcup\mathfrak{S}$ with vertices not in the interior of $\mathfrak{S}$

vertices in the interior of $\mathfrak{S}$

Fig. 3. Type 3

Let $(\mathfrak{S}, C)$ be a block of G and let T be a minimal separator of $H = G(\mathfrak{S}, C)$ such that $T \setminus S \neq \varnothing$ for all $S \in \mathfrak{S}$. In this case let $\{H[D_i] : i \in I\}$ be the set of components of $H - T$. Based on Proposition 3 we define

$$\mathsf{Dec}(\mathfrak{S}, C, T) = \{(\{T\} \cup \{S : S \in \mathfrak{S} \text{ and } S \cap D_i \neq \varnothing\}, C \cap D_i) : i \in I\}.$$

The following theorem summarizes Lemma 5 and Propositions 1, 2 and 3.

Theorem 4. *Let* $(\mathfrak{S}, C)$ *be a block of* G *and let* T *be a minimal separator of* $H = G(\mathfrak{S}, C)$. *Then we have a bijection between the blocks* (T, D) *corresponding with the components of* $H - T$ *and the blocks* $(\mathfrak{T}, D')$ *in* $\mathsf{Dec}(\mathfrak{S}, C, T)$ *such that* $G(\mathfrak{T}, D') = H(\{T\}, D)$.

6   Algorithms

The approach of the previous section enables two different types of algorithms. One type is a dynamic programming algorithm as used in [8, 12, 19, 20].

Here we use another type of algorithm sometimes called recursive algorithm with memoization (see e.g. [9]). First we describe the generic version. The input is a graph $G = (V, E)$. In a preprocessing the algorithm computes $\mathsf{Sep}(G)$ using the listing algorithm given in [17].

The procedure **compute** is the heart of the algorithm. It is recursive via **access**. The macros **compute** and **main** use **collect**, **complete**, **initialize**, **update** and **start**, which are specific to the algorithmic problem.

The algorithm uses a data structure X that can store any block $(\mathfrak{S}, C)$ of a graph $G = (V, E)$ with a value $p(\mathfrak{S}, C)$, and retrieve these values. Suppose $V =$

```
procedure main;
begin
  compute Sep(G);
  p ← start;
  for C ∈ Comp(G) do p ← collect(access(∅, C));
  return(p)
end.

procedure access(S, C);
begin
  if not present(S, C) then compute(S, C);
  return(value(S, C))
end;

procedure compute(S, C);
begin
  p ← complete;
  if G(S, C) is not complete then
    for T ∈ Sep(G) do
      if T is a minimal separator of G(S, C) then
        begin
          q ← initialize;
          for (T, D) ∈ Dec(S, C, T) do q ← update(access(T, D));
          p ← min{p, q};
        end;
  store(S, C, p)
end;
```

$\{1, 2, \ldots, n\}$. Any block $(\mathfrak{S}, C)$ is stored as a set $C \subseteq V$ followed by a sequence of the minimal separators $S_1, S_2, \ldots, S_j$ in $\mathfrak{S}$ that are lexicographically ordered (as subsets of V). The data structure X supports the following operations:

- store$(\mathfrak{S}, C, p)$ stores for the block $(\mathfrak{S}, C)$ the value p,
- present$(\mathfrak{S}, C)$ returns true, if an operation store$(\mathfrak{S}, C, p)$ has been performed before, for any value of p, and false otherwise, and
- value$(\mathfrak{S}, C)$ returns the value p of the (last) operation store$(\mathfrak{S}, C, p)$, if present$(\mathfrak{S}, C) = $ true.

All three operations can be executed by iterated search for a vertex in the universe V. A single search can be done in time $O(\log n)$ by standard techniques. To find a whole block $(\mathfrak{S}, C)$ we need $|C| + |\bigcup \mathfrak{S}|$ single searches if $|\mathfrak{S}| \leq 1$ and $\sum_{S \in \mathfrak{S}} |S|$ single searches if $|\mathfrak{S}| \geq 2$. We refer to [3] for an implementation of a related data structure that can easily be extended to one satisfying our purposes. Notice that our algorithm calls value$(\mathfrak{S}, C)$ only if present$(\mathfrak{S}, C) = $ true. Furthermore if store$(\mathfrak{S}, C)$ is called, then present$(\mathfrak{S}, C) = $ false, i.e., for each block of G, store is called at most once.

	treewidth	minimum fill-in	ranking number
collect(c)	$\max\{p,c\}$	$p+c$	$\max\{p,c\}$
complete	$\|C \cup \bigcup S\| - 1$	fill($C \cup \bigcup S$)	$\|C\|$
initialize	0	fill(T)	$\|T \cap C\|$
update(c)	$\max\{q,c\}$	$q + c - $ fill(T)	$\max\{q, c + \|T \cap C\|\}$
start	0	0	0

We consider the running time of our algorithm on an input graph $G = (V,E)$ with $|V| = n$, $|E| = m$, $|\mathsf{Sep}(G)| = r$ and $\mathsf{an}(G) = k$. First the algorithm in [17] needs $O(n^5 r + m)$ time to list all minimal separators of G.

For the following analysis, we assume that all macros can be evaluated in constant time. (If this is not the case in a particular application, it should be easy to achieve the corresponding time bound with a similar analysis.) To determine the overall running time, we estimate the running time of compute($\mathfrak{S}, C$) for any block $(\mathfrak{S}, C)$ of G without counting the running time of those recursive calls compute($\mathfrak{T}, D$) for which present($\mathfrak{T}, D$) = false when compute($\mathfrak{T}, D$) is called. For any block $(\mathfrak{S}, C)$ of G, access calls compute at most once, namely when present($\mathfrak{S}, C$) = false. In this case, for each minimal separator T of G, compute needs $O(n + m)$ time to test whether T is a minimal separator of $G(\mathfrak{S}, C)$ and, if so, to compute the blocks in Dec($\mathfrak{S}, C, T$). For each of the at most n blocks $(\mathfrak{T}, D)$ in Dec($\mathfrak{S}, C, T$), access($\mathfrak{T}, D$) is executed. If access is called for a block $(\mathfrak{T}, D)$ of G, when present($\mathfrak{T}, D$) = true, then access does not call compute.

Procedure access looks up the value $p(\mathfrak{S}, C)$ in the data structure X. Using an implementation of the data structure X, similar to the one described in [3], one look-up can be done in time $\sum_{S \in \mathfrak{S}} |S| \cdot O(\log n) = O(kn \log n)$.

By Observation 1, the number of different blocks of the input graph G is at most $(r+1)n + \sum_{i=2}^{k} \binom{r}{i}$. Consequently, the total running time of the algorithm is $O(n^5 r + m + kr^{k+1}(n + m)n \log n)$.

Theorem 5. *The generic algorithm runs in $O(n^5 r + m + kr^{k+1}(n + m)n \log n)$ time, where r is the number of minimal separators and k is the asteroidal number of the input graph (under some assumptions on the macros).*

The generic algorithm can be used to compute a graph parameter which can be evaluated via a certain type of recurrence involving the minimal separators of the graph (see e.g. Section 3). In particular, Theorem 5 has the following consequence.

Corollary 1. *For each of the problems TREEWIDTH, MINIMUM FILL-IN and VERTEX RANKING there is an algorithm to compute the corresponding graph parameter for any input graph G in time $O(n^5 r + m + kr^{k+1}(n + m)n \log n)$, where r is the number of minimal separators of G and $k = \mathsf{an}(G)$.*

References

1. Arnborg, S., Efficient algorithms for combinatorial problems on graphs with bounded decomposability—A survey. *BIT* **25** (1985), pp.2–23.
2. Arnborg, S., D. G. Corneil and A. Proskurowski, Complexity of finding embeddings in a *k*-tree, *SIAM J. Alg. Disc. Meth.* **8** (1987), pp. 277–284.
3. Bodlaender, H., Kayles on special classes of graphs - An application of Sprague-Grundy theory. *Proceedings of the 18th International Workshop on Graph-Theoretic Concepts in Computer Science, WG'92*, Springer-Verlag, 1993, LNCS 657, pp. 90–102.
4. Bodlaender, H., A tourist guide through treewidth, *Acta Cybernetica* **11** (1993), pp. 1–23.
5. Bodlaender, H., A linear time algorithm for finding tree-decompositions of small treewidth, *SIAM Journal on Computing* **25** (1996), pp. 1305–1317.
6. Bodlaender, H., J. S. Deogun, K. Jansen, T. Kloks, D. Kratsch, H. Müller, Zs. Tuza, Rankings of graphs, *SIAM Journal on Discrete Mathematics* **11** (1998), pp. 168–181
7. Bodlaender, H. L., J. R. Gilbert, H. Hafsteinsson, T. Kloks, Approximating tree-width, pathwidth and minimum elimination tree height, *Journal of Algorithms* **18** (1995), pp. 238–255.
8. Bodlaender, H., T. Kloks and D. Kratsch, Treewidth and pathwidth of permutation graphs, *SIAM Journal on Discrete Mathematics* **8** (1995), pp. 606–616.
9. Cormen, T. H., C. E. Leiserson and R. L. Rivest, *Introduction to algorithms*, MIT Press, Cambridge, Massachusetts, USA, 1990.
10. Corneil, D. G., S. Olariu and L. Stewart, Asteroidal triple-free graphs, *SIAM Journal on Discrete Mathematics.* **10** (1997), pp. 399–430.
11. Corneil, D. G., S. Olariu and L. Stewart, A linear time algorithm to compute dominating pairs in asteroidal triple-free graphs, *Proceedings of ICALP'95*, Springer-Verlag, LNCS 944, 1995, pp. 292–302.
12. Deogun, J. S., T. Kloks, D. Kratsch and H. Müller, On vertex ranking for permutation and other graphs, *11th Annual Symposium on Theoretical Aspects of Computer Science*, Springer-Verlag, LNCS 775, 1994, pp. 747–758.
13. Dirac, G. A., On rigid circuit graphs, *Abh. Math. Sem. Univ. Hamburg* **25** (1961), pp. 71–76.
14. Golumbic, M. C., *Algorithmic graph theory and perfect graphs*, Academic Press, New York, 1980.
15. Habib, M. and R. H. Möhring, Treewidth of cocomparability graphs and a new order-theoretic parameter, *Order* **11** (1994), pp. 47–60.
16. Kloks, T., *Treewidth – Computations and Approximations*, Springer Verlag, LNCS 842, (1994).
17. Kloks, T. and D. Kratsch, Finding all minimal separators of a graph, *SIAM Journal on Computing* **27** (1998), pp. 605–613.
18. Kloks, T., D. Kratsch and H. Müller, Asteroidal sets in graphs, *Proceedings of the 19th International Workshop on Graph-Theoretic Concepts in Computer Science (WG'97)*, Springer-Verlag, LNCS 1335 (1997), pp. 229–241.
19. Kloks, T., D. Kratsch and J. Spinrad, On treewidth and minimum fill-in of asteroidal triple-free graphs, *Theoretical Computer Science* **175** (1997), pp. 309–335.
20. Kloks, T., H. Müller and C. K. Wong, Vertex ranking of asteroidal triple-free graphs, *Proceedings of the 7th International Symposium on Algorithms and Computation*, pp. 174–182, Springer Verlag, LNCS 1178, 1996.

21. Leiserson, C. E., Area efficient graph layouts for VLSI, *Proceedings of the 21st Annual IEEE Symposium on Foundations of Computer Science*, 1980, pp. 270–281.

22. Lekkerkerker, C. G. and J. Ch. Boland, Representation of a finite graph by a set of intervals on the real line, *Fundamenta Mathematicae* **51** (1962), pp. 45–64.

23. Lin, I. J., T. A. McKee and D. B. West, Leafage of chordal graphs, Manuscript 1994.

24. Liu, J. W. H., The role of elimination trees in sparse factorization, *SIAM Journal of Matrix Analysis and Applications* **11** (1990), pp. 134–172.

25. Möhring, R. H., Triangulating graphs without asteroidal triples, *Discrete Applied Mathematics* **64** (1996), pp. 281–287.

26. Parra, A., Structural and algorithmic aspects of chordal graph embeddings, PhD. thesis, Technische Universität Berlin, 1996.

27. Prisner, E., Representing triangulated graphs in stars, *Abh. Math. Sem. Univ. Hamburg* **62** (1992), pp. 29–41.

28. Rose, D. J., R. E. Tarjan and G. S. Lueker, Algorithmic aspects of vertex elimination on graphs, *SIAM Journal on Computing* **5** (1976), pp. 266–283.

29. Walter, J. R., Representations of chordal graphs as subtrees of a tree, *Journal of Graph Theory* **2** (1978), pp. 265–267.

30. Yannakakis, M., Computing the minimum fill-in is NP-complete, *SIAM Journal on Algebraic and Discrete Methods* **2** (1981), pp. 77–79.

A Polynomial-Time Algorithm for Finding Total Colorings of Partial k-Trees

Shuji Isobe, Xiao Zhou, and Takao Nishizeki

Graduate School of Information Sciences,
Tohoku University, 980-8579, JAPAN.

Abstract. A total coloring of a graph G is a coloring of all elements of G, i.e. vertices and edges, in such a way that no two adjacent or incident elements receive the same color. Many combinatorial problems can be efficiently solved for partial k-trees (graphs of treewidth bounded by a constant k). However, no polynomial-time algorithm has been known for the problem of finding a total coloring of a given partial k-tree with the minimum number of colors. This paper gives such a first polynomial-time algorithm.

1 Introduction

A *total coloring* of a graph G is a coloring of all elements of G, i.e. vertices and edges, so that no two adjacent or incident elements receive the same color. Figure 1 depicts a total coloring of a graph with four colors. This paper deals with the *total coloring problem* which asks to find a total coloring of a given graph G with the minimum number of colors. The minimum number of colors is called the *total chromatic number* $\chi_t(G)$ of G. The total coloring problem arises in many applications, including various scheduling and partitioning problems [Yap96]. The problem is NP-complete [Sán89], and hence it is very unlikely that there exists an algorithm to find a total coloring of a given graph G with $\chi_t(G)$ colors in polynomial time.

It is known that many combinatorial problems can be solved very efficiently for partial k-trees or series-parallel graphs [ACPS93, AL91, BPT92, Cou90, TNS82, ZNN96, ZSN96, ZTN96]. Partial k-trees are the same as graphs of treewidth at most k. In the paper we assume that $k = O(1)$. The class of partial k-trees includes trees (k=1), series-parallel graphs (k=2) [TNS82], Halin graphs (k=3),

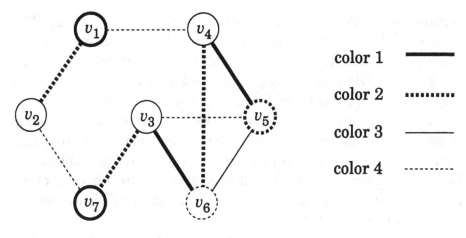

Fig. 1. A total coloring.

and k-terminal recursive graphs. Any partial k-tree can be decomposed into a tree-like structure T of small "basis" graphs, each with at most $k + 1$ vertices. Many problems can be solved efficiently for partial k-trees by a dynamic programming (DP) algorithm based on the tree-decomposition [ACPS93,AL91,BPT92,Cou90]. In particular, it is rather straightforward to design polynomial-time algorithms for vertex-type problems on partial k-trees. For example, the vertex-coloring problem, the maximum independent vertex-set problem, the minimum dominating vertex-set problem, and the vertex-disjoint paths problem can be solved all in linear time for partial k-trees [BPT92,Sch94,TP97]. However, this is not the case for edge-type problems such as the edge-coloring problem and the edge-disjoint paths problem. It needs sophisticated treatment tailored for individual edge-type problems to design efficient algorithms. For example, the edge-coloring problem can be solved in linear time for partial k-trees and series-parallel multigraphs, but very sophisticated algorithms are needed [ZSN97,ZSN96]. On the other hand, the edge-disjoint paths problem is NP-complete even for partial k-trees [ZN98], although the problem can be solved in polynomial time for partial k-trees under a certain restriction on the number of terminal pairs or the location of terminal pairs [ZTN96]. The difficulty of edge-type problems stems from the following facts: the number of vertices in a basis graph (a node of a tree-decomposition T) is bounded by $k + 1$ and hence the size of a DP table required to solve vertex-type problems can be easily bounded by a constant, say 2^{k+1} or $(k+1)^{k+1}$; however, the number of edges incident to vertices in a basis graph is not always bounded and hence it is difficult to bound the size of a DP table for edge-type problems by a constant or a polynomial in the number of vertices in a partial k-tree.

Clearly the mixed type problem like the total coloring problem is more difficult in general than the vertex- and edge-type problems. Both the vertex-coloring problem and the edge-coloring problem can be solved in linear time for partial k-trees. Therefore a natural question is whether the total coloring problem can be efficiently solved for partial k-trees or not.

In this paper we give a polynomial-time algorithm to solve the total coloring problem for partial k-trees G. Our idea is to bound the size of a DP table by $O(n^{2^{2k+3}})$, applying and extending techniques developed for the edge-coloring problem [Bod90,ZN95,ZNN96]. The paper is organized as follows. In section 2 we present some preliminary definitions. In section 3 we give a polynomial-time algorithm for the total coloring problem on partial k-trees. Finally we conclude our result in section 4 with some comments on a parallel algorithm.

2   Terminology and Definitions

In this section we give some definitions. Let $G = (V, E)$ denote a graph with vertex set V and edge set E. We often denote by $V(G)$ and $E(G)$ the vertex set and the edge set of G, respectively. We denote by n the number of vertices in G. The paper deals with *simple undirected* graphs without multiple edges or self-loops. An edge joining vertices u and v is denoted by (u, v). We denote by $\Delta(G)$ the *maximum degree* of G.

The class of k-trees is defined recursively as follows:

(K1) A complete graphs with k vertices is a k-tree.

(K2) If $G = (V, E)$ is a k-tree and k vertices $v_1, v_2, \ldots, v_k$ induce a complete subgraph of G, then $G' = (V \cup \{w\}, E \cup \{(v_i, w) : 1 \le i \le k\})$ is a k-tree where w is a new vertex not contained in G.

(K3) All k-trees can be formed with rules (K1) and (K2).

A graph is a *partial k-tree* if it is a subgraph of a k-tree. Thus a partial k-tree $G = (V, E)$ is a simple graph, and $|E| < kn$. Figure

2 illustrates a process of generating a 3-tree. The graph in Figure 1 is indeed a subgraph of a 3-tree in Figure 2, and hence is a partial 3-tree.

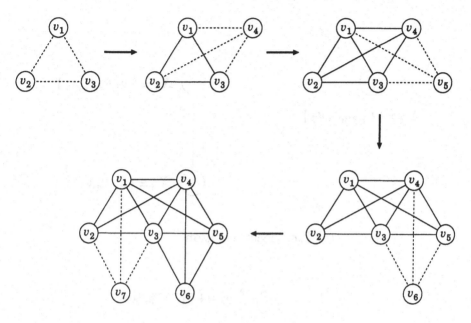

Fig. 2. 3-trees.

A *tree-decomposition* of G is a tree $T = (V_T, E_T)$ where V_T is a family of subset of V with the following three properties (a), (b) and (c):

(a) $\bigcup_{X \in V_T} X = V$;
(b) for each $e = (u, v) \in E$, there is a node $X \in V_T$ such that $u, v \in X$; and
(c) if node X_j lies on the path in T from node X_i to node X_l, then $X_i \cap X_l \subseteq X_j$.

Figure 3(a) illustrates a tree-decomposition of the partial 3-tree in Figure 1. The *width* of a tree-decomposition $T = (V_T, E_T)$ is $\max\{|X| - 1 : X \in V_T\}$. The *treewidth* of graph G is the minimum width of a tree-decomposition of G, taken over all possible tree-decompositions of G. It is known that every graph with treewidth $\leq k$ is a partial k-tree, and conversely, that every partial k-tree has a tree-decomposition with width $\leq k$.

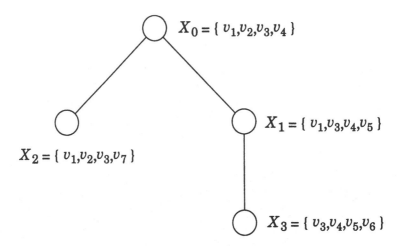

(a) Tree-decomposition

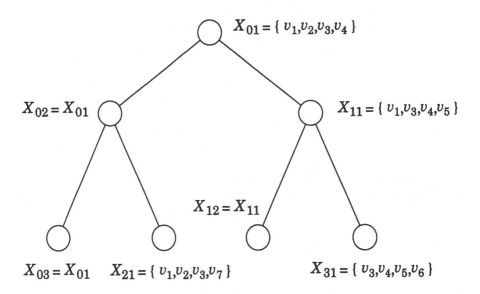

(b) Binary tree-decomposition

Fig. 3. Tree-decompositions of the partial 3-tree in Figure 1.

Bodlaender has given a linear time sequential algorithm to find a tree-decomposition of a given graph with width $\leq k$ for bounded k [Bod96]. We consider a tree-decomposition of a partial k-tree G with width $\leq k$. We transform it to a binary tree T as follows: regard the tree-decomposition as a rooted tree by choosing an arbitrary node as the root X_0 and replace every internal node X_i with r children $X_{j_1}, X_{j_2}, \ldots, X_{j_r}$ by $r+1$ new nodes $X_{i_1}, X_{i_2}, \ldots, X_{i_{r+1}}$ which are copies of X_i, where X_{i_1} has the same father as X_i, X_{i_q} is the father of $X_{i_{q+1}}$ and the q-th child X_{j_q} of X_i ($1 \leq q \leq r$), and $X_{i_{r+1}}$ is a leaf of T. This transformation can be done in linear time and doesn't change width [Bod90]. T is a tree-decomposition of G with the following properties:

(a) The number of nodes in T is $O(n)$.
(b) Each internal node X_i has exactly two children, say X_l and X_r, and either $X_i = X_l$ or $X_i = X_r$.
(c) For each edge $e = (u, v) \in E$, there is at least one leaf X_i with $u, v \in X_i$.

Such a tree T is called a *binary tree-decomposition*. Figure 3(b) illustrates a binary transformation of the tree-decomposition in Figure 3(a). Let T be a binary tree-decomposition with width $\leq k$ of a partial k-tree G. For each edge $e = (u, v) \in E(G)$, we choose an arbitrary leaf X_i with $u, v \in X_i$ and denote it by rep(e). We define a vertex set $V_i \subseteq V(G)$ and an edge set $E_i \subseteq E(G)$ for each node X_i of T as follows: if X_i is a leaf, then let $V_i = X_i$ and $E_i = \{e \in E(G) : \text{rep}(e) = X_i\}$; if X_i is an internal node with children X_l and X_r, then let $V_i = V_l \cup V_r$ and $E_i = E_l \cup E_r$. Note that $V_l \cap V_r \subseteq X_i$ and $E_l \cap E_r = \emptyset$. We denote by G_i the graph with vertex set V_i and edge set E_i. Then graphs G_l and G_r share common vertices only in X_i because of the property (c) of a tree-decomposition.

3   A Polynomial-Time Algorithm

In this section we prove the following theorem.

Theorem 1. *Let $G = (V, E)$ be a partial k-tree of n vertices given by its tree-decomposition with width $\leq k$, let C be a set of colors,* *and let* $\alpha = |C|$. *Then it can be determined in time*

$$O(n(\alpha + 1)^{2^{2k+3}})$$

whether G *has a total coloring:* $V \cup E \to C$.

One can easily know that the following lemma holds.

Lemma 1. *Every partial* k-*tree* G *satisfies*

$$\Delta(G) + 1 \leq \chi_t(G) \leq \Delta(G) + k + 2.$$

Proof: Clearly $\Delta(G) + 1 \leq \chi_t(G)$ for any graph.

Since a partial k-tree G is a simple graph, G has an edge-coloring with $\Delta(G)$ or $\Delta(G)+1$ colors by the classical Vizing theorem [FW77]. On the other hand, one can easily observe that a partial k-tree G has a vertex-coloring with at most $k + 1$ colors. These two colorings immediately yield a total coloring of G with at most $\Delta(G) + k + 2$ colors. Thus $\chi_t(G) \leq \Delta(G) + k + 2$. □

Thus one can compute $\chi_t(G)$ by applying the algorithm in Theorem 1 to G for $k + 2$ distinct values α, $\Delta(G) + 1 \leq |C| = \alpha \leq \Delta(G) + k + 2$. Furthermore, since $\alpha \leq n + k + 2$ and $k = O(1)$, the term $(\alpha + 1)^{2^{2k+3}}$ is bounded by a polynomial in n. Thus we have the following corollary.

Corollary 1. *The total coloring problem can be solved in polynomial time for partial* k-*trees.*

In the remainder of this section we will give a proof of Theorem 1. Although we give an algorithm to decide whether $G = (V, E)$ has a total coloring $f : V \cup E \to C$ for a given set C of colors, it can be easily modified so that it actually finds a total coloring f with colors in C. Our idea is to reduce the size of a DP table to $O((\alpha + 1)^{2^{2k+3}})$ by considering "pair-counts" and "quad-counts" defined below. A similar technique has been used for the ordinary edge-coloring and the f-coloring [Bod90,ZN95,ZNN96].

Let $C = \{1, 2, \ldots, \alpha\}$ be the set of colors. Let $G = (V, E)$ be a partial k-tree, and let X_i be a node of a binary tree-decomposition T of G. We say that a total coloring of graph G_i is *extensible* if it can be extended to a total coloring of $G = G_{01}$ without changing the coloring of G_i, where X_{01} is the root of T. Figure 4 illustrates total colorings of G_{02} and G_{11} for the partial 3-tree of G

in Figure 1 and its binary tree-decomposition T in Figure 3(b), where $X_{02} = \{v_1, v_2, v_3, v_4\}$ is the left child of the root X_{01} and $X_{11} = \{v_1, v_3, v_4, v_5\}$ is the right child. Both of the colorings are extensible because either can be extended to the total coloring of G in Figure 1.

For a total coloring f of G_i and a color $c \in C$, we define subsets $Y(X_i; f, c)$ and $Z(X_i; f, c)$ of X_i as follows:

$$Y(X_i; f, c) = \{v \in X_i : f(v) = c\}, \quad \text{and}$$
$$Z(X_i; f, c) = \{v \in X_i : G_i \text{ has an edge } (v, w) \text{ with } f((v, w)) = c\}.$$

Clearly,

$$Y(X_i; f, c) \cap Z(X_i; f, c) = \emptyset. \tag{1}$$

We call a mapping $\gamma : 2^{X_i} \times 2^{X_i} \rightarrow \{0, 1, 2, \ldots, \alpha\}$ a *pair-count* on a node X_i. A pair-count γ on X_i is defined to be *active* if G_i has a total coloring f such that

$$\gamma(A, B) = |\{c \in C : A = Y(X_i; f, c), B = Z(X_i; f, c)\}|$$

for each pair of $A, B \subseteq X_i$. Such a pair-count γ is called the *pair-count of the total coloring f*. Clearly, for any active pair-count γ,

$$\sum_{A, B \subseteq X_i} \gamma(A, B) = |C| = \alpha.$$

Furthermore, Eq. (1) implies that if $\gamma(A, B) \geq 1$ then $A \cap B = \emptyset$.

Let f be the total coloring f of $G = G_{01}$ for the root $X_{01} = \{v_1, v_2, v_3, v_4\}$ depicted in Figure 4(a), then

$$
\begin{aligned}
&Y(X_{01}; f, 1) = \{v_1\}, &&Z(X_{01}; f, 1) = \{v_3, v_4\}, \\
&Y(X_{01}; f, 2) = \emptyset, &&Z(X_{01}; f, 2) = \{v_1, v_2, v_3, v_4\}, \\
&Y(X_{01}; f, 3) = \{v_2, v_3, v_4\}, &&Z(X_{01}; f, 3) = \emptyset, \\
&Y(X_{01}; f, 4) = \emptyset, &&Z(X_{01}; f, 4) = \{v_1, v_2, v_3, v_4\}.
\end{aligned}
$$

Therefore f has the pair-count $\gamma_{X_{01}}$ such that

$$
\begin{aligned}
&\gamma_{X_{01}}(\{v_1\}, \{v_3, v_4\}) = 1, \\
&\gamma_{X_{01}}(\emptyset, \{v_1, v_2, v_3, v_4\}) = 2, \\
&\gamma_{X_{01}}(\{v_2, v_3, v_4\}, \emptyset) = 1,
\end{aligned}
$$

and $\gamma_{X_{01}}(A, B) = 0$ for any other pair of $A, B \subseteq X_{01}$. On the other hand, the total coloring of G_{02} for the left child $X_{02} = \{v_1, v_2, v_3, v_4\}$ of X_{01} depicted in Figure 4(b) has the pair-count $\gamma_{X_{02}}$ such that

$$\gamma_{X_{02}}(\{v_1\}, \emptyset) = 1,$$
$$\gamma_{X_{02}}(\emptyset, \{v_1, v_2, v_3\}) = 1,$$
$$\gamma_{X_{02}}(\{v_2, v_3, v_4\}, \emptyset) = 1,$$
$$\gamma_{X_{02}}(\emptyset, \{v_1, v_2, v_4\}) = 1,$$

and $\gamma_{X_{02}}(A, B) = 0$ for any other pair of $A, B \subseteq X_{02}$. The total coloring of G_{11} for the right child $X_{11} = \{v_1, v_3, v_4, v_5\}$ of X_{01} depicted in Figure 4(c) has the pair-count $\gamma_{X_{11}}$ such that

$$\gamma_{X_{11}}(\{v_1\}, \{v_3, v_4, v_5\}) = 1,$$
$$\gamma_{X_{11}}(\{v_5\}, \{v_4\}) = 1,$$
$$\gamma_{X_{11}}(\{v_3, v_4\}, \{v_5\}) = 1,$$
$$\gamma_{X_{11}}(\emptyset, \{v_3, v_5\}) = 1,$$

and $\gamma_{X_{11}}(A, B) = 0$ for any other pair of $A, B \subseteq X_{11}$.

We now have the following lemma.

Lemma 2. *Let two total colorings f and g of G_i for a node X_i of T have the same pair-count on X_i. Then f is extensible if and only if g is extensible.*

Thus an active pair-count on X_i characterizes an equivalence class of extensible total colorings of G_i. Since $|X_i| \le k + 1$, there are at most $(\alpha + 1)^{2^{2(k+1)}}$ active pair-counts on X_i. The main step of our algorithm is to compute a table of all active pair-counts on each node of T from leaves to the root X_{01} of T by means of dynamic programming. From the table on the root X_{01} one can easily determine whether G has a total coloring using colors in C, as follows.

Lemma 3. *A partial k-tree G has a total coloring using colors in C if and only if the table on root X_{01} has at least one active pair-count.*

We first compute the table of all active pair-counts on each leaf X_i of T as follows:

(1) enumerate all mappings:

$$V(G_i) \cup E(G_i) \to \{1, 2, \ldots, \min\{\alpha, (k+1)(k+2)/2\}\};$$

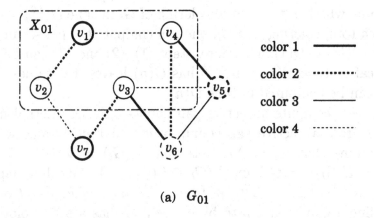

(a) G_{01}

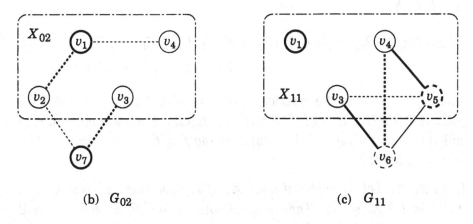

(b) G_{02} (c) G_{11}

Fig. 4. Total colorings of (a) $G = G_{01}$, (b) G_{02}, and (c) G_{11}.

(2) find all total colorings of G_i from the mappings above; and

(3) compute all active pair-counts on X_i from the total colorings of G_i.

Since $|V_i| \le k+1$ and $|E_i| \le k(k+1)/2$ for leaf X_i, the number of distinct mappings $f : V(G_i) \cup E(G_i) \to \{1, 2, \ldots, \min\{\alpha, (k+1)(k+2)/2\}\}$ is at most $O(k^{k^2}) = O(1)$. For each mapping f of G_i, one can determine whether f is a total coloring of G_i in time $O(k^2) = O(1)$. For each total coloring f of G_i, one can compute the pair-count of f in time $O(k^2) = O(1)$. Therefore, steps (1), (2) and (3) can be done for a leaf in time $O(1)$. Since T has $O(n)$ leaves, the tables on all leaves can be computed in time $O(n)$.

We next compute all active pair-counts on each internal noode X_i of T from all active pair-counts of its children X_l and X_r. We may assume that $X_i = X_l$. Note that $V(G_i) = V(G_l) \cup V(G_r)$, $E(G_i) = E(G_l) \cup E(G_r)$ and $E(G_l) \cap E(G_r) = \emptyset$. We call a mapping $\rho : 2^{X_l} \times 2^{X_l} \times 2^{X_r} \times 2^{X_r} \to \{0, 1, 2, \ldots, \alpha\}$ a *quad-count* on X_i. We define a quad-count ρ to be *active* if G_i has a total coloring f such that, for each quadruplet (A_l, B_l, A_r, B_r) with $A_l, B_l \subseteq X_l$ and $A_r, B_r \subseteq X_r$

$$\rho(A_l, B_l; A_r, B_r) = |\{c \in C : A_l = Y(X_l; f_l, c), B_l = Z(X_l; f_l, c),$$
$$A_r = Y(X_r; f_r, c), B_r = Z(X_r; f_r, c)\}|$$

where $f_l = f|G_l$ is the restriction of f to $V(G_l) \cup E(G_l)$ and $f_r = f|G_r$ is the restriction of f to $V(G_r) \cup E(G_r)$. Such a quad-count is called the *quad-count of the total coloring f of G_i*. Then we have the following lemma.

Lemma 4. *Let an internal node X_i of T have two children X_l and X_r, and let $X_i = X_l$. Then a quad-count ρ on X_i is active if and only if ρ satisfies the following conditions (a) and (b):*

(a) *if $\rho(A_l, B_l; A_r, B_r) \ge 1$ then $A_l \cap X_r = A_r \cap X_l$ and $B_l \cap B_r = \emptyset$; and*

(b) *there are two active pair-counts γ_l on X_l and γ_r on X_r such that*
   (i) *for each pair $A_l, B_l \subseteq X_l$,*

$$\gamma_l(A_l, B_l) = \sum_{A, B \subseteq X_r} \rho(A_l, B_l; A, B);$$

(ii) *for each pair* $A_r, B_r \subseteq X_r$,

$$\gamma_r(A_r, B_r) = \sum_{A,B \subseteq X_l} \rho(A, B; A_r, B_r).$$

Using Lemma 4, we compute all active quad-counts ρ on X_i from all pairs of active pair-counts γ_l on X_l and γ_r on X_r. Since there are at most $(\alpha+1)^{2^{2k+3}}$ pairs of active pair-counts on X_l and X_r, there are at most $(\alpha+1)^{2^{2k+3}}$ distinct active quad-counts ρ. For each ρ of them, we determine whether ρ satisfies Conditions (a) and (b) in Lemma 4. For each ρ, one can determine in time $O(1)$ whether ρ satisfies Condition (a), because there are at most $2^{4(k+1)} = O(1)$ distinct quadruplets (A_l, B_l, A_r, B_r). Furthermore, checking Condition (b) for all possible ρ's can be done in time $O((\alpha + 1)^{2^{2k+3}})$ since there are at most $(\alpha + 1)^{2^{2k+3}}$ pairs of γ_l and γ_r. Thus we have shown that all active quad-counts ρ on X_i can be computed in time $O((\alpha+1)^{2^{2k+3}})$.

We now show how to compute all active pair-counts on an internal node X_i from all active quad-counts on X_i.

Lemma 5. *Let an internal node X_i of T have two children X_l and X_r with $X_i = X_l$. A pair-count γ on X_i is active if and only if there exists an active quad-count ρ on X_i such that for each pair $A, B \subseteq X_i$*

$$\gamma(A, B) = \sum \rho(A_l, B_l; A_r, B_r). \qquad (2)$$

The summation above is taken over all quadruplets (A_l, B_l, A_r, B_r) such that $A = A_l$ and $B = (B_l \cup B_r) \cap X_l$.

Using Lemma 5 we compute all active pair-counts γ on X_i from all active quad-counts ρ on X_i. There are at most $(\alpha+1)^{2^{2k+3}}$ distinct active quad-counts ρ. From each ρ of them, we compute γ satisfying Eq. (2) in time $O(1)$ since $|A_l|, |B_l|, |A_r|, |B_r|, |X_i|, |X_l|, |X_r| \leq k+1$. Thus we have shown that all active pair-counts γ on X_i can be computed in time $O((\alpha + 1)^{2^{2k+3}})$. Since T has $O(n)$ internal nodes, one can compute the tables for all internal nodes in time $O(n(\alpha + 1)^{2^{2k+3}})$.

This completes a proof of Theorem 1.

4 Conclusion

In the paper we have given a polynomial-time algorithm to solve the total coloring problem for partial k-trees. One can immediately obtain a parallel algorithm to solve the total coloring problem for partial k-trees, slightly modifying the algorithm as follows. For a given tree-decomposition of a graph G with width at most k, one can obtain a binary tree-decomposition T of G with height $O(\log n)$ and width at most $3k + 2$ in $O(\log n)$ parallel time using $O(n)$ operations on the EREW PRAM [BH95]. Since each leaf of T has at most $3k + 3$ vertices, the tables of all active pair-counts on all leaves of T can be computed in $O(1)$ parallel time using $O(n)$ operations on the common CRCW PRAM. For each internal node X of T, the number of all active pair-counts on X is at most $(\alpha + 1)^{2^{6(k+1)}}$ since $|X| \leq 3k + 3$. Therefore the table on each internal node can be computed from all active pair-counts of the two children in $O(1)$ parallel time using $O((\alpha+1)^{2^{6k+7}})$ operations on the common CRCW PRAM. Since the height of the binary tree-decomposition T is $O(\log n)$, one can compute the table on the root in $O(\log n)$ parallel time using $O(n(\alpha + 1)^{2^{6k+7}})$ operations on the CRCW PRAM. Thus the parallel algorithm runs in $O(\log n)$ parallel time using $O(n(\alpha + 1)^{2^{6k+7}})$ operations on the common CRCW PRAM.

References

[ACPS93] S. Arnborg, B. Courcelle, A. Proskurowski and D. Seese. An algebraic theory of graph reduction, *J. Assoc. Comput. Mach.*, 40(5), pp. 1134–1164, 1993.

[AL91] S. Arnborg and J. Lagergren. Easy problems for tree-decomposable graphs, *Journal of Algorithms*, 12(2), pp. 308–340, 1991.

[BH95] H. L. Bodlaender and T. Hagerup. Parallel algorithms with optimal speedup for bounded treewidth, *Proc. of the 22nd International Colloquium on Automata, Languages and Programming, Lecture Notes in Computer Science,* 944, pp. 268–279, Springer Verlag, 1995.

[Bod90] H. L. Bodlaender. Polynomial algorithms for graph isomorphism and chromatic index on partial k-trees, *Journal of Algorithms*, 11(4), pp. 631–643, 1990.

[Bod96] H. L. Bodlaender. A linear time algorithm for finding tree decompositions of small treewidth, *SIAM Journal on Computing*, 25, pp. 1305–1317, 1996.

[BPT92] R. B. Borie, R. G. Parker and C. A. Tovey. Automatic generation of linear-time algorithms from predicate calculus descriptions of problems on recursively constructed graph families, *Algorithmica*, 7, pp. 555–581, 1992.

[Cou90] B. Courcelle. The monadic second-order logic of graphs I: Recognizable sets of finite graphs, *Information and Computation*, 85, pp. 12–75, 1990.

[FW77] S. Fiorini and R. J. Wilson. Edge-Colourings of Graphs, Pitman, London, 1977.

[Sán89] A. Sánchez-Arroyo. Determining the total colouring number is NP-hard, *Discrete Math.*, 78, pp. 315–319, 1989.

[Sch94]   P. Scheffler. A practical linear time algorithm for disjoint paths in graphs with bounded tree-width, *Technical Report* 396, *Dept. Mathematics, Technische Universität Berlin*, 1994.

[TNS82]   K. Takamizawa, T. Nishizeki and N. Saito. Linear-time computability of combinatorial problems on series-parallel graphs, *J. Assoc. Comput. Mach.*, 29(3), pp. 623–641, 1982.

[TP97]   J. A. Telle and A. Proskurowski. Algorithms for vertex partitioning problems on partial k-trees, *SIAM J. Discrete Math.*, 10, pp. 529–550, 1997.

[Yap96]   H. P. Yap. Total Colourings of Graphs, Lecture Notes in Mathematics, 1623, Springer-Verlag, Berlin, 1996.

[ZN95]   X. Zhou and T. Nishizeki. Algorithms for finding f-coloring of partial k-trees, In *Proc. of the Sixth International Symposium on Algorithms and Computation, Lecture Notes in Computer Science*, 1004, pp. 332–341, Springer-Verlag, 1995.

[ZN98]   X. Zhou and T. Nishizeki. The edge-disjoint paths problem is NP-complete for partial k-trees, submitted to a symposium.

[ZNN96]   X. Zhou, S. Nakano and T. Nishizeki. Edge-coloring partial k-trees, *Journal of Algorithms*, 21, pp. 598–617, 1996.

[ZSN96]   X. Zhou, H. Suzuki and T. Nishizeki. A linear algorithm for edge-coloring series-parallel multigraphs, *Journal of Algorithms*, 20, pp. 174–201, 1996.

[ZSN97]   X. Zhou, H. Suzuki and T. Nishizeki. An NC parallel algorithm for edge-coloring series-parallel multigraphs, *Journal of Algorithms*, 23, pp. 359–374, 1997.

[ZTN96]   X. Zhou, S. Tamura and T. Nishizeki. Finding edge-disjoint paths in partial k-trees, In *Proc. of the Seventh International Symposium on Algorithms and Computation, Lecture Notes in Computer Science, Springer*, 1178, pp. 203–212, 1996.

Rankings of Directed Graphs
(extended abstract)

Jan Kratochvíl [1*] and Zsolt Tuza [2**]

[1] Department of Applied Mathematics
Charles University
Malostranské nám. 25, 118 00 Praha 1
Czech Republic

E-mail: honza@kam.ms.mff.cuni.cz

[2] Computer and Automation Institute
Hungarian Academy of Sciences
H–1111 Budapest, Kende u. 13–17.

Hungary
E-mail: tuza@sztaki.hu

Abstract. A ranking of a graph is a coloring of the vertex set with positive integers such that on every path connecting two vertices of the same color there is a vertex of larger color. We consider the directed variant of this problem, where the above condition is imposed only on those paths in which all edges are oriented in the same direction. We show that the ranking number of a directed tree is bounded by that of its longest directed path plus one, and that it can be computed in polynomial time. Unlike the undirected case, however, deciding whether the ranking number of a directed (and even of an acyclic directed) graph is bounded by a constant is NP-complete. In fact, the 3-ranking of planar bipartite acyclic digraphs is already hard.

1 Introduction

Given an undirected graph G, its *ranking number* $\chi_r(G)$ is the minimum integer k for which there exists a *(vertex) k-ranking*, that is a mapping $f : V(G) \to \{1, 2, \ldots, k\}$ such that every path connecting two vertices u, v of the same rank $f(u) = f(v)$ contains a vertex w with higher rank, $f(w) > f(u)$.

* Research supported in part by the Czech Research Grants GAUK 194 and GAČR 201/1996/0194.

** Research supported in part by the Hungarian Scientific Research Fund, Grant OTKA T–016416.

It is well known and easy to see that for the path P_ℓ of length $\ell - 1$ on ℓ vertices,

$$\chi_r(P_\ell) = \lfloor \log \ell \rfloor + 1$$

holds, and that the longest k-rankable path $P_{2^k-1} = x_1 x_2 \ldots x_{2^k-1}$ admits the *unique* optimal ranking f with

$$f(x_i) = \max \left\{ j : 2^j | i \right\} + 1$$

for all $1 \leq i < 2^k$. (Throughout, log means logarithm of base 2.)

This paper is the first approach to the ranking of *directed* graphs. The ranking number of a digraph G is naturally defined as the minimum k such that there exists a mapping $f : V(G) \to \{1, 2, \ldots, k\}$ with the property that every *directed path* (i.e., path in which all edges are oriented consecutively) connecting two vertices u, v of the same rank $f(u) = f(v)$ contains a vertex w with higher rank, $f(w) > f(u)$. We denote the ranking number of a directed graph G again by $\chi_r(G)$.

Obviously, the ranking number of a directed path equals that of the undirected path of the same length. Directed and undirected rankings, however, have a strikingly different behavior already on trees. For instance, an undirected tree containing no path longer than t can have as large ranking number as $\lceil t/2 \rceil + 1$. This is far from being true in the directed case. We shall prove that the ranking number of a directed tree can exceed that of its longest directed path by at most 1 (Corollary 3), hence it grows just with $\log t$.

We also consider rankings from the computational complexity point of view. The problem RANKING takes as input a graph G and a positive integer k, and asks whether $\chi_r(G) \leq k$. It is known that RANKING on undirected graphs is NP-complete in general, but solvable in polynomial time for every fixed k; see [1] for results and further references. For the analogous problem of DIRECTED RANKING, however, we prove in Theorem 9 that it is NP-complete even if the input is restricted to fixed $k = 3$ and to *acyclic* orientations of *planar bipartite* graphs. On the other hand, the 2-rankable directed graphs can be characterized in several different ways, as shown in Section 5. We also prove that the ranking number of directed *trees* can be determined in polynomial time (Section 3). More generally, a similar approach can be applied to determine in polynomial time the ranking number of digraphs whose underlying graphs have *bounded treewidth*, but we do not include the proof of this stronger assertion here.

2   Upper bound for trees

In this section we prove general bounds on the ranking number of oriented trees and also on that of orientations of a path of given length. We begin with some definitions.

Notation. We write $p(\ell) := \lfloor \log \ell \rfloor + 1 = \chi_r(P_\ell)$ for the ranking number of the (directed or undirected) path with ℓ vertices (i.e., $p(\ell) = k$ if and only if $2^{k-1} \leq \ell \leq 2^k - 1$). Moreover, we define $r_t(\ell)$ and $r_p(\ell)$ as the maximum ranking number of orientations of trees and that of orientations of undirected paths, respectively, under the condition that *no directed subpath has more than ℓ vertices*.

Our results will show that the above three parameters are very close to each other, in the entire range of ℓ. Clearly $r_t(\ell) \geq r_p(\ell) \geq p(\ell)$.

Theorem 1 *For every $k \geq 2$ and ℓ such that $2^{k-2} + 1 \leq \ell \leq 2^{k-1}$,*

$$r_t(\ell) = k.$$

Proof. We first show that $\chi_r(T) \leq k$ holds whenever each directed subpath of a given oriented tree T has at most 2^{k-1} vertices. Consider an infinite directed path with vertices x_i and edges $x_i x_{i+1}$, $i \in \mathbf{Z}$. Define a mapping $\phi : \{x_i : i \in \mathbf{Z}\} \to \{1, 2, \ldots, k\}$ by

$$\phi(x_i) = \begin{cases} k & \text{if } i \equiv 0 \bmod 2^{k-1}, \\ \max\{j : i \equiv 0 \bmod 2^{j-1}\} & \text{if } i \not\equiv 0 \bmod 2^{k-1}. \end{cases}$$

Obviously, any segment of length at most 2^{k-1} is ranked feasibly by ϕ.

Now we consider a directed tree T containing no directed subpath with more than 2^{k-1} vertices. We view such a tree as a Hasse diagram of a partially ordered set, and as such, partition its vertices into levels: we choose an arbitrary vertex and call its level $L(0)$, and then recursively sort the other vertices — a vertex u is placed into level $L(i+1)$ if there is a vertex v already in level $L(i)$ such that $uv \in E(T)$, and a vertex w is placed into level $L(i-1)$ if there is a vertex v already in level $L(i)$ such that $vw \in E(T)$. A mapping f defined by $f(u) = \phi(x_i)$ for $u \in L(i)$ is then a feasible k-ranking of T. (The above procedure partitions T into levels correctly, since T is a tree.) This proves $r_t(\ell) \leq k$.

The lower bound for $r_t(\ell)$ can be found in the full version of the paper. $\square$

Next, we show that the ranking number of directed trees of maximum degree 2 (i.e., orientations of undirected paths) usually equals the ranking number of their longest directed paths.

Theorem 2 *For every $k \geq 3$ and every ℓ such that $2^{k-1} - 1 \leq \ell \leq 2^k - 2$,*

$$r_p(\ell) = k.$$

Proof. We first prove the upper bound, i.e., $r_p(2^k - 2) \leq k$. It is easy to see that every (directed or undirected) path with at most $2^k - 2$ vertices has a feasible k-ranking such that the first vertex is ranked 1 and the last vertex is ranked 2.

Thus, if T is an orientation of a path consisting of several segments of length at most $2^k - 3$ (a segment is a maximal directed subpath), we can k-rank each segment separately so that the sources are ranked 1 and the sinks are ranked 2.

On the other hand, to show the lower bound, we take two vertex-disjoint paths of length $2^k - 2$ each, and orient an arc from the first vertex of one of them to the last vertex of the other one. The resulting graph has no feasible k-ranking, because in every k-ranking of a directed path of length $2^k - 2$, both endvertices are ranked 1, thus the added arc would connect two vertices ranked 1, a contradiction. Therefore $r_p(2^k - 1) \geq k + 1$.   □

Reformulating the results proven above, and relating the ranking number of a directed tree to the ranking number of its longest paths, we obtain:

Corollary 3 *The ranking number of a directed tree is always less than or equal to the ranking number of its longest directed paths plus 1. This bound is best possible, as*

$$r_t(\ell) = \begin{cases} p(\ell) & \text{if } \ell = 2^k, \\ p(\ell) + 1 & \text{if } \ell \neq 2^k. \end{cases}$$

Similarly, for orientations of undirected paths, we have

$$r_p(\ell) = \begin{cases} p(\ell) & \text{if } \ell \neq 2^k - 1, \\ p(\ell) + 1 & \text{if } \ell = 2^k - 1. \end{cases}$$

We illustrate the functions $p(\ell)$, $r_p(\ell)$, and $r_t(\ell)$ in the schematic figure 1.

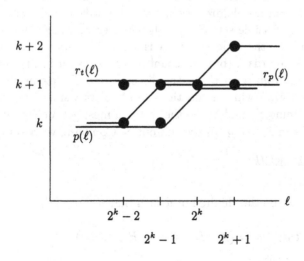

Fig. 1. Trees and paths vs. the undirected path $p(\ell)$

3   Algorithm for trees

In this section we prove that the ranking number of a directed tree can be determined by a polynomial-time algorithm.

Assuming that a natural number k and a tree T with n vertices, rooted at a vertex r, are given, our next goal is to decide by an efficient algorithm if $\chi_r(T) \leq k$.

We shall use the following notation. For a vertex u of T, denote by T_u the subtree rooted at u, that is induced by those vertices — including u as well — from which the path (in the underlying undirected graph of T) to the root of T passes through u. If u is not the root, then u^+ denotes the first vertex on the path from u to the root r. The vertices adjacent to u other than u^+ are called the children of u.

The algorithm described below scans recursively the vertices of T from the leaves to the root and computes a set system $S(u)$ for every $u \in V(T)$. Each $S(u)$ is a family of subsets of $\{1, 2, \ldots, k\}$, storing essential information concerning the feasible rankings of the subtree rooted at u. Namely, $S \in S(u)$ if and only if the subtree rooted in u allows a ranking such that S is the set of colors visible (from u) on directed paths from the inside of T_u to u (when u is not the root of T and the upgoing edge is oriented towards u^+), and otherwise (i.e., if u is the root or if $u^+u \in E(T)$), $S \in S(u)$ if and only if T_u admits a ranking such that S is the set of colors visible (from u) on directed paths leading from u into T_u. Also, the values of auxiliary functions $Up(u)$, $Down(u)$, and $Compose(\mathcal{A}, \mathcal{B})$ are collections of subsets of $\{1, 2, \ldots, k\}$. The meaning of them is as follows: $Up(u)$ will store the unions of all feasible sets S from the children of u such that the edges from the children are directed towards u, and $Down(u)$ will store the unions of all feasible sets S from the children of u such that the edges are directed from u to the children. If S_A is a set of colors visible from u on the paths directed towards u (but not counting the vertex u itself) and S_B is a set of colors visible from u on the paths directed into T_u (again disregarding u), then $Compose(\mathcal{A}, \mathcal{B})$ will contain the sets of colors visible from u in rankings obtained by joining S_A and S_B — the vertex u must get a color large enough to block all colors in $S_A \cap S_B$. In the subroutine $Compose$, we assume $\max \emptyset = 0$.

Algorithm TREE(k)

Function $Up(u)$:
Let $u_1, u_2, \ldots, u_t$ be the children of u such that $u_i u \in E(T)$.
$Up := \{\emptyset\}$;
for $j := 1$ **to** t **do** $Up := \{A \cup B : A \in Up, B \in S(u_j)\}$.

Function $Down(u)$:
Let $u_1, u_2, \ldots, u_t$ be the children of u such that $u u_i \in E(T)$.
$Down := \{\emptyset\}$;
for $j := 1$ **to** t **do** $Down := \{A \cup B : A \in Down, B \in S(u_j)\}$.

Function $Compose(\mathcal{A}, \mathcal{B})$:
$Compose := \emptyset$;
for $A \in \mathcal{A}$ **do**
  for $B \in \mathcal{B}$ **do**
    for $i := \max(A \cap B) + 1$ **to** k **do**
      if $i \notin A \cup B$ **then** $Compose := Compose \cup \{(A \cap \{i+1, i+2, \ldots, k\}) \cup \{i\}\}$.

Function $S(u)$:
if $u \neq r$ **and** $uu^+ \in E(T)$
  then $S := Compose(Up(u), Down(u))$
  else $S := Compose(Down(u), Up(u))$.

Program body:
if $S(r) = \emptyset$
  then $\chi_r(T) > k$
  else $\chi_r(T) \leq k$.

For a vertex u and a path $P = u_1 \ldots u_j$, $u_j = u$, we say that a color i is *visible on P from u* if some vertex u_h on this path receives color i and no vertex u_ℓ, $\ell = h+1, \ldots, j$ is colored with a color higher than i. The correctness of our algorithm follows from the following proposition, whose detailed proof can be found in the full version of the paper.

Proposition 4 *If u is not the root of T and $uu^+ \in E(T)$, then $S \in S(u)$ if and only if T_u admits a ranking such that S is the set of colors visible (from u) on directed paths from the inside of T_u to u. Otherwise (i.e., if u is the root or if $u^+ u \in E(T)$), $S \in S(u)$ if and only if T_u admits a ranking such that S is the set of colors visible (from u) on directed paths leading from u into T_u.*

Corollary 5 *The algorithm TREE(k) gives the correct answer to the question whether $\chi_r(T) \leq k$.*

Proposition 6 *The running time of the algorithm TREE(k) is at most $cnk^2 \, 2^{2k}$, for some absolute constant c independent of k.*

Proof. The function Up (which is a dynamic programming version for computing the set of all unions of type $\bigcup_{j=1}^s A_j$ for $A_j \in S(u_j)$) needs at most 2^{2k} set unions in each of the s steps. Hence, Up on a vertex with s ingoing children runs in $O(sk \, 2^{2k})$ time. The analogous property holds for $Down$ as well. Throughout the entire tree T, there are as many children of processed vertices as the number of edges of T, and therefore Up and $Down$ will consume in total at most $O(nk \, 2^{2k})$ steps.

The procedure $Compose$ requires at most $O(k^2 \, 2^{2k})$ steps, and being performed for every vertex, it requires running time at most $O(nk^2 \, 2^{2k})$. $\square$

In conclusion, we obtain

Theorem 7 *For any directed tree T on n vertices, the directed ranking number of T can be determined in time $O(n\,\ell^2 \log^3 \ell)$, where $\ell \geq 2$ is the length of a longest directed path in T.*

Proof. Assume $n \geq 2$. We know from Theorem 2 that $1 \leq \chi_r(T) - 1 \leq \log \ell$. Therefore, it suffices to run the algorithm TREE(k) for at most $\log \ell$ values of $k \leq \log \ell + 1$, and for each of them, TREE(k) takes at most $O(n \cdot \log^2 \ell \cdot 2^{2\log \ell}) = O(n \cdot \ell^2 \cdot \log^2 \ell)$ time. □

The above results can be extended to the following more general theorem whose proof is omitted. (For the definition of treewidth, see e.g. [1].)

Theorem 8 *For every fixed natural number t, the directed ranking number can be determined in polynomial time for any digraph whose underlying graph has treewidth at most t.*

4   Ranking number of bipartite acyclic digraphs

Here we consider the algorithmic problem on DAGs (directed acyclic graphs).

Theorem 9 *The problem DIRECTED RANKING is NP-complete on DAGs with planar bipartite underlying graphs, even for fixed ranking number $k = 3$.*

Proof. We show a reduction from a variant of the PRECOLORING EXTENSION problem of undirected graphs. It is known [4] that the following problem is NP-complete:

> *Given a planar bipartite graph with some of its vertices properly colored with three colors, does G admit a proper 3-coloring that extends the precoloring?*

Given such a bipartite graph $G = (A \cup B, E)$, observe that we may assume without loss of generality that all the precolored vertices belong to A. Indeed, for each precolored vertex $v \in B$, we create two new precolored vertices of degree 1, adjacent to v and assigned to the two colors different from the one prescribed for v; then v can be made precolorless, as its precolored pendant neighbors force it to get the originally prescribed color.

Given such a bipartite graph $G = (A \cup B, E)$ with precolored vertex set $Z \subseteq A$ and precoloring $\phi : Z \to \{1, 2, 3\}$, we construct a directed graph D with vertex set

$$V(D) = A \cup B \cup \{z_i^j : z \in Z,\ 1 \leq i \leq 7,\ 1 \leq j \leq 2\}$$

and arc set

$$E(D) = \bigcup_{\substack{u \in A,\, v \in B \\ uv \in E}} \{uv\} \cup \bigcup_{\substack{z \in Z \\ 1 \leq i \leq 6 \\ 1 \leq j \leq 2}} \{z_i^j z_{i+1}^j\} \cup \bigcup_{z \in Z} \{z z_{i_1(z)}^1, z z_{i_2(z)}^2\}$$

where

$$i_1(z) = \begin{cases} 6 & \text{if } \phi(z) = 1 \\ 7 & \text{if } \phi(z) = 2 \vee 3 \end{cases} \qquad i_2(z) = \begin{cases} 4 & \text{if } \phi(z) = 1 \vee 2 \\ 6 & \text{if } \phi(z) = 3 \end{cases}$$

Obviously, D is acyclic, and it also remains planar and bipartite because so is G. We claim that D is 3-rankable if and only if G admits a precoloring extension with 3 colors.

Suppose first that D is 3-rankable, and let $f : V(D) \to \{1, 2, 3\}$ be a feasible ranking. Since the paths $P_{z,j} = z_1^j z_2^j \ldots z_7^j$ ($z \in Z$, $j = 1, 2$) are uniquely 3-rankable induced subgraphs of D, we must have $f(z_1^j) = f(z_3^j) = f(z_5^j) = f(z_7^j) = 1$, $f(z_2^j) = f(z_6^j) = 2$, and $f(z_4^j) = 3$. In this way, each $P_{z,j}$ excludes one well-defined color from its neighbor in A, and the total effect is that precisely the two colors distinct from $\phi(z)$ get excluded at each $z \in Z$. It follows that $f(z) = \phi(z)$ holds, and therefore f is a proper 3-coloring of G extending the precoloring ϕ.

On the other hand, any proper precoloring extension of ϕ together with the color sequence 1213121 on each $P_{z,j}$ gives a feasible 3-ranking. $\square$

Corollary 10 *For every fixed ranking number $k \geq 3$, the problem* DIRECTED RANKING *is NP-complete.*

Proof. Take a dag G whose $\chi_r(G) \leq 3$ is questioned, add a tournament on $k-3$ vertices and join every vertex of G to every vertex of the tournament by an arc directed towards the tournament. In any ranking, the vertices of the tournament have to receive distinct colors, and all these colors must differ from the colors used on G. Hence the new graph is k-rankable if and only if G is 3-rankable. $\square$

In fact, one can extend the method of the proof of Theorem 9 to show that for every fixed $k \geq 3$, DIRECTED RANKING remains NP-complete for DAGs with planar bipartite underlying graphs.

5 Directed 2-rankable graphs

Here we investigate directed rankings with $k = 2$ colors. For the structural characterization of 2-rankable digraphs, the following concept will be convenient to introduce. By an *alternating walk of length ℓ* we mean a sequence $P = x_0 x_1 \ldots x_\ell$ of (not necessarily distinct) vertices such that its orientation is $x_0 \to x_1 \leftarrow x_2 \to x_3 \leftarrow \ldots$, i.e., $x_{2i}x_{2i+1} \in E$ for all $0 \leq i < \ell/2$ and $x_{2i}x_{2i-1} \in E$ for all $1 \leq i \leq \ell/2$. An alternating walk is an *alternating path* if its vertices are mutually distinct. Moreover, we say that a vertex v is *starting*, *central*, or *ending*, if there is a *directed path* $P_3 = x_1 x_2 x_3$ with $x_1 = v$, $x_2 = v$, or $x_3 = v$, respectively. In the present context, alternating paths and cycles of *odd* lengths will be crucial. The proof of the following characterization theorem can be found in the full version of the paper.

Theorem 11 *For every digraph $G = (V, E)$, the following conditions are equivalent.*

(1) *G is 2-rankable.*

(2) *G contains no alternating path of odd length from a starting vertex to an ending vertex.*

(3) *G contains no alternating walk of odd length with both endpoints being central vertices.*

(4) *G admits a proper 2-coloring in which the set of central vertices is monochromatic.*

Remarks. 1. Algorithmically it is very easy to decide whether a digraph G is 2-rankable. Indeed, the answer is negative whenever G is not bipartite, and otherwise it suffices to test separately in each connected component if some of the two possible 2-colorings is a 2-ranking. Cf. also condition (4).

2. Similar types of problems have been studied in the framework of precoloring extension in several papers. Good characterizations are known for the existence of k-colorings of trees with any number of prescribed monochromatic independent sets [2, 3], and also for one prescribed monochromatic independent set in *perfect* graphs [5]. (As we have mentioned before, the problem for bipartite graphs with at least three precolored vertices of distinct colors is algorithmically hard [4], and so is for two monochromatic vertex pairs in distinct colors, too.) For an extensive survey on this subject, see [7].

3. Some small subgraphs excluded by the degenerate 'alternating' path of length 1 are:

- the cyclic triangle $y_1 \rightarrow y_2 \rightarrow y_3 \rightarrow y_1$, where any two of the y_i are adjacent central vertices and also each edge joins a starting vertex with an ending vertex,

- the transitive triangle $y_1 \rightarrow y_2 \rightarrow y_3 \leftarrow y_1$, where $y_1 y_3$ is an edge from a starting vertex to an ending vertex (and $y_2 y_3 y_1 y_2$ is an odd alternating walk from the central vertex y_2 to itself),

- the path $y_1 \rightarrow y_2 \rightarrow y_3 \rightarrow y_4$ of length 3, where the edge $y_2 y_3$ joins a starting vertex with an ending vertex, both of which are central as well.

Moreover, chordless odd cycles of lengths ≥ 5 (with any orientation) are also excluded by the longer alternating paths or by the entire cycle as an alternating walk, according to the conditions (2) and (3) for longer paths/walks. Note that the characterization of 2-rankable digraphs in terms of forbidden subgraphs involves an *infinite* family of minimal configurations, which is not the case for undirected rankings.

6   Open problems

There are many interesting related problems arising in the above context in a natural way. Below we mention some of them.

1. Draw a sharper line between the polynomial instances of oriented trees and the NP-complete class of directed acyclic bipartite planar graphs, by describing large subclasses of the latter in which the ranking number still can be determined in polynomial time.
2. What is the complexity of DIRECTED EDGE RANKING for a fixed number of colors? (The undirected version is linear [1], but NP-complete if the number of colors is unrestricted [6].)
3. More generally, which classes of directed graphs admit polynomial-time decision algorithms for k-ranking and/or edge k-ranking, for every fixed k?

Acknowledgement.   We are grateful to Hans Bodlaender for fruitful discussions. Moreover, the second author thankfully acknowledges support from the Konrad-Zuse-Zentrum für Informationstechnik Berlin, where part of this research was carried out.

References

1. H. Bodlaender, J. S. Deogun, K. Jansen, T. Kloks, D. Kratsch, H. Müller and Zs. Tuza: Rankings of graphs. In: Graph Theoretic Concepts in Computer Science (E. W. Mayr et al., eds.), Lecture Notes in Computer Science 903, Springer-Verlag, 1995, 292–304. *Extended version in: SIAM J. Discr. Math.* 11 (1998), 168–181.
2. M. Hujter and Zs. Tuza: Precoloring extension. II. Graph classes related to bipartite graphs. *Acta Math. Univ. Carolinae* 62 (1993), 1–11.
3. M. Hujter and Zs. Tuza: Precoloring extension. III. Classes of perfect graphs. *Combin. Probab. Computing* 5 (1996), 35–56.
4. J. Kratochvíl: Precoloring extension with fixed color bound. *Acta Math. Univ. Carolinae* 62 (1993), 139–153.
5. J. Kratochvíl and A. Sebő: Coloring precolored perfect graphs. *J. Graph Theory* 25 (1995), 207–215.
6. T. W. Lam and F. L. Yue: Egde ranking is NP-complete, to appear in *Discrete Applied Math.*
7. Zs. Tuza: Graph colorings with local constraints—A survey. *Discuss. Math. Graph Theory* 17 (1997), 161–228.

Drawing Planar Partitions II: HH-Drawings

Therese Biedl[1], Michael Kaufmann[2], and Petra Mutzel[3]

[1] School of Computer Science, McGill University, 3480 University St., Montréal,
Québec H3A2A7, Canada. therese@cgm.cs.mcgill.ca[†]
[2] Wilhelm-Schickard Institut für Informatik, Universität Tübingen, D-72116
Tübingen, Germany. mk@informatik.uni-tuebingen.de
[3] Max-Planck-Institut für Informatik, Im Stadtwald, D-66123 Saarbrücken, Germany.
mutzel@mpi-sb.mpg.de

Abstract. Let a planar graph $G = (V, E)$ and a vertex-partition $V = A \cup B$ be given. Can we draw G without edge crossings such that the partition is clearly visible? Such drawings aid to display partitions and cuts as they arise in various applications. In this paper, we study planar drawings of G in which the vertex classes A and B are separated by a horizontal line (so-called HH-drawings). We provide necessary and sufficient conditions for the existence of so-called y-monotone planar HH-drawings, and a linear time algorithm to construct, if possible, a y-monotone planar HH-drawing of area $\mathcal{O}(|V|^2)$ with few bends. Furthermore, we give an exponential lower bound for the area of straight-line planar HH-drawings. Finally, we study planar HH-drawings that are not y-monotone.

1 Introduction

Assume that $G = (V, E)$ is a graph and $V = A \cup B$ is a partition of the vertices of G. How should we draw G such that the partition is clearly visible? Our study of this question was motivated by a competition graph (Graph B) of the Graph Drawing Competition 1996 (see [17]), which is a graph of telephone calls and turned out to be bipartite, hence had a natural partition structure.

Many such partition structures arise in applications, in particular for the special case that G is a bipartite planar graph with the vertex classes A and B. For example, transportation problems are represented by bipartite graphs, with the vertex classes as the sellers and the buyers, respectively. Cuts, and hence partitions, arise in many algorithms for combinatorial optimization, for example maximum flows and Dijkstra's algorithm.

A drawing of G such that the partition is clearly visible aids in understanding the structure of these problems better. Two typical drawings of partitions are shown in Figure 1. Creating such drawings is the subject of this paper.

This paper fits into the larger frame of graph drawing, a relatively new field in computer science, see for example [6]. During the last years, many drawing styles have been developed. In the present paper, we consider *straight-line* and *poly-line grid drawings*, that is, vertices are drawn as points on a grid (i.e., with

---

[†] This paper was written while the author was at RUTCOR, Rutgers University.

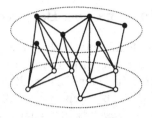

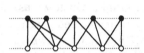

Fig. 1. Drawings of two graphs with a given partition (shown as black and white points), one using two lines and one using two disjoint convex sets.

integer coordinates), and edges are drawn either as straight lines or as sequences of straight-line segments where the bend points lie on a grid as well. We study only *planar partitions*, i.e., the graph $G = (A \cup B, E)$ is planar, and we require that the drawing of G has no crossing.

For planar graphs, many algorithms to construct planar straight-line drawings exist. The first algorithm to achieve an area of $\mathcal{O}(|V|^2)$ was presented in 1988 [11]. Many improvements appeared later, see e.g. [4, 5, 12, 15]. Algorithms for planar poly-line drawings can for example be found in [15, 20]. However, none of these algorithms is designed to pay respect to a partition of the vertices.

On the other hand, two types of drawings that clearly display a partition are demonstrated in Figure 1, but these may have many crossings, even if the graph is planar. Crossings were shown to impede the understanding of the graph [19], and hence should be avoided.

Our goal for this series of papers on drawing planar partitions is to find drawings that display the partition clearly and at the same time have no crossings. There exist various models of what is meant by "displaying the partition"; a unifying naming scheme of those can be found in [1]. For the special case of bipartite graphs, this problem has been solved for a number of models (see e.g. [8, 10, 14, 16]); for arbitrary planar partitions, the only results known to the authors are in the preceding paper of this series [1]. In the present paper, we study so-called *HH-drawings*, i.e., drawings in which the vertex classes A and B are separated by a horizontal line; see for example the right picture of Figure 1. This model has not been studied before. Our results hold for arbitrary partitions.

Not any planar partition has a planar HH-drawing if the edges are required to be drawn y-monotone, i.e., with monotonically increasing y-coordinates. We provide necessary and sufficient conditions for the existence of planar y-monotone HH-drawings. In the third paper of the series [3], it is shown that these conditions can be tested in linear time. One surprising corollary is that any bipartite planar graph has a planar y-monotone HH-drawing.

The proof of sufficiency yields an algorithm for planar y-monotone HH-drawings with area $\mathcal{O}(n^2)$ and at most one bend per edge. We prove that straight-line HH-drawings of polynomial area are not always possible, viz., there exists a graph class for which any planar straight-line HH-drawing requires exponential area. Finally, we drop the monotonicity-requirement, and prove that then every planar partition has a planar HH-drawing with at most three bends per edge.

2   Definitions and Background

We assume familiarity with graphs and algorithms, and refer for example to
[9] for further reference. Let $G = (V, E)$ be a graph, which we assume to be
finite, simple (no loops and multiple edges), and connected. Denote $n = |V|$
and $m = |E|$. The graph G is called *planar* if it can be drawn in the plane
without edge crossings. A planar drawing of G splits the plane into connected
pieces called *faces*, the unbounded face is called the *outer-face*. A specific planar
drawing of a planar graph determines the *planar embedding*, that is, the cyclic
order of edges around each vertex in the incidence list.

Assume that a planar embedding of G is fixed. The *dual graph* G^* is defined
by adding a vertex in G^* for every face in G, and an edge $e^* = (F, F')$ in G^*
for every edge e incident to the two faces F and F'. The edges e^* and e are
called *dual* to each other. The structure of the dual graph depends on the planar
embedding of G. The dual graph G^* may have loops and multiple edges even if
the primal graph G is simple.

A *planar partition* is a planar graph $G = (V, E)$ together with a partition of
the vertices $V = A \cup B$, and will be denoted $G = (A \cup B, E)$. In [1, 10], three
models of displaying a partition were introduced that are relevant to this paper.
A drawing of G is called an *LL-drawing* if the vertices of A have y-coordinate 0
while the vertices of B have y-coordinate 1. A drawing of G is called an *LH-drawing* if the vertices of A have y-coordinate 0 while the vertices of B have
y-coordinate > 0. A drawing of G is called an *HH-drawing* if the vertices of A
have y-coordinate < 0 while the vertices of B have y-coordinate > 0. L stands
for line, and H stands for half-space. See also Figure 2.

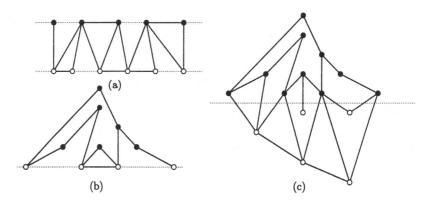

Fig. 2. (a) An LL-drawing, (b) an LH-drawing, and (c) an HH-drawing.

We distinguish between monotone and non-monotone drawings as follows:
An edge (v, w) is drawn *y-monotone* if, after exchanging v and w if necessary,
the y-coordinate does not decrease when walking from v to w. A drawing is
y-monotone if all edges are drawn y-monotone. This concept is closely related
to the one of upward drawings for directed graphs: A drawing of a directed

graph is called *upward* if for each edge the y-coordinate increases as we walk from the tail to the head. However, as opposed to upward drawings, horizontal lines are permitted in y-monotone drawings, hence any straight-line drawing is y-monotone.

2.1 Results for LH-drawings

To construct HH-drawings, we will use the results on LH-drawings presented in [1], to be reviewed briefly here.

Theorem 1. [1] A planar partition $G = (A \cup B, E)$ has a planar straight-line LH-drawing if and only if

1. the vertices of A induce a collection of paths,
2. if we add a vertex v_A and connect it to all vertices in A, the resulting graph G^+_{LH} is planar, and
3. there exists a planar embedding of G^+_{LH} such that any triangle $\{v_A, a', a''\}$ with $a', a'' \in A$ is a face.

If these conditions are satisfied, then we can find a planar straight-line LH-drawing with area $\mathcal{O}(n^2)$ in linear time.

The algorithm to create these LH-drawings works roughly as follows: Let a planar embedding of G^+_{LH} be fixed as demanded in Theorem 1, part 3. Let $a_1, \ldots, a_k$ be the vertices in A, enumerated in clockwise order around v_A (with respect to the planar embedding of G^+_{LH}). Here, choose a_1 such that there is no edge (a_1, a_k) (this exists by Theorem 1, part 1). Let $a_1, \ldots, a_k$ be placed with y-coordinate 0, and with any integer x-coordinates $x_1 < \ldots < x_k$. Then we can add the vertices of B one by one such that the resulting drawing is a planar straight-line LH-drawing of width and height at most $x_k - x_1 + |B|$. See [1] for details.

3   Monotone HH-drawings

Not every planar partition has a y-monotone planar HH-drawing; one counter-example is the graph K_5 with one edge (v_1, v_2) missing, with vertex partition $A = \{v_1, v_2\}$ and $B = \{v_3, v_4, v_5\}$. If this planar partition had a planar y-monotone HH-drawing, then v_3, v_4, v_5 would have y-coordinates > 0. The points and lines of the triangle v_3, v_4, v_5 would also be drawn with y-coordinates > 0, by definition of y-monotone. Since $K_5 - (v_1, v_2)$ is triconnected and has a unique planar embedding, one of $\{v_1, v_2\}$ would be drawn inside the triangle $\{v_3, v_4, v_5\}$, hence also with y-coordinate > 0, which contradicts the definition of an HH-drawing. Note that with another partition, $K_5 - (v_1, v_2)$ would be drawable.

So our goal is develop an algorithm to test whether a given planar partition has a planar y-monotone HH-drawing, and if the answer is yes, create one. To that end, we prove necessary and sufficient conditions for the existence of planar y-monotone HH-drawings, which can be tested in linear time. The sufficiency proof is algorithmic, and yields a linear-time drawing algorithm.

3.1   Necessary condition

Assume that a planar partition $G = (A \cup B, E)$ has a planar y-monotone HH-drawing. Then all edges of the form (a, b), $a \in A$, $b \in B$ cross the x-axis; these edges are called *cut-edges*. Since edges are drawn y-monotonically, each cut-edge crosses the x-axis exactly once, and no other edge crosses it.

Let G^* be the dual graph of G in the embedding induced by the HH-drawing. The *dual cut-edges* are those edges in G^* that are dual to the cut-edges. The *cut-dual graph* G_C^* is the graph formed by the dual cut-edges, i.e., the graph that consists of the dual cut-edges and their endpoints. Note that this is not necessarily the same as the graph induced by the endpoints of the dual cut-edges. For example, in Figure 3, the edge (d_1, d_2) is not in G_C^*, even though both its endpoints are in G_C^*.

Enumerate the cut-edges as $e_1, \ldots, e_k$, from left to right with respect to their intersecting with the x-axis, and let $e_1^*, \ldots, e_k^*$ be their dual edges. Then e_i^* and e_{i+1}^* have a common endpoint, for $i = 1, \ldots, k-1$, therefore G_C^* is connected.

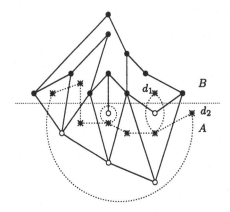

Fig. 3. A planar partition in an HH-drawing and its cut-dual graph (shown with stars and dotted lines).

Lemma 1. *A planar y-monotone HH-drawing of G can exist only if there exists a planar embedding of G such that the cut-dual graph G_C^* is connected.*

3.2   Sufficiency of the necessary condition

To show that the necessary condition of Lemma 1 is also sufficient, we present an algorithm in three steps. First, we transform the input graph, by subdividing the cut-edges and adding some edges. Then we split the graph into two pieces and create an LH-drawing of each. Combining the drawings we arrive at an HH-drawing.

So assume from now on that a planar embedding of G is fixed such that the connected cut-dual graph G_C^* is connected. Choose an arbitrary outer-face of G.

A crossing-free Eulerian circuit of G_C^* The cut-dual graph G_C^* is Eulerian, since any face of G has an even number of incidences to cut-edges. Find a Eulerian circuit $e_1, \ldots, e_k$ for G_C^* which is *crossing-free*, i.e., if a vertex is incident to the edges $e_i, e_{i+1}, e_j, e_{j+1}$ of the Eulerian circuit, for some i, j, then the sets $\{e_i, e_{i+1}\}$ and $\{e_j, e_{j+1}\}$ are not interleaved in the order of edges around v. Such a circuit exists, since we can remove crossings by swapping pieces of the Eulerian circuit, see Figure 4.

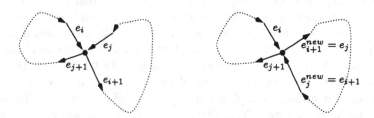

Fig. 4. The left picture shows a crossing of the circuit, which can be removed by reversing a piece of the circuit.

In fact, we can compute a crossing-free Eulerian circuit directly in linear time. The dual graph of any Eulerian planar graph is bipartite, so in particular the dual graph of G_C^* is bipartite. The circuits of those faces of G_C^* that correspond to one of the vertex classes of $(G_C^*)^*$ cover every edge in G_C^* exactly once, and can be merged iteratively, using constant time each, into one crossing-free traversal.

A Jordan-curve crossing the cut-edges Using the crossing-free Eulerian circuit of G_C^*, we define an extension G_{HH}^+ of the original graph G. Subdivide each cut-edge of G and connect the subdivision vertices v_{e_1} and v_{e_2} of two cut-edges e_1 and e_2 whenever their dual edges e_1^* and e_2^* are consecutive in the Eulerian circuit of G_C^*. Since the Eulerian circuit was crossing-free, the resulting graph G_{HH}^+ is planar. See also Figure 5.

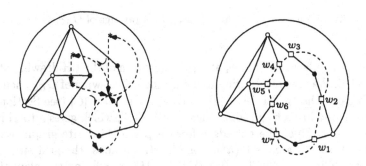

Fig. 5. We subdivide the cut-edges and add edges along the route of the Eulerian circuit. Since the Eulerian circuit is crossing-free, the resulting graph G_{HH}^+ is planar.

The resulting crossing-free cycle C of subdivision vertices mirrors the course of the Eulerian circuit in G_C^*. It defines a closed Jordan-curve in the drawing of G, which intersects each cut-edge exactly once, and no other edges. This, together with the connectivity of G, implies that the vertices of A are outside C and the vertices of B are inside C, or vice versa. Let $w_1, \ldots, w_k$ be the vertices of C.

Splitting G_{HH}^+ into two graphs Let G_{AC} be the planar subgraph of G_{HH}^+ induced by the vertices of A and C, and let w.l.o.g. let the vertices of A be outside C. Then C forms a face of G_{AC}, with $w_1, \ldots, w_k$ on it in this order. If we add a vertex v_C inside C and connect it with the vertices in C, then the resulting graph G_{AC}^+ is planar. Furthermore, C induces a cycle and the triangles containing v_C are faces in this embedding of G_{AC}^+. Therefore, the graph $G_{AC} - (w_1, w_k)$ satisfies the necessary conditions for a planar straight-line LH-drawing (cf. Theorem 1), with respect to the partition $C \cup A$. The same holds for the graph $G_{BC} - (w_1, w_k)$, where G_{BC} is the graph induced by the vertices of B and C.

Creating the drawings Create a planar straight-line LH-drawing of $G_{AC} - (w_1, w_k)$ as described in Section 2.1. Flip this drawing upside down, and use it as a starting point for creating a planar straight-line LH-drawing of $G_{BC} - (w_1, w_k)$ (this is possible since the vertices $w_1, \ldots, w_k$ are in the same order on C in both G_{AC} and G_{BC}).

Deleting the vertices in C, and the edges between them, we arrive at an HH-drawing of G. Edges not in the cut are drawn straight. The cut-edges may have a bend, with y-coordinate 0, but they are drawn y-monotonically.

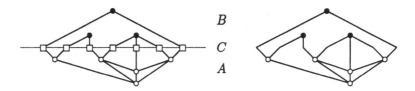

Fig. 6. An HH-drawing is produced using LH-drawings of G_{AC} and G_{BC}.

If we start by placing vertex w_i at $(i, 0)$, the width of the LH-drawing of G_{AC} is $|C| - 1 + |A|$. We use this drawing to create the LH-drawing of G_{BC}, hence the HH-drawing of G has width $|A| + |B| + |C| - 1 = n - 1 + |C|$ (see Section 2.1). Each LH-drawing has a height of at most $n - 1 + |C|$, which gives a total height of $2|C| + 2n - 2$. Since the cut-edges form a planar bipartite graph, we have $|C| \leq 2n - 4$, so the area is $\mathcal{O}(n^2)$. For bipartite graphs, the grid size can be improved to $(|C| - 1) \times (2|C| - 2)$ with $|C| = |E|$ by refining the algorithm to create the LH-drawings suitably.

Theorem 2. *If a planar partition $G = (A \cup B, E)$ has a planar embedding such that the dual edges of the cut-edges form a connected graph, then in linear time we can find a planar y-monotone poly-line HH-drawing in an $\mathcal{O}(n^2)$-grid such that only the cut-edges have at most one bend.*

In [3] we give a constructive linear time test whether a given planar partition has a planar embedding such that the dual edges of the cut-edges form a connected graph. Hence, determining whether a planar partition has a planar y-monotone HH-drawing, and if so, finding one of quadratic area, can be done in linear time.

3.3   The special case of bipartite graphs

Assume that $G = (A \cup B, E)$ is a bipartite planar graph. Then every edge of G is a cut-edge, so for any planar embedding the cut-dual graph is the same as the dual graph, and therefore connected. So any bipartite planar graph satisfies the condition of Theorem 2, and we get the following corollary.

Corollary 1. *Any bipartite planar graph, in any planar embedding, has a y-monotone poly-line HH-drawing in an $\mathcal{O}(n^2)$-grid such that each edge has at most one bend; and it can be found in linear time.*

Figure 7 shows a planar HH-drawing of a subgraph of the bipartite competition graph (mentioned in Section 1) constructed by our algorithm.

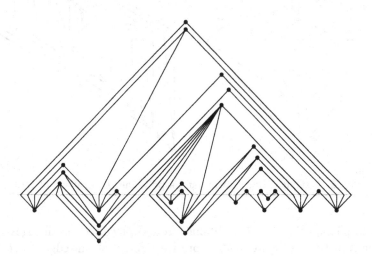

Fig. 7. A planar HH-drawing of a bipartite graph.

4   A lower bound for straight-line drawings

We would like to avoid the bends in the drawing of the algorithm of Section 3 while maintaining similar area-bounds, but as we show in this section, this is impossible. We present a family of graphs G_n together with a partition $A \cup B$ of the vertices that require exponential area in all straight-line planar HH-drawings.

Our graphs are similar to the ones presented by Di Battista et al [7]. They proved that for any positive integer n, there exists a planar acyclic digraph D_n with $2n+2$ vertices such that any planar straight-line upward drawing of D_n that reflects a given planar embedding has area $\Omega(2^n)$. Our graphs G_n are identical to D_n except that G_n is undirected, whereas D_n is a directed acyclic graph.

The precise definition is as follows: Graph G_1 consists of vertices a_0, a_1, b_0 and b_1 and edges $(a_0, b_0), (a_1, a_0), (b_0, b_1), (a_1, b_0)$ and (a_0, b_1). For $n \geq 2$, the graph G_n is constructed by adding vertices a_n and b_n and edges $(a_n, a_{n-1}), (b_{n-1}, b_n)$, $(a_{n-2}, b_n), (a_n, b_{n-2}), (a_n, b_{n-1})$ and (a_{n-1}, b_n) to G_{n-1} (see Figure 8). For $n \geq 2$, the graph G_n is planar and triconnected and thus has a unique planar embedding. Let $A = \{a_0, \ldots, a_n\}$ and $B = \{b_0, \ldots, b_n\}$.

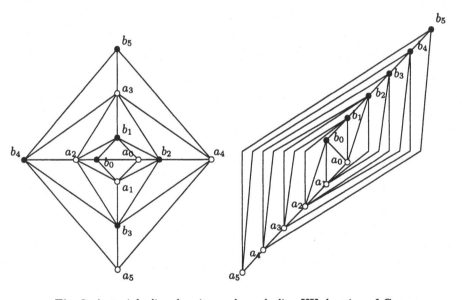

Fig. 8. A straight-line drawing and a poly-line HH-drawing of G_5.

The graph D_n defined in [7] is obtained from G_n by directing all edges upward in the right picture of Figure 8. More precisely, direct all cut-edges (a_i, b_j) from a_i to b_j, all edges (a_i, a_{i+1}) from a_{i+1} to a_i, and all edges (b_i, b_{i+1}) from b_i to b_{i+1}, for $0 \leq i < n$ and $j \in \{i-2, i-1, i+1\}$.

Theorem 3. *Any planar straight-line HH-drawing of G_n with the partition $A \cup B$ has area $\Omega(2^{n/2})$.*

Proof: Consider any planar straight-line HH-drawing of G_n. We show that a large subgraph of G_n is drawn upward with respect to the orientations in D_n, hence the drawing must have a large area. All cut-edges are drawn upward by definition of an HH-drawing.

Since G_n is triconnected, it has a unique planar embedding and therefore a unique cyclic ordering of the incident edges around each vertex. From Figure 8, we can see that for $0 < i < n$, the unique order around a_i contains one or two vertices in B, then one vertex in A, then one or two vertices in B, and finally one vertex in A again.

Let a_j be the vertex with the smallest y-coordinate, breaking ties arbitrarily.

Claim: The chains $a_{j-1}, \ldots, a_0$ and $a_{j+1}, \ldots, a_n$ have strictly increasing y-coordinates.

We prove this claim by induction and only for the second chain. Let $j < l < n$. All neighbors of a_l in B *lie above* a_l, that is, they have a strictly larger y-coordinate. By our observation about the order of incident edges around a_l, one neighbor of a_l in A must also lie above a_l. Vertex a_{l-1} does not lie above a_l (by definition of j for $l = j + 1$ and by induction for $l > j + 1$), so it must be a_{l+1} that lies above a_l. The claim follows.

Similarly one shows that if b_k is the vertex with the largest y-coordinate, then the chains $b_{k+1}, \ldots b_n$ and $b_{k-1}, \ldots, b_0$ have strictly decreasing y-coordinates.

Since a_j and b_k both lie on the outer-face, we must have $|k-j| \leq 2$. If $j \geq n/2$, then $k \geq n/2 - 2$. The chain $a_{n/2-3}, \ldots, a_0$ and the chain $b_0, \ldots b_{n/2-3}$ are drawn upward, therefore the graph induced by these vertices, which is $D_{n/2-3}$, is drawn upward. If $j \leq n/2$, then $k \leq n/2 + 2$, and the graph induced by $a_{n/2+3}, \ldots, a_n$ and $b_n, \ldots, b_{n/2+3}$, which is isomorphic to $D_{n/2-3}$, is drawn upward.

So either way, a subgraph isomorphic to $D_{n/2-3}$ is drawn upward, hence from [7] the area of the drawing must be at least $\Omega(2^{n/2-3}) = \Omega(2^{n/2})$.   □

5   Non-monotone drawings

In this section, we study non-monotone drawings. Halton [13] has shown that any planar graph can be embedded on any given set of points without crossing, using Jordan curves to represent the edges. It follows that any planar partition has a planar LL-drawing, by defining suitable points for the vertices, and then apply the technique by Halton. Since any LL-drawing is an LH-drawing and an HH-drawing, any planar partition has a planar drawing in any of the three models.

The drawings by Halton use Jordan curves to represent edges. It is not too hard to create drawings with poly-lines instead, but no non-trivial upper bounds on the number of bends is known to the authors. In fact, it can be shown that even for such a simple graph as a perfect matching, $\Omega(n)$ bends may be required for one edge [18].

However, for the special case of requiring one of our models, rather than arbitrary points, we can achieve a constant number of bends per edge, since the choice of suitable x-coordinates can be left to the algorithm.

5.1 A non-monotone HH-drawing

First we show how to create a planar HH-drawing of any planar partition $G = (A \cup B, E)$, by turning G into a bipartite graph through subdividing edges. We subdivide every edge (a, a') with $a, a' \in A$, and add the subdivision vertex to B. We subdivide every edge (b, b') with $b, b' \in B$ and add the subdivision vertex to A. The resulting graph is planar and bipartite with vertex classes A and B.

Any bipartite graph has a planar y-monotone HH-drawing, with at most one bend per edge (Corollary 1). Removing the subdivision vertices, we get a planar HH-drawing with at most 3 bends per edge. See also Figure 9.

Theorem 4. *Any planar partition $G = (A \cup B, V)$ has a (not necessarily monotone) planar poly-line HH-drawing in an $\mathcal{O}(n^2)$-grid with at most three bends per edge.*

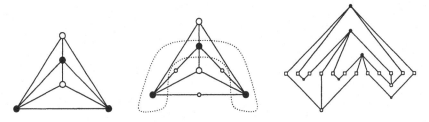

Fig. 9. By subdividing edges, we make G bipartite and get an HH-drawing with at most three bends per edge.

5.2 A non-monotone LL-drawing

By subdividing edges further, we can achieve that every vertex in A is incident to at most one cut-edge, and that no two vertices in A that are incident to a cut-edge are adjacent; furthermore, the corresponding statements hold for vertices in B at the same time. Using this subdivision, one can create a non-monotone LL-drawing with a constant number of bends per edge. Details are omitted and can be found in [2].

Theorem 5. *Any planar partition $G = (A \cup B, V)$ has a (not necessarily monotone) planar poly-line LL-drawing in an $\mathcal{O}(n^2)$-grid with at most seven bends per edge.*

6 Conclusion

In this paper, we studied drawings of planar graphs where a partition of the vertices is given and should be clearly visible in a planar drawing. We focused on so-called HH-drawings where the two vertex classes are drawn separated by a horizontal line.

For the case that edges are required to be drawn y-monotonically, we present necessary and sufficient conditions for the existence of such drawings. In particular, these conditions are always satisfied for bipartite graphs. They can be tested in linear time, and yield a linear-time algorithm to create HH-drawings, if possible. However, the resulting drawings have a bend for each cut-edge. We prove that drawings without bends are impossible unless we allow an exponential area of the underlying grid.

Finally, we study non-monotone drawings. In this case, any planar partition has a planar HH-drawing with up to 3 bends per edge, and a planar LL-drawing with up to 7 bends per edge, and they can be constructed in linear time.

For the future, it would be interesting to consider multiway partitions as well as scenarios where some crossings between the partitions are allowed.

References

1. T. Biedl. Drawing planar partitions I: LL-drawings and LH-drawings. In *14th Annu. ACM Symp. Computational Geometry*, pages 287–296, 1998.
2. T. Biedl, M. Kaufmann, and P. Mutzel. Drawing planar partitions II: HH-drawings. Technical Report RRR 12-98, RUTCOR, Rutgers University, 1998.
3. T. Biedl. Drawing planar partitions III: Two constrained embeddings. Technical Report RRR 13-98, RUTCOR, Rutgers University, 1998.
4. M. Chrobak and G. Kant. Convex grid drawings of 3-connected planar graphs. *Internat. J. Comput. Geom. Appl.*, 7(3):211–223, 1997.
5. M. Chrobak and S. Nakano. Minimum-width grid drawings of plane graphs. In R. Tamassia and I. Tollis, editors, *DIMACS International Workshop, Graph Drawing 94*, volume 894 of *Lecture Notes in Computer Science*, pages 104–110. Springer-Verlag, 1995.
6. G. Di Battista, P. Eades, R. Tamassia, and I. Tollis. Algorithms for drawing graphs: an annotated bibliography. *Comp. Geometry: Theory and Applications*, 4(5):235–282, 1994.
7. G. Di Battista, R. Tamassia, and I. Tollis. Area requirement and symmetry display of planar upward drawings. *Discrete Computational Geometry*, 7(4):381–401, 1992.
8. P. Eades, B. D. McKay, and N. Wormald. On an edge crossing problem. In *ACSC 9, 9th Australian Computer Science Conference*, pages 327–334, 1986.
9. S. Even. *Graph Algorithms*. Computer Science Press, 1979.
10. U. Fößmeier and M. Kaufmann. Nice drawings for planar bipartite graphs. In *3rd Italian Conference on Algorithms and Complexity, CIAC '97*, volume 1203 of *Lecture Notes in Computer Science*, pages 122–134. Springer-Verlag, 1997.
11. H. de Fraysseix, J. Pach, and R. Pollack. How to draw a planar graph on a grid. *Combinatorica*, 10:41–51, 1990.

12. C. Gutwenger and P. Mutzel. Grid embedding of biconnected planar graphs. Extended Abstract, Max-Planck-Institut für Informatik, Saarbrücken, Germany, 1998.
13. J. Halton. On the thickness of graphs of given degree. *Information Sciences*, 54:219–238, 1991.
14. J. Harary and A. Schwenk. A new crossing number for bipartite graphs. *Utilitas Mathematica*, 1:203–209, 1972.
15. G. Kant. Drawing planar graphs using the canonical ordering. *Algorithmica*, 16:4–32, 1996.
16. J. Kratochvil and M. Křivanek. Satisfiability of co-nested formulas. *Acta Informatica*, 30(4):397–403, 1993.
17. S. North, editor. *Symposium on Graph Drawing 96*, volume 1190 of *Lecture Notes in Computer Science*. Springer-Verlag, 1997. For information on the contest, see *http://www.research.att.com/conf/gd96/contest.html*.
18. J. Pach, F. Shahrokhi, and M. Szegedy. Applications of the crossing number. *Algorithmica*, 16:111–117, 1996.
19. H. Purchase. Which aesthetic has the greatest effect on human understanding? In G. Di Battista, editor, *Graph Drawing '97*, volume 1353 of *Lecture Notes in Computer Science*, pages 248–261. Springer-Verlag, 1998.
20. R. Tamassia. On embedding a graph in the grid with the minimum number of bends. *SIAM J. Computing*, 16(3):421–444, 1987.

Triangles in Euclidean Arrangements

STEFAN FELSNER and KLAUS KRIEGEL

Freie Universität Berlin
Fachbereich Mathematik und Informatik
Takustr. 9, 14195 Berlin, Germany
E-mail: {felsner,kriegel}@inf.fu-berlin.de

Abstract. The number of triangles in arrangements of lines and pseudolines has been object of some research. Most results, however, concern arrangements in the projective plane. We obtain results for the number of triangles in Euclidean arrangements of pseudolines. Though the change in the embedding space from projective to Euclidean may seem small there are interesting changes both in the results and in the techniques required for the proofs.

In 1926 Levi proved that a nontrivial arrangement -simple or not- of n pseudolines in the projective plane contains n triangles. To show the corresponding result for the Euclidean plane, namely, that a simple arrangement of n pseudolines contains $n - 2$ triangles, we had to find a completely different proof. On the other hand a non-simple arrangements of n pseudolines in the Euclidean plane can have as few as $2n/3$ triangles and this bound is best possible. We also discuss the maximal possible number of triangles and some extensions.

Mathematics Subject Classifications (1991). 52A10, 52C10.

Key Words. Arrangement, Euclidean plane, pseudoline, strechability, triangle.

1 Introduction, Definitions and Overview

A natural approach to generate an object from a combinatorial class at random is to set up an appropriate Markov chain. Basically this works as follows: The objects in the class are the vertex set of a so called transition graph. The (directed) edges in this graph are defined by some local operations which transform one object into another one.

For arrangements of pseudolines a natural choice for the local operations are triangular flips. The mixing rate of the Markov chain depends on the expansion properties of the transition graph. A necessary condition for good expansion is a sufficiently large degree. In the setting described the degree of a vertex, i.e., an arrangement $\mathcal{A}$, is the number $p_3(\mathcal{A})$ of triangles of the arrangement. In this article we show some results concerning the number of triangles in Euclidean arrangements of pseudolines.

Grünbaum [4] defines an *arrangement $\mathcal{A}$ of lines* as a finite collection $\{L_0, L_1, \ldots, L_n\}$ of lines, i.e., 1–dimensional subspaces in the real projective plane $\mathbb{P}$. Specifying a line L_0 in $\mathcal{A}$ as the "line at infinity" induces the arrangement $\mathcal{A}_{L_0}$ of lines $\{L_1, \ldots, L_n\}$ in the Euclidean plane $\mathbb{E} = \mathbb{P} \backslash L_0$.

With an arrangement we associate the cell complex of vertices edges and cells into which the lines of the arrangement decompose the underlying space $\mathbb{P}$ or $\mathbb{E}$. Arrangements are *isomorphic* provided their cell complexes are isomorphic.

An *arrangement $\mathcal{B}$ of pseudolines* in $\mathbb{P}$ is a collection $\{P_0, P_1, \ldots, P_n\}$ of simple closed curves (we call them *pseudolines*) in $\mathbb{P}$ such that every two curves have exactly one point in common. Specifying a pseudoline P_0 in $\mathcal{B}$ as the line at infinity induces the arrangement $\mathcal{B}_{P_0}$ of pseudolines $\{P_1, \ldots, P_n\}$ in $\mathbb{P} \backslash P_0$. Since $\mathbb{P} \backslash P_0$ is homeomorphic to the Euclidean plane and we are interested in properties of the induced cell complex we may regard $\mathcal{B}_{P_0}$ as an arrangement in $\mathbb{E}$.

Already in early work of Levi [7] and Ringel [8] it has been noted that arrangements of pseudolines are a proper generalization of arrangements of lines. This is due to the existence of incidence laws in plane geometry ,e.g., the Theorem of Pappus. Arrangements of pseudolines have gained attention since they provide a generic model for oriented matroids of rank 3. In this context questions of strechability have attained considerable interest. For more about these connections we refer the reader to the 'bible of oriented matroids' [1].

An arrangement is called *trivial* if all the (pseudo)lines intersect in a single point. If no point belongs to more then two of the (pseudo)lines we call the arrangement *simple*.

Euclidean arrangements of pseudolines will be the main object of investigations in this paper. Work with these objects is simplified by the fact that every arrangement of pseudolines, i.e., of doubly unbounded curves, is isomorphic to an arrangement of x-monotone pseudolines, i.e., of curves that intersect every vertical line in exactly one point. Particularly nice pictures of Euclidean arrangements of pseudolines are given by their *wiring diagrams* introduced in Goodman [3], see Figure 1. In this representation the n x-monotone curves are restricted to n y-coordinates except for some local switches where adjacent lines cross. Knuth [6] points out a connection with 'primitive sorting networks'.

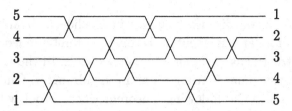

Figure 1. Wiring diagram of a simple arrangement of 5 pseudolines.

We now summarize bounds for the number p_3 of triangles in arrangements.

Theorem 1. *For every arrangement $\mathcal{A}$ of n pseudolines in $\mathbb{P}$:*

(1) *Every pseudoline is incident with at least three triangles. Since every triangle is incident with three lines this implies $p_3 \geq n$.*

(2) *$p_3 \leq \frac{1}{3}n(n-1)$ for $n \geq 10$ with equality for infinitely many values of n.*

Part (1) is due to Levi [7]. The lower bound for p_3 it best possible. To see this take the n supporting lines of the edges of a regular n-gon for $n \geq 4$. The arrangement thus obtained is a simple arrangement of lines with $p_3 = n$.

Part (2) has a more entangled history. In [4] the following easy argument for $p_3 \leq \frac{1}{3}n(n-1)$ in simple arrangements is found. If $\mathcal{A}$ is simple then only one of the cells bounded by an edge can be a triangle. Since there are $n(n-1)$ edges and every triangle uses three of them the bound is established. Grünbaums conjectured the same bound for nonsimple arrangements of lines with sufficiently large n. Several special cases and lower bounds where proved by Strommer, Purdy and others. Finally Roudneff [11] proved the conjectured bound for $n \geq 10$. By perturbing high degree vertices so that suitable arrangements are formed in the neighborhood he shows that p_3 is maximized by what he calls 'reduced arrangements'. In particular these arrangements have no vertices of degree more then four. The crucial part of the proof is to show that if t_i counts vertices of degree i then for $n \geq 9$ every reduced arrangement has

$$3p_3 \leq 2(t_2 + 3t_3 + 6t_4).$$

Since $\sum_k \binom{k}{2}t_k = \binom{n}{2}$ this implies the bound.

Infinite families of simple arrangements with $p_3 = \frac{1}{3}n(n-1)$ have been obtained by Roudneff [9] and Harborth [5]. For stretchable arrangements the best known constructions are due to Füredi and Palásti [2]. Their examples have at least $\frac{1}{3}n(n-3)$ triangles.

In this paper we discuss triangles in Euclidean arrangements. The cell complex of an arrangement in $\mathbb{E}$ consists of unbounded and bounded cells. In our treatment we ignore unbounded cells. In the arrangement of Figure 1 we thus count 3 triangles and 3 quadrangles. Our main results are summarized in the following Theorem whose proof will be given in sections 2 and 3.

Theorem 2. *For every arrangement $\mathcal{B}$ of n pseudolines in $\mathbb{E}$:*

(1) *If $\mathcal{B}$ is simple then $p_3 \geq n - 2$.*

(2) *If $n \geq 6$ then $p_3 \geq \frac{2}{3}n$ with equality for all $n = 0 \pmod 3$.*

(3) *$p_3 \leq \frac{1}{3}n(n-2)$ with equality for infinitely many values of n.*

Part (1) again has a long history. Roberts 1889 claimed that for every simple arrangement $\mathcal{A}$ of $n + 1$ lines in $\mathbb{P}$ and every line L of $\mathcal{A}$ there are $n - 2$ triangles not incident with L. The argument however was considered non-convincing. Ninety years later Shannon [12] proved Roberts theorem, actually, he proved the analog of Roberts theorem for arbitrary dimensions. In particular this implies that every stretchable arrangement $\mathcal{B}$ of n lines in $\mathbb{E}$ has at least $n - 2$ triangles. Add the line at infinity to obtain a projective arrangement and apply Roberts theorem.

Shannon's proof does not require that the arrangement is simple. Therefore, Shannon's theorem together with Theorem 2 (2) gives the following amazing result.

Corollary 1. *The count of triangles can be a certificate for nonstrechability of nonsimple Euclidean arrangements.*

A similar effect in the projective setting was conjectured by Grünbaum and proved by Roudneff [10]. A nonsimple projective arrangement with $p_3 = n$ is nonstrechable. An Example of such an arrangement is due to Canham, see Grünbaum [4, page 55]. In Section 3 we describe a family W_n of arrangements with few triangles. If W_n is considered as an arrangement in the projective plane it is a nonsimple arrangement with n lines and $p_3 = n$.

It is interesting to note that Levi's theorem about the number of triangles incident to a line and Roberts respectively Shannons theorem about the number of triangles avoiding a line both give easy double-counting proofs for $p_3 \geq n$. We elaborate the second:

Corollary 2. *The number of triangles in a simple arrangement $\mathcal{A}$ of n pseudolines in $\mathbb{P}$ is at least n.*

Proof. For each pseudoline P_i consider the Euclidean arrangement $\mathcal{A}_{P_i}$ obtained by taking P_i as line at infinity. Each such arrangement has at least $(n-1)-2$ triangles. Altogether this gives at least $n(n-3)$ triangles. Any fixed triangle Δ in $\mathcal{A}$ is bounded by three pseudolines and hence counted exactly $n-3$ times. This shows that there are at least n different triangles. □

The upper bound on the number of triangles in the Euclidean case claimed in (3) of Theorem 2 can be proved along the lines of Roudneff's upper bound for the projective case. The proof is long and the changes necessary for to adopt it to the Euclidean case obvious. Therefore, we will refrain from elaborating on it and refer to Roudneff's original paper [11].

To show that the bound is best possible again the examples from the same paper [11] do the work. Roudneff shows that there is an infinite family of simple projective arrangements with $n+1$ lines and $(n+1)n/3$ triangles. Each line of such an arrangement is incident to n triangles. Choose an arbitrary line l as line in infinity. The remaining Euclidean arrangement of n lines has $(n+1)n/3 - n = n(n-2)/3$ triangles.

2 Simple Euclidean arrangements

In this section we prove the lower bound for the number of triangles in simple arrangements in $\mathbb{E}$.

Proposition 1. $p_3 \geq n-2$ *for every simple arrangement $\mathcal{B}$ of n pseudolines in $\mathbb{E}$.*

Proof. We consider the finite part of $\mathcal{B}$ as a planar graph. Let V be the number of vertices E be the number of edges and F be the number of (finite!) faces. These statistics can all be expressed as functions of the number of pseudolines.

$$V = \binom{n}{2}, \qquad E = n(n-2), \qquad F = \binom{n-1}{2}$$

Note that in this setting Euler's formula gives $V - E + F = 1$.

We assign labels $\oplus$ or $\ominus$ to each side of every edge. Let f be one of the two (possibly unbounded) faces bounded by e and let e' and e'' be the edge-neighbors of e along f. Let l, l' and l'' be the supporting pseudolines of e, e' and e'' respectively. The label of e on the side of f is $\oplus$ if f is contained in the finite triangle T of the arrangement $\{l, l', l''\}$ otherwise the label is $\ominus$. See Figure 2 for an illustration of the definition and Figure 3 for a complete labeling. With the next lemmas we collect important properties of the edge labeling.

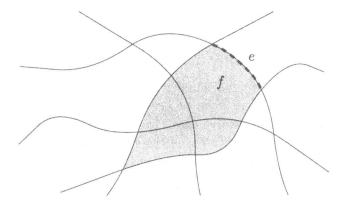

Figure 2. The label of e at f is $\oplus$ since f is contained in the shaded triangle.

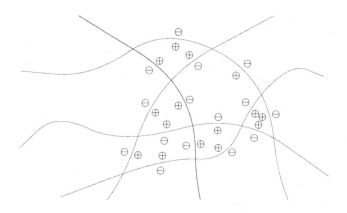

Figure 3. The arrangement of Figure 2 with the completed edge labeling.

Lemma 1. *Every edge e of a simple arrangement has a $\oplus$ and a $\ominus$ label.*

Proof. Let f_1 and f_2 be the two faces bounded by e and let e_1', e_1'' and e_2', e_2'' be the edge-neighbors of e in these two faces. Since the arrangement is simple the supporting lines $\{l_1', l_1''\}$ of both pairs of edges are the same. The finite triangular region T of the arrangement $\{l, l', l''\}$ has edge e on its boundary. Therefore, exactly one of the two faces f_1 and f_2 is contained in T.   △

As seen in the proof of the lemma the triangular region T used to define the edge label of e on the side of f is independent of f. This allows to adopt the notation $T(e)$ for this region.

Lemma 2. *All three edge labels in a triangle are $\oplus$. A quadrangle contains two $\oplus$ and two $\ominus$ labels. For $k \geq 5$ a k sided face contains at most two $\oplus$ labels.*

Proof. If f is a triangle then for each of its edges e the triangular region $T(e)$ is f itself.

Let f be a quadrangle and $e, \bar{e}$ be a pair of opposite edges of f. Both edges have the same neighboring edges, hence, two of the lines bounding the triangles $T(e)$ and $T(\bar{e})$ are equal. It is easy to see that either $T(e) = f \cup T(\bar{e})$ or $T(\bar{e}) = f \cup T(e)$. In the first case e has label $\oplus$ and $\bar{e}$ has label $\ominus$ in the second case the labels are exchanged. The second pair of opposite edges also has one label $\oplus$ and the other $\ominus$.

Let f be a face with $k \geq 5$ sides the lemma immediately follows from the following

Claim. *Any two edges with label $\oplus$ in f are neighbors, i.e., share a common vertex.*

Let $e_1, e_2, \ldots, e_k$ be the edges of f numbered in counterclockwise direction along f and let l_i be the supporting line of e_i. Let e_1 have label $\oplus$ and consider an edge e_i with $3 \leq i \leq k - 1$. We will show that the label of e_i is $\ominus$. The argument as given applies to the case $4 \leq i \leq k - 1$, the remaining case situation $i = 3$, however is symmetric to $i = k - 1$.

Face f is contained in $T(e_1)$ and line l_i has to leave $T(e_1) \setminus f$ through l_k and l_2. Figure 4 is a generic sketch of the situation.

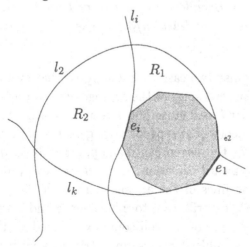

Figure 4. Edge e_1 has label $\oplus$ in f so e_i must have $\ominus$.

Consider line l_{i-1}. This line enters the region R_1 bounded by l_2, l_i and the chain of edges $e_3, e_4, \ldots, e_{i-1}$ at the vertex $e_{i-1} \cap e_i$. To leave region R_1 line l_{i-1} has to cross l_2. Therefore, l_{i-1} has to leave the region R_2 bounded by l_i, l_2 and l_k through l_k. Symmetrically, l_{i+1} has a crossing with l_k to leave the region bounded by l_k, l_i and the chain of edges $e_{i+1}, e_{i+2}, \ldots, e_k$. Therefore, to leave region R_2 line l_{i+1} has to cross l_2. This shows that l_{i-1} and l_{i+1} cross inside region R_2. Hence, $T(e_i)$ is contained in R_2 and e_i has label $\ominus$ in f.

Since e_1 was an arbitrary $\oplus$ labeled edge in F we have shown the claim. $\triangle$

We use the two lemmas to count the number of $\oplus$ labels in different ways:

$$E = \sum_f \#\{+ \text{ labels in } f\} \leq 2F + p_3.$$

With $E = n(n-2)$ and $2F = (n-1)(n-2)$ this implies

$$p_3 \geq n - 2.$$

$\square$

3   Nonsimple Euclidean arrangements

We now come to the lower bound for the number of triangles in the nonsimple case.

Proposition 2. *A Euclidean nonsimple and nontrivial arrangement of $n \geq 6$ pseudolines has at least $2n/3$ triangles with equality for all $n = 0$ (mod 3).*

Proof. We distinguish two cases. First suppose that every line l of the arrangement contains crossings of the arrangement in both open halfspaces it defines. Consider l as a state of a sweepline going across the arrangement. From the theory of sweeps for arrangements of pseudolines (see e.g. [13]) we know that the sweep can make progress both in the forward as well as in the backward direction. A progress-move pulls line l across a crossing c of some lines of the arrangement with the property that the portion of all lines contributing to c between c and l are free of further crossings, i.e. are edges of the cell complex induced by the arrangement. Hence such a move pulls l across some triangles with corner c and an edge on l. This shows that l contributes to at least one triangle on either side. Since we assumed that every line has crossings on either side this accounts for $2n$ triangles each counted at most three times and the claim is proved in this case.

Now assume that there is a line l so that all crossings of the arrangement not on l are on one side of l. If taking away l all lines cross in just one point c then there are $n-2$ triangles in the arrangement and since we assume $n \geq 6$ we are done. Else removing l from the arrangement we still have a nontrivial arrangement which by induction has at least $2(n-1)/3$ triangles. Since l can make a sweep move to one of its sides there is at least one triangle with an edge on l that disappeared after removal of l (it turned into an unbounded region). His makes a total of $2(n-1)/3 + 1 > 2n/3$ triangles in the initial arrangement.

It remains to describe a family W_n of arrangements with $3n$ lines but only $2n$ triangles. A drawing of W_4 is given in Figure 5.

Let P be a regular $2n$-gon with edges $e_1, e_2, \ldots, e_{2n}$ in counterclockwise ordering and barycenter c. Let lines $l_1, \ldots, l_{2n}$ be straight lines such that l_i contains edge e_i of P. Orient the lines such that P is to their left. Note that l_i is crossed by lines $l_{i+n+1}, l_{i+n+2}, \ldots, l_{i-1}, l_{i+1}, l_{i+2}, \ldots l_{i+n-1}$ in this order with indices being taken cyclically. The arrangement $\mathcal{A}$ formed by these $2n$ lines has $2n$ triangles all adjacent to P. All the other faces of the arrangement are quadrangles.

For every pair l_i, l_{i+n} of parallel lines we construct an additional line g_i. We lead g_1 from the unbounded region between the positive end of l_1 and the negative end of l_n to the unbounded region between the positive end of l_{n+1} and the negative end of l_{2n}. The first line crossed by g_1 is l_1. Parallel to l_{n+1} line g_1 crosses $l_2, l_3, \ldots, l_{n-1}$ and splits quadrangles into two. Before entering P line g_1 splits the triangle sitting over edge e_n into a quadrangle and a triangle. From edge e_n line g_1 joins to point c and then to the opposite edge e_{2n} to cross lines $l_{2n}, l_{2n-1}, \ldots, l_{n+1}$ in this order.

Define lines $g_2, \ldots, g_n$ by rotational symmetry and note that $g_1, \ldots, g_n$ all cross in c. The arrangement $\mathcal{A} \cup \{g_1, g_2, \ldots, g_n\}$ has the same number of triangles as $\mathcal{A}$.

So far we still have n pairs of parallel lines. Note however that without increasing the number of triangles we may arbitrarily choose to have the crossing of pair $\{l_i, l_{i+n}\}$ to be on the side of the positive end of either l_i or l_{i+n}. Thus W_n is itself not just one but an exponentially large class of examples. $\square$

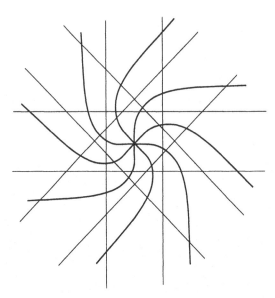

Figure 5. The arrangement W_4 with 12 lines and 8 triangles.

4   Triangles in arrangements with multiple intersections

In his monograph Grünbaum extends the notion of arrangements in several directions. Let an *arrangement of pseudocircles* be a family of closed curves with the property that any two curves cross twice*. A *digon* in such an arrangement is a face bounded by only two of the curves. Grünbaum asks for the relationship between the number of triangles and digons in such arrangements. In particular he conjectures [4, Conjecture 3.7] that every digon-free arrangement of pseudocircles contains $2n - 4$ triangles. The only progress on this conjecture is a result of Snoeyink and Hershberger [13]. They prove $p_3 \geq 4n/3$. The proof is only given for the simple case, i.e., no three curves cross in a single point. However, it is not hard to see that it also applies to the general case.

Based on the arrangements W_n from Section 3 it is possible to construct examples of nonsimple arrangements of pseudocircles in $\mathbb{P}$ with only $4n/3$ triangles. The idea is to glue two copies of W_n together such that all faces generated by gluing are quadrangles, see Figure 6. Hence, the result of Snoeyink and Hershberger is best possible. However, if the arrangement is simple, i.e., no three curves meet in a single point we think that Grünbaum's conjecture should prove correct. For emphasis we restate the conjecture.

---

* Grünbaum calls this an *arrangement of curves*

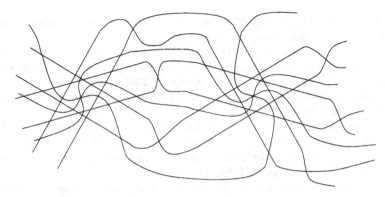

Figure 6. A digon-free arrangement of 9 two-intersecting curves and 12 triangles.

Conjecture 1. Every simple digon-free arrangement of pseudocircles contains at least $2n - 4$ triangles.

We feel that the spirit of Euclidean arrangements is captured well with the following generalization. Call an arrangement of x-monotone curves with the property that any two curves cross exactly k times a k-curve arrangement. Again based on the family W_n it is possible to obtain k-curve arrangements of n curves with only $2kn/3$ triangles. On the other hand we conjecture.

Conjecture 2. Every simple digon-free k-curve arrangement contains at least $k(n - 2)$ triangles.

If true this would obviously be best possible since gluing together k appropriate arrangements of pseudolines with $n - 2$ triangles each gives arrangements with only $k(n - 2)$ triangles.

References

1. A. BJÖRNER, M. LAS VERGNAS, B. STURMFELS, N. WHITE, AND G. ZIEGLER, *Oriented Matroids*, Cambridge University Press, Cambridge, 1993.
2. Z. FÜREDI AND I. PALASTI, *Arrangements of lines with large number of triangles.*, Proc. Am. Math. Soc., 92 (1984), pp. 561–566.
3. J. E. GOODMAN, *Proof of a conjecture of Burr, Grünbaum and Sloane*, Discrete Math., 32 (1980), pp. 27–35.
4. B. GRÜNBAUM, *Arrangements and spreads*, Regional Conf. Ser. Math., Amer. Math. Soc., Providence, RI, 1972.
5. H. HARBORTH, *Some simple arrangements of pseudolines with a maximum number of triangles.*, in Discrete geometry and convexity, Proc. Conf., New York 1982, vol. 440, Ann. N. Y. Acad. Sci., 1985, pp. 31–33.

6. D. E. KNUTH, *Axioms and Hulls*, vol. 606 of Lecture Notes in Computer Science, Springer-Verlag, Heidelberg, Germany, 1992.

7. F. LEVI, *Die Teilung der projektiven Ebene durch Gerade oder Pseudogerade*, in Berichte über die Verhandlungen der sächsischen Akademie der Wissenschaften, Leipzig, Mathematisch-physikalische Klasse 78, 1926, pp. 256–267.

8. G. RINGEL, *Teilungen der Ebenen durch Geraden oder topologische Geraden.*, Math. Z., 64 (1956), pp. 79–102.

9. J.-P. ROUDNEFF, *On the number of triangles in simple arrangements of pseudolines in the real projective plane.*, Discrete Math., 60 (1986), pp. 243–251.

10. ———, *Arrangements of lines with a minimum number of triangles are simple.*, Discrete Comput. Geom., 3 (1988), pp. 97–102.

11. ———, *The maximum number of triangles in arrangements of pseudolines.*, J. Comb. Theory, Ser. B, 66 (1996), pp. 44–74.

12. R. W. SHANNON, *Simplicial cells in arrangements of hyperplanes.*, Geom. Dedicata, 8 (1979), pp. 179–187.

13. J. SNOEYINK AND J. HERSHBERGER, *Sweeping arrangements of curves*, in Proceedings of the 5th Annual Symposium on Computational Geometry (SCG '89), ACM Press, 1989, pp. 354–363.

Internally Typed Second-Order Term Graphs

Wolfram Kahl

Institut für Softwaretechnologie, Fakultät für Informatik
Universität der Bundeswehr München, D-85577 Neubiberg, Germany
`kahl@informatik.unibw-muenchen.de`

Abstract. We present a typing concept for second-order term graphs that does not consider the types as an external add-on, but as an integral part of the term graph structure. This allows a homogeneous treatment of term-graph representations of many kinds of typing systems, including second-order λ-calculi and systems of dependent types. Applications can be found in interactive systems and as typed intermediate representation for example in compilers.

1 Introduction

Term graphs have originally been introduced as efficient representations of terms. The key to this efficiency is the possibility of *sharing* of what on the term side are equal substructures. Linear notation systems for term graphs (such as in some functional programming languages) often use names bound by e.g. `where`-clauses to express sharing; this kind of binding is perceived as different from that introduced by λ-abstractions as represented in term graphs e.g. by Wadsworth, the inventor of graph reduction [12], the difference being that the names bound by `where`-clauses do not appear in the graph, but those bound by λ-abstractions do.

In previous work [3] we took the step to consider *both* uses of bound variable names only as *coding* of structure that can be made *explicit* with an appropriate definition of term graphs. The structure element encoded by `where`-bindings is traditionally explicit in term graphs as the possibility that nodes have several predecessors; the structure element encoded by the names and scopes of λ-bound variables is in the first instance that of *variable binding*, a function that assigns every bound variable its binder, and in the second instance that of *variable identity*, an equivalence relation among variables which makes explicit which variable occurrences belong to the same variable.

This principle of rigorously making structure explicit is now applied to typing in this paper.

In conventional typing systems, types are assigned to *terms*. But when formally reasoning about terms, usually the corresponding *abstract syntax trees* are considered instead. Now trees can not only be considered as a free algebra, where trees are built from constituents (subtrees) via operators, they can also be considered as a special kind of directed graphs, where a node may have successor nodes. Therefore, the types of terms in conventional systems correspond to types of nodes in term graphs, and where conventional systems relate the types of terms with the types of their immediate constituents, a term graph typing system should relate the type of a node with the type of its successor nodes.

Furthermore, types usually are again terms of some language, so when we represent terms and their subterms as nodes in a term graph, we can equally represent types as nodes in a term graph of types. Now there are systems that use (sublanguages of) the same language for programs and types, and that even allow references from programs to types (as in second-order λ-calculi) or from types to program terms (as in dependent types), so it seems only natural to consider a program and its type as parts of *one* term graph that is enriched with an additional *typing function* from program nodes to their type nodes.

After establishing some notation in Sect. 2, we define internally typed second-order term graphs in Sect. 3 and their homomorphisms in Sect. 4. The principles of an appropriate framework for typing systems are laid out in Sect. 5, with some details about the typing of multi-node variables left to Sect. 6. Section 7 shows how our typed term graphs can be considered to result from a kind of algebraic graph grammars. Finally we shortly present a few more complicated typing systems in Sect. 8 for giving an impression of the power of our formalism.

2 Notation

In our formalisation, we frequently use relational operations since this allows very clear and concise formalisations in the context of graphs, see e.g. [9, 2].

For many purposes we use parts of the Z-notation [10], most notably for set comprehensions where Z uses the pattern "{*signature* | *predicate* • *term*}" instead of the otherwise frequently observed "{*term* | *predicate*}". So we have as an example $\{n : \mathbb{N} \mid n < 4 \bullet n^2\} = \{0, 1, 4, 9\}$. If the predicate is constantly true, then we can also write "{*signature* • *term*}", e.g., $\{x : \mathbb{B} \bullet (x, x)\} = \{(\text{True}, \text{True}), (\text{False}, \text{False})\}$; if the *term* is just the tuple of the variables introduced in the *signature*, then another possibility is "{*signature* | *predicate*}", e.g. $\{x, y : \mathbb{B} \mid x \neq y\} = \{(\text{True}, \text{False}), (\text{False}, \text{True})\}$. Quantification uses the same patterns; here most frequently the predicate is omitted, so we have for example $\forall x : \mathbb{N} \bullet x + 1 > x$. The powerset of a set A is written $\mathbb{P}.A$.

The set of relations between two sets A and B is written $A \leftrightarrow B$ and is equal to the power set of the cartesian product: $(A \leftrightarrow B) := \mathbb{P}.(A \times B)$. The set of univalent relations or partial functions from A to B is written $A \nrightarrow B$, and that of total functions or mappings is written $A \rightarrow B$. Application of a function $f : A \nrightarrow B$ to an argument $x : A$ is written "$f.x$" and is only used if the argument is known to be in the domain of the function: $x \in \text{dom}.f$, where for any relation R the domain of R is $\text{dom}.R := \{(x, y) : R \bullet x\}$, and the range of R is $\text{ran}.R := \{(x, y) : R \bullet y\}$.

The set A^* is the set of finite sequences of elements of A; these sequences are considered to be partial functions of type $\mathbb{N} \nrightarrow A$ with contiguous domain which, when nonempty, always includes zero; therefore, if l is a sequence, then $l.i$ denotes the $(i + 1)$-th element of l. For any set A, the function $\text{len} : A^* \rightarrow \mathbb{N}$ calculates the length of sequences.

The identity relation on a set A is $I_A : A \rightarrow A$, and we usually just write I. For two sets A and B, the universal relation is $\mathbb{T}_{A,B} := A \times B$ and the empty relation is $\perp\!\!\!\perp_{A,B} := \emptyset$; again we usually just write $\mathbb{T}$ and $\perp\!\!\!\perp$.

For two relations $R, S : A \leftrightarrow B$, their intersection is $R \cap S$ and their union is $R \cup S$; inclusion is written $R \subseteq S$. The complement of R is $\bar{R}$. The converse of R is the relation $R^{\cup} : B \leftrightarrow A$, defined by $R^{\cup} := \{(x, y) : R \bullet (y, x)\}$.

For two relations $R : A \leftrightarrow B$ and $S : B \leftrightarrow C$, their composition is $R;S : A \leftrightarrow C$ with $R;S := \{(x, y) : R; (u, z) : S \mid y = u \bullet (x, z)\}$. The transitive closure of a homogeneous relation $R : A \leftrightarrow A$ is R^+, and the reflexive transitive closure is R^*.

When a relation $R : A \leftrightarrow A$ is considered as a **graph**, a node $y : A$ is *reachable* from another node $x : A$ if and only if $(x, y) \in R^*$. The relation R is *acyclic* if $R^+ \subseteq \bar{I}$. A node r is a *source* if $r \notin \text{ran}.R$, and r is a *root* if it is the only source (at least in the DAG setting).

3 Term Graph Definition

In comparison with the untyped graphs of [3,5], we present a simplified formalisation; for the sake of brevity we do not fully formalise obvious concepts. However, we immediately present a definition for *typed* term graphs. For this purpose, we first formalise our view of typing without reference to any concrete typing system, just regarding the typing function as another term graph component.

The "lexical material" which we fill our term graph structure with is essentially the same as for second-order terms (or metaterms, introduced by Klop [7]), but we do not introduce a separate class of binders, and what usually is called "function symbol" is called "constant constructor" here, since we want to stress the contrast with variables:

Definition 3.1. A **term graph alphabet** is a tuple $(\mathcal{L}, A, \mathcal{C}, \mathcal{B}, \mathcal{M})$ with the set $\mathcal{L}$ of *node labels*, the *arity function* $A : \mathcal{L} \to \mathbb{N}$, and a partition of $\mathcal{L}$ into the sets $\mathcal{C}$ of labels for *constant constructors*, $\mathcal{B}$ for *bindable variables*, and $\mathcal{M}$ for *metavariables*.   □

In the following we assume a fixed term graph alphabet $(\mathcal{L}, A, \mathcal{C}, \mathcal{B}, \mathcal{M})$.

The main differences between the following definition and those of [3,5] — besides the introduction of typing — are that we here only consider finite acyclic graphs and that the set of edge labels is fixed as the set of natural numbers.

Definition 3.2. A **term graph** is a tuple $G = (\mathcal{N}, L, S, D, B, W, T)$ with
- $\mathcal{N}$, the finite **node set**,
- $L : \mathcal{N} \to \mathcal{L}$, the **node labelling** function,
- $S : \mathcal{N} \to \mathcal{N}^*$, the **successor** function with $L;A = S;\text{len}$, i.e., the length of the successor list of each node has to be the arity of its label,
- $D : \mathcal{N} \leftrightarrow \mathcal{N}$, the **associated relation**, $D := \{(x, l) : S; y : \text{ran}.l \bullet (x, y)\}$; obviously D is not a primitive component but derived from S; it is listed here for its importance,
- $T : \mathcal{N} \twoheadrightarrow \mathcal{N}$, the partial **typing** function, where $(D \cup T)$ has to be acyclic,
- $B : \mathcal{N} \twoheadrightarrow \mathcal{N}$, the **binding** function, where for $(x, b) \in B$, the *bound variable* x has to have a label in $\mathcal{B}$, the *binder* b a label in $\mathcal{C}$, and b *dominates*[1] x in the graph induced by $(D \cup T)$,

---

[1] In graph theory, a node b dominates another node x, if for every node a and every path from a to x either b lies on that path or a is reachable from b. Domination therefore implies reachability.

– $W : \mathcal{N} \leftrightarrow \mathcal{N}$, the **variable identity**, a *partial equivalence relation*[1] defined exactly on *variables*, i.e. on nodes with labels from $\mathcal{B} \cup \mathcal{M}$. The variable identity has to be compatible with the labelling: $W \mathbin{;} L \subseteq L$, and with the binding[2]: $W \mathbin{;} B \subseteq B$.

Roots are considered wrt. $(D \cup T)$, and the **type part** of a typed term graph is the set ran.$(T \mathbin{;} (D \cup T)^{*})$ containing all nodes reachable from typing nodes.   □

At first sight it might seem strange that we did not impose any restriction on the interplay of the typing function with the other term graph components, most notably with variable identity. But as we shall see in Sects. 6 and 8, different typing systems open up very different possibilities and also impose different restrictions, so that it does not make sense to impose restrictions on the level of the term graph definition, especially since we still lack the machinery to formulate most of the useful restrictions.

Terms corresponding to a rooted term graph are easily recovered by unfolding recursively along the D-Paths from the root — creation of a unique name for every variable is the easiest means to ensure preservation of the binding and variable identity structure. Consider the following examples (of untyped graphs, i.e., of graphs with empty typing), where successor edges are black arrows with their sequence indicated by the left-to-right order of their attachment to their source node; binding edges are drawn in red resp. as thick, dark grey, usually curved arrows, and an irreflexive kernel of variable identity is indicated by blue resp. thick medium gray lines:

The first two correspond to the terms "$2 \cdot 2 + 2 \cdot 2 \cdot 1$" and "$\lambda x.\lambda f.fx$" from arithmetic resp. λ-calculus. For the last two, let us assume that A and B are metavariables — in HOPS (see [6]), where the pictures have been produced, arity is part of the node label, so that unary and zero-ary metavariables all are drawn with the label V, but according to their arity they should be considered as different labels $V_0, V_1 : \mathcal{M}$ in the examples. The last two term graphs then correspond to the metaterms "$(\lambda x. B[x]) A$" and "$B[\text{rec } x. B[x]]$", respectively.

The concept of free variables is easy to transfer to term graphs; especially important is the relation between a binder and those nodes below which its bound variable occurs freely:

Definition 3.3. A variable node x is **free below** a node a, if there is a $(D \cup T)$-path from a to x such that no binder of x lies on that path; if in this constellation x is bound by b, then b **encapsulates** a. The **encapsulation** $C : \mathcal{N} \leftrightarrow \mathcal{N}$ relates b with a exactly when b encapsulates a.   □

---

[1] A partial equivalence relation is a symmetric and transitive relation.

[2] This condition, $W \mathbin{;} B \subseteq B$, means that if any node in an equivalence class wrt. W is bound by some binder, then all nodes in that class are bound by the same binder — note the conciseness and elegance of the relational formulation!

4 Homomorphy

In conventional term graph formalisms there are no bound variables, and variables corresponding to our metavariables are always zero-ary. Therefore, when considering homomorphisms based on node mappings, the image of such a variable is the whole subgraph starting at the image node of the variable node. In the case of second-order term graphs however, there are metavariables with successors, and their images have to *stop* before the image nodes of their successors. Therefore we introduce:

Definition 4.1. An **interval** in a graph G is a pair $(t,b) : (\mathcal{N} \times (\text{IN} \nrightarrow \mathcal{N}))$ consisting of a *top node t* and a finite *lower border b*, which should be considered as a partial node sequence. The **inner nodes** of the interval (t,b) are those nodes that are $(D \cup T)$-reachable from t via paths on which there lies no node of ran.b. The interval (t, b) is **coherent** if all nodes in the lower border (i.e. all nodes in ran.b) are $(D \cup T)$-reachable from t. □

Not every interval is a reasonable candidate for being image of metavariables; conditions corresponding to "no capture of variables" have to be fulfilled. Auxiliary concepts for dealing with this issue are:

Definition 4.2. An interval is **consistent**, if all nodes encapsulated by inner nodes are inner nodes. The **encapsulation skeleton** of an interval is the set of those inner nodes, from which a node in the lower border can be reached via a $(B \cup (D \cup T))$-path. □

The most important use of term graph homomorphisms is to serve as matchings from rule sides into application graphs for transformation or rewriting. In the term context, matching is usually defined as "there exist a context and a (second-order) substitution, such that the result of inserting the substituted rule side into the context is α-equivalent to the application term" — with the definitions of substitution application and insertion into contexts taking care of avoiding "variable capture". In term graphs, a more direct approach is necessary, and in second-order term graphs the structure that is not there is almost as important as the structure that is there, so the conditions for structure preservation take on an unusual shape, and for avoiding "variable capture" several special conditions are needed. While in [3] we worked only with total functions and in [5] we went all the way to possibly partial and multivalent relations, here we just present a definition using potentially partial functions. This still allows a reasonably simple treatment of the images of metavariables:

Definition 4.3. If for two graphs G_1 and G_2, a partial function $F : \mathcal{N}_1 \nrightarrow \mathcal{N}_2$ is given, then the **image interval** for a metavariable node $m : \mathcal{N}_1$ with $m \in \text{dom}.F$ and $L_1.m \in \mathcal{M}$ is defined to be the interval $(F.m, (S_1.m)\mathbin{;}F)$. □

The fact that we only consider acyclic graphs and homomorphisms with rooted domain graphs helps considerably to keep the conditions simple:

Definition 4.4. A **metavariable base** in a term graph G is a set v of metavariable nodes that is closed under variable identity. □

Definition 4.5. A v-**fitting** from a term graph G_1 with metavariable base v to a term graph G_2 is a function $F : \mathcal{N}_1 \nrightarrow \mathcal{N}_2$ that (we let $F_0 := F \cap \overline{v \times \mathcal{N}_2}$ be the *constant part* of F relative to v)

- *preserves labels:* $F_0^{\smile}{}_{;}L_1 \subseteq L_2$,
- *preserves successors:* $\forall(n1, n2) : F_0 \bullet S_1.n_1 \subseteq (S_2.n_2){}_{;}F^{\smile}$,
- is *coherent* and *consistent:* the image interval of every metavariable node in dom. F is coherent and consistent,
- *strictly preserves binding:* $F_0^{\smile}{}_{;}B_1 = B_2{}_{;}F_0^{\smile}$,
- *strictly preserves variables:* $W_1{}_{;}F_0 = F_0{}_{;}W_2$,
- *respects equally bound variables:* $F_0{}_{;}W_2{}_{;}F_0^{\smile} \cap B_1{}_{;}B_1^{\smile} \subseteq W_1$,
- *respects free variables:* $F_0{}_{;}W_2{}_{;}F_0^{\smile} \cap \overline{B_1{}_{;}\mathbb{T}} \subseteq W_1$,
- *controls variables:* if for some metavariable node $m : \mathcal{N}_1$ in dom. F, a variable node x_2 in the image interval of m is free below $F.m$, then there is no (variable) node $x_1 : \mathcal{N}_1$ in dom. F_0, such that $(F_0.x_1, x_2) \in W_2$.
- *preserves typing:* $F^{\smile}{}_{;}T_1 \subseteq T_2{}_{;}F^{\smile}$, and *preserves typelessness:* $F{}_{;}T_2 \subseteq T_1{}_{;}F$. $\square$

Note that the conditions for the typing must not be restricted to the constant part of F.

For **linear** term graphs, i.e. where the variable identity restricted to metavariables is trivial, this is already the definition of homomorphisms; for non-linear term graphs we need a concept of "isomorphism up to sharing" between image intervals of metavariables, so we define:

Definition 4.6. A **correspondence** between two intervals (t_1, b_1) and (t_2, b_2) in a graph G is a relation $H : \mathcal{N} \leftrightarrow \mathcal{N}$ on the node set fulfilling the following conditions:
- H exactly covers the inner nodes: dom. H is the set of the inner nodes of (t_1, b_1), and ran. H is the set of the inner nodes of (t_2, b_2),
- H preserves upper borders: if H is non-empty, it relates t_1 exactly to t_2 and vice versa,
- H preserves node labels: $H{}_{;}L \subseteq L$,
- H preserves successors: $(H \parallel I){}_{;}S \subseteq S{}_{;}(H \cup b_1^{\smile}{}_{;}b_2)$,
- H preserves bindings on internally bound nodes:
  $$H{}_{;}B \cap \mathbb{T}{}_{;}H \subseteq B{}_{;}H \subseteq H{}_{;}B \quad \text{and} \quad H^{\smile}{}_{;}B \cap \mathbb{T}{}_{;}H^{\smile} \subseteq B{}_{;}H^{\smile} \subseteq H^{\smile}{}_{;}B,$$
- H respects the distinctness of nodes internally bound by the same constructor:
  $$H^{\smile}{}_{;}W{}_{;}H \cap \mathbb{T}{}_{;}H{}_{;}B \cap B{}_{;}B^{\smile} \subseteq W,$$
  and so does $H^{\smile}$,
- H preserves variables: $H{}_{;}W \cap \mathbb{T}{}_{;}H = W{}_{;}H \cap H{}_{;}\mathbb{T}$,
- H stays within the same variable for variables that are not internally bound:
  $$H \cap \overline{B{}_{;}H{}_{;}\mathbb{T}} \cap W{}_{;}\mathbb{T} \subseteq W,$$
- H preserves typing: $H{}_{;}T \cap \mathbb{T}{}_{;}H = T{}_{;}H \cap H{}_{;}\mathbb{T}$. $\square$

It is relatively easy to see that for a correspondence H, its converse $H^{\smile}$ is a correspondence, too. Also the identity relation restricted to the inner nodes of a consistent interval obviously is a correspondence, and it can be proved that the composition of two correspondences is again a correspondence. Existence of correspondences between consistent image intervals of metavariables is therefore an equivalence relation, and it is natural to define:

Definition 4.7. A **v-homomorphism** is a v-fitting where for every two metavariable nodes in $W_1 \cap (v \times v)$ there is a correspondence between their image intervals.   □

Without further restrictions, these homomorphisms are not composable, but in [3] we have shown that this is not necessary for being able to define a sound rewriting concept (see also [4]).

As an example homomorphism (indicated by the thin, dark grey (violet) arrows) we show an untyped β-redex to the right.

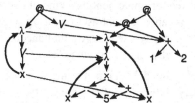

5 Well-Typed Term Graphs

We now introduce a means to distinguish well-typed term graphs. The system we propose is a system that could equally well be employed for untyped term graphs; there it would allow to make distinctions that are not covered by Def. 3.2, such as which node labels are allowed as binders, and below which successors bound variables may occur (a step in the direction of the general "binding structures" of [11]). As shown here, the system still may be used towards these purposes, although its main motivation is to ascertain legal typing.

We define "typing elements" as schema graphs that encode what "locally legal" should mean for a graph:

Definition 5.1. A **typing element** is a typed term graph G which either is rooted or has all its sources related to each other by the variable identity, and where all successors of the source nodes are metavariables and all successors of those metavariables are bound by the root node. Such a typing element is said to be **for** the label of its root node.   □

Here we assume that all bindable variables have zero arity; otherwise the same conditions would have to be enforced for them as for the sources.

We provide three example typing elements for simply-typed λ-calculus and three more for arithmetics — the typing function is denoted by green resp. thin, light grey arrows:

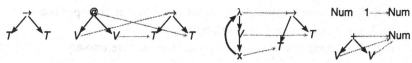

Obviously typing elements closely correspond to typing rules in type derivation systems as they can be found e.g. in the *pure type systems* of [1]. For reasons of space we refrain from formalising this relation, since then we would have to introduce the formalism of pure type systems, too; we only state:

Proposition 5.1. As long as there is no reduction among types, there is a one-to-one correspondence between typing elements and typing rules of pure type systems.   □

Just as only a set of rules defines a language in pure type systems, only a set of typing elements defines a typed term graph language:

Definition 5.2. A **term graph language** is a set $\mathcal{T}$ of typing elements, such that $\mathcal{T}$ contains at most one typing element for each node label $l : \mathcal{C}$. □

The restriction that there is at most one typing element for every node label might be relaxed for implementing a kind of overloading, but we shall not pursue that possibility in this paper. In contrast to most typing systems, that build on some kind of derivation or inference, we use a more simultaneous concept here; we simply define "globally legal" as "locally legal everywhere":

Definition 5.3. A typed term graph G is **well-typed** wrt. a term graph language $\mathcal{T}$ if for every node $n : \mathcal{N}$ of G there is a typing element τ for $L.n$ and a homomorphism from τ into G that maps a source node of τ to n. □

Note that a typing element is necessary for *every* node, and not only for every typed node — this exactly corresponds to the situation in pure type systems, where a trivial type " □ " is given to those terms that correspond to term graph nodes without type. Otherwise this definition would for example not be able to ensure consistency in the presence of dependent types. For graphically conveying the idea behind our definition of well-typedness, we draw a typed version of the β-redex example from below Def. 4.7 once separately and once together with two example homomorphisms from the typing elements for function application and λ-abstraction.

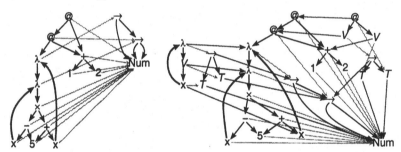

For being able to discuss properties of term graph languages, we first introduce a few classifications of typing elements:

Definition 5.4. A typing element is called
- **separated**, iff it contains no $(D \cup B)$-edges between typed and untyped nodes,
- **well-typed wrt. $\mathcal{T}$**, iff its type part is well-typed wrt. $\mathcal{T}$.
- **first-order**, iff all metavariables contained in the type part are zero-ary. □

For the kind of typing elements we have seen so far, a principal type property is easy to establish:

Theorem 5.1 (Principal Types). If the term graph language $\mathcal{T}$ fulfils the following conditions:
- all typing elements are separated and first-order,
- the relation $Q : \mathcal{L} \leftrightarrow \mathcal{L}$ with $Q := \{G : \mathcal{T}; l : \mathrm{ran}.(T_G; D_G^*; L_G) \bullet (L_G.r_G,\ l)\}$, where r_G denotes the root of graph G, is acyclic,
- if G is a typing element for l in $\mathcal{T}$, then G is well-typed wrt. the sub-language of $\mathcal{T}$ for the labels that are reachable from l in Q,

then there is a **principal type** for every graph in the following sense: Let the program
part of a graph be the untyped subgraph induced by all typed nodes (closing it wrt.
successors, binding, and bound variables); then among all well-typed graphs with
isomorphic program parts there is always one from which there is a homomorphism to
every other.   □

Term graph languages with these restrictions correspond to simple parametric
polymorphism; full Hindley-Milner polymorphism with `let`-polymorphism requires
features of type abstraction, see Sect. 8.2. Other principal type results can also be carried
over to the term graph setting.

6 Typing Elements for Variables

The definition 5.3 of well-typedness implies that there also have to be typing elements
for variables. Since variables may consist of many nodes, and since the condition of
strict variable preservation (Def. 4.5) demands at least as many variable nodes in the
source as in the target, the term graph language (Def. 5.2) which is used as the basis for
well-typedness may severely restrict possible variable occurrences.

For example, if for the bound variable label x there is only one typing element
$x \longrightarrow T$, this implies that on x-nodes the variable identity is trivial. HOPS currently
restricts bound variables in this way, which is only natural in a term DAG setting.

With metavariables, on the other hand, such a restriction is not feasible anymore, at
least not for metavariables with successors. For untyped (i.e., type-level) metavariables
without successors, HOPS also provides only one typing element, since in HOPS the
type part of every term DAG is kept maximally identified.

Obviously, if more than one node per variable is going to be allowed, it makes little
sense to impose other restrictions on the number of those nodes, so we need infinitely
many typing elements for every metavariable label concerned.

For untyped metavariables with untyped successors, this is not a big problem since
their typing elements will have empty typing and the only restriction one might impose
is that the different nodes of one variable shared their successors. However, such a
restriction seems to be difficult to motivate.

For the other cases there are more possibilities. Let us restrict our attention here to
typed metavariables with typed successors, which is also the only kind currently fully
supported by HOPS. In HOPS, for every positive integer n and every natural number i a
typing element is assumed for n i-ary typed metavariable nodes all related by the variable
identity; the successors are all distinct zero-ary typed metavariables with trivial variable
identity, and there are $i + 1$ zero-ary untyped metavariables with trivial variable identity,
one for the sources and one for every successor index. We show the typing elements for
$(n, i) \in \{(2, 0), (2, 1), (3, 1), (2, 2)\}$:

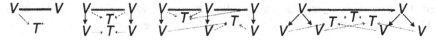

Well-typedness with such a term graph language implies very simple restrictions on the
typing function which can also be expressed directly: All nodes of a variable have to

have the same type: $W \mathbin{;} T \subseteq T$, and the corresponding successors of the different nodes of one variable have the same type: $\forall (x, y) : W \bullet (S.x) \mathbin{;} T = (S.y) \mathbin{;} T$.

We shall see more general typing elements for metavariables in Sect. 8 together with the frameworks that make them necessary.

7 Construction Step Homomorphisms — Towards Typed Term Graph Grammars

In this section we show how stepwise term graph construction — as for example in an interactive system — can be viewed as a sequence of homomorphisms. These "construction step homomorphisms" belong to a few simple classes, which together can be considered to define a special kind of graph grammar which is closely related to algebraic graph grammars, since it relies on category-theoretic concepts.

It is important to note that in this section homomorphisms are restricted to homomorphisms between well-typed graphs wrt. some term graph language.

The simplest kind of homomorphism that is sometimes needed during construction does not instantiate any metavariables:

Definition 7.1. An **identification** is a homomorphism with empty metavariable base; i.e., it is an $\emptyset$-homomorphism. □

Identifications stand in a one-to-one correspondence with congruence relations on term graphs; for every identification F, the equivalence relation $(F \mathbin{;} F^{\cup})^{*}$ is a congruence, and for every congruence, the quotient projection is an identification.

Since we restrict ourselves to finite graphs, every identification can be broken down into a sequence of "primitive identifications", the correspondences of which are the correspondence closure of two-node sets.

Isomorphisms between typed term graphs may instantiate metavariables in that the only visible change that can be brought about by isomorphisms is consistent permutation of the successors of metavariables.

Similar to the identification of two nodes, there is the possibility of unification of two nodes. But since this in general brings about the non-determinism of second-order unification, we have to restrict ourselves to simple deterministic cases:

Definition 7.2. A **replacement** of a metavariable node v with another node r which is not a successor of v is a homomorphism F which unifies v and r and which uniquely factorises any other such homomorphism. □

It is comparatively easy to see that if such a replacement exists, then the replacement is unique up to isomorphism. Replacements can also be regarded as special instances of the following:

Definition 7.3. An **internal instantiation** of an n-ary metavariable node v in G_1 with an interval (t, b) in G_1 is an equaliser of the following two homomorphisms F and G from the Graph G_0 containing a $(1, n)$-typing element for the label of v:

- F is total and maps the source of G_0 to v,
- G maps the source r_0 of G_0 to t and behaves as b on the successors: $(S_0 . r_0) \mathbin{;} G = b$.

(Such an equaliser is a homomorphism such that $F \,;\, H = G \,;\, H$ and which uniquely factorises any other such homomorphism.) □

Those internal instantiations where the interval has an empty lower border are just the replacements of above.

Definition 7.4. An **external instantiation** of an n-ary metavariable node v in G_1 with an interval (t, b) in G_2 is a pushout of the following two homomorphisms F and G from the Graph G_0 containing a one-source typing element for a metavariable with typing compatible to that of v and with arity $m \leq n$:
- F is total and maps the source of G_0 to v,
- G maps the source r_0 of G_0 to t and behaves as b on the successors: $(S_0 . r_0) \,;\, G = b$.

(Such a pushout is a "pushout graph" G_3 together with two homomorphisms H_F from G_1 to G_3 and H_G from G_2 to G_3 such that $F \,;\, H_F = G \,;\, H_G$ and H_F and H_G uniquely factorise any other constellation like that.) □

The pushouts of external and the equalisers of internal instantiations need not exist; this fact usually corresponds to the attempt to introduce a type error.

External instantiations with simple intervals and with the metavariable in G_0 having the same arity as the instantiated metavariable can serve to cut or permute outgoing edges of metavariables; another important instantiation uses an interval where one lower border node is identical to the top node, thus "shrinking" every node of the instantiated metavariable to its corresponding successor.

Besides these language-independent instantiations, the most important are those where G_0 contains a zero-ary metavariable mapped to some typing element as G_2; in these cases the metavariables in the typing element have the arity of the instantiated metavariable added to theirs:

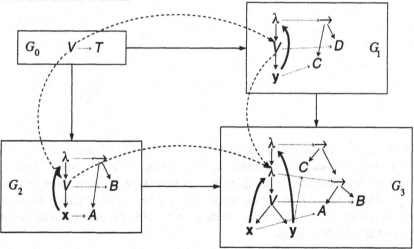

Rewriting of our typed term graphs is defined essentially in the same way as in [3,4] via the fibred approach, a generalisation of the traditional double-pushout approach. For reasons of space we do not present this here.

8 Advanced Instances

In this section we show how three kinds of advanced type systems can be transferred to the term graph setting. For reasons of space we have to assume that the reader is already familiar with the basics of the respective systems. Although the examples can still be constructed and drawn in HOPS, none of them has been implemented so far because of the problems with second-order unification.

8.1 Recursive Datatypes and Polytypic Programming

The essential feature of polytypic programming — see e.g. [8] — is an explicit recursion operator encapsulating a second-order variable on the type side; this variable stands for some appropriate functor, and the whole construct then stands for the (usually) least fixed point of that functor.

The basic functions necessary for programming on these recursive datatypes are the isomorphisms abstr and repr between the fixed-point domain and its images under the functor. Therefore, we need the following three typing elements — they are all separated:

The isomorphism properties can be expressed as rules[1]:

For the first rule, this is indeed the principal type (when limited to maximally identified type parts); the second rule, however, also has another "simplest" type relying on a nested recursion and which is incomparable to the given type — as soon as second-order typing elements are used, automatic unification cannot be used anymore because of these ambiguities of second-order unification. In HOPS, it is planned to make some user-assisted solution available.

---

[1] A term graph rule is a term graph together with two distinguished nodes; these two nodes are indicated by thick long-tipped orange resp. grey arrows in HOPS drawings and are the roots of the rule's *left-hand side* and *right-hand side*, respectively. Since rewriting is not the topic of this paper, we do not explain the rule application mechanism here, let us only mention that for multi-node metavariables with successors, no matter whether typed or untyped, the encapsulation skeleton of their image on the left-hand sides has to be copied for constructing their image on the right-hand side.

Now, the most popular polytypic functions are probably polytypic map and catamorphisms; we give the typing elements:

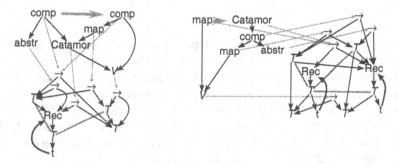

For specific functors, maps can be specialised to known functions, as in the following example rules, which recognise the functor involved by their sensitivity to typing, since their left-hand sides do not have principal types (pupd is — as obvious from its typing — function product):

The basic rule for catamorphisms allows an unfolding of the transferred functor, i.e., a map, to a catamorphism preceded by an abstr; see the drawing to the left:

Together with the map rules for the different type constructors — to the right above, there is the map rule for datatype recursion — it is possible to transform arbitrary specific legal catamorphisms into general recursions just by applying these type-sensitive rules.

8.2 Second-order λ-Calculus

In second-order λ-calculus, there is a different kind of abstraction on the type level, usually denoted "∀"; we shall use the label forall. This then enters the typing elements of a second kind of λ-abstraction and application together with multi-node second-order metavariables:

The last two typing elements are not separated: in the abstraction, there is an untyped variable bound at the typed Lambda node, and the typed Apply node has an untyped second successor. The β-reduction rule for this type abstraction nicely shows the interplay between bound variables and metavariables, and with respect to the discussion in Sect. 6 we also show a typing element for the unary typed metavariable used here:

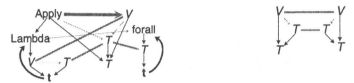

8.3 Dependent Types

Dependent types are to a certain extent "dual" in a syntactic sense to second-order λ-calculus; here the border crossings between program and type part run in the contrary direction, and there are already border crossings in the typing element for the type constructor Pi: a typed variable x (which has a typing element x⸱⸱⸱▸T of its own) is bound at the untyped Pi, and is successor of an untyped unary metavariable:

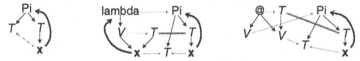

In the typing element of λ-abstraction, the typed bound variable x is also successor to an untyped unary metavariable, and this does not violate the condition that binders should dominate bound variables because of the modified reachability. In the typing element of application, finally, a zero-ary typed metavariable occurs as successor of a unary untyped metavariable — an indication that we have to be very careful in this kind of calculus since now application of rules on the program side may at the same time be on the type side. The pictures of the rule of β-reduction and of the metavariable typing element are astonishingly similar to those for second-order λ-calculus:

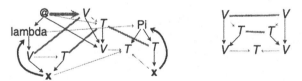

9 Conclusion

We have introduced a formalism that allows a large class of type systems to be translated to the term graph setting in a very homogeneous way. The explicitness with which all the structure including typing is incorporated into our term graphs makes them extremely useful for human interaction with complex formalisms — in fact, our formalisation is the result of long-running efforts to provide the graphically interactive term graph programming system HOPS [6,13] with an appropriate typing system both from the implementation point of view and from theoretical considerations.

Another potential use for our internally typed term graphs is to serve as internal data structure in symbolic computation systems including interpreters and compilers, true to the recent trend to keep typing information until much later phases in the compilation process.

References

1. Hendrik P. Barendregt. Lambda Calculi with Types. In S. Abramsky, Dov M. Gabbay, T.S.E. Maibaum, *Handbook of Logic in Computer Science, Vol. 2*, pages 117–309. Oxford Univertity Press, 1992.

2. Chris Brink, Wolfram Kahl, Gunther Schmidt (eds.). *Relational Methods in Computer Science*. Advances in Computing. Springer-Verlag, Wien, New York, 1997. ISBN 3-211-82971-7.

3. Wolfram Kahl. *Algebraische Termgraphersetzung mit gebundenen Variablen*. Reihe Informatik. Herbert Utz Verlag, München, 1996. ISBN 3-931327-60-4, zugleich Dissertation an der Fakultät für Informatik, Universität der Bundeswehr München.

4. Wolfram Kahl. A Fibred Approach to Rewriting — How the Duality between Adding and Deleting Cooperates with the Difference between Matching and Rewriting. Tech. Rep. 9702 (May 1997), Fakultät für Informatik, Universität der Bundeswehr München.

5. Wolfram Kahl. Relational Treatment of Term Graphs With Bound Variables. *Journal of the IGPL* **6** (2), 259–303 (March 1998).

6. Wolfram Kahl. The Higher Object Programming System — User Manual for HOPS, Fakultät für Informatik, Universität der Bundeswehr München, February 1998. URL http:// diogenes.informatik.unibw-muenchen.de:8080/kahl/HOPS/.

7. Jan Willem Klop. Combinatory Reduction Systems. Mathematical Centre Tracts 127 (1980), Centre for Mathematics and Computer Science, Amsterdam. PhD Thesis

8. Erik Meijer, Maarten Fokkinga, Ross Paterson. Functional Programming with Bananas, Lenses, Envelopes and Barbed Wire. In John Hughes (ed.), *Functional Programming Languages and Computer Architecture, 5th ACM Conference*, pages 124–144. LNCS 523. Springer Verlag, 1991.

9. Gunther Schmidt, Thomas Ströhlein. *Relations and Graphs, Discrete Mathematics for Computer Scientists*. EATCS-Monographs on Theoretical Computer Science. Springer-Verlag, Berlin/Heidelberg/New York, 1993.

10. J. M. Spivey. *The Z Notation: A Reference Manual*. Prentice Hall International Series in Computer Science. Prentice Hall, 1989.

11. Carolyn L. Talcott. A Theory of Binding Structures and Applications to Rewriting. *Theoretical Computer Science* **112**, 68–81 (1993).

12. Christopher Peter Wadsworth. *Semantics and Pragmatics of the Lambda Calculus*. Ph.D. thesis, Oxford University, September 1971.

13. Hans Zierer, Gunther Schmidt, Rudolf Berghammer. An Interactive Graphical Manipulation System for Higher Objects Based on Relational Algebra. In *Proc. 12th International Workshop on Graph-Theoretic Concepts in Computer Science*, pages 68–81. LNCS 246. Springer-Verlag, Bernried, Starnberger See, June 1986.

Compact Implicit Representation of Graphs [*]
(Extended Abstract)

Maurizio Talamo[1] and *Paola Vocca*[2]

[1] Dipartimento di Informatica e Sistemistica, Università di Roma "La Sapienza",
Via Salaria 113, I-00198 Rome, Italy.
talamo@dis.uniroma1.it,
[2] Dipartimento di Matematica, Università di Roma "Tor Vergata",
Via della Ricerca Scientifica, I-00133 Rome, Italy.
vocca@axp.mat.uniroma2.it

Abstract. How to represent a graph in memory is a fundamental data structuring problem. In the usual representations, a graph is stored by representing explicitly all vertices and all edges. The names (labels) assigned to vertices are used only to encode the edges and betray nothing about the structure of the graph itself and hence are a "waste" of space. In this context, we present a general framework for labeling *any* graph so that adjacency between any two given vertices can be tested in constant time. The labeling schema assigns to each vertex x of a general graph a $O(\delta(x) \log^3 n)$ bit label, where n is the number of vertices and $\delta(x)$ is x's degree. The adjacency test can be performed in 5 steps and the schema can be computed in polynomial time. This representation strictly contrasts with usual representations, i.e. adjacency matrix and adjacency list representations, which require $O(n \log n)$ bit label per vertex and constant time adjacency test, and $O(\delta(x) \log n)$ bit label per vertex and $O(\log \delta(x))$ steps to test adjacency, respectively. Additionally, the labeling schema is *implicit*, that is: no pointers are used.

1 Introduction

The representation of graphs has received much attention since the very beginning of the study of computer science theory [2, 3, 18, 6, 21, 11, 16, 22]. What is generally required is a "good" and efficient representation: good for the efficiency of algorithms running on it, and efficient both in terms of the space to store data and the computational time needed to derive the representation.

In this context, we present a general framework for encoding *any* graph which leads to an *implicit* representation, that is: no pointers are used, allowing to test adjacency in a constant number of steps. In particular, the problem considered in this paper can be stated as follows. Given any graph $G = (V, E)$, with $n = |N|$, label the vertices such that, given the names of two vertices, we can determine

[*] Work partially supported by National Authority for Information Technology for the Public Administration and the Italian Project MURST-"Algorithms and Data Structure"

adjacency in $O(1)$ steps. The representation proposed assigns to each vertex x a $O(\delta(x) \log^3 n)$ bit label, where $\delta(x)$ is the degree of x, allows to test adjacency in 5 steps, and can be computed in $O(n^3)$ time.

Implicit representations are widely accepted as good representations since implicit data structures lend themselves to a sequential storage scheme which requires no pointers, thus providing for a compact way to store data with no waste of space for pointers [13, 4, 12, 15]. Moreover, the algorithms are easier to implement and are often more efficient.

Additionally, the representation we proposed satisfies the property of "locality", that is data are stored on a per–node basis, thus providing an efficient representation for distributed computation [14] and secondary memory storage [23].

Several authors worked on this problem. Breuer [2] and Breuer and Folkman [3] considered the problem of labeling vertices such that adjacency would be determined by the Hamming distance of the labels. Their schema is very restricted and for general graphs the length of the labels can be $O(n \log n)$. Turan [18] and Kannan et al. [11] considered the problem of representing a graph as succinctly as possible. However, they gave an efficient implicit representation for the adjacency list of the graph only for restricted classes of graphs: trees, graphs with bounded arboricity, intersection graphs, and c–decomposable graphs [8]. Additionally, the class of bounded treewidth can be represented with $O(b \log n)$ bit labels, where b is the treewidth (see [19]).

The same problem has been studied in a different context, namely for routing messages in a distributed network. In [16, 20, 21, 5–9] the problem considered is how to store routing information at the vertices of a distributed network so as to computer near–optimal routes. Again, the problem is optimally solved only for restricted class of graphs, as trees, rings, complete graphs, planar st–graphs, interval graphs, or for specific network topologies as hypercube, meshes.

In this paper, we propose a k–step labeling schema based on a set of k mutually composable labeling functions, each one evaluable in 1 step. In order to test adjacency, the k functions are evaluated sequentially, thus adjacency can be tested in k steps.

In this scenario we obtain the following results:

1. a 3–step labeling schema for regular bipartite graph of degree δ. The schema assigns $O(\delta \log^2 n)$ bit label to each vertex x;
2. a 5–step labeling schema for general graph. The schema assigns $O(\delta(x) \log^3 n)$ bit label to each vertex x, where $\delta(x)$ is the degree of x.

It is worth noting that if we compare our results with usual representation strategies, the amount of space on a per–node basis is increased of a $\log n$ factor for the former and a $\log^2 n$ factor for the latter. On the counterpart, the adjacency test can be performed in only 3 steps and 5 steps, respectively.

The proof strategy proceeds as follows: i) the adjacency test problem for a general graph is reduced to an equivalent problem on a general bipartite graph (Section 5); ii) the same problem for a general bipartite graph is reduced to an equivalent problem for a collection of regular bipartite graphs (Section 5);

iii) the problem for a regular bipartite graph is then reduced to an equivalent problem for a collection of bounded degree bipartite graphs (Section 4) which we show to be efficiently representable in Section 3.

The technique used in this paper is based on the one described in [17] for directed acyclic graphs.

Without loss of generality, we will prove all theorems under the assumption that $G = (V, E)$ is connected. In the case of non-connected graphs, all results can be applied to each connected component without any change in complexity bounds (see [1]).

2   Preliminaries

In this section we introduce the notation used and some basic definitions. More definitions on graphs can be found in textbooks as [10].

We denote a *graph* by $G = (V, E)$, where V is the *vertex* set, with $n = |V|$, and E is the *edge set*, with $m = |E|$. Given a vertex x, the set of vertices *adjacent to* x is defined as $adj(x) = \{y \in V | (x, y) \in E\}$. The *degree* of a vertex x is $\delta(x) = |adj(x)|$. For any set of vertices $X \subseteq V$, the set of vertices *adjacent to* X is defined as $adj(X) = \{y \in V | (x, y) \in E \land x \in X\}$.

$K = (A \cup B, E)$ denotes a *bipartite graph*, where A is the set of the *upper vertices*, B is the set of the *lower vertices*, and $E \subseteq A \times B$ is the set of non-directed edges. Moreover, we assume $n_a = |A|$, $n_b = |B|$, and $m = |E|$.

Given a bipartite $K = (A \cup B, E)$ and a set C such that either $C \subseteq A$ or $C \subseteq B$, *the subgraph induced by* C, denoted $K_{/C}$, is the subgraph induced by the set of vertices $\{C \cup adj(C)|\}$ or $\{adj(C) \cup C\}$, respectively, that is $K_{/C} = (C \cup adj(C), E')$, where $E' = \{(x, y) \in E | x \in C \text{ and } y \in adj(C)\}$ or $K_{/C} = (adj(C) \cup C, E')$, where $E' = \{(x, y) \in E | x \in adj(C) \text{ and } y \in C\}$. Let $x \in adj(C)$. We denote $\delta_{/C}(x)$ the degree of x in the subgraph $K_{/C}$.

A *regular bipartite graph* $K_{n,\delta}$ is a bipartite graph $K = (A \cup B, E)$, where $|A| = |B| = n$ and for any $x \in A \cup B$ then $\delta(x) = \delta$.

Given a graph $G = (V, E)$, we define a *k–steps labeling schema for* G, $\mathcal{L} = \{f_1 \ldots f_k\}$, where f_i is a partial function computable in 1 step and such that a composition between f_i and f_{i-1} can be defined, for $2 \leq i \leq k$, and the composition is well defined.

Given a graph $G = (V, E)$ and two vertices x and y, a *k–step labeling schema* $\mathcal{L}$ for G is *valid* iff:

$$y \in adj(x) \iff (f_k \circ f_{k-1} \circ \ldots \circ f_1(x, y) = y) \text{ or } (f_k \circ f_{k-1} \circ \ldots \circ f_1(y, x) = x).$$

3   Basic Labeling Schema

In this section, we describe the first two functions of the labeling schema for general graphs. Even though the result we show is not satisfying in terms of the length of the labels assigned to vertices, it represents a basic step for the overall strategy. As an intermediate step, we present a 2–step labeling schema for bipartite graphs of bounded degree.

First we need some more definitions.

Let $K = (A \cup B, E)$ be a bipartite graph, two vertices $x_1, x_2 \in B$, with $x_1 \neq x_2$, are *independent* if $\{adj(x_1) \cap adj(x_2)\} = \emptyset$.

Let $\mathcal{L}_1 : (x) \in B \longmapsto c \in I\!\!N$ be a labeling function of the lower vertices, called *cluster labeling function*. Given $x_1, x_2 \in B$ then $\mathcal{L}_1(x_1) = \mathcal{L}_1(x_2) = c$ iff x_1 and x_2 are independent vertices.

We define the *size* of the cluster labeling function $\mathcal{L}_1$ as $|\mathcal{I}m(\mathcal{L}_1)|$, that is the number of different c values necessary to encode all lower vertices. In other words, function $\mathcal{L}_1$ labels each lower vertex such that two lower vertices receive the same label if and only if they are independent.

The following Lemma allows to bound the size of the cluster labeling function $\mathcal{L}_1$ in terms of the maximum degrees of the upper and lower vertices.

Lemma 1. *Let $K = (A \cup B, E)$ be a bipartite graph, and d_a, d_b two positive integer values such that for any $x \in A$, $d_a/2 \leq \delta(x) \leq d_a$ and for any $x \in B$, $\delta(x) \leq d_b$. There exists a cluster labeling function of size $O(d_a d_b)$.*

Proof.(Skecth) Let $n_a = |A|$ and $n_b = |B|$. Consider the following greedy strategy. Choose any vertex $x \in B$ and set $\mathcal{L}_1(x) = 1$. Set $\mathcal{L}_1(y) = 1$ for any other vertex $y \in B$ such y is independent with all vertices x such that $\mathcal{L}_1(x) = 1$, until there are no other independent vertices. The choice criteria is not relevant. Values i, for $2 \leq i \leq p$, are assigned in a similar way choosing vertices not already labeled until all vertices in B are considered.

When label i is assigned to a vertex x at most $\delta(x)(d_a - 1)$ cannot receive the same label. Hence, the number t_1 of vertices with label equal to 1 is satisfies the following inequality:

$$\sum_{j=1}^{t_1} \delta(x_j) d_a \geq n_b \qquad \forall x \text{ s.t } \mathcal{L}_1(x_j) = 1. \qquad (1)$$

Observe that if a vertex x has not been labeled yet this implies that one of its adjacent vertices is adjacent to one of the already labeled vertex, thus inequality (1) holds for all value i of the label, $2 \leq i \leq p$. Hence,

$$|\mathcal{L}_1^{-1}(i)| \geq \sum_{j=1}^{t_i} \delta(x_j) \geq \frac{n_b}{d_a} \qquad \forall x \text{ s.t } \mathcal{L}_1(x_j) = i. \qquad (2)$$

Observing that the labeling process ends when all the edges have been considered, that is when:

$$\sum_{i=1}^{p} \sum_{j=1}^{t_i} \delta(x_j) \geq m \qquad \forall x \text{ s.t } \mathcal{L}_1(x_j) = i, \qquad (3)$$

where $m = |E|$. Additionally, $\frac{n_a d_a}{2} \leq m \leq n_a d_a$. Hence, we have:

$$n_a d_a \geq m = \sum_{i=1}^{p} \sum_{j=1}^{t_i} \delta(x) \geq p\frac{n_b}{d_a} \qquad \forall x \text{ s.t } \mathcal{L}_1(x) = i \qquad (4)$$

that is, $p \leq \frac{n_a d_a^2}{n_b}$. The proof follows observing that, by hypothesis, $n_b \geq \frac{n_a d_a}{2d_b}$.

It is interesting noting that the above bound is tight [17].

The cluster labeling function allows to derive a 2–steps valid labeling schema for bounded degree bipartite graph, as shown in the following theorem:

Theorem 1. *Let $K = (A \cup B, E)$ be a bipartite graph, and d_a, d_b two positive integer values such that for any $x \in A$, $d_a/2 \leq \delta(x) \leq d_a$ and for any $x \in B$, $\delta(x) \leq d_b$. There is 2–step valid labeling schema that assigns to each upper vertex a $O(d_a d_b \log n)$ bit label and can be computed in $O(m d_a)$ time.*

Proof.(Sketch) Let $\mathcal{L}_2 : (x) \in A \longmapsto B^p$, where p is the size of the cluster labeling function $\mathcal{L}_1$, be a labeling function which assigns to each vertex x in A, for each value $i \in |\mathcal{I}m(\mathcal{L}_1)|$, the unique vertex $y \in B$ with label i adjacent to x.

The labeling schema easily follows from the definition of the following two functions:

$$\begin{cases} f_1 : (x, y) \in A \times B \longmapsto (x, \mathcal{L}_1(y)) \in A \times \mathbb{N}; \\ f_2 : (x, c) \in A \times \mathbb{N} \longmapsto y \in B, \end{cases}$$

where $f_2(x, c) = y$ iff $\mathcal{L}_1(y) = c$ and y is adjacent to x.

It is easy to verify that, using the two labelings $\mathcal{L}_1$ and $\mathcal{L}_2$, the adjacency test between $x \in A$ and $y \in B$ can be computed in 2 steps evaluating the two composable functions $f_2 \circ f_1(x, y)$. More precisely, $y \in adj(x)y \Longleftrightarrow f_2 \circ f_1(x, y) = y$. By Lemma 1, $\mathcal{L}_2$ assigns to each vertex $x \in A$ a $O(d_a d_b \log n)$ bit label. Finally, the time bound on the computation of the labeling schema easily derives observing that once a label is assigned to vertex $x \in B$, $\delta(x)d_a$ updates are necessary.

Obviously, although the labeling schema proposed is valid for general bipartite graphs the length of the labels assigned to upper vertices can be as greater as $\delta(x)n \log n$, and hence, it is not competitive with classical representations, unless the degree of the lower vertices is bounded by a constant value.

4   Labeling Schema for Regular Bipartite Graphs

In this section, we describe a 3–step valid labeling schema for regular bipartite graph of degree δ, which assigns $O(\delta \log^2 n)$ bit label to each vertex. This not only represents a result by itself but it is an intermediate step for our strategy which reduces the problem of testing adjacency in general graph to the same problem first for a general bipartite graph and then for a collection of regular bipartite graphs.

The implicit representation of regular bipartite graph is obtained by showing that any regular bipartite graph can be decomposed into a $O(\delta \log n)$ collection of special bounded degree bipartite graphs, called *log–graphs* (Lemma 2 and Corollary 2). Log–graphs are bipartite graphs such that the degree of the lower vertices isbounded by $\lfloor \log n \rfloor$, and, hence, they admit a vertex labeling function with a suitable label length for vertices (see Theorem 1).

First we give a set of technical lemmas already described in [17], and presented here again for clarity of exposition.

Lemma 2. *Let $K_{n,\delta}$ be a regular bipartite graph. A set $A' \subseteq A$ exists such that the following two conditions hold: i) $|A'| = \left\lfloor \frac{n}{6\delta+1} \right\rfloor$; ii) for any $x \in adj(A')$, $\delta_{/A'}(x) < \lfloor \log n \rfloor$.*

Proof. Let $h = \lfloor \log n \rfloor$ and $k = \left\lfloor \frac{n}{6\delta+1} \right\rfloor$.

For any $i \geq 0$ we define a collection of sets $A(i)$ as follows:

$$\begin{cases} A(0) = \emptyset; \\ A(i) = A(i-1) \cup \{x_i\}, \end{cases}$$

where $x_i \in A - A(i-1)$ is a vertex belonging to a set $S_h(i)$ to be defined later.

We claim that $A(k)$ is a set satisfying the statement of Lemma. For each set $A(i)$ consider the sub-bipartite $K_{/A(i)}$. With reference to $K_{/A(i)}$, for any $1 \leq j \leq i$ we define the following sets:

$$C_j(i) = \{x \in adj(A(i)) | \delta_{/A(i)}(x) = j \text{ and } \delta_{/A(i-1)}(x) = j-1\}; \quad (5)$$

$$C_j^T(i) = \{x \in adj(A(i)) | \delta_{/A(i)}(x) = j\}; \quad (6)$$

$$Out(C_j^T(i)) = \{(x,y) \in E | y \in C_j^T(i) \text{ and } x \in A - A(i)\} \quad (7)$$

Loosely speaking, set $C_j(i)$ is the set of vertices in $adj(A(i))$ having degree j only after the addition of x_i to the set $A(i-1)$; $C_j^T(i)$ is the set of all vertices in $adj(A(i))$ of degree j; finally, $Out(C_j^T(i))$ is the set of edges leaving vertices in $C_j^T(i)$ towards vertices in $A - A(i)$.

Sets $S_h(i)$ to which x_i belongs are, for $j = h$, the sub-collection of $S_j(i)$ defined recursively as follows:

$$S_1(i) = A - A(i-1);$$

$$S_j(i) = \begin{cases} \{x \in S_{j-1}(i) | \ |C_{j-1}^T(i-1) \cap adj(x)| \leq \frac{t|Out(C_{j-1}^T(i-1))|}{|S_{j-1}(i)|} \} & \text{for } 1 \leq j \leq i \\ S_i(i) & \text{for } j > i \end{cases}$$

where $t > 1$ is a parameter whose proper evaluation will allow to derive the Lemma. Sets $S_h(i)$ is composed by those vertices x_i which do not increase the degrees of the lower vertices of $K_{/A(i)}$ more than t times the average.

The proof of Lemma proceeds in two steps. In the first step, we prove that the construction of sets $A(i)$ always halts finding at least one x_i; in the second step, we show that $K_{/A(k)}$ satisfies conditions *i)* and *ii)* of the statement of Lemma, which, in terms of the notation above introduced, is equivalent to show that the inequality $|C_h^T(k)| < 1$ holds.

Claim. $|S_h(i)| \geq 1$ for any $t \geq \lfloor \log n \rfloor$ and $1 \leq i \leq k$.

Proof.(Claim) The proof of Claim proceeds by showing that:

$$|S_h(i)| \geq |S_{h-1}(i)| \left(1 - \frac{1}{t}\right), \tag{8}$$

Solving the recurrence (8) with the initial condition $|S_1(i)| = n - i + 1$, we obtain:

$$|S_h(i)| \geq (|A| - |A(i-1)|) \left(1 - \frac{1}{t}\right)^{h-1} \tag{9}$$

hence, the proof follows by choosing $t \geq \lfloor \log n \rfloor$ and $1 \leq i \leq n - 4$.

Let us define:

$$\mu = \frac{|Out(C_{h-1}^T(i-1))|}{|S_{h-1}(i)|}, \tag{10}$$

and,

$$\overline{S}_h(i) = S_{h-1}(i) - S_h(i) = \{x \in S_{h-1}(i) | |C_{h-1}^T(i-1) \cap adj(x)| > \frac{t|Out(C_{h-1}^T(i-1))|}{|S_{h-1}(i)|}\}.$$

Hence, for any $x \in \overline{S}_h(i)$, $|C_{h-1}^T(i-1) \cap adj(x)| > \mu t$. Moreover,

$$\sum_{x \in \overline{S}_h(i)} |C_{h-1}^T(i-1) \cap adj(x)| + \sum_{x \in S_h(i)} |C_{h-1}^T(i-1) \cap adj(x)| = |Out(c_{h-1}^T(i-1))|.$$

Hence,

$$\mu t |\overline{S}_h(i)| \leq |Out(C_{h-1}^T(i-1))| \tag{11}$$

Recurrence (8) derives from equations (10) and (11).

Claim. $|C_h^T(k)| < 1$.

Proof.(Claim) First observe that, by definition in (6), we have:

$$C_h^T(k) \subseteq \bigcup_{i=1}^{k} C_h(i),$$

hence,

$$|C_h^T(k)| \leq \sum_{i=1}^{k} |C_h(i)| = \sum_{i=1}^{k} |C_{h-1}^T(i) \cap adj(x_i)|.$$

Due to the choice of x_i and the definition of $S_h(i)$, it follows:

$$|C_h^T(k)| \le \sum_{i=1}^{k} \frac{t|Out(C_{h-1}^T(i-1))|}{|S_{h-1}(i)|} \tag{12}$$

Moreover, by definition

$$|Out(C_{h-1}^T(i-1))| < |C_{h-1}^T(i-1)|\delta.$$

Hence, the following recurrence holds:

$$\begin{cases} |C_1^T(i)| \le i\delta \\ |C_j^T(i)| < \sum_{i=1}^{k} \frac{t\delta|C_{h-1}^T(i-1)|}{|S_{h-1}(i)|} \text{ for } j > 1 \end{cases} \tag{13}$$

From the Inequality (9) and using basic algebra, the evaluation of recurrence (13) for $j = h$, is given by:

$$|C_h^T(k)| < \lfloor \gamma_h k^h \rfloor. \tag{14}$$

where,

$$\gamma_h = \frac{t^{h-1}\delta^h}{h!(n-k)^{h-1}(1-\frac{1}{t})^{\frac{(h-1)(h-2)}{2}}}. \tag{15}$$

The Claim is proved substituting the values of h and k and choosing $t = h$ (see Claim 4). The proof of Lemma follows from Claims 4 and 4.

Given a bipartite $K_{n,\delta}$, we define a *special set* any set $A' \subseteq A$ satisfying condition *ii*) in the statement of Lemma 2. A *log-graph* is $K_{/A'}$, and we denote it by $K_{/A'}^*$.

It is worth noting that Lemma 2 holds for a wider class of bipartite. In fact, in the proof it is required that the degree of the upper vertices is regular, while, for the lower vertices, it suffices to be bounded. More precisely,

Corollary 1. Let $K = (A \cup B, E)$ be a bipartite graph such that $|A| = |B| = n$, and for any $x \in A$ then $\delta(x) = \delta$, while for any $x \in B$, $\delta(x) \le \delta$. A set $A' \subseteq A$ exists such that the following two conditions hold: i) $|A'| = \lfloor \frac{n}{6\delta+1} \rfloor$; ii) for any $x \in adj(A')$, $\delta_{/A'}(x) < \lfloor \log n \rfloor$.

Lemma 3. Let $K_{n,\delta}$ be a regular bipartite graph. There is a subset of A of size at least $\lceil \frac{n}{2} \rceil$ that can be partitioned in a collection of at most $12\delta+2$ special sets.

Proof.(Sketch) Let us denote by $\{A_1, \ldots, A_s\}$ the collection of special sets. A_i, $1 \le i \le s$, is obtained by applying the strategy described in Lemma 2 to bipartite $K_{/(A-\bigcup_{j=0}^{i-1} A_j)}$, where $A_0 = \emptyset$, until $\sum_{i=0}^{s} |A_i| \ge \lceil \frac{n}{2} \rceil$ (see Corollary 1).

The Lemma derives by showing that $|A_i| \ge \lfloor \frac{n}{12\delta+2} \rfloor$, for $1 \le i \le s$.

The proof proceeds by induction. The base case easily follows from Lemma 2 applied to $K_{n,\delta}$. Further, applying Lemma 2 to $K_{/(A-\bigcup_{j=0}^{s-1} A_j)}$ and observing that $\sum_{i=0}^{s-1} |A_i| < \lceil \frac{n}{2} \rceil$, we have $|A_s| \ge \lfloor \frac{n}{12\delta+2} \rfloor$.

An immediate consequence of Lemma 2 is:

Corollary 2. *Let $K_{n,\delta}$ be a regular bipartite graph. The set of vertices A can be partitioned in a $O(\delta \log n)$ sequence of special sets.*

The main consequence of the above technical lemmas is that, applying Theorem 1 to each log–graph induced by the special set collection, it is possible to find a suitable labeling schema for the representation of the whole regular bipartite graph (see Theorem 2).

Theorem 2. *Let $K_{n,\delta}$ be a regular bipartite graph. There is 3–step valid labeling schema that assigns to each vertex a $O(\delta \log^2 n)$ bit label and can be computed in $O(m\delta + n^2)$ time.*

Proof.(Sketch) Let $\mathcal{A} = \{A_i\}$ be the special set collection which partition the upper vertices set A (see Lemma 3 and Corollary 2), and let $\{K^*_{/A_i}\}$ be the corresponding collection of log-graphs.

The following labeling functions can be defined.

i. $\mathcal{L}_1 : x \in A \longmapsto s \in \mathbb{N}$ is a labeling function of the upper vertices which assigns to each vertex $x \in A$ an integer value $s \in \{1, \ldots, (12\delta + 2)\lceil \log n \rceil\}$ representing the unique special set to which x belongs according to the decomposition in special sets $\mathcal{A} = \{A_i\}$.

ii. $\mathcal{L}_2 : x \in B \longmapsto \mathbb{N}^{|\mathcal{A}|}$ is a labeling schema function which maps each lower vertex x to a $|\mathcal{A}|$–tuple of integers. The i-th integer, for $1 \leq i \leq |\mathcal{A}|$, is the label assigned to x according to the cluster labeling function of the log-graph $K^*_{/A_i}$. In fact, the log-graph $K^*_{/A_i}$ satisfies the conditions of Theorem 1, where $d_a = \delta$ and $d_b = \lfloor \log n \rfloor$.

iii. $\mathcal{L}_3 : x \in A \longmapsto B^{p_i}$ is a labeling schema function which maps each upper vertex x, with respect to the unique special set A_i to which x belongs, for each value $i \in \{1, \ldots, p_i\}$, the unique vertex $y \in B$ adjacent to x, where p_i is the size of the cluster labeling function $\mathcal{L}_2$ restricted to $K^*_{/A_i}$.

The 3–step labeling schema easily follows from the definition of the following three functions:

$$\begin{cases} f_1 : (x,y) \in A \times B & \longmapsto (x, y, \mathcal{L}_1(x)) \in A \times B \times \mathbb{N}; \\ f_2 : (x, y, \mathcal{L}_1(x)) \in A \times B \times \mathbb{N} & \longmapsto (x, \mathcal{L}_1(x), \mathcal{L}_2(y)) \in A \times \mathbb{N}^2; \\ f_3 : (x, \mathcal{L}_1(x), \mathcal{L}_2(y)) \in A \times \mathbb{N}^2 & \longmapsto \mathcal{L}_3(x) = y \in B, \end{cases}$$

More precisely, $f_1(x, y)$ returns the tuple $(x, y, \mathcal{L}_1(x))$, where $\mathcal{L}_1(x)$ is the index of the unique special set to which x belongs to; $f_2(x, y, \mathcal{L}_1(x))$ returns the tuple $(x, \mathcal{L}_1(x), \mathcal{L}_2(y))$, where $\mathcal{L}_2(y)$ is the unique label of y according to the cluster labeling function restricted to the log–graph $K^*_{/A_i}$, with $i = \mathcal{L}_1(x)$; finally, $f_3(x, \mathcal{L}_1(x), \mathcal{L}_2(y))$ returns the unique lower vertex y to which x is connected with respect to the log-graph $K^*_{/A_i}$ and the cluster labeling function $\mathcal{L}_2(y)$.

Thus $f_3 \circ f_2 \circ f_1(x, y) = y$ iff y is adjacent to x. From Lemma 2, Corollary 2, and Theorem 1 the labeling functions assign to vertices in $O(\delta \log^2 n)$ as required.

The analysis of the time complexity to compute the data structure can be divided into two parts: i.) the time required to compute the special set collection;

ii.) the time complexity to compute the cluster labeling function. The latter derives from Theorem 1. For the former, observe that the insertion of a vertex x into a special set implies δ^2 updates. It is easy to design a data structure which allows to compute the special sets collection $\mathcal{A} = \{A_i\}$ in the time required by the Theorem.

5   Labeling Schema for General Bipartite Graph and General Graphs

In the first part of this section we show how to reduce adjacency testing problem for general bipartite graph to the same problem for a collection of regular bipartite graph and, finally, we show how to reduce the same problem for general graphs to general bipartite graphs. Actually, the collection of bipartite graphs considered is slightly more general than that of regular bipartite graphs. The bipartite graphs considered are what we call *bounded bipartite graphs*.

A *bounded bipartite graph*, denoted by $K_{(n_a,n_b,\delta_a,\delta_b)}$, is a bipartite graph $K_{(n_a,n_b,\delta_a,\delta_b)} = (A \cup B, E)$, such that $|A| = n_a$, $|B| = n_b$, $|E| = m$, and for any $x \in A$ then $\delta_a/2 \le \delta(x) \le \delta_a$, while for any $x \in B$, $\delta(x) \le \delta_b$.

It is trivial to extend the results in Section 4 to bounded bipartite graphs. More precisely they can be restated as follows:

Lemma 4. *Let* $K_{(n_a,n_b,\delta_a,\delta_b)}$ *be a bounded bipartite graph. There exists* $A' \subseteq A$ *such that the following two conditions hold:* i) $|A'| = \left\lfloor \frac{n_a}{6\delta_b+1} \right\rfloor$; ii) *for any* $x \in adj(A')$, $\delta_{/A'}(x) < \lfloor 2\log n_a \rfloor$.

Lemma 5. *Let* $K_{(n_a,n_b,\delta_a,\delta_b)}$ *be a bounded bipartite graph. There is a subset of* A *of size at least* $\lceil \frac{n_a}{2} \rceil$ *that can be partitioned in a collection of at most* $12\delta_b + 2$ *special sets.*

Corollary 3. *Let* $K_{(n_a,n_b,\delta_a,\delta_b)}$ *be a bounded bipartite graph. The set of vertices* A *can be partitioned in a* $O(\delta_b \log n_a)$ *sequence of special sets.*

Theorem 3. *Let* $K_{(n_a,n_b,\delta_a,\delta_b)}$ *be a bounded bipartite graph. There is 3–step valid labeling schema that assigns to each upper vertex a* $O(\delta_a \log^2 n_a)$ *bit label and to each lower vertex* $O(\delta_b \log^2 n_a)$ *and can be computed in* $O(m\delta_a + n_a^2)$ *time.*

It is now possible to study general bipartite graphs.

Theorem 4. *Let* $K = (A \cup B, E)$, *where* $|A \cup B| = n$ *and* $|E| = m$, *be a bipartite graph. There is 5–step labeling schema that assigns to a vertex* x *a* $O(\delta(x) \log^3 n)$ *bit label and can be computed in* $O(m\delta + n^2)$ *time, where* δ *is the maximum degree of the lower vertices.*

Proof.(Skecth) Let δ_a and δ_b be the maximum degrees of the upper and lower vertices, respectively. Let us consider a partition of the set of upper vertices A into sets $\{D_i\}$, for $1 \le i \le \lceil \log \delta_a \rceil + 1$, such that $D_i = \{x \in A | 2^{i-1} \le \delta(x) \le 2^i\}$.

With reference to $K_{/D_i}$, define a partition of the set of lower vertices $adj(D_i)$ in sets $\{B_{i,j}\}$, for $1 \leq j \leq \lceil \log \delta_b \rceil + 1$, such that $B_{i,j} = \{x \in adj(D_i) | 2^{j-1} \leq \delta(x) \leq 2^j\}$.

Moreover, let $K_{i,j} = (D_i \cup B_{i,j}, E_{i,j})$, where $E_{i,j} = \{(x,y) \in E | x \in D_i$ and $y \in B_{i,j}\}$. By construction each $K_{i,j}$ is a bounded bipartite graph, hence, Theorem 3 holds.

It is trivial to see that $(x,y) \in E$ if and only if there are $1 \leq i \leq \lceil \log \delta_a \rceil + 1$ and $1 \leq j \leq \lceil \log \delta_b \rceil + 1$, such that $(x,y) \in E_{i,j}$.

Hence, the following labeling functions can be defined.

i. $\mathcal{L}_1 : x \in A \longmapsto i \in \{1, \ldots, \lceil \log \delta_a \rceil\}$ is a labeling function of the upper vertices that assigns to each vertex $x \in A$ an integer value i representing the unique set D_i to which x belongs.

ii. $\mathcal{L}_2 : x \in B \longmapsto \{1, \ldots, \lceil \log \delta_b \rceil\}^q$, where $q = \{1, \ldots, \lceil \log \delta_a \rceil\}$, is a labeling function which maps each lower vertex x to a q–tuple of integers. The i-th integer is the value j of the unique set $B_{i,j}$ to which x belongs.

iii. $\mathcal{L}_3 : x \in A \longmapsto h \in \mathbb{N}^t$, where $t = \lceil \log \delta_b \rceil$, is a labeling function of the upper vertices which assigns to each vertex $x \in A$ an integer value $h \in \{1, \ldots, (12 \cdot 2^i + 2) \lceil \log n_a \rceil\}$, representing the unique special set to which x belongs according to the decomposition in special sets $\mathcal{A}_i = \{A_{i,h}\}$ of the bipartite $K_{/D_i}$ to which x belongs.

iv. $\mathcal{L}_4 : x \in B \longmapsto \{1, \ldots, \lceil \log \delta_b \rceil\}^q \times \mathbb{N}^{|\mathcal{A}_i|}$, with $q = \lceil \log \delta_a \rceil$, is a labeling function that maps each lower vertex x, with respect to $\mathcal{L}_2(x)$, to a $|\mathcal{A}_i|$–tuple of integers. The h-th integer of the tuple, for $1 \leq h \leq |\mathcal{A}_i|$, is the label assigned to x according to the cluster labeling function of the bipartite graph $K_{i,j/A_{i,h}}$ which satisfies the conditions in Theorem 1, where $d_a = 2^i$ and $d_b = h$.

v. $\mathcal{L}_5 : x \in A \longmapsto \mathbb{N}^t \times B^{p_{j,k}}$, where $t = \lceil \log \delta_b \rceil$, is a labeling function which maps each upper vertex x, with respect to $\mathcal{L}_3(x)$, for each value $i \in \{1, \ldots, p_{j,k}\}$, the unique vertex $y \in B$ adjacent to x, where $p_{j,k}$ is the size of the cluster labeling function $\mathcal{L}_4$ restricted to the bipartite graph $K_{i,j/A_{i,h}}$.

The 5–step valid labeling schema easily follows from the definition of the following five functions:

$$
\begin{cases}
f_1 : (x,y) \in A \times B \longmapsto (x,y,\mathcal{L}_1(x)) \in A \times B \times \mathbb{N}; \\
f_2 : (x,y,\mathcal{L}_1(x)) \longmapsto (x,y,\mathcal{L}_1(x),\mathcal{L}_2(y)) \in A \times B \times \mathbb{N}^2; \\
f_3 : (x,y,\mathcal{L}_1(x),\mathcal{L}_2(y)) \longmapsto (x,y,\mathcal{L}_1(x),\mathcal{L}_2(y),\mathcal{L}_3(x)) \in A \times B \times \mathbb{N}^3; \\
f_4 : (x,y,\mathcal{L}_1(x),\mathcal{L}_2(y),\mathcal{L}_3(x)) \longmapsto (x,y,\mathcal{L}_1(x),\mathcal{L}_2(y),\mathcal{L}_3(x),\mathcal{L}_4(y)) \in A \times B \times \mathbb{N}^4; \\
f_5 : (x,y,\mathcal{L}_1(x),\mathcal{L}_2(y),\mathcal{L}_3(x),\mathcal{L}_4(y)) \longmapsto \mathcal{L}_5(x) = y \in B.
\end{cases}
$$

It is easy to verify that $f_5 \circ f_4 \circ f_3 \circ f_2 \circ f_1(x,y) = y$ iff y is adjacent to x. By Lemma 2, Corollary 2, and Theorem 1, the labeling functions assign to each vertex x a $O(\delta(x) \log^3 n)$ bit label, as required.

The time complexity bound to compute the data structure follows from Theorem 3 and observing that the two partitions in sets $\{D_i\}$ and $B_{i,j}$ can be computed in $O(n^2)$.

The extension to general graphs is now straightforward, as shown in following theorem.

Theorem 5. *Let $G = (V, E)$ be a graph. There is a 5–step valid labeling schema that assigns to each vertex x a $O(\delta(x) \log^3 n)$ bit label and can be computed in $O(m\delta + n^2)$ time, where δ is the maximum degree of vertices.*

Proof.(Skecth) Consider the bipartite $K = (A \cup B, E')$ associate to G defined as follows: $A = B = V$, $E' = \{(x,y) | x \in A$ and $y \in B$ and $(x,y) \in E\}$. The proof follows from Theorem 4.

6   Conclusions and Open Problems

In this paper, a 5–steps labeling schema for an almost optimal graph implicit representation has been presented. The adjacency test can be performed in 5 steps, evaluating 5 mutually composable functions, each one evaluable in 1 step. The labeling functions assign a $O(\delta(x) \log^3 n)$ bit label to each vertex x, where $\delta(x)$ is the degree of x.

The proposed schema favorably compares with usual representations. In fact, even though the amount of space on a per–node basis is increased of a $\log^2 n$ factor, the adjacency test can be performed in only 5 steps instead of $\log \delta(x)$, which, for many applications (i.e distributed computation [14] and secondary memory storage [23]) represents an important improvement

A natural direction for further work is to improve the bound on the length of the labels.

Another interesting research direction, is to apply our approach for coping with secondary memory management problems. In fact, as said before, the "locality" property of the data storage and the limited number of accesses to secondary memory for testing adjacency is an important feature for designing efficient solutions to external memory graph problems.

References

1. A. V. Aho, J. E. Hopcroft, and J. D. Ullman. *Design and Analysis of Computer Algorithms*. Addison-Wesley, 1974.
2. M. Breuer. Coding vertices of a graph. *IEEE Trans. Infor. Theory*, 12:148–153, 1966.
3. M. Breuer and J. Folkman. An unexpected result on coding vertices. *J. Math. Anal. Appl.*, 20:583–600, 1967.
4. G. N. Frederickson. Implicit data structures for the dictionary problem. *J. of the ACM*, 30(1):80–94, 1980.
5. G. N. Frederickson and R. Janardan. Optimal message routing without complete routing tables. In *Proc. 5th Annual ACM Symposium on Principles of Distributed Computing*, pages 88–97, Calgary, August 1986.
6. G. N. Frederickson and R. Janardan. Designing networks with compact routing tables. *Algorithmica*, 3:171–190, 1988.

7. G. N. Frederickson and R. Janardan. Efficient message routing in planar networks. *SIAM Journal on Computing*, 18:843–857, 1989.
8. G. N. Frederickson and R. Janardan. Space efficient message routing in c-decomposable networks. *SIAM Journal on Computing*, 19:164–181, 1990.
9. G. Gambosi and P. Vocca. Topological Routing. In Özalp Babaoğlu and Keith Marzullo, editor, *Distributed Algorithms-WDAG'96*, pages 206–219. LNCS 1151, Springer-Verlag, 1996.
10. F. Harary. *Graph Theory*. Addison-Wesley, 1972.
11. S. Kannan, M. Naor, and S. Rudich. Implicit representation of graphs. *SIAM Journal of Discrete Mathematics*, 5(4):596–603, 1992.
12. J. Ian Munro. An implicit data structure supporting inserting, deletion, and search in $o(\log^2 n)$ time. *J. of Comp. and Syst. Science*, 33:66–74, 1986.
13. J. Ian Munro and H. Suwanda. Implicit data structures for fast search and update. *J. of Comp. and Syst. Science*, 21:236–250, 1980.
14. M. Naor and L. Stockmeyer. What can be computed locally? *SIAM J.Comput.*, 24(6):1259–1277, 1995.
15. M. Ouksel and P. Scheurmann. Implicit data structure for linear hashing schemes. *Inf. Proc. Letters*, 29:183–189, 1988.
16. N. Santoro and R. Khatib. Labelling and implicit routing in networks. *Computer Journal*, 28(1):5–8, 1985.
17. M. Talamo and P. Vocca. A time optimal digraph browsing on a sparse representation. Technical Report 8, Math Department, University of Rome "Tor Vergata", 1997. Submitted to JGAA.
18. G. Turan. Succint representation of graphs. *Discrete Appl. Math.*, 8:289–284, 1984.
19. J. van Leeuwen. Graph algorithms. In J. van Leeuwen, editor, *Handbook of Theoretical Computer Science*, volume A, pages 525–631. Elsevier Science Publisher B.V., Amsterdam, 1990.
20. J. van Leeuwen and R. B. Tan. Computer networks with compact routing tables. In *G. Rozenberg and A. Salomaa (Eds.) The Book of L*, volume 790. Springer-Verlag, 1986.
21. J. van Leeuwen and R. B. Tan. Interval routing. *Computer Journal*, 30:298–307, 1987.
22. J. van Leeuwen and R. B. Tan. Compact routing methods: A survey. In *Proc. Colloquium on Structural Information and Communication Complexity (SICC'94)*. Carleton University Press, 1994.
23. J.S. Vitter and E.A.M. Shriver. Algorithms for parallel memory, i: Two–level memories. *Algorithmica*, 12(6):110–147, 1994.

Graphs with Bounded Induced Distance

Serafino Cicerone and Gabriele Di Stefano*

Dipartimento di Ingegneria Elettrica
Università degli Studi di L'Aquila
I-67040 Monteluco di Roio - L'Aquila - Italy
{cicerone,gabriele}@infolab.ing.univaq.it

Abstract. In this work we introduce *graphs with bounded induced distance* of order k (BID(k) for short). In any graph belonging to BID(k), the length of every induced path between every pair of nodes is at most k times the distance between the same nodes. In communication networks modeled by these graphs any message can be always delivered through a path whose length is at most k times the best possible one, even if some nodes fail.

In this work we first provide a characterization of graphs in BID(k) by means of cycle-chord conditions. After that, we investigate classes with order $k \leq 2$. In this context, we note that the class BID(1) is the well known class of distance-hereditary graphs, and show that 3/2 is a lower bound for the order k of graphs that are not distance-hereditary. Then we characterize graphs in BID(3/2) by means of their minimal forbidden induced subgraphs, and we also show that graphs in BID(2) have a more complex characterization. We prove that the recognition problem for the generic class BID(k) is Co-NP-complete. Finally, we show that the split composition can be used to generate graphs in BID(k).

1 Introduction

In communication networks, nodes are connected by point-to-point communication links for exchanging messages between neighbors. Consequently, messages from a sender to a destination are delevered through intermediate node(s). Some networks are not reliable, that is, at some time certain nodes can *fail* and, consequently, these nodes cannot cooperate to the communication process. In case of failures, whenever the sender and the destination are still connected, the messages are always delivered within unknown but some finite delay due to the fact that the distance between sender and destination might increase. We assume that transmitting a message incurs a constant cost for each link, that is the distance between any pair of nodes is given by the minimum number of links that a message must traverse to reach the destination.

Delivering messages is done according to a well defined *routing* strategy. Naturally, it is desirable that the strategy routes messages through *shortest paths*.

---

* Authors are partially supported by the Italian MURST Project "Teoria dei Grafi ed Applicazioni".

The usefulness of routing messages through shortest paths is extensively discussed in [19, 22], and also used by the Internet Protocol [18]. In networks using this protocol, some nodes, also called gateways, use shortest paths information about the current state of the network to recalculate shortest paths quickly in the face of node failures.

In this network model, to deliver a message when node failures have occurred corresponds to deliver the same message in a subnetwork modeled by a subgraph induced by the unfailed nodes. Moreover, shortest paths in this subgraph correspond to induced paths in the graph that models the whole network.

In this work we investigate network topologies in which, in case of node failures, messages will eventually reach the destination within a bounded delay, that is, each message traverses a path whose length is at most k times the length of a shortest path computed in absence of node failures. To this end, we introduce graphs having *bounded induced distance of order k*. To define this kind of graphs, let us consider two nodes x and y in a graph G, and the set of all the induced paths joining them. Let us consider the length of a shortest one, that is $d(x, y)$ (i.e., the distance between x and y in G), and the length of a longest one in this set of paths, that is $D(x, y)$. Now, the following definition can be given.

Definition 1. *Let G be a graph, and $\{x, y\}$ a pair of distinct connected nodes in G. The* stretch number *of the pair $\{x, y\}$, denoted by $s(x, y)$, is given by $s(x, y) = \frac{D(x, y)}{d(x, y)}$. The* stretch number *of G, denoted by $s(G)$, is the maximum stretch over all the possible pairs, that is $s(G) = \max_{\{x, y\}} s(x, y)$.*

In a graph G in which the stretch number is $s(G) \leq k$, the length of every induced path is at most k times the length of the shortest one between the same pair of nodes, and hence the delay ratio in case of node failures is always less or equal to k. This fact holds in any bounded induced distance graph of order k. Then, we can define all the following new classes:

Definition 2. *Let k be a real number. A graph G is a* bounded induced distance graph of order k *if and only if $s(G) \leq k$. The class of all the bounded induced distance graph of order k is denoted by* BID(k).

In communication network design, the efficiency of a routing scheme is measured in terms of its *stretch factor* [13, 23], that is the maximum ratio between the length of a route computed by the scheme and that of a shortest path connecting the same pair of nodes. If a network G has a k-spanner [21], then a routing scheme having stretch factor k exists for G. A k-spanner of a graph G is a spanning subgraph G' of G in which every pair of nodes that are adjacent in G are at distance no greater than k in G'. Neither the concept of spanners nor the concept of stretch factor can be used to measure the efficiency decrease in message routing due to node failures. On the other hand, if $G \in$ BID(k), any routing scheme delivering messages along induced paths has a stretch factor no greater than k.

The class BID(1) corresponds to the well known class of distance-hereditary graphs: a graph is distance-hereditary if and only if the lengths of any two induced paths joining the same pair of nodes are equal. Distance-hereditary graphs [2, 17] have already been investigated to design interconnection network topologies [7, 10, 12], and many papers have been devoted to them (e.g., see [1, 5, 6, 9, 11, 15, 20, 24]).

Given the relevance of graphs with bounded induced distance, our purpose is to provide a first characterization of graphs belonging to BID(k) of any order $k \geq 1$. To this end, we first prove that the class BID(k) is hereditary for each k, that is it is closed under induced subgraphs. Then, we give a characterization of graphs in BID(k) based on cycle-chord conditions.

In particular, we study the classes with small order, that is BID(k), $k \leq 2$. In this context we show that, in networks that are not distance-hereditary and in which node failures have occurred, $k = 3/2$ represent a lower bound on the transmission delay ratio. Then we characterize graphs in BID(3/2) by means of their forbidden induced subgraphs, and we also show that graphs in BID(2) have a more complex characterization than graphs belonging to BID(k), $k < 2$.

We also investigate the complexity of the recognition problem for the generic class BID(k), proving that this problem is Co-NP-complete. Moreover, in order to define operations to yield graphs in the class under consideration, we study the relationships between the stretch number of the graph obtained by split composition [8] and the stretch number of the composed graphs. By using these relations we can show that the split composition can be applied to generate graphs in the class BID(k).

The remainder of this paper is organized as follows. Notations and basic concepts used in this work are given in Section 2. In Section 3, the new characterization of graphs in BID(k) is shown. In Section 4, we investigate the graph classes BID(k) when $k \leq 2$. In Section 5 we give the complexity result for the recognition problem for the generic class BID(k). Section 6 shows when the split composition can be used as an operation to generate graphs in BID(k). Finally, in the last section we give the conclusions of this work and list some open problems.

2   Preliminaries

In this work we consider finite, simple, loopless, undirected and unweighted graphs $G = (V, E)$ with node set V and edge set E. We use standard terminologies from [16], some of which are briefly reviewed here.

A *subgraph* of G is a graph having all its nodes and edges in G. Given a subset S of V, the *induced subgraph* $\langle S \rangle$ of G is the maximal subgraph of G with node set S. S is *independent* if $\langle S \rangle$ has no edges. $|G|$ denotes the cardinality of V.

A sequence of pairwise distinct nodes $(x_0, \ldots, x_n)$ is a *path* in G if $(x_i, x_{i+1}) \in E$ for $0 \leq i < n$. The *length* of path $(x_0, \ldots, x_n)$ is n, and $|p|$ denotes the number of nodes in the path p. A path $(x_0, \ldots, x_n)$ is an *induced path* iff $\langle \{x_0, \ldots, x_n\} \rangle$ has n edges. A graph G is *connected* iff for each pair of nodes x and y of G there is a path from x to y in G. By $N(x)$ we denote the *neighbors* of x, that is, the set of vertices in G that are adjacent to x.

The notion of cycle has great relevance in this paper. A *cycle* C_n in G is a path $(x_0, \ldots, x_{n-1})$ where also $(x_0, x_{n-1}) \in E$. Two nodes x_i and x_j are *consecutive* in C_n when $j = (i+1) \bmod n$ or $i = (j+1) \bmod n$. A *chord* of a cycle is an edge joining two non-consecutive nodes in the cycle. Particular cycles are the *fan* (a cycle C_5 with two chords both incident the same node), the *house* (obtained by removing a chord from a fan), and the *domino* (a cycle C_6 with only one chord that divides the cycle into two chordless C_4).

Given a cycle C_n its *chord distance* is denoted by $cd(C_n)$, and it is the minimum number of consecutive nodes in C_n such that every chord of C_n is incident to some of such nodes. We assume $cd(C_n) = 0$ when C_n is chordless.

The set $\mathcal{S}(G)$ contains all the pairs of nodes inducing the stretch number of G, that is, $\mathcal{S}(G) = \{\{x, y\} \mid s(x, y) = s(G)\}$. We use the symbols $P(x, y)$ and $p(x, y)$ to denote a longest and a shortest induced path between x and y, respectively. Sometimes, when no ambiguity occurs, we use $P(x, y)$ and $p(x, y)$ to denote the sets of nodes belonging to the corresponding paths. When $\langle P(x, y) \cup p(x, y) \rangle$ forms a cycle, we use $C_{(x,y)}$ to denote it.

In this paper, we use the *split composition* graph operation, the inverse of the decomposition operation introduced by Cunningham [8]. In the following we recall the split composition terminology.

Let G_1, G_2 be graphs having node sets $V_1 \cup \{m_1\}$, $V_2 \cup \{m_2\}$ and edge sets E_1, E_2, respectively, where $\{V_1, V_2\}$ is a partition of V and $m_1, m_2 \notin V$. The split composition of G_1 and G_2 is the graph $G = G_1 * G_2$ having node set V and edge set $E = E_1' \cup E_2' \cup \{(x, y) \mid x \in N(m_1), y \in N(m_2)\}$, where $E_i' = \{(x, y) \in E_i \mid x, y \in V_i\}$ for $i = 1, 2$. Nodes which the operation $*$ is applied to, i.e. m_1 and m_2, are called *marked nodes* of the split composition. In this paper we also use the version of Bouchet [3] where marked nodes m_1 and m_2 are joined by a marked edge (e.g., see Fig. 2).

3   A characterization of graphs in BID(k)

In this section we investigate about the relationships between class BID(k) and graph classes already known. We also provide a cycle-chord characterization for graphs belonging to BID(k).

Distance-hereditary graphs represent a well known graph class; it has been introduced and studied by Howorka [17] and further characterized in terms of the distance function, forbidden isometric subgraphs, generative operations, cycle-chord conditions, and others [2, 4, 9, 15].

The metric characterization of distance-hereditary graphs is: **a graph is distance-hereditary if and only if the lengths of any two induced paths joining the same pair of vertices are equal.** Hence, the following proposition is straightforward.

Proposition 1. *A graph G is distance-hereditary if and only if $G \in$ BID(1).*

The following proposition states the relationship between different classes of bounded induced distance graphs of distinct orders.

Proposition 2. $\mathrm{BID}(k_1) \subseteq \mathrm{BID}(k_2)$, *for each* $k_1 < k_2$.

The above two propositions suggest that bounded induced distance graphs can be also thought as a parametric extension of distance-hereditary graphs. Moreover, given a graph G with n nodes, it is easy to see that $G \in \mathrm{BID}((n-2)/2)$. In particular, if a cycle C_n is chordless, then $s(C_n) = (n-2)/2$. Notice that, if G is not distance-hereditary then $s(x,y) = s(G)$ implies $d(x,y) \geq 2$.

The following technical results will be used to prove the characterization of graphs belonging to the class $\mathrm{BID}(k)$ given in Theorem 2.

Theorem 1. *The class* $\mathrm{BID}(k)$ *is closed under induced subgraphs.*

Proof. Let G be a graph in $\mathrm{BID}(k)$, $x, y \in G$, and G' an induced subgraph of G such that $x, y \in G'$. Let $D'(x,y)$ and $d'(x,y)$ be the length of a longest and a shortest induced path between x and y in G', respectively. Since G belongs to $\mathrm{BID}(k)$ then $D(x,y) \leq k \cdot d(x,y)$, whereas relationships $D'(x,y) \leq D(x,y)$ and $d'(x,y) \geq d(x,y)$ hold in every graph. Hence, $D'(x,y) \leq k \cdot d'(x,y)$ and the theorem follows. $\square$

Lemma 1. *Let* G *be a graph, and* $\{x,y\} \in S(G)$. *If* $z \in P(x,y) \cap p(x,y)$, *and* $z \notin \{x,y\}$, *then* $\{x,z\} \in S(G)$ *or* $\{z,y\} \in S(G)$.

Proof. Let $k = s(G)$. By contradiction, let us suppose that $\{x,z\}, \{z,y\} \notin S(G)$. Then $D(x,z) < k \cdot d(x,z)$ and $D(z,y) < k \cdot d(z,y)$. Hence, $\frac{D(x,y)}{d(x,y)} = \frac{D(x,z)+D(z,y)}{d(x,z)+d(z,y)} < \frac{k \cdot d(x,z)+k \cdot d(z,y)}{d(x,z)+d(z,y)} < k$. This contradicts the hypothesis $\{x,y\} \in S(G)$. $\square$

Corollary 1. *Let* $G \in \mathrm{BID}(k)$, *and* $s(G) > 1$. *Then, there exists a pair* $\{x,y\} \in S(G)$ *such that* $\langle P(x,y) \cup p(x,y) \rangle$ *is a cycle.*

Proof. Let $\{x,y\} \in S(G)$. If $\langle P(x,y) \cup p(x,y) \rangle$ is not a cycle, then there exists a node z belonging both to $P(x,y)$ and to $p(x,y)$. Now, by Lemma 1, another pair of nodes having a distance less than $d(x,y)$ is in $S(G)$. If this pair determines a cycle we are done, otherwise we apply the previous lemma recursively until either we find a pair of nodes $\{u,v\} \in S(G)$ which determines the requested cycle or $d(u,v) = 1$. The latter case implies that $s(G) = 1$ and this contradicts the hypothesis $s(G) > 1$. $\square$

Corollary 2. *Let* $G_1, G_2, \ldots, G_n$ *be the subgraphs induced by the maximal biconnected components of a graph* G. *Then* $s(G) = \max_{1 \leq i \leq n} s(G_i)$.

Proof. Let $\{x,y\} \in S(G)$ a pair of nodes belonging to two different maximal biconnected components of G. If G is connected, then there exists an articulation point z belonging both to $P(x,y)$ and to $p(x,y)$. In this case, by Lemma 1, it follows that $\{x,z\} \in S(G)$ or $\{z,y\} \in S(G)$. Let us suppose $\{x,z\} \in S(G)$. If x and z are in the same component, the theorem is proved. Otherwise, we can apply the Lemma 1 recursively until we find a pair of nodes $\{u,v\}$ belonging to the same component and such that $\{u,v\} \in S(G)$.

If G is not connected, we can apply the same argumentations to any connected component having a pair of nodes in $\mathcal{S}(G)$. $\qquad\square$

The following theorem gives a characterization of graphs belonging to BID(k) for any given order k.

Theorem 2. *Let G be a graph and $k \geq 1$ a real number. Then, $G \in$ BID(k) if and only if $cd(C_n) > \left\lceil \frac{n}{k+1} \right\rceil - 2$ for each cycle C_n, $n > 2k + 2$, of G*

Proof. Only if case. By contradiction, let us suppose that a cycle C_n, $n > 2k+2$, exists in G such that $cd(C_n) \leq \left\lceil \frac{n}{k+1} \right\rceil - 2$. Let $(x, v_1, v_2, \ldots, v_q, y, u_1, u_2, \ldots u_p)$, $p + q + 2 = n$, be the cycle C_n, and $\{v_1, v_2, \ldots, v_q\}$ the set of nodes giving the chord distance of C_n; then, $d(x, y) \leq \left\lceil \frac{n}{k+1} \right\rceil - 1$. It follows that $d(x, y) + D(x, y) \leq$

$$d(x,y) + k \cdot d(x,y) \leq d(x,y)(k+1) \leq (k+1)(\left\lceil \tfrac{n}{k+1} \right\rceil - 1) < (k+1)\tfrac{n}{k+1} = n.$$

Hence, $D(x, y) < n - d(x, y)$, and this implies that there must exist chords in C_n not incident to any node in $\{v_1, v_2, \ldots, v_q\}$, contradicting the definition of chord distance.

If case. By contradiction, let us suppose $G \notin$ BID(k), and $\{x, y\} \in \mathcal{S}(G)$. By Corollary 1 it follows that $C_n \equiv C_{(x,y)}$, and by contradiction hypothesis we have that $D(x, y) > k \cdot d(x, y)$. Then, $n = D(x, y) + d(x, y) > k \cdot d(x, y) + d(x, y) = (k+1)d(x, y)$. This implies that

$$cd(C_n) > \left\lceil \tfrac{n}{k+1} \right\rceil - 2 \geq \left\lceil \tfrac{n}{k+1} \right\rceil - 1 \geq \frac{n}{k+1} - 1 \geq \frac{(k+1)d(x,y)}{k+1} - 1 = d(x,y) - 1$$

Hence, $cd(C_n) \geq d(x, y)$. As $d(x, y) - 1$ is the number of nodes between x and y in the cycle C_n, then there must exists a chord not incident to any node in $p(x, y)$; this chord joins nodes in $P(x, y)$, and this is a contradiction because $P(x, y)$ is an induced path. $\qquad\square$

The following result represents a reformulation of the well known crossing-chord characterization of distance-hereditary graphs [17]: G is distance-hereditary if and only if every cycle C_n, $n \geq 5$, in G has two crossing chords.

Corollary 3. *$G \in$ BID(1) if and only if $cd(C_n) > 1$ for every cycle C_n, $n \geq 5$, in G.*

Proof. Only if case. By contradiction, let us suppose that a cycle C_n, $n \geq 5$, such that $cd(C_n) \leq 1$. If $cd(C_n) = 0$ there are no chords in C_n, and trivially $G \notin$ BID(1). Otherwise, if $cd(C_n) = 1$ then there exists a node x incident to all the chords of the cycle. Let u and v the nodes adjacent to x in C_n. In this case $d(u, v) = 2$ and $D(u, v) > 2$, which is a contradiction because G is distance-hereditary.

If case. By contradiction, let us suppose $G \notin$ BID(1). Then, by Theorem 2 in [2] or by Theorem 4.2 in [15], G must contain, as induced subgraph, a cycle C corresponding to either a chordless cycle C_n, $n \geq 5$, or a fan, or a house, or a domino. For all these cycles, either $cd(C) = 0$ or there exists a node x such that all the chords are incident to it. This contradicts the hypothesis $cd(C) > 1$. $\qquad\square$

4   Graphs in BID(k) with small k

In this section we investigate the structure of graphs belonging to BID(k) when k is close to 1. The motivation is quite natural: we want to relax the restriction for which all the induced paths between two nodes have the same length (as happens in any distance-hereditary graph) and maintain a small ratio between the lengths of the longest and shortest path connecting any pair of nodes. In other words we are investigating graphs that are not distance-hereditary, but with a stretch number very small (e.g., graphs belonging to BID(k), $k \leq 2$).

Theorem 3. *Let G be a graph. If G is not distance-hereditary, then $s(G) \geq 3/2$.*

Proof. If G is not distance-hereditary, by Theorem 2 in [2] or by Theorem 4.2 in [15] G must contain, as induced subgraph, a chordless cycle C_n, $n \geq 5$, or a fan, or a house, or a domino. If G contains a chordless cycle C_n with at least 5 nodes, then $s(G) \geq (n-2)/2 \geq 3/2$. Fan and house are cycles with 5 nodes and chord distance equal to 1: then, their stretch number is exactly $3/2$. The domino graph is a cycle with 6 nodes and a chord: then its stretch number is 2. In any case, $s(G) \geq 3/2$.   □

Remark 1. Despite of the simpleness of its proof, Theorem 3 implies that in case of node failures no routing scheme on computer networks not based on distance-hereditary graphs is able to assure the delivery of messages through paths whose length is less than 1.5 times the best possible, that is, when no node has failed.

At this point, the class BID(3/2) assumes a great relevance. In the following, we give a characterization of this class based on forbidden induced graphs.

Theorem 4. *Let G be a graph. $G \in$ BID(3/2) if and only if the following graphs are not induced subgraphs of G: chordless cycles $C_n, n \geq 6$, cycles C_6 and C_7 with chord distance equal to 1, and cycles C_8 with chord distance equal either to 1 or 2.*

Proof. Only if part. All the considered cycles have a stretch number greater than $3/2$, and hence they are forbidden induced subgraphs for every graph belonging to BID(3/2).

If part. By contradiction, let us assume that $G \notin$ BID(3/2). This implies there are two nodes $x, y \in S(G)$ such that $s(x, y) > 3/2$. Let us assume that $C_{(x,y)}$ has n nodes. In the following we analyze three different cases: *(i)* $d(x, y) = 2$, *(ii)* $d(x, y) = 3$, and *(iii)* $d(x, y) \geq 4$.

(i) If $d(x, y) = 2$, then $D(x, y) > 3$ and $n \geq 6$. Being $d(x, y) = 2$, then $p(x, y) = (x, w, y)$ and any chord (if any) in $C_{(x,y)}$ is incident to w. If there are no chords in $C_{(x,y)}$, this contradicts the hypothesis. In the case of chords incident to w, n cannot be equal to 6, 7 or 8, by hypothesis. If $n > 8$ then any chord in $C_{(x,y)}$ generates an induced cycle with at least 6 nodes and $cd(C_{(x,y)}) \leq 1$. By recursively applying this property to this cycle, we reach either an empty cycle with more than 5 nodes or a forbidden graph with 6, 7 or 8 nodes, a contradiction of the hypothesis.

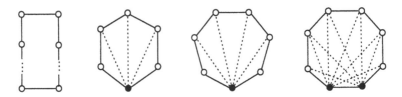

Fig. 1. *The forbidden induced subgraphs for* BID(3/2)*. In each cycle there must exist at least one of the dotted chords.*

(ii) If $d(x,y) = 3$, then $D(x,y) > 9/2$ and $n \geq 8$. Since $d(x,y) = 3$ then $p(x,y) = (x,v,w,y)$, and any chord in $C_{(x,y)}$ is incident to either v or w. Moreover, since $cd(C_{(x,y)}) \leq 2$ then $n \neq 8$. Assuming $n > 8$, $C_{(x,y)}$ must have at least one chord. Let us assume that v is incident to some chord in $C_{(x,y)}$.

Assuming $P(x,y) = (x,u_1,u_2,\ldots,u_t,y)$, let us consider the chord (v,u_i) when $i = \max_{1 \leq j \leq t}\{j \mid (v,u_j) \text{ is a chord in } C_{(x,y)}\}$. Chord (v,u_i) cuts $C_{(x,y)}$ into two cycles. The one containing x cannot be formed by more than 5 nodes, otherwise the corresponding induced graph is either forbidden or it is a graph considered in case *(i)*. Hence, the other cycle has $n-3 \geq 6$ nodes with every chord (if any) incident only to w. This cycle is an induced subgraph already considered in case *(i)*.

(iii) When $d(x,y) \geq 4$ we show that the graph induced by the nodes of $C_{(x,y)}$ contains an induced subgraph already considered in the previous cases. Let $p(x,y) = (x,v_1,v_2,\ldots,v_l,y)$ and $P(x,y) = (x,u_1,u_2,\ldots,u_t,y)$. As in case *(ii)*, let us consider the chord (v_2,u_i) when $i = \max_{1 \leq j \leq t}\{j \mid (v_2,u_j) \text{ is a chord in } C_{(x,y)}\}$.

This chord cuts $C_{(x,y)}$ into two cycles: the first containing x with m nodes, and the other one with m' nodes. In this case, m must be less or equal to 7, otherwise we are in the case *(ii)*. As a consequence we have that $m' \geq n-5$. By Theorem 2 we have that $l \leq \left\lceil \frac{n}{5/2} \right\rceil - 2$, and then $n > \frac{5}{2}(l+1)$. Now, by showing that $cd(C_{m'}) = l - 2 \leq \left\lceil \frac{m'}{5/2} \right\rceil - 2$, we prove that the cycle with m' nodes does not belong to BID(3/2). It follows that:

$$\left\lceil \frac{m'}{5/2} \right\rceil - 2 \geq \left\lceil \frac{n-5}{5/2} \right\rceil - 2 > \left\lceil \frac{(5/2)(l+1)-5}{5/2} \right\rceil - 2 = \lceil (l+1) - 2 \rceil - 2 = l - 3$$

Hence, $\left\lceil \frac{m'}{5/2} \right\rceil - 2 \geq l - 2$ implies that the subgraph induced by the cycle $C_{m'}$ is not in BID(3/2). Then, either it is a graph considered in the previous two cases or we can recursively apply the same argumentations to it.

If v_2 has no incident chords, then trivially $C_{(x,y)}$ has an induced cycle with at least m' nodes and chord distance less or equal to $l - 2$. Then we can apply to this cycle the argumentations of the previous three cases.

This concludes the proof.   □

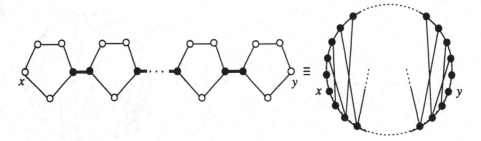

Fig. 2. *A graph $C_5 * C_5 * \cdots * C_5$, obtained by $i - 1$ chordless cycles C_5 obtained by split composition, having stretch equal to $2 - 1/i$.*

By computing the stretch number of every minimal forbidden subgraph for class BID(3/2), we can state the following:

Corollary 4. *Let G be a graph. If $G \notin \text{BID}(3/2)$, then $s(G) \geq 5/3$.*

According to all the previous results, as there exists no graph with stretch number between 1 and 3/2, and between 3/2 and 5/3, we are confident that the following conjecture holds.

Conjecture 1. There exists no graph G such that $2 - \frac{1}{i} < s(G) < 2 - \frac{1}{i+1}$, for each $i = 1, 2, 3, \ldots$.

Moreover, we are able to prove the following theorem:

Theorem 5. *There exists a graph G_i such that $s(G_i) = 2 - \frac{1}{i}$, for each $i = 1, 2, \ldots$.*

Proof. Assuming that C_5 denotes here the chordless cycle with five nodes, let us define the graph G_i.

$$\overbrace{\qquad\qquad\qquad}^{i-1 \text{ cycles}}$$

If $i = 1$, then $G_1 \equiv C_4$. If $i > 1$, then $G_i \equiv C_5 * C_5 * \cdots * C_5$, where marked nodes, belonging to the same cycle C_5, have distance 2. Figure 2 shows the graph G_i.

Now we have to prove that $s(G_i) = 2 - \frac{1}{i}$. For $i = 1$ the claim is trivially true, whereas for $i = 2$, $G_2 \equiv C_5$, and $s(G_2) = \frac{3}{2}$. When $i > 2$, let us consider the nodes x and y in Fig. 2. By definition of split composition, since there are exactly $i - 1$ cycles, it is easy to see that $\{x, y\} \in S(G_i)$. In fact, $d(x, y) = i$, $D(x, y) = 2i - 1$, and hence $s(G_i) = 2 - \frac{1}{i}$. $\qquad \square$

In the sequel of this section, we prove that the class BID(2) is quite different from classes BID(k), $k < 2$. In fact, assuming that the Conjecture 1 holds, likely all the classes properly included in BID(2) can be characterized by listing all their minimal forbidden subgraphs. For instance, for BID(1): chordless C_n with $n \geq 5$, cycles C_5 and C_6 with chord distance equal to 1; for BID(3/2): chordless C_n with $n \geq 6$, cycles C_6 and C_7 with chord distance equal to 1, and cycles

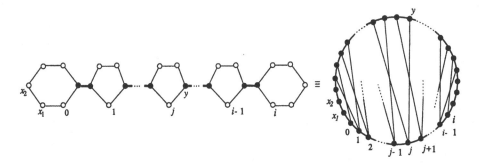

Fig. 3. *A minimal forbidden induced subgraph C_{7+3i} in class* BID(2) *having chord distance* $cd(C_{7+3i}) = i + 1$, $i = 1, 2, \ldots$

C_8 with chord distance equal to 1 or 2. Conversely, the following theorem shows that in BID(2) there are infinitely many different forbidden subgraphs with chord distance greater than zero.

Theorem 6. *For graphs in the class* BID(2) *there exist minimal forbidden induced cycles* C_{7+3i}, $i = 1, 2, 3, \ldots$, *with chord distance equal to* $i + 1$.

Proof. Let $i \geq 1$ be an integer. We build a cycle $G_i \equiv C_{7+3i}$, with $cd(G_i) = i+1$ by using the split composition:

$$G_i \equiv C_6 * \overbrace{C_5 * C_5 * \cdots * C_5}^{i-1 \text{ cycles}} * C_6$$

where C_6 and C_5 are chordless cycles, and the two marked nodes belonging to the same cycle C_5 have distance 2 (see Fig. 3). The resulting graph G_i has $7+3i$ nodes (the non-marked nodes) and chord distance $cd(G_i) = i + 1$. According to the characterization of graphs in BID(2) given in Theorem 2, since $\lceil \frac{7+3i}{3} \rceil - 2 = i+1$, then G_i is forbidden in BID(2).

To prove that G_i is minimal, we have to prove that $s(x, y) \leq 2$ for each pair $\{x, y\}$ of distinct and connected nodes of G_i such that the subgraph induce by $P(x, y) \cup p(x, y)$ does not coincide with G_i.

We have to consider three different cases: *(i)* neither x nor y belongs to the same cycle C_6, *(ii)* only x is in a cycle C_6, and *(iii)* x is in the first cycle C_6 and y in the second one.

(i) Since the longest and the shortest induced paths cannot include nodes of the C_6 cycles (otherwise they are not induced), we are in the same case of proof of Theorem 5 (see also Fig. 2), and then $s(x, y) < 2$.
(ii) In this case we have to consider just the pairs of nodes $\{x_1, y\}$ and $\{x_2, y\}$, where y belongs to the j-th cycle C_5, and x_1 and x_2 belong to the same cycle C_6. These nodes are pointed out in Fig. 3, and, since all the other possible pairs are symmetrical, they are not considered. As regard $\{x_1, y\}$, $d(x_1, y) = j + 2$ (one edge to reach node 0, one edge for each marked edge,

and one edge to reach y), and $D(x_1, y) = 2j + 4$ (3 edges in the first C_6 cycle, one edge for each marked edge, one edge for each C_5 cycle, and one edge to reach y) then $s(x_1, y) = 2$. As regard $\{x_2, y\}$, $d(x_2, y) = d(x_1, y) + 1 = j + 3$ and $D(x_2, y) = D(x_1, y) - 1 = 2j + 3$, then $s(x_1, y) < 2$.

(iii) In this case, the longest and shortest induced paths between x and y always form the same cycle of $7 + 3i$ nodes, which coincides with G_i.

This concludes the proof.   □

5   Recognition problem for the class BID(k)

Although Theorem 2 provides a characterization for graphs with bounded induced distance of a generic order k, it cannot be used to devise an efficient algorithm to solve the recognition problem for the class BID(k). Moreover, the following complexity result can be shown.

Definition 3. Stretch Number *Problem:*
INSTANCE: *A graph $G = (V, E)$, a rational number $q \geq 1$.*
QUESTION: *Is the stretch number of G greater than q?*

The NP-completeness of this problem can be shown by providing a polynomial transformation from the NP-complete problem *Induced Path* [14], that can be formally defined as follows:

INSTANCE: A graph $G = (V, E)$, a positive integer $k \leq |V|$.
QUESTION: Is there a subset $P \subseteq V$ with $|P| \geq k$ such that the subgraph induced by P is an induced path on $|P|$ nodes ?

Theorem 7. *Stretch Number is NP-complete.*

Proof. It is easy to see that the Stretch Number problem belongs to NP, as given a pair of paths joining two nodes in V it is possible to check in polynomial time whether the ratio of their lengths is greater than q.

Given a graph $G = (V, E)$ and a positive integer k representing an instance of Induced Path, we construct in polynomial time a graph G' and define a rational number q such that there is the required induced path in G if and only if $s(G')$ is greater than q.

The reduction graph $G' = (V', E')$ is obtained as follows: add a pendant node $\bar{v}$ to each node $v \in V$, forming an independent set $W = \{\bar{v} \mid v \in V\}$. Then connect all the nodes in $V \cup W$ to a new node u. Formally, $V' = V \cup W \cup \{u\}$, V, W and $\{u\}$ are pairwise disjoint sets with $|W| = |V|$, and $E' = E \cup \{(v, \bar{v}) \mid v \in V, \bar{v} \in W\} \cup \{(u, v) \mid v \in V \cup W\}$ (see Fig. 4). Concerning the rational number q, it is given by $q = (k + 1)/2$.

Now we prove that the instance of Induced Path has a positive answer if and only if $s(G') > q$.

Only if case. Let us assume that the instance of Induced Path has a positive answer. This implies that an induced path $p = (v_1, v_2, \ldots, v_n)$ exists in $\langle V \rangle$ such

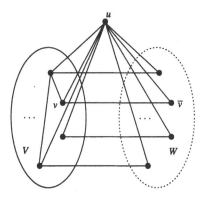

Fig. 4. *The graph G' built using the instance $G = (V, E)$ of the Induced Path problem. W is an independent set containing a "copy" $\bar{v}$ for each node $v \in V$.*

that $|p| \geq k$. Then the path $\bar{p} = (\bar{v}_1, v_1, \ldots, v_n, \bar{v}_n)$ is also an induced path in G' and $|\bar{p}| \geq k + 2$.

By definition of G', nodes $\bar{v}_1$ and $\bar{v}_n$ are not connected, and since they are both adjacent to u, then $d(\bar{v}_1, \bar{v}_n) = 2$. Hence, the following relation holds:

$$\frac{D(\bar{v}_1, \bar{v}_n)}{d(\bar{v}_1, \bar{v}_n)} \geq \frac{k+2}{2} > \frac{k+1}{2} = q$$

This implies that the instance of *Stretch Number* has a positive answer.

If case. Let us assume that *Stretch Number* has a positive answer, that is $s(G') > q$. By definition of stretch number there exist two nodes $x, y \in G'$ such that $s(x, y) > q$. Vertices x and y cannot be adjacent otherwise $s(x, y) = 1$. For the same reason, neither x nor y can coincide with u, being u adjacent to each other node in G'. Then, $d(x, y) = 2$ and, as consequence, $D(x, y) > 2q = k + 1$.

Let $p = (x, v_1, \ldots, v_n, y)$ be an induced path between x and y whose length is equal to $D(x, y)$. This path p cannot include u, since the only induced path including u is (x, u, y); but in this case $D(x, y) = 2$, contradicting the relation $D(x, y) > 2q$ just found. Hence, x, y and v_i, $1 \leq i \leq n$, are elements of $V \cup W$. Moreover, since the elements of W are pendant nodes in $\langle V \cup W \rangle$, then $v_i \notin W$, $1 \leq i \leq n$.

Now, three different cases arise, according to the membership of x and y to W.

- Both x and y are in V. In this case p is an induced path in G, and since $|p| > k + 1$, then p itself is a solution for the instance of the *Induced Path* problem.
- $x \in V$ and $y \in W$. In this case $p' = (x, v_1, \ldots, v_n)$ is an induced path in G, and since $|p| > k + 1$, then $|p'| > k$ and p' is a solution for the instance of the *Induced Path* problem.
- Both x and y are in W. In this case $p'' = (v_1, \ldots, v_n)$ is an induced path in G, and since $|p| > k + 1$, then $|p''| \geq k$ and p'' is a solution for the instance of the *Induced Path* problem.

This implies that the instance of *Induced Path* has a positive answer.   □

Since the recognition problem for the class $\mathrm{BID}(k)$ is exactly the complementary problem of *Stretch Number*, we can state the following complexity result.

Corollary 5. *The recognition problem for the class* $\mathrm{BID}(k)$ *is Co-NP-complete.*

6 Building graphs in BID(k)

The complexity result of the recognition problem for the class $\mathrm{BID}(k)$ makes difficult to devise operations able to build *all* the graphs with bounded induced distance of order k. In this section, we introduce simple operations able to yield graphs representing a part of the class under consideration.

The following theorem gives an upper bound to the order of the class which a graph belongs to when it is obtained by a split composition.

Theorem 8. *Let* G_1, G_2 *be two graphs belonging to* $\mathrm{BID}(k_1)$ *and* $\mathrm{BID}(k_2)$, *respectively. Then,*

$$G_1 * G_2 \in \mathrm{BID}(k), \ \ where \ k = \max\{k_1, k_2, \frac{2(k_1 + k_2) - 1}{3}\}$$

Proof. Let m_1 and m_2 the marked nodes in G_1 and G_2, respectively. If $x \in G_1 - N(m_1)$ and $y \in G_2 - N(m_2)$, then $D(x, y) = D(x, m_1) + D(m_2, y) - 1$ and $d(x, y) = d(x, m_1) + d(m_2, y) - 1$. Since $D(x, m_1) \leq k_1 \cdot d(x, m_1)$ and $D(m_2, y) \leq k_2 \cdot d(m_2, y)$, then $D(x, m_1) + D(m_2, y) - 1 \leq k_1 \cdot d(x, m_1) + k_2 \cdot d(m_2, y) - 1 \leq k_1(d(x, m_1) + d(m_2, y) - 1) + k_2(d(x, m_1) + d(m_2, y) - 1) + k_1(1 - d(m_2, y)) + k_2(1 - d(x, m_1)) - 1$. Then, $s(x, y) \leq k_1 + k_2 + (k_1(1 - d(m_2, y)) + k_2(1 - d(x, m_1)) - 1)/(d(x, m_1) + d(m_2, y) - 1) \leq k_1 + k_2 + (-k_1 - k_2 - 1)/3 = (2(k_1 + k_2) - 1)/3$. Hence, $s(G_1 * G_2) \leq \max\{k_1, k_2, \frac{2(k_1+k_2)-1}{3}\}$.   □

Notice that this theorem provides a tight upper bound to the stretch number of the graph $G_1 * G_2$. In fact, if the cycle C_5 is chordless, then $s(C_5) = 3/2$ and $s(C_5 * C_5) = \frac{2(\frac{3}{2} + \frac{3}{2}) - 1}{3} = \frac{5}{3}$.

Corollary 6. *Let* G *be a graph in* $\mathrm{BID}(k)$. *If* G' *belongs to* $\mathrm{BID}(\frac{k+1}{2})$, *then* $G * G' \in \mathrm{BID}(k)$.

This corollary can be used to build graphs belonging to $\mathrm{BID}(k)$ by split composition with graphs having stretch not greater then $(k + 1)/2$.

Corollary 7. *The class of distance-hereditary graphs is closed under split composition. The class* $\mathrm{BID}(k)$ *is closed under extension by distance-hereditary graphs via split composition.*

7 Conclusions

In this paper we have introduced graph classes that represent a parametric extension of the class of distance-hereditary graphs. In any graph belonging to the generic new class $BID(k)$, the length of every induced path between any pair of nodes is at most k times the length of the shortest one between the same nodes. These graphs can model communication networks with shortest paths routing, and in which node failures can occur: any message is always delivered through a path (that, due to node failures, could be longer than the shortest one) whose length is at most k times the best possible.

In spite of the results provided in this work, many questions are left open:

1. Can the Conjecture 1 be proven?
2. The recognition problem for $BID(1)$ can be solved in linear time [2, 15], whereas it is Co-NP-complete for the generic case (Theorem 5). What is the largest constant k such that the recognition problem for $BID(k)$ can be solved in polynomial time?
3. Many other combinatorial problems are solvable in polynomial time for $BID(1)$. Can some of these results be extended to $BID(k)$, $k > 1$?
4. Can the characterization of graphs in $BID(3/2)$ given in Theorem 4 be extended to other classes $BID(k)$, $k < 2$?
5. Is it possible to characterize graphs in $BID(k)$, $k < 2$, by split decomposition?
6. Can some results of this work be extended to graphs such that $D(x,y) \leq f(d(x,y))$ (e.g., $f(d(x,y)) = d(x,y) + k$)?
7. Is it possible to define compact routing schemes (or other kinds of routing schemes) for networks based on graphs in $BID(k)$?

References

1. H. J. Bandelt, A. D'Atri, M. Moscarini, H. M. Mulder, and A. Schultze. Operations on distance hereditary graphs. Technical Report 226, CNR, Istituto di Analisi dei Sistemi e Informatica del CNR (Rome Italy), 1988.
2. H. J. Bandelt and M. Mulder. Distance-hereditary graphs. *Journal of Combinatorial Theory, Series B*, 41(2):182–208, 1986.
3. A. Bouchet. Transforming trees by successive local complementations. *Journal of Graph Theory*, 4:196–207, 1988.
4. A. Brandstädt. Special graph classes. Technical Report SM-DU-199, University Duisburg, 1993.
5. A. Brandstädt and F. F. Dragan. A linear time algorithm for connected r-domination and Steiner tree on distance-hereditary graphs. Technical Report SM-DU-261, University Duisburg, 1994.
6. S. Cicerone and G. Di Stefano. Graph classes between parity and distance-hereditary graphs. In *1st Discrete Mathematics and Theoretical Computer Science (DMTCS'96)*, Combinatorics, Complexity, and Logic, pages 168–181. Springer-Verlag, 1996.
7. S. Cicerone and G. Di Stefano. Port and node support to compact routing. Technical Report R.97-17, Dipartimento di Ingegneria Elettrica, Università di L'Aquila (L'Aquila, Italy), 1997.

8. W. H. Cunningham. Decomposition of directed graphs. *SIAM Journal on Alg. Disc. Meth.*, 3:214–228, 1982.

9. A. D'Atri and M. Moscarini. Distance-hereditary graphs, steiner trees, and connected domination. *SIAM Journal on Computing*, 17:521–530, 1988.

10. G. Di Stefano. A routing algorithm for networks based on distance-hereditary topologies. In *3rd Int. Colloquium on Structural Information and Communication Complexity (SIROCCO'96)*, 1996.

11. F. F. Dragan. Dominating cliques in distance-hereditary graphs. In *SWAT: Scandinavian Workshop on Algorithm Theory*, volume 824, pages 370–381. Lecture Notes in Computer Science, 1994.

12. A. H. Esfahanian and O. R. Oellermann. Distance-hereditary graphs and multidestination message-routing in multicomputers. *Journal of Comb. Math. and Comb. Computing*, 13:213–222, 1993.

13. G. N. Frederickson and R. Janardan. Designing networks with compact routing tables. *Algorithmica*, 3:171–190, 1988.

14. M.R. Garey and D.S. Johnson. *Computers and Intractability. A Guide to the Theory of NP-completeness*. W.H. Freeman, 1979.

15. P. L. Hammer and F. Maffray. Completely separable graphs. *Discrete Applied Mathematics*, 27:85–99, 1990.

16. F. Harary. *Graph Theory*. Addison-Wesley, 1969.

17. E. Howorka. Distance hereditary graphs. *Quart. J. Math. Oxford*, 2(28):417–420, 1977.

18. J. M. McQuillan, I. Richer, and E. C. Rosen. The new routing algorithm for the ARPANET. *IEEE Trans. Commun.*, COM-28:711–719, 1980.

19. P. M. Merlin and A. Segall. A failsafe distributed routing protocol. *IEEE Trans. Commun.*, COM-27:1280–1289, 1979.

20. F. Nicolai. Hamiltonian problems on distance-hereditary graphs. Technical Report SM-DU-264, University Duisburg, 1994.

21. D. Peleg and A. Schaffer. Graph spanners. *Journal of Graph Theory*, 13:99–116, 1989.

22. M. Schwartz and T. E. Stern. Routing techniques used in computer communication networks. *IEEE Trans. Commun.*, COM-28:539–552, 1980.

23. J. van Leeuwen and R.B. Tan. Interval routing. *The Computer Journal*, 30:298–307, 1987.

24. H. G. Yeh and G. J. Chang. Weighted connected domination and steiner trees in distance-hereditary graphs. In *Franco-Japanese and (n-4th) Franco-Chinese Conference Combinatorics and Computer Science*, volume 1120 of *Lecture Notes in Computer Science*, pages 48–52. Springer-Verlag, 1996.

Diameter Determination on Restricted Graph Families

Derek G. Corneil[1], Feodor F. Dragan[2], Michel Habib[3], and Christophe Paul[3]

[1] Dpt. of Computer Science, University of Toronto, Toronto M5S 3G4, Canada
[2] Dpt. of Computer Science, University of Rostock, D-18051 Rostock, Germany
[3] LIRMM, UMR CNRS-Universit de Montpellier II, 161 rue Ada, 34392 Montpellier Cedex 5, France

Abstract. Determining the diameter of a graph is a fundamental graph operation, yet no efficient (*i.e.* quadratic time) algorithm is known. In this paper, we examine the diameter problem on chordal and AT-free graphs and show that a very simple (linear time) 2-sweep Lex-BFS algorithm identifies a vertex of maximum eccentricity unless the given graph has a specified induced subgraph (it was previously known that a single Lex-BFS algorithm is guaranteed to end at a vertex that is within 1 of the diameter for chordal and AT-free graphs). As a consequence of the forbidden induced subgraph result on chordal graphs, our algorithm is guaranteed to work optimally for directed path graphs (it was previously known that a single LexBFS algorithm is guaranteed to work optimally for interval graphs).

1 Introduction

Recently considerable attention has been given to the problem of developing fast and simple algorithms for various classical graph problems. The motivation for such algorithms stems from our need to solve these problems on very large input graphs, thus the algorithms must be not only fast, but also easily implementable. Determining a graph diameter is a classical and well-known problem. For arbitrary graphs, as well as for various restricted graph families, the current fastest algorithm for this problem achieves the time bound of $O(nm)$ (see for example [12]) which is too slow to be practical for very large graphs.

In this paper, we study the problem of determining a vertex of high eccentricity for chordal and AT-free graphs. The *eccentricity* of a vertex x is $ecc(x) = max_{y \in V} d(x, y)$ where $d(x, y)$ denotes the distance between x and y. The *diameter* of a graph equals the maximum eccentricity achieved by any vertex in the graph. Given v, a vertex of maximum eccentricity, it is trivial to determine the set of vertices whose distance from v equals the diameter of G (*i.e.* these vertices constitute the last layer of a Breadth First Search, (BFS) from v). A graph is *chordal* iff there is no chordless cycle of length more than 3. It is well-known that chordal graphs are exactly the intersection graphs of subtrees in trees [1, 7]. Interval graphs can be defined as the intersection graphs of subpath in paths (see [11]). A natural generalization of interval graphs is the concept of directed path graphs. A graph is a *directed path graph* iff it is the intersection graph of a collection of directed paths in a rooted directed tree [8]. Three vertices u, v, w are an *asteroidal triple* (AT) if between any two of them, there exists a path that avoids the neighbourhood of the remaining vertex. A graph is *AT-free* if it does not contain an AT.

The algorithm that we present involves two sweeps of the well-known Lexicographic Breadth First Search (Lex-BFS) introduced by Rose, Tarjan and Lueker [14] (see algorithm 1) for the recognition of chordal graphs. An example of a Lex-BFS sweep is presented in figure 8. It is somewhat surprising that Lex-BFS seems to play a fundamental role for both chordal and AT-free graphs, two families that exhibit very little structural similarity (see for example [3, 4, 7, 13, 14]). Furthermore Dragan et al [5] and Dragan [6] have shown that the eccentricity of the vertex visited last in a Lex-BFS is within 1 of the diameter for chordal and AT-free graphs respectively, and is the diameter for interval graphs.

Algorithm 1: Lexicographic Breadth First Search (Lex-BFS(z)) [14]

Input: A graph $G = (V, E)$ with $z \in V$

Output: An ordering σ of the vertices of V, with z the last element.

begin

> assign the label $\{n + 1\}$ to vertex z and $\emptyset$ to every other vertex ;
> **for** $i = n$ *to 1* **do**
>> pick an unnumbered vertex x with the largest label in the lexicographic order (number x by i);
>> **for** *each unnumbered neighbour y of x* **do**
>>> add i to $label(y)$;
>> $\sigma(i) \leftarrow x$;

end

Note that Lex-BFS can be started from any vertex of the graph G. We will denote by Lex-BFS(w) a Lex-BFS started from vertex w. In this paper, we examine the following very simple 2-sweep Lex-BFS algorithm and study its performance on chordal and AT-free graphs.

In particular, we examine conditions when $ecc(v) = diam(G) - 1$, where v is the vertex returned by the 2-sweep algorithm. These conditions include forbidden subgraph results for both chordal and AT-free graphs. The forbidden subgraph result for chordal graphs immediately shows that the algorithm works optimally for directed path graphs.

Before presenting these results, we show that it is unlikely that the diameter problem on either chordal or AT-free graphs can be solved in quadratic time. To do this we introduce the disjoint sets problem.

Algorithm 2: 2-sweep Lex-BFS

Input: A graph G
Output: A vertex v
begin
  Let w be an arbitrary vertex;
  $u \leftarrow$ the last vertex numbered by Lex-BFS(w);
  $v \leftarrow$ the last vertex numbered by Lex-BFS(u);
  return v;
end

2   Disjoint Sets Problem

Given $\mathcal{S} = \{S_1, S_2, \ldots, S_n\}$ sets over the base set $\mathcal{X}$, the Disjoint Sets Problem (DSP) asks whether there exist i and j such that $S_i \cap S_j = \emptyset$. To our knowledge, there is no algorithm for this problem with a better running time than the order of Boolean Matrix Multiplication (BMM). As pointed out by Chepoi [2], a fast algorithm (*i.e.* in quadratic time or better) for determining whether a split graph (and thus a chordal graph) has diameter 2 or 3 would imply a fast algorithm (*i.e.* faster than BMM) for the DSP (see figure 1).

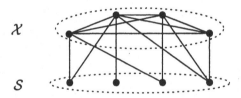

Fig. 1. The set $\mathcal{X}$ is represented by a clique and $\mathcal{S}$ by an independent set. A set S_i is adjacent to its elements in $\mathcal{X}$. The diameter of this graph is 3 iff there exist two disjoint sets in $\mathcal{S}$.

In figure 2, a similar transformation is presented to show that the diameter equals 2 or 3 problem on co-comparability graphs (and thus AT-free graphs) would have the same impact on the DSP.

Thus it seems unlikely that a linear or quadratic time algorithm exists for the diameter problem on either chordal or AT-free graphs. We now present the main results of our paper.

3   Results

The proofs of the results will be included in the journal version of this paper. First we note that the algorithm is guaranteed to find the diameter for arbitrary graphs, if the diameter equals 2.

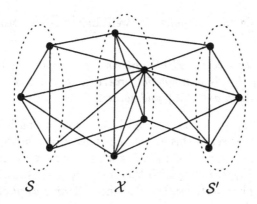

Fig. 2. The set $\mathcal{X}$ is represented by a clique. Two copies of S are also represented by two cliques. A set S_i (resp. S_i') is adjacent to its elements in $\mathcal{X}$. The diameter of this graph is 3 iff there exist two disjoint sets in S.

Proposition 1 *Let G be an arbitrary graph and u be the vertex of G visited last by a Lex-BFS. If $diam(G) = 2$, then $ecc(u) = diam(G)$.*

Although the algorithm does not guarantee a maximum eccentricity vertex for chordal and AT-free graphs, the following lemma shows that such a vertex is in the last BFS layer from v, the vertex returned by algorithm 2.

Lemma 1. *If G is chordal or AT-free and algorithm 2 returns vertex v, then there exists a vertex x of maximum eccentricity such that $d(v, x) = ecc(v)$.*

The distance between a vertex x and a set of vertices S, denoted by $d(S, x)$, is the minimum distance between x and a vertex of S. The following easy property of Lex-BFS holds for arbitrary graphs.

Lemma 2. *Let S_i be the numbered vertices at step i of Lex-BFS. If $x \notin S_i$ and $y \notin S_i$ are two vertices such that $d(S_i, x) < d(S_i, y)$, then x will be numbered before y.*

3.1   Chordal Graphs

Let $\sigma = (v_1, v_2, \ldots, v_n)$ be an ordering of the vertex set of a graph G. We write $a < b$ whenever in a given ordering σ vertex a has a smaller number than vertex b. Moreover, $\{a_1, \cdots, a_l\} < \{b_1, \cdots, b_k\}$ is an abbreviation for $a_i < b_j$ $(i = 1, \cdots, l; j = 1, \cdots, k)$. An ordering of the vertex set of a graph G generated by LexBFS is called a *LexBFS–ordering*.

In what follows we will often use the following property (cf.[10]) :

(P1)   If $a < b < c$ and $ac \in E$ and $bc \notin E$ then there exists a vertex d such that $c < d$, $db \in E$ and $da \notin E$.

It is well–known that any LexBFS–ordering has property (P1) [9]. Moreover, any ordering fulfilling (P1) can be generated by LexBFS [5].

The following lemma presents the well known characterization of chordal graphs.

Lemma 3. *[14] Let σ be a LexBFS–ordering of a graph G. Then G is a chordal graph if and only if σ is a simplicial ordering, that is : $bc \in E$ holds for all vertices a, b and c with $a < \{b, c\}$ and $ab, ac \in E$.*

Let $P = (x_0 - x_1 - \cdots - x_{k-1} - x_k)$ be an arbitrary path of G and σ be an ordering of the vertex set of this graph. The path P is *monotonic* (with respect to σ) if $x_0 < x_1 < \cdots < x_{k-1} < x_k$ holds whenever $x_0 < x_k$, and P is *convex* if there is an index i ($1 \leq i < k$) such that $x_0 < x_1 < \cdots < x_{i-1} < x_i > x_{i+1} > \cdots > x_{k-1} > x_k$. Then vertex x_i is called the *switching point* of the convex path P.

In the sequel of this subsection we assume that G is a chordal graph and σ is a LexBFS–ordering of G.

By lemma 3 no induced path $P = (x_0 - \cdots - x_k)$ of G can contain a vertex x_j ($1 \leq j < k$) with $x_{j-1} > x_j < x_{j+1}$. Hence, we have the following.

Lemma 4. *Every induced path of G is either monotonic or convex.*

Now let $P = (x_0 - \cdots - x_k)$ be a shortest path of G connecting x_0 and x_k. We say that P is a *rightmost shortest path* if the sum $x_0 + x_1 + \cdots + x_k$ of the positions of $x_0, \cdots, x_k$ in σ is largest among all shortest paths connecting x_0 and x_k.

Lemma 5. *Let $P = (x_0 - \cdots - x_{2k})$ be a shortest path in G such that the subpath $P' = (x_k - \cdots - x_{2k})$ of P is a rightmost shortest path connecting x_k and x_{2k}. If $x_0 < x_{2k}$ and x_k is the switching point of P, then $x_{k+j} > x_{k-j}$ holds for each j ($1 \leq j \leq k$).*

Lemma 6. *Let $P = (x_0 - \cdots - x_k)$ be a rightmost shortest path in G which is convex and x_i be the switching point of P. Then $d(x_0, x_i) \geq d(x_k, x_i)$ whenever $x_0 < x_k$.*

Lemma 7. *Let u be the vertex of a chordal graph G visited last by a LexBFS. For every two vertices x and y of G such that $d(x, u) \leq d(y, u)$, $d(x, y) \leq d(y, u) + 1$ holds. Moreover, if $d(x, y) = d(y, u) + 1$ then $d(y, u) = d(x, u)$ and $d(y, u)$ is even.*

Theorem 1 *Let u be the vertex of a chordal graph G visited last by a LexBFS, and x, y be a pair of vertices such that $d(x, y) = diam(G)$. If $ecc(u) < diam(G)$ then $ecc(u)$ is even, $d(u, x) = d(u, y) = ecc(u)$ and $ecc(u) = diam(G) - 1$.*

We continue with rather surprising results concerning the parity of the diameter of the graph and the parity of the eccentricity of the vertex visited last by LexBFS.

Corollary 1 *If the diameter of a chordal graph G is even, then the vertex visited last by a LexBFS has eccentricity equal to $diam(G)$.*

Corollary 2 *If the vertex u of a chordal graph G visited last by a LexBFS has odd eccentricity, then ecc(u) = diam(G).*

The main result of this subsection is the following.

Theorem 2 *If G is a chordal graph and if v, the vertex returned by algorithm 2, is not of maximum eccentricity, then G contains either an induced 3-sun or an induced 4-sun (see figure 3) or one of the graphs in figure 4 as an induced subgraph.*

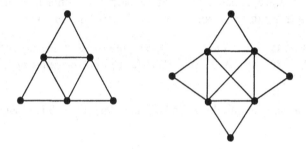

Fig. 3. The 3-sun and the 4-sun.

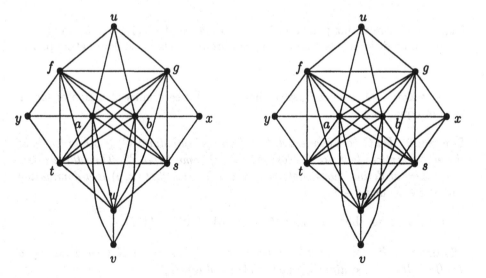

Fig. 4. Strongly chordal graphs where 2 sweeps of LexBFS are not enough to find the diameter.

Since none of the graphs from figure 3 and figure 4 is a directed path graph, this theorem immediately yields the following corollary.

Corollary 3 *Algorithm 2 finds a vertex of maximum eccentricity for directed path graphs.*

3.2   AT-free Graphs

We now turn our attention to AT-free graphs and start by recalling some known results. A pair of vertices (x, y) is said to be a *dominating pair* if for every x, y path P and every vertex $z \in V$, $N(z) \cap P \neq \emptyset$. In [4], it was shown that every connected AT-free graph has a dominating pair. For sufficiently high diameter AT-free graphs this can be strengthened to the "polar lemma".

Lemma 8. *[4] If G is a connected AT-free graph with $diam(G) > 3$ then there exists disjoint vertex sets X, Y such that (x, y) is a dominating pair of G iff $x \in X$ and $y \in Y$.*

The fact that this lemma does not hold for $diam(G) = 3$ is illustrated by the graph in figure 5.

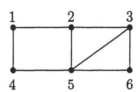

Fig. 5. The dominating pairs are $(1, 3), (1, 6), (4, 3), (4, 6), (1, 5), (2, 5)$ and $(4, 5)$. There are no disjoint sets whose cartesian product defines all dominating pairs.

A weaker version does however hold for AT-free graphs of diameter larger than or equal to 3.

Lemma 9. *Let G be a connected AT-free graph with $diam(G) \geq 3$ and let V_1 be the set of vertices that are the last vertices of some Lex-BFS. Then there exists a partition of V_1 into non-empty sets X and Y such that (x, y) is a dominating pair if $x \in X$ and $y \in Y$.*

For the graph shown in figure 5, $X = \{6\}$ and $Y = \{1, 4\}$.

Theorem 3 *[6] Let u be the vertex of an AT-free graph G visited last by a LexBFS, If $ecc(u) < diam(G)$ then $ecc(u) = diam(G) - 1$.*

The next proposition presents further facts about the structure of the AT-free graphs.

Proposition 2 *Let G be an AT-free graph with $diam(G) = k > 2$. If $ecc(v) = k - 1$, where v is the vertex returned by algorithm 2, and u', v' achieve the diameter, then :*

1. *every u', v' shortest path is vertex disjoint from every u, v shortest path,*
2. *$d(u, v) = d(u, v') = d(u', v) = k - 1$,*
3. *$uu' \in E$ and $vv' \in E$.*

Before presenting the final result on AT-free graphs, we introduce the notion of an h-ladder and an h-*ladder.

Definition 1. *An h-ladder consists of a chain of h 4-cycles where the 4-cycles are attached as shown in figure 6. In an h-*ladder the first 4-cycle has a diagonal.*

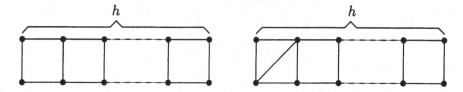

Fig. 6. The h-ladder and the h-*ladder.

Theorem 4 *If G is an AT-free graph with $diam(G) = k \geq 3$ and $ecc(v) = k-1$, where v is the vertex returned by algorithm 2, then G contains an induced $(k-1)$-ladder or an induced $(k-1)$-*ladder.*

Note that this theorem considerably strengthens the following result by Dragan [6]. An HHD-free graph does not contain an induced house (complement of P_5) or an induced hole (a cycle of length at least 5) or an induced domino (a 2-ladder).

Theorem 5 *[6] If G is an HHD-free, AT-free graph, then the vertex visited last by a Lex-BFS has maximum eccentricity.*

4   Concluding Remarks

First of all, the reader should note a kind of duality in the results when algorithm 2 finds a vertex whose eccentricity is not maximum. For chordal graphs, eacg of the forbidden subgraphs (the 3-sun and the 4-sun) has an AT. For AT-free graphs, the h-ladder and the h-*ladder are built with 4-cycles, the smallest non-chordal graph.

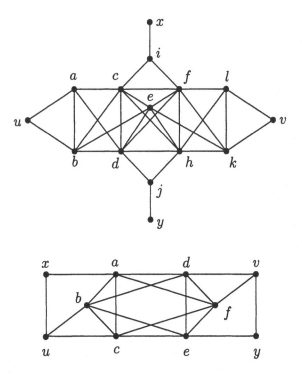

Fig. 7. A chordal graph and an AT-free graph where an infinite number of Lex-BFS sweeps never end at a maximum eccentricity vertex.

Having seen the power of the 2-sweep Lex-BFS algorithm, it is natural to ask whether significant improvements can be achieved by performing c sweeps for some $c > 2$. In particular, can we find a vertex of maximum eccentricity, although in light of the results of section 2, this is highly unlikely for c a constant? As shown by the graphs in figure 7, for no c, is the c-sweep algorithm guaranteed to find a vertex of maximum eccentricity. The first graph is chordal, the second AT-free. In both graphs any Lex-BFS starting at u must end at v and vice versa. Thus if the initial choice of vertex is either u or v, a multi-sweep Lex-BFS algorithm will forever alternate between u and v, thereby missing x and y, the two vertices of maximum eccentricity.

A second obvious question concerns the power of the 2-sweep algorithm on arbitrary graphs. Unfortunately, the answer again is negative. In particular, for any $i > 1$, there is a graph G_i where $ecc(v) = diam(G_i) - 2^{i-1} + 1$, where v is the vertex returned by algorithm 2. We construct G_i as follows: Let T_1 be a 2-leaf tree with root r_1. T_i, $i > 1$, is formed from two copies of T_{i-1} by making r_i, the root of T_i, adjacent to the two r_{i-1} roots. Each $r_i r_{i-1}$ edge then has $2^{i-2} - 1$ new vertices inserted. Finally G_i is formed from T_i by creating a path on the leaves of T_i in the obvious way. G_4 is shown in figure 8. If w is the leftmost leaf of the right T_{i-1} and the next vertex choosen in the Lex-BFS from w is the rightmost leaf of the left T_{i-1}, then the Lex-BFS will end at $u = r_i$ (see figure 8). If the second Lex-BFS starts at u and breaks ties by choosing the last eligible vertex in the previous sweep, then v, the last vertex, is the same as w (see figure 9). It is easy to see that $ecc(v) = 2^{i-1}$ and $diam(G_i) = 2^i - 1$ as witnessed by the extreme leaves.

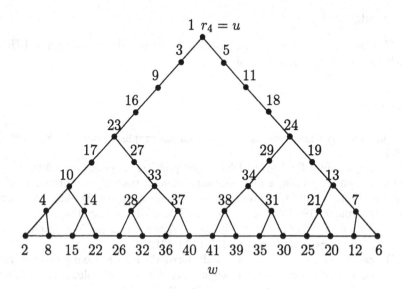

Fig. 8. G_4 together with the first Lex-BFS.

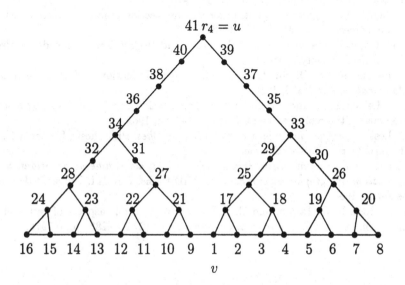

Fig. 9. G_4 and the second Lex-BFS.

As a final comment, we note that the results in this paper add to the growing evidence of the similar roles played by Lex-BFS for chordal and AT-free graphs. It would be interesting to find a structural result to explain this surprising phenomenon.

Acknowledgments

Derek G. Corneil wishes to thanks NSERC for financial assistance and LIRMM for their hospitality during his visit.

References

1. P. Buneman. A characterization of rigid circuit graphs. *Discr. Math.*, 9:205–212, 1974.
2. V.D. Chepoi and F.F. Dragan. Disjoint set problem. unpublished, 1992.
3. D.G. Corneil, S. Olariu, and L. Stewart. Linear time algorithms for dominating pairs in asteroidal triple-free graphs. Technical Report 294–95, University of Toronto, January 1995. To appear in SIAM Journal of Computing.
4. D.G. Corneil, S. Olariu, and L. Stewart. Asteroidal triple-free graphs. *SIAM Journal on Discrete Mathematics*, 10(3):399–431, 1997.
5. F. Dragan, F. Nicolai, and A. Brandstädt. Lex-BFS orderings and powers of graphs. In *Graph-Theoretic Concepts in Computer Science, WG'96*, volume 1197 of *Lecture Notes in Computer Science*, pages 166–180. WG'96, 1997.
6. Feodor Dragan. Almost diameter in hhd-free graphs in linear time via lex-bfs. In *Optimal Discrete Structures and Algorithms*. University of Rostock, Germany, 1997.
7. F. Gavril. The intersection graphs of a path in a tree are exactly the chordal graphs. *Journ. Comb. Theory*, 16:47–56, 1974.
8. F. Gavril. A recognition algorithm for the intersection graph of directed paths in directed trees. *Discr. Math.*, 13:337–349, 1975.
9. M.C. Golumbic. *Algorithms Graph Theory and Perfect Graphs*. Academic Press, New York University, 1980.
10. B. Jamison and S. Olariu. On the semi-perfect elimination. *Advances in Applied Mathematics*, 9:364–376, 1988.
11. C.G. Lekkerkerker and J.C Boland. Representation of a finite graph by a set of intervals on the real line. *Fund. Math.*, 51:45–64, 1962.
12. M. Lesk. *Couplages maximaux et diamtre de graphes*. PhD thesis, Universit Pierre et Marie Curie, Paris 6, October 1984.
13. C. Paul. *Parcours en largeur lexicographique : un algorithme de partitionnement, application aux graphes et gnralisation*. PhD thesis, LIRMM, Universit de Montpellier II, 1998.
14. D. J. Rose, R. E. Tarjan, and G. S. Lueker. Algorithmic aspects of vertex elimination on graphs. *SIAM Journal of Computing*, 5(2):266–283, June 1976.

Independent Tree Spanners[*]

Fault-Tolerant Spanning Trees with Constant Distance Guarantees

(Extended Abstract)

Dagmar Handke

Universität Konstanz, Fakultät für Mathematik und Informatik,
78457 Konstanz, Germany.
email: Dagmar.Handke@uni-konstanz.de

Abstract. For any fixed parameter $t \geq 1$, a *tree t–spanner* of a graph G is a spanning tree T of G such that the distance between every pair of vertices in T is at most t times their distance in G. In this paper, we incorporate a concept of fault-tolerance by examining *independent tree t–spanners*. Given a root vertex r, this is a pair of tree t–spanners, such that the two paths from any vertex to r are edge (resp., internally vertex) disjoint. It is shown that a pair of independent tree 2–spanners can be found in linear time, whereas the problem for arbitrary $t \geq 4$ is $\mathcal{NP}$–complete.

As a less restrictive concept, we treat *tree t–root-spanners*, where the distance constraint is relaxed. Here, we show that the problem of finding an *independent* pair of such subgraphs is $\mathcal{NP}$–complete for all t. As a special case, we then consider *direct tree t–root-spanners*. These are tree t–root-spanners where paths from any vertex to the root have to be detour-free. In the *edge* independent case, a pair of these can be found in linear time for all t, whereas the *vertex* independent case remains $\mathcal{NP}$–complete.

1 Introduction

A *t–spanner* of an unweighted graph G is a spanning subgraph T in which the distance between every pair of vertices is at most t times their distance in G. Throughout this paper, $t \geq 1$ will be an arbitrary, but fixed integer. The main idea of this concept is to find a subgraph of a given graph G that is sparse, but still guarantees a so–called *stretch factor* on the vertex-to-vertex distances of G that is bounded by a constant independent of the size of G.

The concept of spanners has been introduced by Peleg and Ullman in [10], where they used spanners to synchronize asynchronous networks. One of the many other applications for spanners are communication networks, where one is interested in finding a sparse subnetwork that nevertheless guarantees a constant delay factor. Further results and discussions concerning t–spanners and variants

---

[*] Research supported by *Deutsche Forschungsgemeinschaft* under grant Wa 654/10-1.

thereof can be found in [12]. The sparsest t–spanners are *tree t–spanners* (i.e. t–spanners that are trees, see [3]). Apart from their sparseness, the tree property is of particular interest in several applications. As an example consider communication networks where trees result in small sized routing tables [11].

One major drawback of tree subnetworks is their vulnerability: A fault of one single link or node results in disconnecting the subnetwork. In this paper, we consider *independent tree t–spanners*: a pair of tree t–spanners with the additional property that the unique pair of tree paths from any vertex of the graph to a specified root vertex is disjoint. We examine the *edge* as well as the *vertex* disjoint case, thus guaranteeing fault-tolerance in either one link or node, such that the distance conditions are still maintained. The subgraph consisting of the union of two independent tree t–spanners is also sparse and has good structural properties. Thus this sort of subgraphs may serve for example as a reliable means for doing message-efficient broadcast on a distributed network.

Previous work has concentrated on only one of the two aspects of fault-tolerance and distance guarantee at a time. Independent spanning trees are treated in [7, 8, 13, 6], for example. For a general survey on *tree t–spanners*, see [3]. Very recently, in [9], Levcopoulos et al. examined fault-tolerant spanners in *geometric* graphs (i.e. complete graphs with Euclidean edge weights) using a slightly different definition.

In this paper, we give a constructive, efficient characterization for graphs admitting a pair of independent tree 2–spanners. This results in a linear-time algorithm. For $t \geq 4$, however, the corresponding decision problem is $\mathcal{NP}$–complete. Both results hold for the edge and vertex disjoint case. Additionally, the first result can be easily extended to more than two independent tree 2–spanners. The case $t = 3$ remains open.

Unfortunately, the concept of independent tree t–spanners is very restrictive in the sense that the class of graphs admitting a pair of independent tree t–spanners for a fixed t is quite small. In many settings in the design of subnetworks, however, it is not strictly necessary that the distance guarantee holds for all pairs of vertices, but only for the distances between each vertex and the specified root vertex. This leads to the definition of t–*root-spanners*, and we are interested in finding a pair of independent spanning trees such that the distance between any vertex and the root in each tree is at most t times the distance in the graph.

Annexstein et. al. [1, 2] have considered related concepts, but the stretch factors considered there may well be non-constant. Here, we show that the problem of deciding the existence of independent spanning trees with fixed constant stretch factors is hard by showing its $\mathcal{NP}$–completeness for all possible values of t.

Motivated by the hardness of finding independent tree t–root-spanners, we also consider restricted versions of the problem. For example, it is a natural constraint that paths are not allowed to 'run uphill'. This means that on the path from a vertex to the root, no deviations to vertices that have a longer distance to the root are allowed. We call these spanners *direct*. For the decision problem of finding such a pair of independent direct tree t–root-spanners, the situation for the edge and vertex disjoint case differs significantly: We can decide the existence of a pair of edge independent direct tree t–root-spanners (and its construction) in linear time for all t, whereas the problem of finding vertex independent direct tree t–root-spanners remains $\mathcal{NP}$–complete.

This extended abstract concentrates on the algorithmic aspects of the work. We give only one exemplary $\mathcal{NP}$–completeness proof, the details of the other two can be found in the full paper [5].

2 Notation and Previous Results

In what follows, $G = (V, E)$ denotes a simple, unweighted, undirected graph with vertex set V and edge set E. The *root* is a specified vertex $r \in V$. $V(G)$ and $E(G)$, respectively, is the vertex set and edge set, respectively, of a graph G. $G[R]$, where $R \subseteq V$, denotes the subgraph of G induced by R. Since spanners of each connected component can be determined independently, we only consider *connected* graphs. Thus, in the following, G denotes a connected, unweighted graph and r is the specified root vertex.

The *length* of a path P is the number of its edges and is denoted by $|P|$. The *distance* between two vertices u and v in G, i.e. the length of the shortest path, is denoted by $d_G(u, v)$. For a vertex v, denote by $N(v)$ the set of all *neighbors* of v, i.e. vertices that are adjacent to v. A vertex v of a graph G is called *universal* w.r.t. G if $N(v) = V(G)$. Let v be universal w.r.t. G, then the *star centered at* v is the graph consisting of all vertices of G and all edges incident to v.

Given a root vertex r, we partition V into disjoint *levels* $L_\ell := \{v \in V : d_G(v, r) = \ell\}$ for $\ell \in \{0, \ldots, \max_v\{d_G(v, r)\}\}$. The level index of a vertex is indicated by $l(v) := d_G(v, r)$. The level $L_{l(v)-1}$ is called *parent level* of vertex v. A vertex in $L_{l(v)-1} \cap N(v)$ is called a *parent vertex* of v, and a vertex in $L_{l(v)} \cap N(v)$ is called a *sibling vertex* of v. Note that every vertex $v \neq r$ has at least one parent vertex. For a tree T rooted at r, denote the unique path from a vertex v to the root r (called *root path*) by $rp(v, T)$. Sometimes we abuse the notation $rp(v, T)$ to indicate the set of internal vertices of the root path.

For a connected graph, a k–*separator* is a set of k vertices the deletion of which disconnects the graph. A graph is *biconnected* (or *2–vertex–connected*) if it has no 1–separator, and k–*vertex–connected* if it has no $(k-1)$–separator. A *block* of a graph is a maximal biconnected subgraph. A graph is *2–edge–connected* if no deletion of a single edge disconnects it.

Definition 1 *For any integer $t \geq 1$, a spanning subgraph $T = (V, E')$ with $E' \subseteq E$ is a t–spanner of a graph $G = (V, E)$, if for all $u, v \in V$: $d_T(u, v) \leq t \cdot d_G(u, v)$.*

The parameter t is called *stretch factor*. We say that an edge $e \in E$ is *covered* if in T there exists a path of length at most t that connects the endpoints of e. In order to prove that a given spanning subgraph is a t–spanner, we do not have to consider all pairwise distances of the vertices. It is sufficient to only check whether all edges of the original graph that are not part of the spanning subgraph are covered (cf. [3]). A graph always admits a t–spanner, since G itself is a t–spanner.

Tree t–spanners are t–spanners that are trees. As opposed to t–spanners, there are graphs that, for a fixed stretch factor t, do not admit a tree t–spanner as a subgraph. In [3], Cai and Corneil show that the corresponding decision problem, the Tree t–Spanner Problem, is $\mathcal{NP}$–complete for all $t \geq 4$. They also give linear-time algorithms to find a tree t–spanner for $t = 1$ or 2 if it exists. The case $t = 3$ is still open.

3   Independent Tree t–Spanners

3.1   Definitions and Results

Definition 2 *Two spanning trees T_1 and T_2 of a graph G, both rooted at r, are called* edge independent *(resp.,* vertex independent*) w.r.t. r, if for every vertex $v \in V(G)$ the unique paths $rp(v, T_1)$ and $rp(v, T_2)$ are edge disjoint (resp., internally vertex disjoint).*

In most cases considered here, the situation for vertex independence and edge independence is similar. Thus in the following, if not stated explicitly, the terms *disjointness* and *independence* always stand for both cases.

There is a strong relationship, though not equivalence, between the connectivity of a graph and the existence of independent trees. As shown in [7], every biconnected (resp., 2–edge–connected) graph admits a pair of vertex (resp., edge) independent trees. On the other hand, a graph admitting a pair of *edge* independent trees is certainly 2–edge–connected. But note that a graph admitting a pair of *vertex* independent trees does not necessarily have to be biconnected: $\{r\}$ may be a 1–separator (but no other vertex than r). Observe that independent trees do not necessarily have to be edge disjoint. The two trees may share an edge as long as the root paths are disjoint.

In the rest of this section, we are interested in finding a pair of tree t–spanners that are independent:

Problem 3 (Independent Tree t–Spanners Problem, ITS$_t$)

    Given: *A graph G and a root vertex r.*

    Problem: *Does G admit a pair of independent tree t–spanners w.r.t. r?*

Figure 1 shows examples of graphs with different behavior concerning admissibility of independent tree 2–spanners. Observe that the number of edges of a pair of independent tree t–spanners cannot exceed $2 \cdot |V| - 2$. Clearly, for a fixed t, the class of graphs admitting a pair of *vertex* independent tree t–spanners is a proper subclass of the class of graphs admitting a pair of *edge* independent tree t–spanners, which is itself a proper subclass of the class of graphs admitting one tree t–spanner.

Considering either edge or vertex disjointness, the case for $t = 1$ is trivial: A pair of independent tree 1–spanners cannot exist, since for a vertex v incident

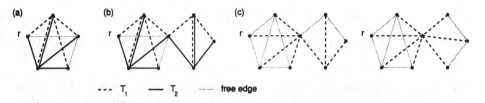

--- T_1 —— T_2 .... free edge

Fig. 1. (a) a graph admitting a pair of vertex (and edge) independent tree 2–spanners, (b) a graph admitting a pair of *edge* (but not *vertex*) independent tree 2–spanners, and (c) two graphs admitting one tree 2–spanner, but no pair of independent tree 2–spanners.

to the root r edge $\{v, r\}$ has to be contained in both trees. For the other values of t, in the following two subsections, we prove the following theorem:

Theorem 4

1. *A pair of independent tree 2–spanners can be found in linear time (if it exists), thus $ITS_2 \in \mathcal{P}$.*
2. *ITS_t is $\mathcal{NP}$–complete for $t \geq 4$.*

Since the main property of independent tree 2–spanners (Lemma 2) can be extended in the obvious way to more than two independent tree 2–spanners, part 1 of Theorem 4 also holds for k independent tree 2–spanners ($k \geq 2$).

3.2   Finding Independent Tree 2–Spanners

In this subsection, we prove part 1 of Theorem 4. For this aim, we first consider the *edge* independent case and describe the structure of the two tree 2–spanners. The strict characterization of graphs that admit a pair of *edge* independent tree 2–spanners and the algorithm to find such a pair then follow directly. The case for *vertex* independent tree 2–spanners is treated subsequently.

Edge independent tree 2–spanners. First, observe that if G is 2–edge–connected then G is not necessarily biconnected. G may consist of several blocks. For each block B of G there is a unique 1–separator $\{s\}$ such that all root paths from vertices in B have to include s. This vertex s (called *root separator*) takes over the role of r in B. Let r be the root separator of the blocks containing r. The first lemma describes the form that two edge independent tree 2–spanners may have. This observation then leads to the characterization lemma 2.

Lemma 1. *If G admits a pair T_1 and T_2 of edge independent tree 2–spanners w.r.t. r then for each block B of G, $T_i[B]$ is a star ($i = 1, 2$).*

Proof. Let T_1 and T_2 be two edge independent tree 2–spanners of G and consider a block B of G with root separator s. From [3] we know that either

1. B is 3–vertex–connected and T_i is a star centered at a vertex u_i ($i = 1, 2$) or

2. for every 2–separator $\{x,y\}$ of G, edge $\{x,y\}$ is in $T_i(i=1,2)$.

In the first case, we are done. The second case can be seen by contradiction (see [5] for details).   □

Lemma 2. *G admits a pair of edge independent tree 2–spanners w.r.t. r if and only if each block B of G contains two distinct universal vertices w.r.t. B that are not root separators of B.*

Proof. For the if-part, observe that if G fulfills the given conditions then we can construct two edge independent tree 2–spanners by combining the stars centered at the universal vertices. To show the opposite direction, assume that G admits a pair of edge independent tree 2–spanners T_1 and T_2. Then by Lemma 1, for each block B, $T_i[B]$ is a star centered at a vertex $u_i^{(B)}$, $i = 1, 2$. Since T_1 and T_2 are edge independent it follows that $u_1^{(B)}$ and $u_2^{(B)}$ are disjoint and not root separators of B.   □

Using the characterization, an algorithm to decide the existence of a pair of edge independent tree 2–spanners can then be implemented in linear time.

Vertex independent tree 2–spanners. The situation for the *vertex* independent case is similar. We get the following characterization:

Lemma 3. *G admits a pair of vertex independent tree 2–spanners w.r.t. r if and only if G admits a pair of edge independent tree 2–spanners w.r.t. r and G is either biconnected or G is 2–edge–connected and $\{r\}$ is the only 1–separator.*

Proof. As stated in Section 3.1, if G admits a pair of vertex independent trees w.r.t. r then G is at least 2–edge–connected and $\{r\}$ is the only 1–separator. Thus, G consists only of blocks containing r. The proof is complete by observing that the edge independent tree 2–spanners found in Lemma 2 for blocks containing r are also vertex independent.   □

3.3   The Hardness of Finding Independent Tree t–Spanners

The membership of ITS$_t$ in $\mathcal{NP}$ is immediate. To prove part 2 of Theorem 4, we use a reduction from the Not-All-Equal Satisfiability Problem with three literals per clause (NAE-3SAT, cf. [4] (LO3)): Given a set U of variables, a collection C of clauses over U such that each clause $c \in C$ has $|c| = 3$, the problem is to find a truth assignment for U such that each clause in C has at least one true literal and at least one false literal. Given an instance (U, C) of NAE-3SAT, we construct the graph G for ITS$_t$ as follows:

– For each variable $x \in U$, create a *variable component* T_x consisting of
  • a *root vertex* r_x and two *literal vertices* x and $\bar{x}$,
  • *literal edges* $\{r_x, x\}$ and $\{r_x, \bar{x}\}$, and
  • a simple path P of length $t - 3$ connecting x and $\bar{x}$, where all internal vertices of P are new vertices.

- Identify all root vertices r_x to form one distinguished root vertex r.
- For each clause $c \in C$ create a *clause vertex* c and connect it by a *clause edge* to all its corresponding literal vertices.

G can be constructed in linear time. The following lemmas are immediate:

Lemma 4. *Let P be the path of length $t - 3$ connecting x and $\overline{x}$ in an isolated variable component T_x. Then T_1 consisting of edges $\{r, x\} \cup P$ and T_2 consisting of edges $\{r, \overline{x}\} \cup P$ are two unique independent tree t–spanners of T_x.*

Lemma 5. *Let G be the graph constructed from an instance of NAE-3SAT. If T_1 and T_2 are independent tree t–spanners (w.r.t. r) of G then*

1. *for all $x \in U$, no literal edge belongs to both T_1 and T_2,*
2. *for all $x \in U$, at least one literal edge belongs to T_1 or T_2, and*
3. *$d_{T_1}(c, r) = d_{T_2}(c, r) = 2$.*

It remains to show that given a not-all-equal truth assignment of (U, C) we can in polynomial time construct a pair of independent tree t–spanners for G and vice versa. Firstly, starting with a not-all-equal truth assignment θ of (U, C), construct T_1 and T_2 according to the following rules:

- If $\theta(x) = true$, put $\{r, x\} \in T_1$ and $\{r, \overline{x}\} \in T_2$; otherwise put $\{r, x\} \in T_2$ and $\{r, \overline{x}\} \in T_1$.
- For each variable x, put path $\{x, \overline{x}\}$ in both T_1 and T_2.
- For each clause vertex c, choose a clause edge for each T_1 and T_2 such that $d_{T_1}(c, r) = d_{T_2}(c, r) = 2$. Such an edge exists since θ is a not-all-equal truth assignment.

It can be easily seen that for $t \geq 4$, T_1 and T_2 are trees, independent, and t–spanners.

For the other direction, it follows from part 1 and 2 of Lemma 5 that we can define θ in the following way: For every $x \in U$, if $\{r, x\} \in T_1$ or $\{r, \overline{x}\} \in T_2$ then $\theta(x) := true$ else $\theta(x) := false$. As a corollary from part 3 of Lemma 5, θ is a not-all-equal truth assignment. This completes the proof of part 2 of Theorem 4.

4 Independent Tree t–Root-Spanners

In many settings, it is sufficient for a subgraph that the distance guarantee just holds for the distances between a vertex and a specified root. We therefore introduce the concept of t–root-spanners:

Definition 5 *For any integer $t \geq 1$, a spanning subgraph $T = (V, E')$ with $E' \subseteq E$ is a t-root-spanner of a graph $G = (V, E)$ w.r.t. r, if for all $v \in V$: $d_T(v, r) \leq t \cdot d_G(v, r)$.*

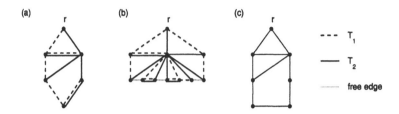

Fig. 2. (a) A graph admitting a pair of vertex (and edge) independent tree 2–root-spanners, (b) a graph admitting a pair of *edge* (but not *vertex*) independent tree 2–root-spanners, (c) a graph not admitting a pair of independent tree 2–root-spanners.

As above, a *tree t–root-spanner* is a *t*–root-spanner that is a tree. Observe that the shortest path tree is a tree 1–root-spanner and thus every graph admits a tree *t*–root-spanner for arbitrary *t*. We are interested in finding a pair of tree *t*–root-spanners that are independent. The corresponding *Independent Tree t–Root-Spanners Problem* ITRS$_t$ is defined analogously to Problem 3.

Figure 2 shows examples of graphs (not) admitting two independent tree *t*–root-spanners. The vertices of the graphs are drawn in a way that reflects the distance to the root, according to their levels. To check whether a given subgraph is a *t*–root-spanner each vertex has to be considered separately. It is not possible to just check edges. For example, edges within a level do not have to be covered.

It can be easily seen that, for every fixed *t*, the class of graphs admitting a pair of *vertex* independent tree *t*–root-spanners is a proper subclass of the class of graphs admitting a pair of *edge* independent tree *t*–root-spanners. On the other hand, the class of graphs admitting a pair of independent tree *t–root-spanners* is a superclass of the class of graphs admitting a pair of independent tree *t*–spanners. As before, the case for $t = 1$ is trivial, since two independent tree 1–root-spanners never exist. Thus the Independent Tree *t*–Root-Spanners Problem is fully characterized by the following theorem:

Theorem 6 *ITRS$_t$ is $\mathcal{NP}$–complete for $t \geq 2$.*

Proof. The proof uses a reduction from NAE-3SAT similar to the one of the preceding section using paths of length $t - 1$ to connect the literal vertices and paths of length $\lceil \frac{t}{2} \rceil + 1$ to connect clause vertices with their corresponding literal vertices. Details can be found in [5]. □

5   Independent Direct Tree *t*–Root-Spanners

As proved in the previous section, it is hard to decide the existence of a pair of general independent tree *t*–root-spanners in a given graph. In this section, we consider the case where the paths from a vertex to the root within the tree *t*–root-spanner have to be direct. No deviations to vertices that have a longer distance to the root are allowed.

5.1   Definitions and Results

Definition 7 *For any integer $t \geq 1$, a tree t-root-spanner T of G (w.r.t. r) is called* direct *if for each vertex $v \in V$ the following holds:for every $w \in rp(v, T)$ $d_G(w, r) \leq d_G(v, r)$.*

We are interested in finding two independent direct tree t-root-spanners. The corresponding *Edge* (resp., *Vertex*) *Independent Direct Tree t-Root-Spanners Problem* EIDTRS$_t$ (resp., VIDTRS$_t$) is defined analogously to Problem 3. For a better handling of direct tree t-root-spanners, we treat the vertices of G according to their level $L_{l(v)}$. Thus, paths from a vertex to the root in a *direct* tree t-root-spanner always stay in the same level or lead to a level with smaller index.

For an example consider again Figure 2 of the previous section: The graph in (a) admits a pair of independent tree 2-root-spanners, but not two *direct* ones, whereas the trees in (b) are direct. It can be easily seen that, for every fixed t, the class of graphs admitting a pair of *vertex* independent *direct* tree t-root-spanners is a proper subclass of the class of graphs admitting a pair of *edge* independent *direct* tree t-root-spanners, which itself is a proper subclass of the class of graphs admitting a pair of edge independent tree t-root-spanners.

As opposed to the Independent Tree t-Root-Spanners Problem, the complexity of the decision problem for the direct version significantly differs for the vertex and edge independent case. We fully characterize the *Independent Direct Tree t-Root-Spanners Problem* by the following theorem.

Theorem 8

1. *A pair of edge independent direct tree t-root-spanners can be found in linear time (if it exists), thus EIDTRS$_t \in \mathcal{P}$.*
2. *VIDTRS$_t$ is $\mathcal{NP}$-complete for $t \geq 2$.*

The proof of part 2 again uses a reduction from NAE-3SAT. This time, the variable components consist of three triangles whereas the clause components consist of several stars. Details are deferred to the full paper [5]. The proof for part 1 is given in the following subsection.

5.2   Finding Edge Independent Direct Tree t-Root-Spanners

In the following, assume that G admits a pair of edge independent direct tree t-root-spanners. The first lemma establishes that there exists also a pair of edge independent direct tree t-root-spanners such that every vertex is connected to a parent vertex in at least one of the two trees. If a vertex has more than one parent vertex it is always possible to choose a direct connection to the parent level in both trees.

Lemma 6. *If G admits a pair of edge independent direct tree t-root-spanners then there exists a pair T_1 and T_2 of edge independent direct tree t-root-spanners w.r.t. r such that*

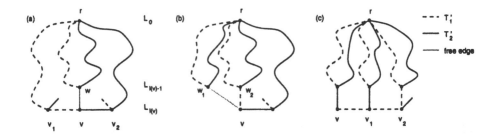

Fig. 3. Illustrations for (a) part 1 of Lemma 6, (b) part 2 of Lemma 6, and (c) Lemma 7.

1. for all $v \in V - \{r\}$ there is an $i \in \{1,2\}$ and a $w \in L_{l(v)-1}$ such that $\{v, w\} \in T_i$.
2. for all $v \in V - \{r\}$ with $w_1, w_2 \in N(v) \cap L_{l(v)-1}, w_1 \neq w_2 : \ \{v, w_i\} \in T_i$.

Proof.
1. The proof of part 1 is by induction. Consider an arbitrary pair T_1' and T_2' of edge independent direct tree t–root-spanners. For vertices at level L_1 the statement is immediate. Now let v be a vertex with minimal level index $l(v)$ such that the constraint given in the lemma is not fulfilled. Let w be a parent vertex of v, and v_i, $i \in \{1,2\}$ be the sibling vertices of v on $rp(v, T_i')$, respectively. See Figure 3(a) for an illustration. From the induction hypothesis, we know that $d_{T_i'}(w, r) \leq t \cdot (l(v) - 1)$, and that the root paths $rp(w, T_i')$ are direct and edge disjoint.
   Now, obtain T_i from T_i' for $i \in \{1,2\}$ as follows: w.l.o.g. delete $\{v, v_1\}$ from T_1' and add $\{v, w\}$. For the other tree, let $T_2 := T_2'$.
   Note that the distance and directness constraints keep trivially valid, and both T_1 and T_2 are trees. To see that $rp(v, T_1)$ and $rp(v, T_2)$ are edge disjoint observe the following: If $rp(v, T_2')$ and $rp(w, T_1')$ are vertex disjoint we are done. Otherwise, take the first common vertex x in $rp(v, T_2')$ and $rp(w, T_1')$. By induction, since $l(x) < l(v)$, the root paths $rp(x, T_1')$ and $rp(x, T_2')$ are edge disjoint and thus are $rp(v, T_1)$ and $rp(v, T_2)$.
2. Using part 1 of the lemma, suppose that T_1' and T_2' fulfill the constraint given there. The proof for part 2 then follows along the same lines. Figure 3(b) illustrates the situation.   □

The characterization can be further enhanced by examining the stretch factor of the two edge independent direct tree t–root-spanners. As shown in the following lemma, the directness constraint dominates the stretch factor, such that the class of graphs admitting a pair of edge independent direct tree t–root-spanners for arbitrary t and the class of graphs admitting a pair of edge independent direct tree 2–root-spanners collapse.

Lemma 7. *If G admits a pair of edge independent direct tree t–root-spanners w.r.t. r then G also admits a pair of edge independent direct tree 2–root-spanners w.r.t. r.*

Proof. To prove the lemma, observe that if G admits a pair T_1' and T_2' of edge independent direct tree t–root-spanners then there exist also two edge independent direct tree t–root-spanners T_1 and T_2 such that

$$|\{w : w \in rp(v, T_i) \cap L_{l(v)}, w \neq v\}| = 1, \quad (*)$$

thus indicating that for each v the distance to the parent level in each T_i is at most 2. This can be seen by similar arguments as in Lemma 6 starting with T_1' and T_2' as indicated in part 2 of that lemma. See Figure 3(c) for an illustration of the situation. Accumulating this over all levels results in a pair of edge independent direct tree 2–root-spanners.   □

As a corollary from the previous lemmas, we get the following characterization of graphs that admit a pair of edge independent direct tree t–root-spanners:

Corollary 1. *A graph G does* not *admit a pair of edge independent direct tree t–root-spanners if and only if there is a vertex v such that $|N(v) \cap L_{l(v)-1}| = 1$ and $N(v) \cap L_{l(v)} = \emptyset$.*

As a consequence of property $(*)$ of Lemma 7, the connection of a vertex to its parent level in both tree 2–root-spanners is independent of the connections in lower levels. Thus, using the given characterization, we obtain a top–down algorithm as described in Figure 4. It starts from the top level vertices and constructs the edge independent direct tree 2–root-spanners level by level, if possible. We begin with level L_1, the direct neighbors of the root. Once we have fixed the tree edges for all vertices of a level $L_{\ell-1}$, we continue with the vertices of level L_ℓ.

The correctness of the algorithm follows directly from Corollary 1. Since every vertex is treated exactly once it can be implemented in linear time. Thus, part 1 of Theorem 8 is proved.

Since the main properties of edge independent direct tree t–root-spanners (Lemma 6 and Corollary 1) can be extended to more than two edge independent direct tree t–root-spanners in the obvious way, part 1 of Theorem 8 also holds for k edge independent direct tree t–root-spanners ($k \geq 2$).

References

1. F. S. Annexstein, K. A. Berman, and R. Swaminathan. Independent spanning trees with small stretch factors. Technical Report 96-13, DIMACS, June 1996.
2. F. S. Annexstein, K. A. Berman, and R. Swaminathan. On computing nearly optimal multi-tree paths and s, t-numberings. In D. Krizanc and P. Widmayer, ed., *Proc. 4th Int. Colloquium on Structural Information and Communication Complexity, SIROCCO'97*, pages 12–23. Carleton Scientific, 1998.
3. L. Cai and D. Corneil. Tree spanners. *SIAM J. Discrete Math.*, 8(3):359–387, 1995.
4. M. R. Garey and D. S. Johnson. *Computers and Intractability: A Guide to the Theory of $\mathcal{NP}$-Completeness.* W H Freeman & Co Ltd, 1979.
5. D. Handke. Independent tree spanners: Fault-tolerant spanning trees with distance guarantees. *submitted*, 1998.

Procedure **Process-Level(ℓ)**

Given: A graph G (all vertices of level L_ℓ unmarked) with root r, and two trees T_1 and T_2 that are edge independent direct tree 2–root-spanners for vertices of level $L_{\ell-1}$ or lower.

Output: Two trees T_1 and T_2 that are edge independent direct tree 2–root-spanners for vertices of level L_ℓ or lower, if they exist.

1. Arbitrarily choose an unmarked $v \in L_\ell$, mark v, let $N := N(v) \cap L_{\ell-1}$.
   (* N is the set of parent vertices of v. *)

2. If $|N| \geq 2$:   (* v can be connected to level $L_{\ell-1}$ in both trees. *)
   arbitrarily choose two vertices $v'_1, v'_2 \in N$,
   add $\{v, v'_1\}$ to T_1 and $\{v, v'_2\}$ to T_2.

3. If $N = \{v'\}$:   (* v has one distinguished parent vertex. *)
   let $M = N(v) \cap L_\ell$.   (* M is the set of sibling vertices of v. *)
   (a) If $M = \emptyset$, then stop: G does not admit a pair of edge independent direct tree 2–root-spanners.

   (b) Else, if all $w_i \in M$ are unmarked, then arbitrarily choose a w, let w' be an arbitrary parent vertex of w, and add edges $\{v, v'\}$ and $\{v, w\}$ to T_1 and edges $\{w, w'\}$ and $\{v, w\}$ to T_2; mark w.

   (c) Else, arbitrarily choose a marked w; add $\{v, w\}$ to the tree in which w is connected to its parent level, and add $\{v, v'\}$ to the other tree.

Fig. 4. The algorithm for EIDTRS$_2$ for vertices of level L_ℓ

6. A. Huck. Independent trees in graphs. *Graphs and Combinatorics*, 10:29–45, 1994.
7. A. Itai and M. Rodeh. The multi-tree approach to reliability in distributed networks. *Information and Computation*, 79(1):43–59, 1988.
8. S. Khuller and B. Schieber. On independent spanning trees. *Information Processing Letters*, 42:321–323, 1992.
9. C. Levcopoulos, G. Narasimhan, and M. Smid. Efficient algorithms for constructing fault-tolerant geometric spanners. Preprint 16, Fakultät für Informatik, Otto-von-Guericke-Universität Magdeburg, Germany, 1997.
10. D. Peleg and J. D. Ullman. An optimal synchronizer for the hypercube. In *Proc. 6th ACM Symposium on Principles of Dist. Comp., Vancouver*, pages 77–85, 1987.
11. N. Santoro and R. Khatib. Labelling and implicit routing in networks. *The Computer Journal*, 28(1):5–8, 1985.
12. J. Soares. Graph spanners: a survey. *Congressus Numerantium*, 89:225–238, 1992.
13. A. Zehavi and A. Itai. Three tree-paths. *J. of Graph Theory*, 13:175–188, 1989.

Upgrading Bottleneck Constrained Forests

S. O. Krumke[1], M. V. Marathe[2], H. Noltemeier[3], S. S. Ravi[4], and H.-C. Wirth[3]

[1] Konrad-Zuse-Zentrum für Informationstechnik Berlin,
Department Optimization, Takustr. 7, 14195 Berlin-Dahlem, Germany.
krumke@zib.de
[2] Los Alamos National Laboratory,
P.O.Box 1663, MS B265, Los Alamos, NM 87545, USA.
madhav@c3.lanl.gov
[3] Department of Computer Science,
University of Würzburg, Am Hubland, 97074 Würzburg, Germany.
{noltemei,wirth}@informatik.uni-wuerzburg.de
[4] Department of Computer Science,
University at Albany – SUNY, Albany, NY 12222, USA.[‡]
ravi@cs.albany.edu

Abstract. We study *bottleneck constrained network upgrading problems*. We are given an edge weighted graph $G = (V, E)$ where node $v \in V$ can be upgraded at a cost of $c(v)$. This upgrade reduces the delay of each link emanating from v. The goal is to find a minimum cost set of nodes to be upgraded so that the resulting network has a good performance. The performance is measured by the bottleneck weight of a constrained forest defined by a proper function [GW95]. These problems are a generalization of the node weighted constrained forest problems studied by Klein and Ravi [KR95].

The main result of the paper is a polynomial time approximation algorithm for this problem with performance guarantee of $2\ln(\sqrt{e}/2 \cdot |K|)$, where $K := \{ v : f(\{v\}) = 1 \}$ is the set of terminals given by the proper function f. We also prove that the performance bound is tight up to small constant factors by providing a lower bound of $\ln |K|$. Our results are obtained by extending the elegant solution based decomposition technique of [KR95] for approximating node weighted constrained forest problems. The results presented here extend those in [KR95, KM+97].

1 Introduction

Several problems arising in the area of communication networks can be expressed in the following general form: Given a network, enhance the performance of that network by modifying parts of the network. Such *network upgrade problems*, as opposed to network reconstruction problems, are convenient for investigating cases where the cost of implementing a new network from scratch exceeds the cost of modifying an already installed network.

[‡] Research supported by NSF Grant CCR-97-34936.

There are two main models for network upgrading problems: the *edge upgrading model* [Ber92, Phi93, KN+96], where upgrading an edge reduces the delay on the upgraded edge, and the *node upgrading model* [PS95, KM+97], where upgrading a node reduces the delay on all the edges incident on the upgraded node. In communication networks, upgrading a node corresponds to installing faster communication equipment at an exchange point which results in a speedup of all links leading through that node. In this paper, we concentrate on the node upgrading problem UPGRADING-CFP: Find a minimum cost set of vertices, such that after upgrading those vertices, the resulting graph contains a constrained forest (as specified by a proper function f) of bottleneck weight at most a given threshold. The problem UPGRADING-CFP generalizes a number of node weighted network design problems, such as node weighted Steiner Trees.

In this paper, we provide the first approximation algorithm for the above problem. The performance guarantee of the algorithm is $2\ln(\sqrt{e}/2 \cdot |K|)$, where $K := \{\, v : f(\{v\}) = 1 \,\}$ is the set of terminals given by the proper function f. We also establish a lower bound of $\ln|K|$ that matches the upper bound to within a constant factor. These results generalize the results in [KM+97], where similar problems for the special case of spanning trees were investigated.

2   Preliminaries

The *node based network upgrading model* used in this paper generalizes the model from [PS95] and was introduced in [KM+97]. It can be formally described as follows. Let $G = (V, E)$ be a connected undirected graph with $n := |V|$ vertices and $m := |E|$ edges. For each edge $e \in E$, we are given three integers $d_0(e) \geq d_1(e) \geq d_2(e) \geq 0$. The value $d_i(e)$ represents the *weight* or *delay* of the edge e if exactly i of its endpoints are upgraded.

For each node $v \in V$, the value $c(v)$ specifies how expensive it is to upgrade the node. For a subset W of V, the cost of upgrading all the nodes in W, denoted by $c(W)$, is equal to $\sum_{v \in W} c(v)$. The edge weight function resulting from an upgrade of the node set W is denoted by d_W. The *bottleneck graph* Bottleneck(G, d_W, D) contains all edges $e \in E$ with $d_W(e) \leq D$.

Given a bound D, we partition the set of edges into four sets according to how many of the endpoints must be upgraded in order to decrease the delay of an edge below the threshold D. An edge of delay $d_0(e) \leq D$ is called an *uncritical* edge. An edge e is said to be *1-critical*, if $d_0(e) > D \geq d_1(e)$, and *2-critical*, if $d_1(e) > D \geq d_2(e)$. Finally, if $d_2(e) > D$, the edge e is called *useless*. Without loss of generality, we can assume that the graph does not contain any useless edges.

2.1   Constrained Forest Problems

Constrained forest problems were introduced by Goemans and Williamson (see [GW95, GG+94]). For a certain family of cuts in a graph, one searches for a subgraph intersecting all the cuts in the family. Fix a graph G. For any nonempty node subset $U \subset V$ with $U \neq V$, there is a corresponding cut $\delta(U)$ in G, namely the cut which contains those edges that have exactly one endpoint in U. Thus, we can use a function $f: 2^V \to \{0, 1\}$ to define a family of cuts: $f(U) = 1$ if and only if $\delta(U)$ is in the family. Such a function f is termed *proper* if it satisfies the following two conditions:

(i) **Symmetry**: $f(U) = f(V \setminus U)$ for all $U \subseteq V$; and
(ii) **Disjointness**: If $A \cap B = \emptyset$, then $f(A) = f(B) = 0$ implies that $f(A \cup B) = 0$.

Any subset F of the edges such that

$$|F \cap \delta(U)| \geq f(U), \quad \text{for all } \emptyset \neq U \subset V$$

is termed a *constrained forest* with respect to f. Any vertex $v \in V$ such that $f(\{v\}) = 1$ is called a *terminal* and $K := \{v : f(v) = 1\}$ is the *set of terminals* given by the proper function f.

Many interesting families of problems can be formulated as constrained forest problems with proper functions. For example, if we define a proper function by

$$f(U) = 1 \quad \text{for all } \emptyset \neq U \subset V, U \neq V,$$

any constrained forest must contain at least one edge of each cut in the graph. Therefore, the corresponding subgraph is connected and the inclusion-wise minimal constrained forests are the spanning trees of the input graph. As a second example, let a set $K \subseteq V$ of terminals be given and define $f(U) = 1$ if and only if $\delta(U)$ separates the set K of terminals. Then any constrained forest must span all terminals, and the minimal constrained forests are Steiner trees with terminal set K.

Lemma 1 ([GW95]). *Let f be a proper function. If $f(U) = 0$ and $f(B) = 0$ for some $B \subseteq U$, then $f(U \setminus B) = 0$.*

2.2   Problem Formulation

We are now ready to formulate the problem UPGRADING-CFP under study: Given a graph $G = (V, E)$ with edge weights $d_0 \geq d_1 \geq d_2$, node weights c as before, a bound D and a proper function f, find a minimum cost set $W \subseteq V$ of nodes such that the resulting graph with edges weights given by d_W has a constrained forest (with respect to f) of bottleneck delay at most D. Notice that the condition just stated is equivalent to saying that after the upgrade the set of all edges of weight at most D forms a constrained forest.

Given a vertex set $W \subseteq V$ it can be easily checked in polynomial time whether W is a valid upgrading set. This can be achieved by computing the bottleneck graph $G' := \text{Bottleneck}(G, d_W, D)$ and evaluating the proper function for each connected component of G'. In fact, we claim that W is valid if and only if f evaluates to zero on each component of G'. Clearly, if $f(C) = 1$ for a connected component C of G' then G' cannot contain a constrained forest, since any such forest must have an edge with exactly one endpoint in C. Assume conversely that $f(C) = 0$ for each connected component C of G'. If C' contained no constrained forest, there would be a set $\emptyset \neq U \subset V$ with $f(U) = 1$ and $\delta(U) \cap H = \emptyset$. But then U is the disjoint union of components of H and from the disjointness property of f we obtain the contradiction that $f(U) = 0$.

The problem UPGRADING-CFP generalizes the problem of finding a node-weighted Steiner Tree of minimum cost. An instance of the node-weighted Steiner Tree problem is given by a graph $G = (V, E)$ with edge weights l and node weights w. For a subset $K \subseteq V$ of terminals, the problem consists of finding a connected subgraph of G of minimum (edge- and node-) weight spanning all the terminals. This problem was studied by Klein and Ravi [KR95] who obtained an approximation with performance $2 \ln |K|$.

Let an instance of the Steiner Tree Problem be given. Notice that without loss of generality we can assume that all edge weights are zero: If not, we can replace each edge (u, v) by two new edges (u, x) and (x, v), where x is a new vertex of weight $l(u, v)$. We can now construct an instance of UPGRADING-CFP by taking the graph G specified in the Steiner Tree instance and defining edge weights $d_0(e) := d_1(e) := 2$ and $d_2(e) := 1$. We set the bottleneck threshold D to be 1. The cost of upgrading a vertex v is set to $c(v) := w(v)$. The proper function f is defined as above to reflect Steiner Trees for the terminal set K.

For each solution of the Steiner Tree Problem the vertices in the tree induce a feasible upgrading set whose cost equals that of the tree. Conversely, it is easy to see that each upgrading set can be used to obtain a solution of the Steiner Tree Problem of at most the same cost.

3   The Algorithm

We first give a brief overview of our algorithm. The set W of upgraded nodes is initially empty. Our algorithm maintains the connected components of the edge subgraph of G consisting of those edges whose delays do no longer exceed the threshold D. Such a connected component C is called *active*, if $f(C) = 1$. In each iteration the algorithm merges at least two active components by upgrading nodes in the network. Notice that from the properties of f it is impossible that there is only one active component remaining. The algorithm terminates when no active components remain.

The basic rule for selecting the nodes to upgrade in an iteration is the following: select a set that gives the best improvement ratio. This ratio is measured by the quotient of the cost of the vertices and the decrease in the number of active components. A formal definition of the *quotient cost* is given in Sect. 3.2.

We will need the following notation for stating our algorithm. Assume that at some stage during the execution of the algorithm $\mathcal{C} = \{C_1 \ldots, C_q\}$ is the set of active components. Then, for a vertex $v \in V$ we define $c^-(v, C_j)$ to be the minimum cost of an upgrading set that does not include v such that v and C_j are connected by a path of bottleneck weight at most D in the upgraded graph. If no such upgrading set exists, we define $c^-(v, C_j) := +\infty$. Moreover, if $v \in C_j$,

---

Algorithm 1 Node upgrading for constrained forests.

Input: A graph $G = (V, E)$ with three edge weight functions d_0, d_1, d_2, a node
   weight function c, a threshold number D,
   a proper function $f: 2^V \rightarrow \{0, 1\}$.

1 $W \leftarrow \emptyset$ and $G' \leftarrow \text{Bottleneck}(G, d_W, D)$
2 **while** G' contains at least one active connected component **do**
3   Assume that $\mathcal{C} = \{C_1 \ldots, C_q\}$ is the set of active components.
4   **for all** $v \in V$, $C \in \mathcal{C}$ **do**
5          $c^-(v, C) \leftarrow$ minimum upgrading cost to obtain a path of bottleneck delay at
            most D from v to C where v is *not* upgraded.
6          $c^+(v, C) \leftarrow$ minimum upgrading cost to obtain a path of bottleneck delay at
            most D from v to C where v is upgraded. This cost does *not* include the
            upgrading cost of v.
7   {Comment: If $v \in C$, then $c^+(v, C) = c^-(v, C) = 0$.}
8   **end for**
9   Find a node $v \in V$ in the graph G with min. quotient cost $q(v)$ as defined in (1).
10   Let $C_1, \ldots, C_r$ be the components in $\mathcal{C}$ chosen in Step 9.
       Let U be the upgraded vertices on the paths from v to the clusters $C_1, \ldots, C_r$.
11     $W \leftarrow W \cup U$
12   Recompute the edge weights d_W, the graph $G' = \text{Bottleneck}(G, d_W, D)$ and its
       connected components.
13 **end while**
14 **return** W

---

then $c^-(v, C_j) = 0$. Similarly, we define $c^+(v, C_j)$ to be the minimum cost of an
upgrading set containing v but not counting the cost $c(v)$ of v itself.

3.1   Computing the Best Upgrading Paths

In this section we show that for each vertex v and active component C we can
compute the values $c^-(v, C)$ and $c^+(v, C)$ in polynomial time. This is done by
two single source shortest path computations on an auxiliary graph H, where
the length of a path is defined to be the costs of the nodes on the path excluding
the source vertex.

For each vertex $v \in V$ the auxiliary graph H contains two vertices v^+ and
v^- representing the upgraded and the untouched version of v. The vertex set is
augmented by one node for each active component.

For each edge $(u, v) \in E$, we insert the edge (u^+, v^+) into H. For each 1-
critical edge (u, v), H contains the edges (u^+, v^-), (u^-, v^+). Finally, for each
uncritical edge (u, v), the auxiliary graph H contains additionally the edges
(u^+, v^-), (u^-, v^+) and (u^-, v^-). The construction is illustrated in Fig. 1.

Each active cluster C is joined to all the vertices v^+ and v^- where $v \in C$.

Let W be the set of nodes already upgraded. For $v \in W$, we set the cost of
vertex v^+ to zero and remove vertex v^-. For $v \notin W$, the cost of vertex v^+ is set
to $c(v)$ and the cost of vertex v^- is zero. Also, the cluster nodes have zero cost.

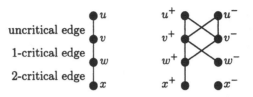

Fig. 1. The graph G (left) and the constructed auxiliary graph H (right).

For a vertex v and an active cluster C let $c(v^-, C)$ and $c(v^+, C)$ denote the length of shortest paths with respect to node weights from v^- and v^+ to C in H, respectively, not including the cost of the source vertex. Thus, the cost $c(v^+, C)$ does *not* contain the cost of v.

The following two lemmas show how to compute $c^-(v, C)$ and $c^+(v, C)$ for a node v and an active cluster C.

Lemma 2. *For each vertex $v \in V \setminus W$ and each active cluster C the minimum cost $c^-(v, C)$ of an upgrading set not containing v such that the resulting graph has a path of bottleneck delay at most D from v to a node in C equals the node weighted distance $c(v^-, C)$ in the auxiliary graph H.* □

Lemma 3. *For each $v \in W$ and each active cluster C, $c^+(v, C) = c(v^+, C)$.* □

3.2 Quotient Costs

Let $v \in V$ be a vertex and $\mathcal{C} = \{C_1, \ldots, C_p\}$ the set of active components. Let

$$q^+(v) := \min_{2 \leq r \leq p} \min_{\substack{\mathcal{C}' \subseteq \mathcal{C} \\ |\mathcal{C}'| = r}} \frac{c(v) + \sum_{C' \in \mathcal{C}'} c^+(v, C')}{r}$$

$$q^-(v) := \min_{2 \leq r \leq p} \min_{\substack{\mathcal{C}' \subseteq \mathcal{C} \\ |\mathcal{C}'| = r}} \frac{\sum_{C' \in \mathcal{C}'} c^-(v, C')}{r}.$$

Then we define the *quotient cost* of v by

$$q(v) := \min \left\{ q^+(v), q^-(v) \right\}. \tag{1}$$

Notice that for each node, its quotient cost can be computed in polynomial time: By ordering the active components such that $c^+(v, C_1) \leq \cdots \leq c^+(v, C_p)$, we can compute $q^+(v)$ by considering only the p subsets of $\mathcal{C}$ of the form $\{C_1, \ldots, C_r\}$, $r = 1, \ldots, p$. The value $q^-(v)$ can be computed similarly.

3.3 Running Time

We briefly argue that our algorithm can be implemented to run in polynomial time. For a terminal set K the number of iterations is at most $|K|$, since we start with at most $|K|$ active components. In each iteration we must solve $\mathcal{O}(n)$ single-source shortest-path problems to compute the best upgrading paths. Each of these shortest-path trees can be computed by Dijkstra's algorithm in time $\mathcal{O}(n \log n + m)$. The quotient cost of any node can then be determined in $\mathcal{O}(n \log n)$ time. This leads to a total time of $\mathcal{O}(n^2 \log n)$ per iteration neglecting the time needed to update the weights and the bottleneck graph. The latter task needs total time $\mathcal{O}(m)$ over all iterations, since each edge weight is updated at most twice. Thus, the algorithm can be implemented to run in time $\mathcal{O}(|K| n^2 \log n)$.

4 Performance Guarantee

The proof of the performance guarantee uses the notion of a spider covering, which extends the definitions given in [KR95].

4.1 Spider Decompositions and Coverings

We first recall the definition of a spider and a spider decomposition.

Definition 4 (Spider). *A spider is a tree with at most one node of degree greater than two. A* center *of a spider is a vertex from which there are edge-disjoint paths to the leaves of the spider. If a spider has at least three leaves, then its center is unique. A* foot *is a leaf, and the path from the center to a non-center foot is called a* leg *of the spider. A* nontrivial spider *is a spider with at least two leaves.*

Notice that our notion of a foot is slightly different from the original definition in [KR95].

Definition 5 (Spider Decomposition). *Let $G = (V, E)$ be a graph and $M \subseteq V$. A spider decomposition of M in G is a set of node-disjoint nontrivial spiders in G such that the union of the feet and the centers of the spiders contains M.*

Lemma 6 ([KR95]). *Let G be a connected graph, and let M be a subset of its nodes such that $|M| \geq 2$. Then G contains a spider decomposition of M.*

Definition 7 (Spider Covering). *Let G be a connected graph, and suppose there is a collection $\{M_1, \ldots, M_p\}$ of disjoint node sets, where each M_i induces a connected subgraph of G. A spider covering of $\{M_1, \ldots, M_p\}$ in G is a collection of node disjoint nontrivial spiders in G such that:*

1. *each set M_i contains a foot or a center of a spider, and*
2. *if a set M_i contains a foot, then M_i does not contain any other foot or center.*

Lemma 8. *Let G be a connected graph and $\{M_1, \ldots, M_p\}$ be a collection of disjoint node sets each inducing a connected subgraph in G. Then there is a spider covering of $\{M_1, \ldots, M_p\}$ in G.*

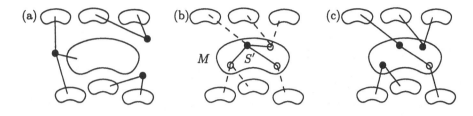

Fig. 2. Illustration on the proof of Lemma 8. (a) No super node appearing as a body. (b) One super node as body. Solid lines represent second spider decomposition. (c) Resulting set of spiders.

Proof. Let $\tilde{G} = (\tilde{V}, \tilde{E})$ be the graph created from G by aggregating each node set M_i to a super node M_i. By Lemma 6 there is a spider decomposition of the set $\{M_1, \ldots, M_p\}$ in $\tilde{G}$.

All super nodes appear as feet or centers in the spider decomposition. Let us first assume that none of the super nodes is a center of a spider (see Fig. 2–a). Then, unfold each super node M_i and choose a node $m_i \in M_i$ to connect as a foot to the corresponding leg of the spider in G. The modified spider decomposition then forms a collection of spiders with the desired properties.

Now we consider the case that there is a super node M which is the center of a spider S in the decomposition. Unfold the super node M and replace it by the corresponding subgraph. Denote by M' the set of nodes in the subgraph $G[M]$ in which the legs of spider S are rooted. Then, perform a second spider decomposition of M' in $G[M]$ (see Fig. 2–b).

Let S' be one of the spiders of that decomposition. For each foot $m' \in M'$ of S' in which paths to two or more super nodes are rooted, disconnect m' from S', and declare m' as the body of a new spider. After this procedure, we are left over with the body of S' and a set of feet each rooting the path to exactly one super node.

If this remaining part of S' connects zero or more than one super nodes, then it can be discarded or it forms a nontrivial spider, respectively. Otherwise, the single remaining super node can be connected through edges of S' to any of the just constructed new spiders (Fig. 2–c).

The given construction can be performed for each spider in M and again for each super node appearing as a center of a spider in the first spider decomposition. The resulting set of spiders is then a spider covering as desired.   □

4.2   An Averaging Lemma

Lemma 9. *Let v be a node chosen in Step 9 of Algorithm 1 and let $c(U)$ denote the total cost of the nodes added to the solution set W in this iteration. Suppose there are p active clusters before v is chosen and assume that in this iteration r clusters are merged. Let* OPT *be the total upgrade cost in an optimal solution.*

Then

$$\frac{c(U)}{r} \leq \frac{\text{OPT}}{p}.$$

Proof. Let W^* be an optimal upgrading set of cost $\text{OPT} := c(W^*)$ and F^* be a constrained forest of bottleneck delay at most D after upgrading the vertices in W^*. Let $C_1, \ldots, C_p$ be the active components at the beginning of the iteration. From the symmetry of f it follows that $p \geq 2$. Also, let W be the upgrading set constructed by the algorithm so far and $F \subset E$ be the set of edges whose delay has already been decreased to be at most D.

Assume in the first case that the graph F' consisting of the edges of $F \cup F^*$ is connected.

We now apply Lemma 8 to the graph F' with $M := \{C_1, \ldots, C_p\}$ to obtain a spider covering of M in F'. Let $P_1, \ldots, P_k$ be the spiders in the decomposition. We define the cost $c(P_i)$ of spider P_i to be the sum of the cost of the vertices from $W^* \setminus W$ that are contained in P_i, i.e., the cost of the vertices from the optimum solution that have not been upgraded yet. Since the spiders are node disjoint we have

$$\sum_{i=1}^{k} c(P_i) \leq c(W^*) - c(W) \leq \text{OPT}. \qquad (2)$$

Let $M' \subset M$ denote those clusters which are not covered by the feet of the spiders. Notice that for each such cluster $C \in M'$ we have at least one spider that contains a node from M' as a center. Denote the number of feet in spider P_i by f_i. Then by Lemma 8 we have

$$|M'| + \sum_{i=1}^{k} f_i \geq p. \qquad (3)$$

We will now show the following: If v_i is the center of spider P_i and is contained in an active component C which is not covered by the feet of the spiders in the cover, then the quotient cost of v_i is at most $c(P_i)/(f_i + 1)$. Otherwise we show the slightly weaker estimate that the quotient cost is bounded by $c(P_i)/f_i$.

Let $C_1, \ldots, C_{f_i}$ be the active clusters covered by the feet of the spider centered at v_i. In the first case we have $v_i \notin W^* \setminus W$. Then, the upgraded vertices from $W^* \setminus W$ on the path from v_i to the foot covering C_j are an upgrading set resulting in a bottleneck path of delay at most D from v_i to some node in C_j. Thus, their costs are at least $c^-(v, C_j)$. Since the legs are node disjoint (except for the center v_i which by assumption does not belong to $W^* \setminus W$), we get that

$$\sum_{j=1}^{f_i} c^-(v_i, C_j) \leq c(P_i). \qquad (4)$$

Since the quotient cost of v_i is at most $\sum_{j=1}^{f_i} c(v_i^-, C_j)/f_i$, it follows that the quotient cost of v_i is bounded from above by $c(P_i)/f_i$. Moreover, if $v_i \in C$ and

C is not covered by the feet of the spider in our collection, then in particular C does not occur in the sum on the left hand side of (4). Since $c^-(v_i, C) = 0$ we get

$$c^-(v_i, C) + \sum_{j=1}^{f_i} c^-(v_i, C_j) \le c(P_i).$$

and, consequently, the quotient cost of v_i is at most $c(P_i)/(f_i + 1)$.

In the second case the vertex v_i is in $W^* \setminus W$. In this case, the upgrading vertices from $W^* \setminus W$ on the leg to C_j excluding v_i have cost at least $c^+(v_i, C_j)$. Again, by the node disjointness of the legs we get that

$$c(v_i) + \sum_{j=1}^{f_i} c^+(v_i, C_j) \le c(P_i).$$

Since the quotient cost of v_i is also at most $c(v_i) + \sum_{j=1}^{f_i} c^+(v_i, C_j)$ divided by f_i, we obtain again that the quotient cost of v_i is at most $c(P_i)/f_i$.

The same arguments as above show that if $v_i \in C$ and C is not covered by the feet of the spiders then the quotient cost of v_i can be bounded by $c(P_i)/(f_i + 1)$.

Let v be the node chosen in Step 9 in the current iteration. Then the quotient cost $q(v)$ of v satisfies $q(v) \le q(v_i)$ for $i = 1, \ldots, k$. Thus we get

$$q(v) \cdot f_i' \le c(P_i), \quad \text{for } i = 1, \ldots, k, \tag{5}$$

where $f_i' \in \{f_i, f_i + 1\}$ is chosen as above such that the quotient cost of the center v_i of spider P_i is bounded by $c(P_i)/f_i'$.

Summing up the inequalities in (5) and using (2) and (3) the claim of the lemma follows in this case.

It remains to consider the case that the graph F' consisting of the edges from $F \cup F^*$ is not connected. Notice that if we show that each connected component of F' contains either none or at least two active clusters, we can apply our arguments from above to each of the connected components and the claim of the lemma will follow by summing up over those components that contain active clusters.

Let $C_1, \ldots, C_p$ be the active components and $Z_1, \ldots, Z_t$ be the inactive components at the beginning of the iteration. Notice that each connected component of F' is the disjoint union of some components C_i and Z_j. Assume for the sake of a contradiction that component Z of F' contains exactly one active cluster, say C_1. As noted above, F' can be written as the disjoint union of the connected components at the beginning of the current iteration, so $F' = C_1 \cup Z_1 \cup \cdots \cup Z_{t'}$ for some inactive components Z_j.

Clearly $f(Z) = 0$, since otherwise one connected component of F^* (and thus of F') would contain vertices from Z as well as from $V \setminus Z$ which is not possible. Moreover, $f(Z_j) = 0$, by the definition of an inactive component. By the disjointness of the Z_j we have for $B := Z_1 \cup \cdots \cup Z_{t'}$ that $f(B) = 0$. Now applying Lemma 1 for $U := Z$ and B as defined above yields that $f(C) = 0$ which contradicts the fact that C is an active component.   □

4.3   Potential Function Argument

We are now ready to prove the main result on the performance of our approximation algorithm.

Theorem 10. *Algorithm 1 is an approximation algorithm for* UPGRADING-CFP *with a performance guarantee of* $2\ln(\sqrt{e}/2 \cdot |K|)$. *Here* $K := \{v : f(\{v\}) = 1\}$ *is the set of terminals given by the proper function* f.

Proof. Let the algorithm use l iterations. Notice that $l \leq n$. We let the potential function ϕ_j denote the number of active components at the end of iteration j. Then $\phi_{l-1} \geq 2$ since the algorithm does not terminate before iteration l and $\phi_l = 0$. Now we have

$$\phi_j \leq \phi_{j-1} - (r_j - 1) \leq \phi_{j-1} - \frac{1}{2}r_j \overset{\text{Lemma 9}}{\leq} \phi_{j-1}\left(1 - \frac{c_j}{2\,\text{OPT}}\right),$$

where r_j denotes the number of active components merged in iteration j, and c_j is the cost spent in that iteration. Consequently,

$$\phi_{l-1} \leq \phi_0 \prod_{i=1}^{l-1}\left(1 - \frac{c_j}{2\,\text{OPT}}\right).$$

Taking natural logarithms on both sides and using $\ln(1 - \tau) \leq -\tau$, we obtain

$$\sum_{j=1}^{l-1} c_j \leq 2\,\text{OPT}\cdot\ln\frac{\phi_0}{\phi_{l-1}} \leq 2\,\text{OPT}\cdot\ln\frac{|K|}{2}$$

as a bound for the cost of the upgraded vertices in all but the last iteration. Also, by Lemma 9 the cost of the vertices upgraded in the last iteration is at most $\text{OPT}/(\phi_{l-1} - \phi_l)\cdot\phi_{l-1} = \text{OPT}$ (since $\phi_l = 0$). Thus, the total cost is bounded by

$$\text{OPT}\cdot\left(2\ln\frac{|K|}{2} + 1\right) = \text{OPT}\cdot\left(2\ln(\sqrt{e}/2\cdot|K|)\right)$$

as claimed.   □

5   Hardness Result

This section contains our hardness result for the node upgrading problem under study.

Theorem 11. *Let* $\varepsilon > 0$ *be arbitrary and the proper function* f *specify Steiner Trees. Unless* $\text{NP} \subseteq \text{DTIME}(N^{\mathcal{O}(\log\log N)})$, *there is no approximation algorithm for* UPGRADING-CFP *(with proper function* f) *with performance* $(1 - \varepsilon)\ln|K|$, *where* K *is the set of terminals. This result continues to hold even if* $c(v) = 1$ *for all vertices* $v \in V$.

Proof. We use a reduction from the MIN SET COVER problem. Given an instance with element set Q and subset collection $R \subseteq 2^Q$, we set up a bipartite graph with node set $Q \cup R$. For each $q \in Q$, we add an edge e between node q and Q of weight $d_0(e) := 2$ and $d_1(e) := d_2(e) := 1$. We add a root node connected to all set nodes through edges e of weight $d_0(e) := d_1(e) := d_2(e) := 1$. The proper function f is chosen to reflect a Steiner Tree with terminal set R. The bottleneck constraint is chosen to be 1. All nodes have an upgrade cost of 1.

It is easy to see that a set cover of some size implies a valid upgrade set of the same cost. For the converse, notice that it is of no use to prefer the root node or element nodes over the adjacent set nodes for upgrading. Hence we can assume that a minimum cost upgrade set consists only of set nodes. Consequently, we obtain a set cover of size equal to the upgrade cost.

Since the reduction is approximation preserving, we can apply the non-approximability result of Feige [Fei96] and the claim follows.   □

References

[Ber92]   O. Berman, *Improving the location of minisum facilities through network modification*, Annals of Operations Research **40** (1992), 1–16.

[Fei96]   U. Feige, *A threshold of* ln *n for approximating set cover*, Proceedings of the 28th Annual ACM Symposium on the Theory of Computing (STOC'96), 1996, pp. 314–318.

[GG+94]   M. X. Goemans, A. V. Goldberg, S. Plotkin, D. B. Shmoys, E. Tardos, and D. P. Williamson, *Improved approximation algorithms for network design problems*, Proceedings of the 5th Annual ACM-SIAM Symposium on Discrete Algorithms (SODA'94), January 1994, pp. 223–232.

[GW95]   M. X. Goemans and D. P. Williamson, *A general approximation technique for constrained forest problems*, SIAM Journal of Computing **24** (1995), 296–317.

[KM+97]   S. O. Krumke, M. V. Marathe, H. Noltemeier, R. Ravi, S. S. Ravi, R. Sundaram, and H. C. Wirth, *Improving spanning trees by upgrading nodes*, Proceedings of the 24th International Colloquium on Automata, Languages and Programming (ICALP'97), Lecture Notes in Computer Science, vol. 1256, 1997, pp. 281–291.

[KN+96]   S. O. Krumke, H. Noltemeier, S. S. Ravi, M. V. Marathe, and K. U. Drangmeister, *Modifying networks to obtain low cost trees*, Proceedings of the 22nd International Workshop on Graph-Theoretic Concepts in Computer Science, Cadenabbia, Italy, Lecture Notes in Computer Science, vol. 1197, June 1996, pp. 293–307.

[KR95]   P. Klein and R.Ravi, *A nearly best-possible approximation for node-weighted Steiner trees*, Journal of Algorithms **19** (1995), 104–115.

[Phi93]   C. Phillips, *The network inhibition problem*, Proceedings of the 25th Annual ACM Symposium on the Theory of Computing (STOC'93), May 1993, pp. 288–293.

[PS95]   D. Paik and S. Sahni, *Network upgrading problems*, Networks **26** (1995), 45–58.

Routing in Recursive Circulant Graphs: Edge Forwarding Index and Hamiltonian Decomposition

G. Gauyacq, C. Micheneau, and A. Raspaud *

LaBRI U.M.R. 5800, Université Bordeaux I,
351, cours de la Libération
33405 Talence Cedex, France
raspaud@labri.u-bordeaux.fr

Abstract. The recursive circulant graphs $G(2^m, 4)$ were described in [13] as a concurrent to the hypercube considered as topology for multicomputer networks. In this paper we give the exact value of the edge forwarding index and bisection width of the generalize recursive circulant graphs $G(cd^m, d)$ with $d > c > 0$. Moreover we prove that they admit a Hamiltonian decomposition.

1 Introduction

In all the following we use the definitions and notation of [2]. The graph we deal with here is the circulant graph $G(cd^m, d)$. This graph belongs to a family of circulant graphs denoted by $G(N, d)$ with $N, d, \in \mathbb{N}$. Let us define $G(N, d)$. The set of vertices is $V = \{0, 1, \ldots, N - 1\}$, and the set of edges is $E = \{\{v, w\}, v \in V, w \in V /$ there exists i, $0 \le i \le \lceil log_d N \rceil - 1$, such that $v \pm d^i \equiv w (mod N)\}$.

By definition $G(cd^m, d)$ has cd^m vertices ($0 < c < d$). $V = \{0, \ldots, cd^m - 1\}$ is the set of vertices, and $E = \{\{v, w\}, v \in V, w \in V / \exists i, 0 \le i \le \lceil log_d cd^m \rceil - 1, v \pm d^i \equiv w (mod \ cd^m)\}$. An edge between v and $w = v \pm d^i$ will have the label d^i. It is easy to see that $G(cd^m, d)$ is a Cayley graph defined on the abelian group $(\mathbb{Z}/cd^m \mathbb{Z}, +)$.

The circulant graphs $G(cd^m, d)$ were described in [13] as a new concurrent to the hypercube. They have similar properties, and some of these graphs contains binary and binomial trees which is important for communications.

A regular graph G is said to be Hamiltonian decomposable if it is possible to find k edge-disjoint Hamiltonian cycles if the degree of G is equal to $2k$, or

---

* This research was supported by ALTEC-KIT, Project no INCO-COP 96-0195 and by CEFIPRA-PROJECT No. 1602-1, Indo-French Cooperation.

k edge-disjoint hamiltonian cycles and a perfect matching if the degree of G is equal to $2k + 1$. In [1] Alspach asked the following question :*Is-it the case that every connected Cayley graph $X(G; H)$ on an abelian group G admits a Hamiltonian decomposition.* So we give the answer for the graphs $G(cd^m, d)$. Moreover it is possible to use these cycles to make communication more efficient, or to be fault tolerant. For the known results concerning this question of the Hamiltonian decomposition, see for example [3, 4, 6, 10]. The vertex and edge forwarding index, defined in [5] and [8], are two parameters characterizing the congestion of routings in an interconnection network.

A Hamilton decomposition of the family of graphs $G(2^m, 4)$, and the edge forwarding index of these graphs have already be found [7, 11]. In this paper, we give an exact value for the edge forwarding index of $G(cd^m, d)$, and in a second part we give a Hamilton decomposition of this graph (see [7, 12] for the version in french).

2   Recursive construction

We recall firts that a Hamiltonian cycle in a graph G is a cycle passing through each vertex of G exactly one.

Proposition 1 ([13]) *For $m \geq 1$ and $1 \leq c < d$, the graph $G(cd^m, d)$ contains d vertex-disjoint copies of the graph $G(cd^{m-1}, d)$. The unused edges, labeled 1, form a Hamiltonian cycle.*

Proof : The set of vertices in the graph $G(cd^m, d)$ is $V = \{0, 1, \ldots, cd^m - 1\}$
Let's denote $G_i(cd^m, d)$ the subgraph of $G(cd^m, d)$ induced by $V_i = \{v/v \equiv i (mod\ d)\}$. $G_i(cd^m, d)$ has the same number of vertices as $G(cd^{m-1}, d)$.
Consider now the application from $G(cd^{m-1}, d)$ to $G_i(cd^m, d)$ defined by $u \longrightarrow du + i$. It is easy to see that this application is an isomorphism, and we can find d copies of the graph $G(cd^{m-1}, d)$ in the graph $G(cd^m, d)$.

Now, the set of unused edges is the set of edges labeled 1 which form a hamiltonian cycle. We will call this Hamiltonian cycle the basic cycle of the graph $G(cd^m, d)$.   □

Remark : The edges which do not belong to a subgraph $G_i(cd^m, d)$ are labeled 1; such an edge has one end in V_i, the other in V_{i+1} for some i, $0 \leq i < cd^m$ (the subscripts are computed modulo cd^m). In that way, $G(cd^m, d)$ is a cycle of d copies of $G(cd^{m-1}, d)$; the edges linking the copies i and $i + 1$ constitute a perfect matching between V_i and V_{i+1}. For example, see figure 1. If $d = 2$, we have two disjoint perfect matchings between V_0 and V_1,see figure 2.

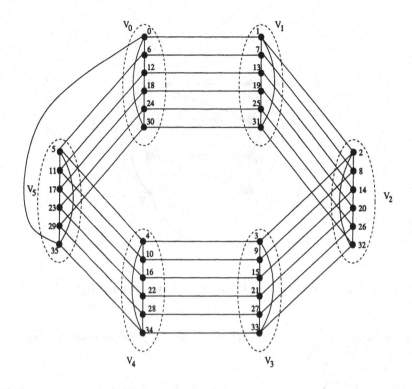

Fig. 1. $G(6^2, 6)$ is a cycle of 6 copies of $G(6, 6)$

3 Forwarding indices

3.1 Definitions

Vertex-forwarding index. Chung, Coffman, Reiman, and Simon introduced in [5] the notion of forwarding index. A *routing* R of a graph Γ of order n is a set of $n(n-1)$ elementary paths $R(u, v)$ specified for all (ordered) pairs (u, v) of vertices of Γ. If all the paths $R(u, v)$ are shortest paths from u to v the routing R is said to be a *routing of shortest paths*.

Let us call the *load of a vertex* v in a given routing R of a graph Γ, denoted by $\xi(\Gamma, R, v)$, the number of paths of R going through v (where v is not an end vertex). A routing for which the load of all vertices is the same will be called a *vertex-uniform routing*. The *vertex-forwarding index* of a network (Γ, R), denoted by $\xi(\Gamma, R)$, is the maximum number of paths of R going through any vertex v in $\Gamma : \xi(\Gamma, R) = \max_{v \in V} \xi(\Gamma, R, v)$. The minimum vertex-forwarding index over all possible routings of a graph Γ will be denoted by $\xi(\Gamma)$ and be called the *vertex-forwarding index* of $\Gamma : \xi(\Gamma) = \min_{R} \xi(\Gamma, R)$. We recall the following result which gives a lower bound for the vertex-forwarding index.

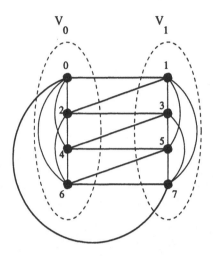

Fig. 2. Construction of $G(2^3, 2)$ with two copies of $G(2^2, 2)$

Proposition 2 ([5]) *Let Γ be a simple connected graph of order n. Then*

(i) $\xi(\Gamma) \geq \frac{1}{n} \sum_{u \in V} \sum_{v \neq u} (d(u, v) - 1)$

(ii) $\xi(\Gamma) = \frac{1}{n} \sum_{u \in V} \sum_{v \neq u} (d(u, v) - 1)$ *if and only if there exists a vertex-uniform routing of shortest paths in Γ.*

M.C. Heydemann, J.C. Meyer and D. Sotteau have proved that in any Cayley graph there exists a vertex-uniform routing of shortest paths ([8]). Since $G(cd^m, d)$ is a Cayley graph, we can calculate its vertex-forwarding index.

Proposition 3 *For $m \geq 3$, the vertex-forwarding index of $G(2^m, 4)$ is given by*

$$\xi(G(2^m, 4)) = \begin{cases} 2^m(\frac{9}{20}m - 0, 92) - 0, 08 \times (-1)^{\frac{m}{2}} + 1 & \text{if } m \text{ is even} \\ 2^m(\frac{9}{20}m - 0, 97) + 0, 04 \times (-1)^{\frac{m-1}{2}} + 1 & \text{if } m \text{ is odd} \end{cases}$$

Proof : For any graph $G(cd^m, d)$, the total distance from vertex 0 to all other vertices in $G(cd^m, d)$ is defined to be $td(G(cd^m, d)) = \sum_{v \in V(G(cd^m, d))} d_G(0, v)$ where $d_G(0, v)$ denotes the distance from 0 to v in $G(cd^m, d)$. Theorem 3 in [13] gives the exact value of $td(G(cd^m, d))$ in any graph $G(cd^m, d)$. For the graph $G(2^m, 4)$ we have :

$$td(G(2^m, 4)) = \begin{cases} 2^m(\frac{9}{20}m + 0, 08) - 0, 08 \times (-1)^{\frac{m}{2}} & \text{if } m \text{ is even} \\ 2^m(\frac{9}{20}m + 0, 03) + 0, 04 \times (-1)^{\frac{m-1}{2}} & \text{if } m \text{ is odd} \end{cases}$$

Since $G(2^m, 4)$ is vertex-transitive, $\frac{1}{|V|} \sum_{u \in V} \sum_{v \neq u} (d(u,v) - 1)$ simplifies to

$\frac{1}{|V|} |V| \sum_{v \neq 0} (d(0,v) - 1)$. This number is equal to $td(G(2^m, 4)) - (2^m - 1)$ and
the result follows.

Edge-forwarding index. M.C. Heydemann, J.C. Meyer and D. Sotteau introduced in [8] the same concepts for the edges of a graph. The *load of an edge*
e in a given routing R of a graph Γ, denoted by $\Pi(\Gamma, R, e)$, is the number of
paths of R going through e. A routing for which the load of all edges is the same
will be called an *edge-uniform routing*. The *edge-forwarding index* of (Γ, R),
denoted by $\Pi(\Gamma, R)$, is the maximum number of paths of R going through any
edge of Γ : $\Pi(\Gamma, R) = \max_{e \in E} \Pi(\Gamma, R, e)$ and the edge-forwarding index of Γ is
defined as $\Pi(\Gamma) = \min_R \Pi(\Gamma, R)$. The following result gives a lower bound for
the edge-forwarding index.

Proposition 4 ([8]) *Let $\Gamma = (V, E)$ be a simple connected graph of order n.*
(i) $\Pi(\Gamma) \geq \frac{1}{|E|} \sum_{(u,v) \in V \times V} d(u,v)$
(ii) $\Pi(\Gamma) = \frac{1}{|E|} \sum_{(u,v) \in V \times V} d(u,v)$ *if and only if there exists in Γ an edge-uniform*
routing of shortest paths.

In the next section we shall give the edge-forwarding index of recursive circulant graphs. Since it is different from the lower bound of proposition 4 we shall
conclude that there is no edge-uniform routing of shortest paths in recursive
circulant graphs.

3.2   Edge-forwarding index of recursive circulant graphs

Let c and d be two positive integers with $1 \leq c < d$. We first give some notations
to be used throughout this section.
Let V be the vertex set of $G(cd^m, d)$, that is the set of integers between 0 and
$cd^m - 1$. For any i, $0 \leq i \leq d - 1$, let $V_i = \{v \in V/ \ v \equiv i \pmod{d}\}$.
If x is an integer modulo cd^m, $|x|_d$ denotes the only integer between 0 and $d/2$
such that $|x|_d \equiv x \pmod{d}$ or $|x|_d \equiv -x \pmod{d}$.
If $u \in V_i$ and $v \in V_j$ let $\delta_{mod\,d}(u,v) = |i - j|_d$ (i.e. the distance in a d-cycle
between vertices i and j).
Let π_m denote the edge-forwarding index of $G(cd^m, d)$.

Inequalities between π_m and π_{m-1}

Proposition 5 *Let $m \geq 1$ be an integer. We have*

$$cd^{m-1}\left\lfloor\frac{d^2}{4}\right\rfloor \leq \pi_m \leq \max\left\{d\pi_{m-1}, cd^{m-1}\left\lfloor\frac{d^2}{4}\right\rfloor\right\}.$$

Proof : Let R be an arbitrary routing of $G(cd^m, d)$. A path of R from a vertex u to a vertex v contains at least $\delta_{mod\,d}(u, v)$ edges having the label 1. Thus the sum of the loads induced on the edges of label 1 by all the paths of R is at least

$$\sigma = \sum_{(u,v)\in V\times V} \delta_{mod\,d}(u, v).$$

$\sum_{v\neq u}\delta_{mod\,d}(u, v)$ is the same for each vertex u, so we have $\sigma = cd^m\sum_{v\neq 0}\delta_{mod\,d}(0, v)$.

Now we note that, if $1 \leq i < d/2$, there exist $2cd^{m-1}$ vertices such that $\delta_{mod\,d}(0, v) = i$: those in V_i and those in V_{d-i}. For d even, there exist cd^{m-1} vertices such that $\delta_{mod\,d}(0, v) = d/2$: those in $V_{d/2}$. Therefore,

$$\sigma = \begin{cases} cd^m(2cd^{m-1} \times 1 + 2cd^{m-1} \times 2 + \ldots + 2cd^{m-1}\lfloor\frac{d}{2}\rfloor) & \text{if } d \text{ is odd} \\ cd^m(2cd^{m-1} \times 1 + 2cd^{m-1} \times 2 + \ldots + 2cd^{m-1}(\frac{d}{2}-1) + cd^{m-1}\frac{d}{2}) & \text{if } d \text{ is even} \end{cases}$$

Combining these equalities yields

$$\sigma = cd^m\left(cd^{m-1}\left\lfloor\frac{d^2}{4}\right\rfloor\right).$$

In $G(cd^m, d)$ there are cd^m edges carrying the label 1, and the maximum number of paths passing through an edge is more than the average number, so we get the lower bound :

$$\pi_m \geq \frac{\sigma}{cd^m} = cd^{m-1}\left\lfloor\frac{d^2}{4}\right\rfloor.$$

For the upper bound we are going to define a routing R_m of $G(cd^m, d)$ for which the load of an edge is either less than $d\pi_{m-1}$ or less than $cd^{m-1}\left\lfloor\frac{d^2}{4}\right\rfloor$.

Let R_{m-1} be a routing of $G(cd^{m-1}, d)$ with a minimum edge-forwarding index, that is $\Pi(G(cd^{m-1}, d), R_{m-1}) = \pi_{m-1}$.
Let $u \in V_i$ and $v \in V_j$. If $i = j$, i.e u and v are in the same copy of $G(cd^{m-1}, d)$ the path $R_m(u, v)$ is simply the path $R_{m-1}(u, v)$ in this copy.
If $i \neq j$ we choose a nearest vertex of v in V_i. If d is even and $|i - j| = d/2$ two choices exist : $v - d/2$ and $v + d/2$; we take $u' = v - d/2$. In the other cases there is only one nearest vertex of v in V_i :

$$u' = \begin{cases} v - |i - j| & \text{if } |i - j| < d/2 \\ v + d - |i - j| & \text{if } |i - j| > d/2 \end{cases}.$$

The path $R_m(u, v)$ is defined as the path $R_{m-1}(u, u')$ in copy i of $G(cd^{m-1}, d)$, which is followed by the shortest path from u' to v, which consists only of edges carrying the label 1.

Let us find the load of an edge e which is not labeled 1. Assume the ends of e are in V_i. All the paths going through e begin in V_i. For fixed j, the load of e induced by the paths $R_m(u, v)$, $u \in V_i$, $v \in V_j$ is $\Pi(G(cd^{m-1}, d), R_{m-1}, e)$ which is less than π_{m-1}. Since there exist d choices for j we get $\Pi(G(cd^m, d), R_m, e) \le d\pi_{m-1}$.

Let us find the load of an edge e which is labeled 1. Let $e = \{x, x+1\}$ with $x \in V_i$. The paths passing through e are :
- the paths going through e from x to $x + 1$, i.e. the paths $R_m(u, x + k)$ for any $k, 1 \le k \le d/2$ and any u in $V_{i-\ell}$ for $0 \le \ell \le d/2 - k$; their number is $cd^{m-1}(1 + 2 + \ldots + \lfloor d/2 \rfloor)$.
- the paths going through e from $x + 1$ to x, i.e. the paths $R_m(u, x - k)$ for any $k, 0 \le k < d/2 - 1$ and any u in $V_{i+\ell}$ for $1 \le \ell < d/2 - k$; their number is
$$\begin{cases} cd^{m-1}(1 + 2 + \ldots + \lfloor d/2 \rfloor) & \text{if } d \text{ is odd} \\ cd^{m-1}(1 + 2 + \ldots + d/2 - 1) & \text{if } d \text{ is even} \end{cases}$$
By combining these equalities we find the load of e : $cd^{m-1} \left\lfloor \frac{d^2}{4} \right\rfloor$.

In [7], this proof is generalized to find lower and upper bounds for the edge-forwarding index of compound graphs.

The case $(c, d) \ne (2, 3)$

Theorem 1 If $m \ge 1$, $d \ge 2$, $(m, d) \ne (1, 2)$ and $(c, d) \ne (2, 3)$ then the edge-forwarding index of $G(cd^m, d)$ is $cd^{m-1} \left\lfloor \frac{d^2}{4} \right\rfloor$.

Proof : 1) If $d = 2$, then $c = 1$ since $1 \le c < d$. The graph $G(2^1, 2)$ is the complete graph on 2 vertices thus $\pi_1 = 2$ and $G(2^2, 2)$ is the complete graph on 4 vertices thus $\pi_2 = 2$. If we suppose $\pi_{m-1} = 2^{m-2}$, by proposition 5 we get $\pi_m = 2^{m-1}$.

2) Now, we assume $d \ge 3$. The proof is also by induction on m. Since $G(cd^0, d)$ is the c-cycle its edge-forwarding index is $\pi_0 = \left\lfloor \frac{c^2}{4} \right\rfloor$ if $c \ne 2$ and $\pi_0 = 2$ if $c = 2$ ([8]). By proposition 5, we get

$$c \left\lfloor \frac{d^2}{4} \right\rfloor \le \pi_1 \le \max \left\{ c \left\lfloor \frac{d^2}{4} \right\rfloor, d\pi_0 \right\}.$$

By the definition, $c \leq d-1$ thus, if $c \neq 2$, $d\pi_0 = d \left\lfloor \frac{c^2}{4} \right\rfloor \leq \frac{c^2 d}{4} \leq c\frac{d(d-1)}{4} < c \left\lfloor \frac{d^2}{4} \right\rfloor$.

If $c = 2$, we have $d\pi_0 = 2d \leq 2 \left\lfloor \frac{d^2}{4} \right\rfloor$ (Recall that $(c,d) \neq (2,3)$). Therefore

$$c \left\lfloor \frac{d^2}{4} \right\rfloor \leq \pi_1 \leq c \left\lfloor \frac{d^2}{4} \right\rfloor$$

and the theorem holds for $m = 1$.

Suppose, now, that $\pi_{m-1} = cd^{m-2} \left\lfloor \frac{d^2}{4} \right\rfloor$. By proposition 5, we get

$$cd^{m-1} \left\lfloor \frac{d^2}{4} \right\rfloor \leq \pi_m \leq \max \left\{ d\pi_{m-1}, cd^{m-1} \left\lfloor \frac{d^2}{4} \right\rfloor \right\}.$$

It follows from the induction hypothesis that $\pi_m = cd^{m-1} \left\lfloor \frac{d^2}{4} \right\rfloor$.

Corollary 1 *The edge-forwarding index of $G(2^m, 4)$ is 2^m.*

Proof : We apply theorem 1 for $c = 1, d = 4$ when m is even and for $c = 2, d = 4$ when m is odd.

Corollary 2 *For $m \geq 3$ there is no edge-uniform routing of shortest paths in $G(2^m, 4)$.*

Proof : By proposition 4, $\Pi(G(2^m, 4)) = \frac{1}{|E|} \sum_{(u,v) \in V \times V} d(u, v)$ if and only if there exists in $G(2^m, 4)$ an edge-uniform routing of shortest paths. We are going to compute $\frac{1}{|E|} \sum_{(u,v) \in V \times V} d(u, v)$ in $G(2^m, 4)$. Let Σ be this number.

As said in the proof of proposition 3 $td(G(cd^m, d)) = \sum_{v \in V(G(cd^m, d))} d_G(0, v)$ is given by theorem 3 in [13]. We have :

$$td(G(2^m, 4)) = \begin{cases} 2^m(\frac{9}{20}m + 0,08) - 0,08 \times (-1)^{\frac{m}{2}} & \text{if } m \text{ is even} \\ 2^m(\frac{9}{20}m + 0,03) + 0,04 \times (-1)^{\frac{m-1}{2}} & \text{if } m \text{ is odd} \end{cases}$$

Since $G(2^m, 4)$ is vertex-transitive, $\sum_{(u,v) \in V \times V} d(u, v)$ simplifies to $|V| \, td(G(2^m, 4))$.

Since $G(2^m, 4)$ has degree m, it has $\frac{m}{2}|V|$ edges and $\Sigma = \frac{2 \, td(G(2^m, 4))}{m}$.

For $m \geq 3$, $\Sigma \neq 2^m$; thus we can conclude. In figure 3, we compare $\lceil \Sigma \rceil$ with 2^m for $2 \leq m \leq 8$.

m	2	3	4	5	6	7	8
2^m	4	8	16	32	64	128	256
$\lceil \Sigma \rceil$	4	8	15	30	60	117	236

Fig. 3. Comparison between 2^m and Σ

Remark : In [13] the graph $G(2^m, 4)$ is compared with the hypercube Q_m. We can notice that they have the same edge-forwarding index 2^m, but unlike in $G(2^m, 4)$, the routing of Q_m for which the edge-forwarding index is 2^m is an edge-uniform routing of shortest paths (for edge-forwarding index of Q_m, see [8]).

The case $c = 2, d = 3$

Proposition 6 *If $m \geq 1$, then*

$$\left\lceil \frac{8m + 6}{2m + 1} 3^{m-1} \right\rceil \leq \Pi(G(2 \times 3^m, 3)) \leq 5 \times 3^{m-1}.$$

Proof : 1) By proposition 4, $\Pi(G(2 \times 3^m, 3)) \geq \frac{1}{|E|} \sum\limits_{(u,v) \in V \times V} d(u, v)$. Let us determine this lower bound.

By theorem 3 in [13] we know the exact value of $td(G(cd^m, d))$ in any graph $G(cd^m, d)$. For odd d, we have

$$td(G(cd^m, d)) = cd^m \left(\frac{d^2 - 1}{4d} m + \frac{\lfloor c^2/4 \rfloor}{c} \right).$$

Applying this theorem to $c = 2, d = 3$ yields

$$td(G(2 \times 3^m, 3)) = (4m + 3)3^{m-1},$$

thus $\sum\limits_{(u,v) \in V \times V} d(u, v) = |V| \, td(G(2 \times 3^m, 3)) = |V| \, (4m + 3)3^{m-1}$.

Since $G(2 \times 3^m, 3)$ has degree $2m + 1$, it has $|V| \frac{2m+1}{2}$ edges and the lower bound is

$$\frac{|V| \, td(G(cd^m, d))}{|V| \frac{2m+1}{2}} = \frac{2 \, td(G(cd^m, d))}{2m + 1} = \frac{8m + 6}{2m + 1} 3^{m-1}.$$

Hence

$$\pi_m = \Pi(G(2 \times 3^m, 3)) \geq \frac{8m + 6}{2m + 1} 3^{m-1} > 4 \times 3^{m-1}. \qquad (1)$$

2) To find the upper bound we shall proceed by induction on m.

π_1 : The graph $G(2 \times 3^1, 3)$ is shown in figure 4. Let $R(x, x+2) = (x, x+1, x+2)$ and $R(x, x+4) = (x, x+3, x+4)$; the other paths are single edges. It is not difficult to verify that the edge-forwarding index of this routing is 5, thus $\pi_1 \leq 5$. On the other hand, by inequality (1), $\pi_1 \geq 5$ hence $\pi_1 = 5$ and the proposition is proved for $m = 1$.

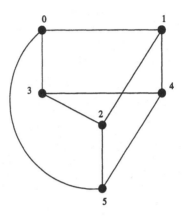

Fig. 4. The graph $G(2 \times 3^1, 3)$

Suppose now that $\pi_{m-1} \leq 5 \times 3^{m-2}$. By proposition 5 :

$$\pi_m \leq \max\left\{3\pi_{m-1}, 2 \times 3^{m-1} \left\lfloor \frac{3^2}{4} \right\rfloor\right\}$$

i.e. $\pi_m \leq \max\left\{3\pi_{m-1}, 4 \times 3^{m-1}\right\}$.

It follows from the induction hypothesis and the inequality (1) that $4 \times 3^{m-1} < 3\pi_{m-1} \leq 5 \times 3^{m-1}$ and then $\pi_m \leq 5 \times 3^{m-1}$.

3.3   Bisection width of $G(cd^m, d)$

The *bisection width* of a graph Γ, denoted by $Bisw(\Gamma)$, is the minimum number of edges that have to be removed in order to disconnect Γ into two subgraphs with identical (within one) number of vertices. Let N be the order of Γ. It is known that

$$\Pi(\Gamma)Bisw(\Gamma) \geq \left\lfloor \frac{N^2}{2} \right\rfloor \quad \text{(see [14] th. 3.9.6 p.119)}.$$

Theorem 2 *If $d \geq 2$ is even, the bisection width of $G(cd^m, d)$ is $2cd^{m-1}$.*

Proof : If d is even, $\left\lfloor \frac{d^2}{4} \right\rfloor = \frac{d^2}{4}$. Since $\Pi(G(cd^m, d)) = cd^{m-1} \left\lfloor \frac{d^2}{4} \right\rfloor$ we have

$$Bisw(G(cd^m, d)) \geq \frac{(cd^m)^2}{2cd^{m-1}\frac{d^2}{4}} = 2cd^{m-1}.$$

It is easy to see that the deletion of $2cd^{m-1}$ edges can disconnect $G(cd^m, d)$ into two halves with identical number of vertices. Just remove the edges having one end in V_0 and one end in V_1 and those having one end in $V_{d/2}$ and one end in $V_{d/2+1}$. For example, see figure 5.

Thus $Bisw(G(cd^m, d)) \leq 2cd^{m-1}$ and so $Bisw(G(cd^m, d)) = 2cd^{m-1}$.

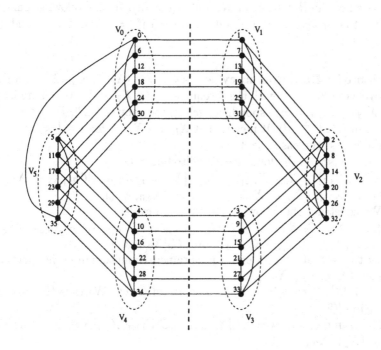

Fig. 5. Minimum bisection of $G(6^2, 6)$

Corollary 3 *The bisection width of* $G(2^m, 4)$ *is* 2^{m-1}.

4   Hamilton decomposition

In the following, we will prove that the graphs $G(cd^m, d)$ are Hamiltonian decomposable. This result is a generalization of the one obtained for the graphs $G(2^m, 4)$ published in [11].

4.1 Disjoint Hamiltonian cycles

Definitions. If $C = c_1 c_2 \ldots c_p c_1$ is a cycle of a graph G, we let $C[c_i, c_j]$ be the subpath $c_i c_{i+1} \ldots c_j$ and $\bar{C}[c_j, c_i]$ the subpath $c_j c_{j-1} \ldots c_i$, where the indices are taken modulo p. We will say that a Hamiltonian cycle of $G(cd^m, d)$ contains a $3 - sequence$, if there exist three consecutive vertices of the Hamiltonian cycle labeled by three consecutive integers (the integers are taken modulo cd^m).

By using the fact that $G(cd^{m+1}, d)$ contains d copies of $G(cd^m, d)$, we will construct a Hamiltonian cycle of $G(cd^{m+1}, d)$ containing a 3-sequence from a Hamiltonian cycle of $G(cd^m, d)$ containing a 3-sequence. We will say that we have made *an extension* of the Hamiltonian cycle of $G(cd^m, d)$. In the following paragraph we assume that $m \geq 2$ if $c = 1$ and $d = 2$, $m \geq 1$ if $c = 2$ and if $c = 1$ and $d \geq 3$, $m \geq 0$ if $c \geq 3$.

Extension of a Hamiltonian cycle. Let $C = X_1 X_2 \ldots X_{cd^m} X_1$ be a Hamiltonian cycle of $G(cd^m, d)$ with the 3-sequence $\{X_1, X_2, X_3\}$. We denote by $C_i = X_{1,i} \ldots X_{cd^m, i} X_{1,i}$ the corresponding cycle in the copy $G_i(cd^m, d)$ $(0 \leq i \leq d-1)$ contained in $G(cd^{m+1}, d)$. It means that $X_{l,i} = dX_l + i$ for $1 \leq l \leq cd^m$.

We consider first the case $d \geq 3$.

d even. We construct the following Hamiltonian cycle:

$X_{1,0} \bar{C}_1 [X_{1,1}, X_{3,1}] C_2 [X_{3,2}, X_{1,2}] \ldots \bar{C}_{2k-1} [X_{1,2k-1}, X_{3,2k-1}] C_{2k} [X_{3,2k}, X_{1,2k}] \ldots$
$\bar{C}_{d-1} [X_{1,d-1}, X_{3,d-1}] X_{2,d-1} X_{2,d-2} \ldots X_{2,0} C_0 [X_{2,0}, X_{1,0}].$

d odd. We construct the following Hamiltonian cycle:

$X_{1,0} \bar{C}_1 [X_{1,1}, X_{3,1}] C_2 [X_{3,2}, X_{1,2}] \ldots \bar{C}_{2k-1} [X_{1,2k-1}, X_{3,2k-1}] C_{2k} [X_{3,2k}, X_{1,2k}] \ldots$
$C_{d-1} [X_{3,d-1}, X_{1,d-1}] X_{2,d-1} X_{2,d-2} \ldots X_{2,0} C_0 [X_{2,0}, X_{1,0}].$

It is easy to see that in each case the obtained Hamiltonian cycle contains the 3-sequence $\{X_{2,0}, X_{2,1}, X_{2,2}\}$.

If $d = 2$, $G(2^{m+1}, 2)$ contains only two copies of $G(2^m)$. We take the 2-sequence $\{X_1, X_2\}$ in $G(2^m)$.

Then the Hamiltonian cycle $C_0 [X_{2,0}, X_{1,0}] \bar{C}_1 [X_{1,1}, X_{2,1}] X_{2,0}$ contains the 2-sequence $\{X_{2,0}, X_{2,1}\}$.

We leave the proof of the following proposition to the reader:

Proposition 7 *The not used edges by the extension which belong to the basic cycle in the subpath linking $X_{1,0}$ to $X_{3,d-1}$ and the not used edges of the C_i's form a path from $X_{1,0}$ to $X_{3,d-1}$ containing all the vertices $X_{i,j}$ $(1 \leq i \leq 3)$, $0 \leq j \leq d - 1$.*

In the figure 6 we have an example for $d = 4$ and 5 of an extended Hamiltonian cycle. The dotted lines form the path linking $X_{1,0}$ to $X_{3,d-1}$ composed of the not used edges of the C_i's and the basic cycle. Now we can establish the following easy proposition:

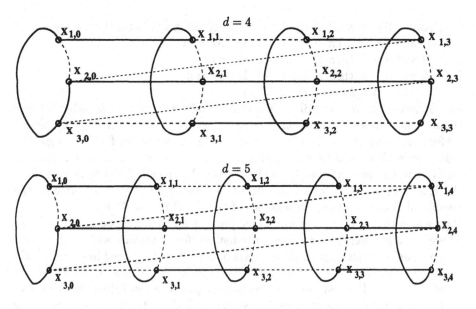

Fig. 6. Extension of a Hamiltonian cycle

Proposition 8 *Let* $C = X_1 X_2 \ldots X_{cd^m}$ *and* $D = Y_1 Y_2 \ldots Y_{cd^m}$ *be two edge-disjoint Hamiltonian cyles of* $G(cd^m, d)$, *each of them containing respectively a 3-sequence* $\{X_1, X_2, X_3\}$ *and* $\{Y_1, Y_2, Y_3\}$ *such that* $\{X_1, X_2, X_3\} \bigcap \{Y_1, Y_2, Y_3\} = \emptyset$. *Then the corresponding extended Hamiltonian cycles in* $G(cd^{m+1}, d)$ *are edge-disjoint and each of them contains respectively the 3-sequence* $\{X_{2,0}, X_{2,1}, X_{2,2}\}$ *and* $\{Y_{2,0}, Y_{2,1}, Y_{2,2}\}$ *such that* $\{X_{2,0}, X_{2,1}, X_{2,2}\} \bigcap \{Y_{2,0}, Y_{2,1}, Y_{2,2}\} = \emptyset$.

Proof : The edges of C_i and D_i $(0 \leq i \leq d-1)$ are distinct by hypothesis. For the two extensions we use the edges labeled 1 having as extremities $\{X_{1,i}, X_{2,i}, X_{3,i}\}$ for the cycle C and the ones that have as extremities $\{Y_{1,i}, Y_{2,i}, Y_{3,i}\}$ for the cycle D. These vertices are different by hypothesis. It is then clear that the used edges are different. This completes the proof.   □

In the following, if 3-sequences have an empty intersection, we will say that they are *disjoint*.

Hamiltonian decomposition. We are now ready to prove the main theorem:

Theorem 3 $G(cd^m, d)$ *is Hamiltonian decomposable* $(1 \leq c < d)$.

Proof : We recall that $G(cd^m, d)$ is a regular graph, let Δ be its degre.

If $c = 1$ and $d = 2$, then $\Delta = 2m - 1$. Let $M = m - 1$.

If $c = 1$ and $d \geq 3$, then $\Delta = 2m$. Let $M = m$.

If $c = 2$ and $d \geq 3$, then $\Delta = 2m + 1$. Let $M = m$.

If $c \geq 3$, then $\Delta = 2m + 2$. Let $M = m$.

By induction on m, we will prove that $G(cd^m, d)$ contains M edge-disjoint Hamiltonian cycles with disjoint 3-sequences (2-sequences, if $d = 2$) plus a perfect matching if $c = 1$, $d = 2$, or $c = 2$. As a matter of fact we begin the induction with $m = 0$ for $c \geq 3$, and with $m = 1$ for $c = 1$ or 2, it is easy to verify that the property holds in these cases. We recall that the basic cyle is the Hamiltonian cycle, which has its edges labeled by 1. It is for us now the Hamiltonian cycle linking the d copies of $G(cd^m, d)$ in $G(cd^m, d)$.

We assume that it is true for m and we prove it for $m + 1$. If we are in the case where there is a perfect matctching, the new perfect matching will be the union of the d perfect matchings of the d copies of $G(cd^m, d)$ contained in $G(cd^{m+1}, d)$. Hence we extend the M edge-disjoint Hamiltonian cycles of $G(cd^m, d)$ in $G(cd^{m+1}, d)$. By Proposition 2 the extensions give M edge-disjoint Hamiltonian cycles with disjoint 3-sequences (2-sequences, if $d = 2$). It remains to prove that the not used edges by the matching and by the M edge-disjoint extended Hamiltonian cycles form a Hamiltonian cycle with a 3-sequence different from the others. We recall that, since the 3-sequences (2-sequences) are disjoint, the edges of the basic cyle, which are used in the M extensions, are all different. We have to consider two cases:

(i) $d = 2$.

It remains the edges labeled 1 of the basic cycle except for the edges of type $\{X_{1,0}, X_{1,1}\}$, $\{X_{2,0}, X_{2,1}\}$ and $\{X_{1,0}, X_{2,0}\}$, $\{X_{1,1}, X_{2,1}\}$ for each cycle. In he basic cycle we replace the subpath $X_{1,0}X_{1,1}X_{2,0}X_{2,1}$ by the path $X_{1,0}X_{2,0}X_{1,1}X_{2,1}$. Since the 2-sequences are disjoint we get a Hamiltonian cycle. There is at least one vertex S of V_0 which is not in a 2-sequence of $G(2^m, 2)$. Then $\{S, S + 1\}$ is a 2-sequence of the new Hamilonian cycle disjoint from the other 2-sequences.

(ii) $d \geq 3$.

- *d even.* In the basic cyle we replace for each 3-sequence the subpath
$X_{1,0}X_{1,1}\ldots X_{1,d-1}X_{2,0}X_{2,1}\ldots X_{2,d-1}X_{3,0}X_{3,1}\ldots X_{3,d-1}$ by the path
$X_{1,0}X_{2,0}X_{1,d-1}X_{2,d-1}X_{3,0}X_{3,1}X_{2,1}X_{1,1}X_{1,2}X_{2,2}X_{3,2}\cdots$
$\ldots X_{1,d-2}X_{2,d-2}X_{3,d-2}X_{3,d-1}.$

- *d odd.* In the basic cyle we replace for each sequence the subpath
$X_{1,0}X_{1,1}\ldots X_{1,d-1}X_{2,0}X_{2,1}\ldots X_{2,d-1}X_{3,0}X_{3,1}\ldots X_{3,d-1}$ by the path
$X_{1,0}X_{2,0}X_{1,d-1}X_{1,d-2}X_{2,d-2}X_{3,d-2}X_{3,d-3}X_{2,d-3}X_{1,d-3}\cdots$
$\ldots X_{1,1}X_{2,1}X_{3,1}X_{3,0}X_{2,d-1}X_{3,d-1}.$

Since the 3-sequences are disjoint we get a Hamiltonian cycle formed by the not used edges of the basic cycle and the M extended Hamiltonian cycles. Moreover there is at least one vertex S of V_0 which is not in a 3-sequence of $G_0(cd^m, d)$. Then $\{S, S+1, S+2\}$ is a 3-sequence of the new Hamilonian cycle disjoint from the others.

This completes the proof.   □

5   Conclusion

The study of the recusive circulant graphs $G(cd^m, d)$ and more particularly of $G(4^m, 4)$ was motivated by the fact that these graphs have good structural properties inducing good routing capabilities. All these properties are of interest from the interconnection networks point of view, for more informations see [9, 14].

References

1. B. Alspach, *Discrete Mathematics, problems*, 50 (1984) pp. 115.
2. C. Berge, Graphs and Hypergraphs (North-Holland, Amsterdam, 2nd ed., 1976).
3. D. Barth and A. Raspaud, Two disjoint Hamiltonian cycles in the butterfly graph, *Inf. Proc. Letters*, 51 (1994) pp. 175-179.
4. J.C. Bermond, O. Favaron and M. Maheo, Hamiltonian decomposition of Cayley graphs of degree 4, *J. Comb. Th. B*, 46 (1989) pp. 142-153.
5. F.R.K. Chung, E.G. Coffman, M.I. Reiman, B. Simon, *The forwarding index of communication Networks*, IEEE Transactions on Information Theory 39 $n^o 2(1987)$, pp. 224-232
6. S.J. Curran, J. A. Gallian, Hamiltonian cycles and paths in Cayley graphs and digraphs, *Discrete Math.* 156, No.1-3,(1996) pp. 1-18 .
7. G. Gauyacq, Routages uniformes dans les graphe sommet-transitifs, Phd Thesis (1995), LaBRI, Université Bordeaux I, France.
8. M.C. Heydemann, J.C. Meyer, D. Sotteau, *On forwarding indices of networks*, *Discrete Applied Maths* 23 (1989) pp. 103-123
9. F.T. Leighton, Introduction to Parallel Algorithms and Architecture: arrays, trees, hypercubes, *Morgan Kaufman Publisher*, (1992).
10. J. Liu, Hamiltonian decompositions of Cayley graphs on abelian groups, *Discrete Math.*, 131 (1994) pp. 163-171.
11. C. Micheneau, Disjoint Hamiltonian cycles in recursive circulant graphs, *Inf. Proc. Letters*, 61 (1997) pp. 259-264.
12. C. Micheneau, Graphes récursifs circulants, communication vagabondes et simulation, Phd Thesis (1996), LaBRI, Université Bordeaux I, France.
13. Jung-Heum Park, Kyung-Yong Chwa, Recursive circulant : A new topology for multicomputer networks, *Proc. Int. Int. Symp. Parallel Architectures, Algorithms and Networks ISPAN'94*, Japan, IEEE (1994) 73-80.
14. J. de Rumeur, *Communications dans les réseaux de processeurs*, Masson (1994)

Improved Compressions of Cube-Connected Cycles Networks *
(Extended Abstract)

Ralf Klasing

Department of Computer Science
University of Warwick
Coventry CV4 7AL, England
rak@dcs.warwick.ac.uk

Abstract. We present a new technique for the embedding of large cube-connected cycles networks (CCC) into smaller ones, a problem that arises when algorithms designed for an architecture of an ideal size are to be executed on an existing architecture of a fixed size. Using the new embedding strategy, we show that the CCC of dimension l can be embedded into the CCC of dimension k with *dilation* 1 and optimum *load* for any $k, l \in I\!N$, $k \geq 8$, such that $\frac{5}{3} + c_k < \frac{l}{k} \leq 2$, $c_k = \frac{4k+3}{3 \cdot 2^{2/3k}}$, thus improving known results. Our embedding technique also leads to improved dilation 1 embeddings in the case $\frac{3}{2} < \frac{l}{k} \leq \frac{5}{3} + c_k$.

1 Introduction

Over the past few years, a lot of research has been done in the field of interconnection networks for parallel computer architectures (for an overview, cf. [19]). Much of the work has been focused on the capability of certain networks to simulate other network or algorithm structures, in order to execute parallel algorithms of a special structure efficiently on different processor networks (see e.g. [5, 17, 25]). One problem that is of specific interest in this context is that many existing algorithms are designed for arbitrarily large networks (see e.g. [19]), whereas, in practice, the processor network will be fixed and of smaller size. Thus, the larger network must be simulated in an efficient way on the smaller target network. There is an enormous literature on this problem (see e.g. [3, 8, 14, 15, 21, 23, 24, 26, 30]).

Customarily, the *simulation* problem is formalized as the *emdedding* problem of one graph in another (for a formal definition of the *embedding* problem, see Section 2). The "quality" of an embedding is measured by the parameters *load*, *dilation*, and *congestion*. The importance of the different parameters becomes apparent through the following result.

---
* This work was partially supported by EU ESPRIT Long Term Research Project ALCOM-IT under contract no. 20244.

Proposition 1 [20]:

> *If there is an embedding of G into H with load ℓ, dilation d, and conges-*
> *tion c, then there is a simulation of G by H with slowdown $O(\ell + d + c)$.*

As a consequence, the load ℓ, dilation d, and congestion c have been investigated for embeddings between many common network structures like hypercubes, binary trees, meshes, shuffle-exchange networks, deBruijn networks, cube-
connected cycles, butterfly networks, etc. Most of the work was done on one-to-
one embeddings (for an overview, see e.g. [25, 29]), but results on many-to-one
embeddings can also be found (see e.g. [2, 6, 7, 9, 12, 13, 16, 18, 22, 26, 27]). In this
paper, we focus on many-to-one embeddings of the cube-connected cycles network (CCC). The CCC was introduced as a network for parallel processing in
[28]. It has fixed degree, small diameter, and good routing capabilities [19]. It
can execute the important class of *normal* hypercube algorithms very efficiently
(see e.g. [19]). In addition, there is also a strong structural relationship to the
deBruijn, shuffle-exchange, and butterfly networks [1, 10]. Hence, the efficient
implementation of algorithms on CCC networks (of fixed size) is of importance.
According to Proposition 1, one way of executing algorithms designed for a CCC
network of arbitrary size efficiently on a CCC network of realistic (fixed) size,
is to find embeddings of large CCC's into small CCC's minimizing the parameters *load, dilation,* and *congestion.* In this paper, we focus on *load* and *dilation.*
Using our embedding strategy, many important algorithms for large CCC's can
be implemented very efficiently on a CCC network of realistic size.

Many-to-one embeddings of the CCC network have been investigated in [2, 6, 12,
16, 27]. In [6, 12, 27], embeddings with optimum dilation and load are presented
in the case of embedding CCC's of dimension l into k where $k|l$. The authors also
restrict themselves to special kinds of embeddings of a very regular structure, like
coverings [6], homogeneous emulations [12], and homomorphisms [27]. Because
of the very restricted nature, Bodlaender [6] and Peine [27] are also able to
classify their embeddings completely. In [2], a general procedure is described
for mapping parallel algorithms into parallel architectures. This procedure is
applied to the CCC network achieving dilation 1, but very high load. Also,
only special kinds of embeddings, so-called contractions, are considered. In [16],
the embedding problem for CCC's is investigated taking into account general
embedding functions and any possible network dimension. More precisely, it is
proved that the *cube-connected cycles* network of dimension l, $CCC(l)$, can be
embedded into $CCC(k)$, $l > k$, with

1.) *dilation 2 and optimum load* $\left\lceil \dfrac{l}{k} \cdot 2^{l-k} \right\rceil$.

2.) *dilation 1 and load*

$$\begin{cases} \left\lceil \dfrac{l}{k} \cdot 2^{l-k} \right\rceil & \text{for } \dfrac{l}{k} \geq 2, \\[3mm] \left\lceil \dfrac{2p-1}{p} \cdot 2^{l-k} \right\rceil & \text{for } p \in \{2, 3, \ldots\} \text{ such that } \dfrac{2p-3}{p-1} < \dfrac{l}{k} \leq \dfrac{2p-1}{p}. \end{cases}$$

In this paper, we present a new technique for the embedding of large cube-connected cycles networks into smaller ones. Using the new embedding strategy, we show:

Let $k, l \in \mathbb{N}$, $k \geq 8$, such that $\frac{5}{3} + c_k < \frac{l}{k} \leq 2$, $c_k = \frac{4k + 3}{3 \cdot 2^{2/3k}}$.
Then, there is a dilation 1 embedding of $CCC(l)$ into $CCC(k)$ with load
$$\left\lceil \frac{l}{k} \cdot 2^{l-k} \right\rceil.$$

This is optimal, and improves the results from [16]. Our embedding technique also leads to improved dilation 1 embeddings in the case $\frac{3}{2} < \frac{l}{k} \leq \frac{5}{3} + c_k$.

The general strategy of the embeddings is the same as in [16], namely to map 2^{l-k} cycles in $CCC(l)$ of length l onto one cycle in $CCC(k)$ of length k and to allocate the nodes of the guest cycles as balancedly as possible on the host cycle. But in order to improve the results from [16], a completely different way of allocating the guest nodes on the host cycle is introduced.

The paper is organized as follows. Section 2 contains the definitions of the terms used in the paper. Section 3 presents the new embedding strategy. Section 4 presents the derived results. The Conclusion gives an outlook on further consequences of the new embedding technique.

2   Definitions

(Most of the terminology is taken from [19, 25].) For any graph $G = (V, E)$, let $V(G) = V$ denote the set of vertices of G, and $E(G) = E$ denote the set of edges of G. Let $\bar{a}$ denote the binary complement of $a \in \{0, 1\}$. For $\alpha = a_0 a_1 ... a_{m-1} \in \{0, 1\}^m$, let $\alpha(i) = a_0 ... a_{i-1} \bar{a}_i a_{i+1} ... a_{m-1}$.

Cube-Connected Cycles Network. The *(wrapped) cube-connected cycles network of dimension m*, denoted by $CCC(m)$, has vertex-set $V_m = \{0, 1, ..., m - 1\} \times \{0, 1\}^m$, where $\{0, 1\}^m$ denotes the set of length-m binary strings. For each vertex $v = (i, \alpha) \in V_m$, $i \in \{0, 1, ..., m - 1\}$, $\alpha \in \{0, 1\}^m$, we call i the *level* and α the *position-within-level (PWL) string* of v. The edges of $CCC(m)$ are of two types: For each $i \in \{0, 1, ..., m - 1\}$ and each $\alpha = a_0 a_1 ... a_{m-1} \in \{0, 1\}^m$, the vertex (i, α) on level i of $CCC(m)$ is connected

- by a *cycle-edge* with vertex $((i + 1) \bmod m, \alpha)$ on level $(i + 1) \bmod m$ and
- by a *cross-edge* with vertex $(i, \alpha(i))$ on level i.

For each $\alpha \in \{0, 1\}^m$, the cycle
$$(0, \alpha) \leftrightarrow (1, \alpha) \leftrightarrow ... \leftrightarrow (m - 1, \alpha) \leftrightarrow (0, \alpha)$$
of length m will be denoted by $C_\alpha(m)$ or C_α.

$CCC(m)$ has $m2^m$ nodes, $3m2^{m-1}$ edges and degree 3. An illustration of $CCC(3)$ is shown in Figure 1.

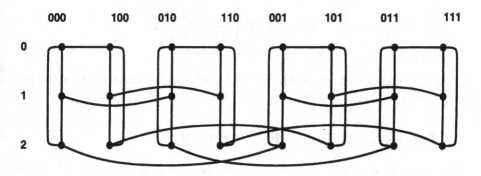

Fig. 1. The cube-connected cycles $CCC(3)$

Graph Embeddings. Let G and H be finite undirected graphs. An *embedding* of G into H is a mapping f from the nodes of G to the nodes of H. G is called the *guest* graph and H is called the *host* graph of the embedding f. The *load* of the embedding f is the maximum number of vertices of the guest graph G that are mapped to the same host graph vertex. [The *optimum load* achievable is the ratio $\lceil |V(G)|/|V(H)| \rceil$ of the number of nodes in G and H.] The *dilation* of the embedding f is the maximum distance in the host between the images of adjacent guest nodes. A *routing* is a mapping r of G's edges to paths in H, $r(v_1, v_2) = $ a path from $f(v_1)$ to $f(v_2)$ in H. The *congestion* of the embedding f is the maximum number of edges that are routed through a single edge of H.

Lexicographic Orderings. Let Lex $: \{0,\ldots,m-1\} \times \{0,1\}^n \to \mathbb{N}_0$, $\text{Lex}(i, a_0 \ldots a_{n-1}) = i2^n + a_0 2^{n-1} + a_1 2^{n-2} + \ldots + a_{n-1} 2^0$. Then, the *lexicographic order* on $\{0, 1, \ldots, m-1\} \times \{0,1\}^n$ is defined by

$$(i, \alpha) < (j, \beta) \iff \text{Lex}(i, \alpha) < \text{Lex}(j, \beta) ,$$

and the *lexicographic distance* between (i, α) and (j, β) is defined as $|\text{Lex}(i, \alpha) - \text{Lex}(j, \beta)|$.

Balanced Allocations. Let $a_1, b_1, a_2, b_2 \in \mathbb{N}_0$ such that $b_1 \geq a_1$, $b_2 \geq a_2$, $b_1 - a_1 \geq b_2 - a_2$. Let $r \in \mathbb{N}$. A function

$$d : \{a_1, a_1 + 1, \ldots, b_1\} \times \{0,1\}^r \to \{a_2, a_2 + 1, \ldots, b_2\}$$

is called a *balanced allocation of* $\{a_1, \ldots, b_1\} \times \{0,1\}^r$ *among* $\{a_2, \ldots, b_2\}$ *according to the lexicographic order on* $\{a_1, \ldots, b_1\} \times \{0,1\}^r$ if d satisfies the following properties:

- $d(a_1, 0^r) = a_2$, $d(b_1, 1^r) = b_2$,
- d is monotonic nondecreasing in the lexicographic ordering of the arguments [i.e., $d(i, \beta) \leq d(i', \beta')$, if $(i, \beta) \leq (i', \beta')$ according to the lexicographic order on $\{a_1, \ldots, b_1\} \times \{0,1\}^r$],

$$-\left\lceil \frac{b_1 - a_1 + 1}{b_2 - a_2 + 1} \cdot 2^r \right\rceil - 1 \le |d^{-1}(j)| \le \left\lceil \frac{b_1 - a_1 + 1}{b_2 - a_2 + 1} \cdot 2^r \right\rceil$$

for all $j \in \{a_2, \ldots, b_2\}$.

[Note that such an allocation function d can always be constructed for the parameters a_1, b_1, a_2, b_2, r as above.]

3   The General Embedding Strategy

The basic idea of the embeddings presented here is to map 2^{l-k} cycles $C_{\alpha_1}, C_{\alpha_2}, \ldots, C_{\alpha_{2^{l-k}}}$ in $CCC(l)$ of length l onto one cycle C_β of length k in $CCC(k)$ and to allocate the $l \cdot 2^{l-k}$ nodes of $C_{\alpha_1}, \ldots, C_{\alpha_{2^{l-k}}}$ appropriately among the k nodes of C_β.

FORMAL CONSTRUCTION:

Consider numbers $\pi(0), \pi(1), \ldots, \pi(k-1)$, where each $\pi(i) \in \{0, 1, \ldots, l-1\}$, and each $\pi(i) < \pi(i+1)$. Let $\bar\pi(0), \bar\pi(1), \ldots, \bar\pi(l-k-1) \in \{0, 1, \ldots, l-1\} \setminus \{\pi(0), \pi(1), \ldots, \pi(k-1)\}$ such that $\bar\pi(0) < \bar\pi(1) < \ldots < \bar\pi(l-k-1)$. [Note that $\{\pi(0), \pi(1), \ldots, \pi(k-1)\} \,\dot\cup\, \{\bar\pi(0), \bar\pi(1), \ldots, \bar\pi(l-k-1)\} = \{0, 1, \ldots, l-1\}$.]

Let $a_{\pi(0)}, a_{\pi(1)}, \ldots, a_{\pi(k-1)} \in \{0, 1\}$. The cycles $\{C_{a_0 a_1 \ldots a_{l-1}} \mid a_{\bar\pi(0)}, a_{\bar\pi(1)}, \ldots, a_{\bar\pi(l-k-1)} \in \{0, 1\}\}$ of $CCC(l)$ are mapped onto the cycle $C_{a_{\pi(0)} a_{\pi(1)} \ldots a_{\pi(k-1)}}$ in $CCC(k)$ such that the nodes $0, 1, \ldots, l-1$ of each $C_{a_0 a_1 \ldots a_{l-1}}$ are allocated appropriately among the nodes of $C_{a_{\pi(0)} a_{\pi(1)} \ldots a_{\pi(k-1)}}$.

The exact allocation of the nodes of $\{C_{a_0 a_1 \ldots a_{l-1}} \mid a_{\bar\pi(0)}, a_{\bar\pi(1)}, \ldots, a_{\bar\pi(l-k-1)} \in \{0, 1\}\}$ on $C_{a_{\pi(0)} a_{\pi(1)} \ldots a_{\pi(k-1)}}$ is determined by an allocation function

$$d : \{0, 1, \ldots, l-1\} \times \{0, 1\}^{l-k} \to \{0, 1, \ldots, k-1\}$$

which specifies, for each node number $\in \{0, 1, \ldots, l-1\}$ on the guest cycle $C_{a_0 a_1 \ldots a_{l-1}}$ and each cycle index $a_{\bar\pi(0)} a_{\bar\pi(1)} \ldots a_{\bar\pi(l-k-1)}$, the position on the host cycle $C_{a_{\pi(0)} a_{\pi(1)} \ldots a_{\pi(k-1)}}$. [On each host cycle $C_{a_{\pi(0)} a_{\pi(1)} \ldots a_{\pi(k-1)}}$, $a_{\pi(0)}, a_{\pi(1)}, \ldots, a_{\pi(k-1)} \in \{0, 1\}$, the same allocation function is used.] Formally, the embedding $f : V(CCC(l)) \to V(CCC(k))$ is of the form

$$f(i, a_0 a_1 \ldots a_{l-1}) := (d(i, a_{\bar\pi(0)} \ldots a_{\bar\pi(l-k-1)}), a_{\pi(0)} \ldots a_{\pi(k-1)})$$

for all $0 \le i \le l-1$, $a_0 a_1 \ldots a_{l-1} \in \{0, 1\}^l$.

The load of f is determined by the allocation function d. Therefore, d should allocate the guest nodes as balancedly as possible on each host cycle. In the sequel, d will be chosen such that

$$\boxed{d(\pi(i), \beta) = i}   \text{ for all } 0 \le i \le k-1, \beta \in \{0, 1\}^{l-k}.$$

This guarantees that all the cross-edges

$$(i, \alpha) \leftrightarrow (i, \alpha(i)), \quad i \in \{\pi(0), \pi(1), \ldots, \pi(k-1)\},$$

of $CCC(l)$ are mapped onto a corresponding cross-edge in $CCC(k)$. All the other edges of $CCC(l)$ are mapped onto a path on a single cycle C_β in $CCC(k)$. So, in this case the dilation is directly dependent on the allocation d of the guest nodes on the host cycle and stands partly in contrast to the desired balancedness of the allocation as explained above.

For low dilation, the values of $\pi(0), \pi(1), \ldots, \pi(k-1)$ should be allocated relatively balancedly among $0, 1, \ldots, l-1$, and the nodes $(i, a_0 a_1 \ldots a_{l-1})$ and $(j, b_0 b_1 \ldots b_{l-1})$ of the cycles $C_{\alpha_1}, C_{\alpha_2}, \ldots, C_{\alpha_{2^{l-k}}}$ of $CCC(l)$ with a small lexicographical distance between $(i, a_{\bar{\pi}(0)} \ldots a_{\bar{\pi}(l-k-1)})$ and $(j, b_{\bar{\pi}(0)} \ldots b_{\bar{\pi}(l-k-1)})$ should be mapped close together on the cycle C_β in $CCC(k)$.

In [16], for $1 < l/k \le 2$, it was shown that the values of $\pi(0), \pi(1), \ldots, \pi(k-1)$ can be specified such that the following holds:

1.) $\pi(i+1) - \pi(i) \le 2$ for all $0 \le i < k-1$.

2.) The nodes $\{(\pi(i), a_0 a_1 \ldots a_{l-1}) \mid a_{\bar{\pi}(0)}, a_{\bar{\pi}(1)}, \ldots, a_{\bar{\pi}(l-k-1)} \in \{0,1\}\}$ are mapped onto $(i, a_{\pi(0)} a_{\pi(1)} \ldots a_{\pi(k-1)})$ for $0 \le i \le k-1$, $a_{\pi(0)}, a_{\pi(1)}, \ldots, a_{\pi(k-1)} \in \{0,1\}$.

3.) The nodes $\{(\bar{\pi}(i), a_0 a_1 \ldots a_{l-1}) \mid 0 \le i \le l-k-1, a_{\bar{\pi}(0)}, a_{\bar{\pi}(1)}, \ldots, a_{\bar{\pi}(l-k-1)} \in \{0,1\}\}$ can be allocated balancedly in certain sections of the host cycle $C_{a_{\pi(0)} a_{\pi(1)} \ldots a_{\pi(k-1)}}$, $a_{\pi(0)}, a_{\pi(1)}, \ldots, a_{\pi(k-1)} \in \{0,1\}$, while maintaining dilation 1 at the same time.

Here, for $\dfrac{5}{3} + c_k < \dfrac{l}{k} \le 2$, $c_k = \dfrac{4k+3}{3 \cdot 2^{2/3k}}$, we show that $\pi(0), \pi(1), \ldots, \pi(k-1)$ can be specified such that the following holds:

1.) $\pi(i+1) - \pi(i) \le 3$ for all $0 \le i < k-1$.

2.) The nodes $\{(\pi(i), a_0 a_1 \ldots a_{l-1}) \mid a_{\bar{\pi}(0)}, a_{\bar{\pi}(1)}, \ldots, a_{\bar{\pi}(l-k-1)} \in \{0,1\}\}$ are mapped onto $(i, a_{\pi(0)} a_{\pi(1)} \ldots a_{\pi(k-1)})$ for $0 \le i \le k-1$, $a_{\pi(0)}, a_{\pi(1)}, \ldots, a_{\pi(k-1)} \in \{0,1\}$.

3.) The nodes $\{(\bar{\pi}(i), a_0 a_1 \ldots a_{l-1}) \mid 0 \le i \le l-k-1, a_{\bar{\pi}(0)}, a_{\bar{\pi}(1)}, \ldots, a_{\bar{\pi}(l-k-1)} \in \{0,1\}\}$ can be allocated balancedly on the complete host cycle $C_{a_{\pi(0)} a_{\pi(1)} \ldots a_{\pi(k-1)}}$, $a_{\pi(0)}, a_{\pi(1)}, \ldots, a_{\pi(k-1)} \in \{0,1\}$, while maintaining dilation 1 at the same time.

The main new technical contribution will be to show that the guest nodes $\{(\pi(i) + 1, a_0 a_1 \ldots a_{l-1}), (\pi(i) + 2, a_0 a_1 \ldots a_{l-1}) \mid a_{\bar{\pi}(0)}, a_{\bar{\pi}(1)}, \ldots, a_{\bar{\pi}(l-k-1)} \in \{0,1\}\}$ can be allocated in an appropriate way among the host nodes $\{(j, a_{\pi(0)} a_{\pi(1)} \ldots a_{\pi(k-1)}) \mid j \in \{i-1, i, i+1, i+2\}\}$ for $0 \le i < k-1$ such that $\pi(i+1) - \pi(i) = 3$, while maintaining dilation 1 at the same time.

4   Improved Dilation 1 Embedding of the CCC

Theorem 1:

Let $k, l \in \mathbb{N}$, $k \geq 8$, such that $\dfrac{5}{3} + c_k < \dfrac{l}{k} \leq 2$, $c_k = \dfrac{4k+3}{3 \cdot 2^{2/3k}}$.
Then, there is a dilation 1 embedding of $CCC(l)$ into $CCC(k)$ with load
$\left\lceil \dfrac{l}{k} \cdot 2^{l-k} \right\rceil$.

Proof:

(A) $\boxed{l - k \text{ even}}$

We show that the construction of Section 3 can be adapted to yield an embedding
of $CCC(l)$ into $CCC(k)$ with dilation 1 and load $\left\lceil \dfrac{l}{k} \cdot 2^{l-k} \right\rceil$.

For this, we specify the allocation d and the indices $\pi(i)$ for the embedding f in
the construction of Section 3.

For $0 \leq i \leq \dfrac{l-k}{2} - 1$, let

$$h(i) := \left\lceil \frac{i \cdot 2l}{l-k} - \frac{3k}{2^{l-k}} \right\rceil + 1.$$

[Then, $h(0) = 1$, $h\left(\dfrac{l-k}{2} - 1\right) = l - 3$.] For $0 \leq i \leq l - k - 1$, let

$$\bar{\pi}(i) := \begin{cases} h\left(\dfrac{i}{2}\right) & \text{if } i \text{ even,} \\ h\left(\left\lfloor \dfrac{i}{2} \right\rfloor\right) + 1 & \text{if } i \text{ odd.} \end{cases}$$

Let $\pi(0), \pi(1), \ldots, \pi(k-1) \in \{0, 1, \ldots, l-1\} \setminus \{\bar{\pi}(0), \bar{\pi}(1), \ldots, \bar{\pi}(l-k-1)\}$
such that $\pi(0) < \pi(1) < \cdots < \pi(k-1)$. [Note that $\{\pi(0), \pi(1), \ldots, \pi(k-1)\} \cup \{\bar{\pi}(0), \bar{\pi}(1), \ldots, \bar{\pi}(l-k-1)\} = \{0, 1, \ldots, l-1\}$.]

For the time being, we only construct the allocation $d : \{0, 1, \ldots, l-1\} \times \{0, 1\}^{l-k} \rightarrow \{0, 1, \ldots, k-1\}$ partially, namely we specify $d(i, \beta)$ for $i \in \{\pi(0), \pi(1), \ldots, \pi(k-1)\}$. Let

$$\boxed{d(\pi(i), \beta) := i} \quad \text{for all } 0 \leq i \leq k-1, \ \beta \in \{0, 1\}^{l-k}. \tag{*}$$

[Later on, $d(i, \beta)$ is specified for $i \in \{\bar{\pi}(0), \bar{\pi}(1), \ldots, \bar{\pi}(l-k-1)\}$. For the moment, $d(i, \beta)$ may have an arbitrary value for $i \in \{\bar{\pi}(0), \bar{\pi}(1), \ldots, \bar{\pi}(l-k-1)\}$.]

Now, the embedding f of $CCC(l)$ into $CCC(k)$ is defined as in the construction
of Section 3:

$$f(i, a_0 a_1 \ldots a_{l-1}) := (d(i, a_{\bar{\pi}(0)} \ldots a_{\bar{\pi}(l-k-1)}), a_{\pi(0)} \ldots a_{\pi(k-1)})$$

for all $0 \leq i \leq l-1$, $a_0 a_1 \ldots a_{l-1} \in \{0,1\}^l$.

Note that (*) guarantees that all the cross-edges

$$(i, \alpha) \leftrightarrow (i, \alpha(i)), \quad i \in \{\pi(0), \pi(1), \ldots, \pi(k-1)\},$$

of $CCC(l)$ are mapped onto a corresponding cross-edge in $CCC(k)$ [the proof is omitted in this Extended Abstract].

Now, we construct $d(i, \beta)$ for $i \in \{\bar{\pi}(0), \bar{\pi}(1), \ldots, \bar{\pi}(l-k-1)\}$. Let $a_{\pi(0)}$, $a_{\pi(1)}, \ldots, a_{\pi(k-1)} \in \{0,1\}$. For the time being, we allocate the guest nodes $\{(i, a_0 a_1 \ldots a_{l-1}) \mid i \in \{\bar{\pi}(0), \bar{\pi}(1), \ldots, \bar{\pi}(l-k-1)\}, a_{\bar{\pi}(0)}, a_{\bar{\pi}(1)}, \ldots, a_{\bar{\pi}(l-k-1)} \in \{0,1\}\}$ balancedly on the host cycle $C_{a_{\pi(0)} a_{\pi(1)} \ldots a_{\pi(k-1)}}$ according to the lexicographical order on $\{0, 1, \ldots, l-1\} \times \{0,1\}^{l-k}$, i.e. we use an allocation function $\tilde{d} : \{\bar{\pi}(0), \bar{\pi}(1), \ldots, \bar{\pi}(l-k-1)\} \times \{0,1\}^{l-k} \rightarrow \{0, 1, \ldots, k-1\}$ such that

- $\tilde{d}(\bar{\pi}(0), 0^{l-k}) = 0$, $\tilde{d}(\bar{\pi}(l-k-1), 1^{l-k}) = k-1$,

- $\tilde{d}$ is monotonic nondecreasing in the lexicographic ordering of the arguments [i.e., $\tilde{d}(i, \beta) \leq \tilde{d}(i', \beta')$, if $(i, \beta) \leq (i', \beta')$ according to the lexicographical order on $\{0, 1, \ldots, l-1\} \times \{0,1\}^{l-k}$],

- $\left\lceil \dfrac{l-k}{k} \cdot 2^{l-k} \right\rceil - 1 \leq |\tilde{d}^{-1}(j)| \leq \left\lceil \dfrac{l-k}{k} \cdot 2^{l-k} \right\rceil$ for all $j = 0, 1, \ldots, k-1$.

[At this point, we are not concerned with the obtained dilation. We will see later on that the allocation $\tilde{d}$ can be changed into an allocation d : $\{\bar{\pi}(0), \bar{\pi}(1), \ldots, \bar{\pi}(l-k-1)\} \times \{0,1\}^{l-k} \rightarrow \{0, 1, \ldots, k-1\}$ which guarantees dilation 1, while maintaining the balancedness of the allocation.]

Let $r(i) := h(i) - 2i - 1$ for all $0 \leq i \leq \dfrac{l-k}{2} - 1$. Then,

1.) $\tilde{d}(h(i) - 1, \beta) = r(i)$

for all $0 \leq i \leq \dfrac{l-k}{2} - 1$, $\beta \in \{0,1\}^{l-k}$,

2.) $\tilde{d}(h(i) + 2, \beta) = r(i) + 1$

for all $0 \leq i \leq \dfrac{l-k}{2} - 1$, $\beta \in \{0,1\}^{l-k}$.

[the proof is omitted in this Extended Abstract]. Also,

1.) $r(i) - 1 \leq \tilde{d}(h(i), \beta) \leq \tilde{d}(h(i) + 1, \beta) \leq r(i) + 2$

for all $0 \leq i \leq \dfrac{l-k}{2} - 1$, $\beta \in \{0,1\}^{l-k}$,

2.) $|\tilde{d}^{-1}(r(i) - 1) \cap \{(h(i), \beta), (h(i)+1, \beta) \mid \beta \in \{0,1\}^{l-k}\}| \leq \left\lceil \dfrac{l-k}{k} \cdot 2^{l-k} \right\rceil - 1$

for all $0 \leq i \leq \dfrac{l-k}{2} - 1$,

3.) $|\tilde{d}^{-1}(r(i)+2) \cap \{(h(i),\beta),(h(i)+1,\beta) \mid \beta \in \{0,1\}^{l-k}\}| \leq \left\lceil \dfrac{l-k}{k} \cdot 2^{l-k} \right\rceil - 1$

for all $0 \leq i \leq \dfrac{l-k}{2} - 1$.

[the proof is omitted in this Extended Abstract].

[As $h(i)-1, h(i)+2 \in \{\pi(0),\pi(1),\ldots,\pi(k-1)\}$, $h(i),h(i)+1 \in \{\bar\pi(0),\bar\pi(1),\ldots,\bar\pi(l-k-1)\}$, the dilation of the embedding f (using the allocation $\tilde{d}$ for $d(i,\beta)$, $i \in \{\bar\pi(0),\bar\pi(1),\ldots,\bar\pi(l-k-1)\}$) would be 2.]

Now, $\tilde{d}$ can be changed to an allocation d such that:

1.) Let $0 \leq i \leq \dfrac{l-k}{2} - 1$. For $1 \leq j \leq 4$, let

$$n_j := |d^{-1}(r(i)-2+j) \cap \{(h(i),\beta),(h(i)+1,\beta) \mid \beta \in \{0,1\}^{l-k}\}|,$$

$$\tilde{n}_j := |\tilde{d}^{-1}(r(i)-2+j) \cap \{(h(i),\beta),(h(i)+1,\beta) \mid \beta \in \{0,1\}^{l-k}\}|.$$

Then,
$$n_1 = \tilde{n}_1,$$
$$n_2 \leq \max\{\tilde{n}_2, \tilde{n}_1 + 1\} \quad \text{if } \tilde{n}_1 > 0,$$
$$n_2 = \tilde{n}_2 \quad \text{if } \tilde{n}_1 = 0,$$
$$n_3 \leq \max\{\tilde{n}_3, \tilde{n}_4 + 1\} \quad \text{if } \tilde{n}_4 > 0,$$
$$n_3 = \tilde{n}_3 \quad \text{if } \tilde{n}_4 = 0,$$
$$n_4 = \tilde{n}_4.$$

2.) For $0 \leq i \leq \dfrac{l-k}{2} - 1$, $\beta = b_{\bar\pi(0)}b_{\bar\pi(1)} \ldots b_{\bar\pi(l-k-1)} \in \{0,1\}^{l-k}$:

$$r(i)-1 \leq d(h(i),\beta) \leq r(i)+1,$$
$$r(i) \leq d(h(i)+1,\beta) \leq r(i)+2,$$
$$|d(h(i)+1,\beta) - d(h(i),\beta)| \leq 1,$$
$$|d(h(i), b_{\bar\pi(0)} \ldots b_{\bar\pi(l-k-1)})$$
$$\qquad -d(h(i), b_{\bar\pi(0)} \ldots b_{\bar\pi(2i-1)}\bar{b}_{\bar\pi(2i)}b_{\bar\pi(2i+1)} \ldots b_{\bar\pi(l-k-1)})| \leq 1,$$
$$|d(h(i)+1, b_{\bar\pi(0)} \ldots b_{\bar\pi(l-k-1)})$$
$$\qquad -d(h(i)+1, b_{\bar\pi(0)} \ldots b_{\bar\pi(2i)}\bar{b}_{\bar\pi(2i+1)}b_{\bar\pi(2i+2)} \ldots b_{\bar\pi(l-k-1)})| \leq 1.$$

[the proof is omitted in this Extended Abstract]. It follows that the final embedding f (using the allocation d) has dilation 1 and load $\left\lceil \dfrac{l}{k} \cdot 2^{l-k} \right\rceil$. $\square$

(B) $\boxed{l-k \text{ odd}}$

We show that the construction of Section 3 can be adapted to yield an embedding of $CCC(l)$ into $CCC(k)$ with dilation 1 and load $\left\lceil \dfrac{l}{k} \cdot 2^{l-k} \right\rceil$.

For this, we specify the allocation d and the indices $\pi(i)$ for the embedding f in the construction of Section 3.

For $0 \leq i \leq \dfrac{l-k-1}{2} - 1$, let

$$h(i) := \left\lceil \frac{i \cdot 2l}{l-k} - \frac{3k}{2^{l-k}} \right\rceil + 1 \, .$$

[Then, $h(0) = 1$, $h\left(\dfrac{l-k-1}{2} - 1\right) \in \{l-6, l-5\}$.] For $0 \leq i \leq l-k-2$, let

$$\bar{\pi}(i) := \begin{cases} h\left(\dfrac{i}{2}\right) & \text{if } i \text{ even,} \\[2mm] h\left(\left\lfloor \dfrac{i}{2} \right\rfloor\right) + 1 & \text{if } i \text{ odd.} \end{cases}$$

Let

$$\bar{\pi}(l-k-1) := l-2 \, .$$

Let $\pi(0), \pi(1), \ldots, \pi(k-1) \in \{0, 1, \ldots, l-1\} \setminus \{\bar{\pi}(0), \bar{\pi}(1), \ldots, \bar{\pi}(l-k-1)\}$ such that $\pi(0) < \pi(1) < \cdots < \pi(k-1)$. [Note that $\{\pi(0), \pi(1), \ldots, \pi(k-1)\} \mathbin{\dot{\cup}} \{\bar{\pi}(0), \bar{\pi}(1), \ldots, \bar{\pi}(l-k-1)\} = \{0, 1, \ldots, l-1\}$.]

For the time being, we only construct the allocation $d : \{0, 1, \ldots, l-1\} \times \{0,1\}^{l-k} \to \{0, 1, \ldots, k-1\}$ partially, namely we specify $d(i, \beta)$ for $i \in \{\pi(0), \pi(1), \ldots, \pi(k-1)\}$. Let

$$\boxed{d(\pi(i), \beta) := i} \qquad \text{for all } 0 \leq i \leq k-1, \ \beta \in \{0,1\}^{l-k}. \tag{$*$}$$

[Later on, $d(i, \beta)$ is specified for $i \in \{\bar{\pi}(0), \bar{\pi}(1), \ldots, \bar{\pi}(l-k-1)\}$. For the moment, $d(i, \beta)$ may have an arbitrary value for $i \in \{\bar{\pi}(0), \bar{\pi}(1), \ldots, \bar{\pi}(l-k-1)\}$.]

Now, the embedding f of $CCC(l)$ into $CCC(k)$ is defined as in the construction of Section 3:

$$f(i, a_0 a_1 \ldots a_{l-1}) := (d(i, a_{\bar{\pi}(0)} \ldots a_{\bar{\pi}(l-k-1)}), a_{\pi(0)} \ldots a_{\pi(k-1)})$$

$$\text{for all } 0 \leq i \leq l-1, \ a_0 a_1 \ldots a_{l-1} \in \{0,1\}^l.$$

Note that $(*)$ guarantees that all the cross-edges

$$(i, \alpha) \leftrightarrow (i, \alpha(i)), \quad i \in \{\pi(0), \pi(1), \ldots, \pi(k-1)\},$$

of $CCC(l)$ are mapped onto a corresponding cross-edge in $CCC(k)$ [the proof is omitted in this Extended Abstract].

Now, we construct $d(i, \beta)$ for $i \in \{\bar{\pi}(0), \bar{\pi}(1), \ldots, \bar{\pi}(l-k-1)\}$. Let $a_{\pi(0)}$, $a_{\pi(1)}, \ldots, a_{\pi(k-1)} \in \{0,1\}$. For the time being, we allocate the guest nodes $\{(i, a_0 a_1 \ldots a_{l-1}) \mid i \in \{\bar{\pi}(0), \bar{\pi}(1), \ldots, \bar{\pi}(l-k-1)\}, \ a_{\bar{\pi}(0)}, a_{\bar{\pi}(1)}, \ldots, a_{\bar{\pi}(l-k-1)} \in \{0,1\}\}$ balancedly on the host cycle $C_{a_{\pi(0)} a_{\pi(1)} \ldots a_{\pi(k-1)}}$ according to the lexicographical order on $\{0, 1, \ldots, l-1\} \times \{0,1\}^{l-k}$, i.e. we use an allocation function $\tilde{d} : \{\bar{\pi}(0), \bar{\pi}(1), \ldots, \bar{\pi}(l-k-1)\} \times \{0,1\}^{l-k} \to \{0, 1, \ldots, k-1\}$ such that

- $\tilde{d}(\bar{\pi}(0), 0^{l-k}) = 0, \quad \tilde{d}(\bar{\pi}(l-k-1), 1^{l-k}) = k-1,$
- $\tilde{d}$ is monotonic nondecreasing in the lexicographic ordering of the arguments [i.e., $\tilde{d}(i, \beta) \leq \tilde{d}(i', \beta')$, if $(i, \beta) \leq (i', \beta')$ according to the lexicographical order on $\{0, 1, \ldots, l-1\} \times \{0, 1\}^{l-k}$],
- $\left\lceil \dfrac{l-k}{k} \cdot 2^{l-k} \right\rceil - 1 \leq |\tilde{d}^{-1}(j)| \leq \left\lceil \dfrac{l-k}{k} \cdot 2^{l-k} \right\rceil \quad$ for all $j = 0, 1, \ldots, k-1$.

[At this point, we are not concerned with the obtained dilation. We will see later on that the allocation $\tilde{d}$ can be changed into an allocation d : $\{\bar{\pi}(0), \bar{\pi}(1), \ldots, \bar{\pi}(l-k-1)\} \times \{0, 1\}^{l-k} \to \{0, 1, \ldots, k-1\}$ which guarantees dilation 1, while maintaining the balancedness of the allocation.]

Let $r(i) := h(i) - 2i - 1$ for all $0 \leq i \leq \dfrac{l-k-1}{2} - 1$. Then,

1.) $\tilde{d}(h(i) - 1, \beta) = r(i)$
   for all $0 \leq i \leq \dfrac{l-k-1}{2} - 1, \beta \in \{0, 1\}^{l-k}$,

2.) $\tilde{d}(h(i) + 2, \beta) = r(i) + 1$
   for all $0 \leq i \leq \dfrac{l-k-1}{2} - 1, \beta \in \{0, 1\}^{l-k}$,

3.) $\tilde{d}(\bar{\pi}(l-k-1) - 1, \beta) = k - 2$
   for all $\beta \in \{0, 1\}^{l-k}$,

4.) $\tilde{d}(\bar{\pi}(l-k-1) + 1, \beta) = k - 1$
   for all $\beta \in \{0, 1\}^{l-k}$.

[the proof is omitted in this Extended Abstract]. Also,

1.) $r(i) - 1 \leq \tilde{d}(h(i), \beta) \leq \tilde{d}(h(i) + 1, \beta) \leq r(i) + 2$
   for all $0 \leq i \leq \dfrac{l-k-1}{2} - 1, \beta \in \{0, 1\}^{l-k}$,

2.) $|\tilde{d}^{-1}(r(i) - 1) \cap \{(h(i), \beta), (h(i) + 1, \beta) \mid \beta \in \{0, 1\}^{l-k}\}| \leq \left\lceil \dfrac{l-k}{k} \cdot 2^{l-k} \right\rceil - 1$
   for all $0 \leq i \leq \dfrac{l-k-1}{2} - 1$,

3.) $|\tilde{d}^{-1}(r(i) + 2) \cap \{(h(i), \beta), (h(i) + 1, \beta) \mid \beta \in \{0, 1\}^{l-k}\}| \leq \left\lceil \dfrac{l-k}{k} \cdot 2^{l-k} \right\rceil - 1$
   for all $0 \leq i \leq \dfrac{l-k-1}{2} - 1$,

4.) $k - 2 \leq \tilde{d}(\bar{\pi}(l-k-1), \beta) \leq k - 1$
   for all $\beta \in \{0, 1\}^{l-k}$.

[the proof is omitted in this Extended Abstract].

[As $h(i) - 1, h(i) + 2, \bar{\pi}(l-k-1) - 1, \bar{\pi}(l-k-1) + 1 \in \{\pi(0), \pi(1), \ldots, \pi(k-1)\}$, $h(i), h(i) + 1, \bar{\pi}(l-k-1) \in \{\bar{\pi}(0), \bar{\pi}(1), \ldots, \bar{\pi}(l-k-1)\}$, the dilation of the embedding f (using the allocation $\tilde{d}$ for $d(i, \beta), i \in \{\bar{\pi}(0), \bar{\pi}(1), \ldots, \bar{\pi}(l-k-1)\}$) would be 2.]

Now, $\tilde{d}$ can be changed to an allocation d such that:

1.) Let $0 \leq i \leq \dfrac{l-k-1}{2} - 1$. For $1 \leq j \leq 4$, let

$$n_j := |d^{-1}(r(i) - 2 + j) \cap \{(h(i), \beta), (h(i) + 1, \beta) \mid \beta \in \{0,1\}^{l-k}\}|,$$

$$\tilde{n}_j := |\tilde{d}^{-1}(r(i) - 2 + j) \cap \{(h(i), \beta), (h(i) + 1, \beta) \mid \beta \in \{0,1\}^{l-k}\}|.$$

Then,

$$n_1 = \tilde{n}_1,$$
$$n_2 \leq \max\{\tilde{n}_2, \tilde{n}_1 + 1\} \quad \text{if } \tilde{n}_1 > 0,$$
$$n_2 = \tilde{n}_2 \quad \text{if } \tilde{n}_1 = 0,$$
$$n_3 \leq \max\{\tilde{n}_3, \tilde{n}_4 + 1\} \quad \text{if } \tilde{n}_4 > 0,$$
$$n_3 = \tilde{n}_3 \quad \text{if } \tilde{n}_4 = 0,$$
$$n_4 = \tilde{n}_4.$$

2.) For $0 \leq i \leq \dfrac{l-k-1}{2} - 1$, $\beta = b_{\bar{\pi}(0)} b_{\bar{\pi}(1)} \ldots b_{\bar{\pi}(l-k-1)} \in \{0,1\}^{l-k}$:

$$r(i) - 1 \leq d(h(i), \beta) \leq r(i) + 1,$$
$$r(i) \leq d(h(i) + 1, \beta) \leq r(i) + 2,$$
$$|d(h(i) + 1, \beta) - d(h(i), \beta)| \leq 1,$$
$$|d(h(i), b_{\bar{\pi}(0)} \ldots b_{\bar{\pi}(l-k-1)})$$
$$\quad -d(h(i), b_{\bar{\pi}(0)} \ldots b_{\bar{\pi}(2i-1)} \bar{b}_{\bar{\pi}(2i)} b_{\bar{\pi}(2i+1)} \ldots b_{\bar{\pi}(l-k-1)})| \leq 1,$$
$$|d(h(i) + 1, b_{\bar{\pi}(0)} \ldots b_{\bar{\pi}(l-k-1)})$$
$$\quad -d(h(i) + 1, b_{\bar{\pi}(0)} \ldots b_{\bar{\pi}(2i)} \bar{b}_{\bar{\pi}(2i+1)} b_{\bar{\pi}(2i+2)} \ldots b_{\bar{\pi}(l-k-1)})| \leq 1.$$

3.) For $k - 2 \leq j \leq k - 1$:

$$|d^{-1}(j) \cap \{(\bar{\pi}(l - k - 1), \beta) \mid \beta \in \{0,1\}^{l-k}\}|$$
$$= |\tilde{d}^{-1}(j) \cap \{(\bar{\pi}(l - k - 1), \beta) \mid \beta \in \{0,1\}^{l-k}\}|.$$

For $\beta \in \{0,1\}^{l-k}$:

$$k - 2 \leq d(\bar{\pi}(l - k - 1), \beta) \leq k - 1.$$

[the proof is omitted in this Extended Abstract]. It follows that the final embedding f (using the allocation d) has dilation 1 and load $\left\lceil \dfrac{l}{k} \cdot 2^{l-k} \right\rceil$.   □

5   Conclusion

In this paper, we have presented a new technique for the embedding of large cube-connected cycles networks into smaller ones. Using the new embedding strategy, we showed:

Let $k, l \in \mathbb{N}$, $k \geq 8$, such that $\dfrac{5}{3} + c_k < \dfrac{l}{k} \leq 2$, $c_k = \dfrac{4k + 3}{3 \cdot 2^{2/3k}}$.
Then, there is a dilation 1 embedding of $CCC(l)$ into $CCC(k)$ with load $\left\lceil \dfrac{l}{k} \cdot 2^{l-k} \right\rceil$.

This is optimal, and improves the results from [16]. In the case that $\frac{l}{k} \cdot 2^{l-k} \in I\!N$, the embedding technique can be adapted to yield an even stronger result:

> Let $k, l \in I\!N$ such that $\frac{5}{3} < \frac{l}{k} \leq 2$, $\frac{l}{k} \cdot 2^{l-k} \in I\!N$. Then, there is a dilation 1 embedding of $CCC(l)$ into $CCC(k)$ with load $\left\lceil \frac{l}{k} \cdot 2^{l-k} \right\rceil$.

The embedding technique can also be applied in the case $\frac{3}{2} < \frac{l}{k} \leq \frac{5}{3} + c_k$ yielding:

1. Let $k, l \in I\!N$, $k \geq 8$, such that $\frac{5}{3} < \frac{l}{k} \leq \frac{5}{3} + c_k$, $c_k = \frac{4k+3}{3 \cdot 2^{2/3k}}$. Then, there is a dilation 1 embedding of $CCC(l)$ into $CCC(k)$ with load $\left\lceil \left(\frac{5}{3} + c_k \right) \cdot 2^{l-k} \right\rceil$.

2. Let $k, l \in I\!N$ such that $\frac{3}{2} < \frac{l}{k} < \frac{5}{3}$. Let $p \in \{1, 2, \ldots\}$ such that $\frac{5p-4}{3p-2} < \frac{l}{k} \leq \frac{5p+1}{3p+1}$. Then, there is a dilation 1 embedding of $CCC(l)$ into $CCC(k)$ with load $\left\lceil \frac{5p+1}{3p+1} \cdot 2^{l-k} \right\rceil$.

This also improves results from [16].

Unfortunately, the new embedding technique does not lead to any improvement in the case $1 < \frac{l}{k} \leq \frac{3}{2}$. Hence, it is still of interest to improve the load of the non-optimal dilation 1 embeddings when $1 < \frac{l}{k} \leq \frac{5}{3} + c_k$ (or to prove their optimality). Finally, a further study should also consider the congestion of the embeddings.

Acknowledgement

The author would like to thank the anonymous referees for their helpful comments.

References

1. F. Annexstein, M. Baumslag, A.L. Rosenberg: Group action graphs and parallel architectures. *SIAM Journal on Computing* 19 (1990), 544–569.
2. F. Berman, L. Snyder: On mapping parallel algorithms into parallel architectures. *Journal of Parallel and Distributed Computing* 4 (1987), 439–458.
3. S.N. Bhatt, J.-Y. Cai: Taking random walks to grow trees in hypercubes. *Journal of the ACM* 40 (1993), 741–764.

4. S.N. Bhatt, F.R.K. Chung, F.T. Leighton, A.L. Rosenberg: Efficient embeddings of trees in hypercubes. *SIAM Journal on Computing* 21(1) (1992), 151–162.

5. S.N. Bhatt, F.R.K. Chung, J.-W. Hong, F.T. Leighton, B. Obrenić, A.L. Rosenberg, E.J. Schwabe: Optimal Emulations by Butterfly-Like Networks. *Journal of the ACM* 43 (1996), 293–330.

6. H.L. Bodlaender: The classification of coverings of processor networks. *Journal of Parallel and Distributed Computing* 6 (1989), 166–182.

7. H.L. Bodlaender, J. van Leeuwen: Simulation of large networks on smaller networks. *Information and Control* 71 (1986), 143–180.

8. S.H. Bokhari: On the mapping problem. *IEEE Transactions on Computers* C-30 (1981), 207–214.

9. M.-Y. Fang and W.-T. Chen: Embedding large binary trees to hypercube multiprocessors. *Proc. International Conference on Parallel Processing*, 1991, I:714–715.

10. R. Feldmann, W. Unger: The Cube-Connected-Cycle is a subgraph of the Butterfly network. *Parallel Processing Letters* 2 (1992), 13–19.

11. M.R. Fellows: *Encoding graphs in graphs*. Ph.D. Dissertation, University of California at San Diego, 1985.

12. J.P. Fishburn, R.A. Finkel: Quotient networks. *IEEE Transactions on Computers* C-31 (1982), 288–295.

13. A.K. Gupta, S.E. Hambrusch: Embedding large tree machines into small ones. *Proc. 5th MIT Conference on Advanced Research in VLSI*, 1988, 179–198.

14. A. Heirich: A scalable diffusion algorithm for dynamic mapping and load balancing on networks of arbitrary topology. *International Journal on Foundations of Computer Science* 8 (1997), 329–346.

15. S.J. Kim, J.C. Browne: A general approach to mapping of parallel computations upon multiprocessor architectures. *Proc. International Conference on Parallel Processing*, 1988, III:1–8.

16. R. Klasing, R. Lüling, B. Monien: Compressing cube-connected cycles and butterfly networks. *Proc. 2nd IEEE Symposium on Parallel and Distributed Processing* (SPDP'90), 1990, 858–865. *Networks*, to appear.

17. R.R. Koch, F.T. Leighton, B.M. Maggs, S.B. Rao, A.L. Rosenberg, E.J. Schwabe: Work-preserving emulations of fixed-connection networks. *Journal of the ACM*, 44(1) (1997), 104–147.

18. S.R. Kosaraju, M.J. Atallah: Optimal simulations between mesh-connected arrays of processors. *Journal of the ACM* 35 (1988), 635–650.

19. F.T. Leighton: *Introduction to Parallel Algorithms and Architectures: Arrays, Trees, Hypercubes*. Morgan Kaufmann Publishers, San Mateo, California, 1992.

20. F.T. Leighton, B.M. Maggs, S.B. Rao: Packet routing and job-shop scheduling in O(congestion + dilation) steps. *Combinatorica* 14 (1994), 167–186.

21. F.T. Leighton, M.J. Newman, A.G. Ranade, E.J. Schwabe: Dynamic tree embeddings in butterflies and hypercubes. *SIAM Journal on Computing* 21 (1992), 639–654.

22. Z. Miller, I.H. Sudborough: Compressing Grids into Small Hypercubes. *Networks* 24 (1994), 327–358.

23. D.I. Moldovan, J.A.B. Fortes: Partitioning and mapping algorithms into fixed size systolic arrays. *IEEE Transactions on Computers* C-35 (1986), 1–12.

24. B. Monien: Simulating binary trees on X-trees. *Proc. 3rd ACM Symposium on Parallel Algorithms and Architectures* (SPAA'91), 1991, 147–158.

25. B. Monien, I.H. Sudborough: Embedding one Interconnection Network in Another. *Computing Supplementum* 7 (1990), 257–282.

26. P.A. Nelson, L. Snyder: Programming solutions to the algorithm contraction problem. *Proc. International Conference on Parallel Processing*, 1986, 258–261.
27. R. Peine: Cayley-Graphen und Netzwerke. *Master Thesis*, Universität-GH Paderborn, Fachbereich 17 – Mathematik/Informatik, Germany, 1990.
28. F.P. Preparata, J.E. Vuillemin: The cube-connected cycles: a versatile network for parallel computation. *Communications of the ACM* 24 (1981), 300–309.
29. A.L. Rosenberg: Graph embeddings 1988: Recent breakthroughs, new directions. *Proc. 3rd Aegean Workshop on Computing (AWOC): VLSI Algorithms and Architectures*, Lecture Notes in Computer Science 319, Springer-Verlag 1988, 160–169.
30. V. Sarkar: *Partitioning and Scheduling Parallel Programs for Multiprocessors*. MIT Press, Cambridge, Massachusetts, 1989.

Efficient Embeddings of Grids into Grids*
(Extended Abstract)

Markus Röttger and Ulf-Peter Schroeder

Department of Computer Science, University of Paderborn,
Fürstenallee 11, D-33102 Paderborn, Germany
{roettger, ups}@uni-paderborn.de

Abstract. In this paper we explore one-to-one embeddings of 2-dimensional grids into their ideal 2-dimensional grids. The presented results are optimal or considerably close to the optimum.

For embedding grids into grids of smaller aspect ratio, we prove a new lower bound on the dilation matching a known upper bound. The edge-congestion provided by our matrix-based construction differs from the here presented tight lower bound by at most one. For embedding grids into grids of larger aspect ratio, we establish five as an upper bound on the dilation and four as an upper bound on the edge-congestion, which are improvements of previous results.

1 Introduction

Let $G = (V, E)$ and $H = (V', E')$ be finite graphs. An *embedding* (ϕ, R_ϕ) of the *guest* graph G into the *host* graph H is a function $\phi : V \longrightarrow V'$ together with a routing scheme R_ϕ which assigns to each edge $e = \{v_1, v_2\} \in E$ a path in H from $\phi(v_1)$ to $\phi(v_2)$. If ϕ is an injective function, we call the embedding *one-to-one*, otherwise *many-to-one*. The *congestion of an edge* $e' \in E'$ is the number of paths in $\{R_\phi(e) \mid e \in E\}$ containing e' as an edge. The *edge-congestion* of (ϕ, R_ϕ) is the maximum congestion of the edges of E'. The *dilation of an edge* $e \in E$ is the length of the path $R_\phi(e)$, the *dilation* of (ϕ, R_ϕ) is the maximum length of the paths in $\{R_\phi(e) \mid e \in E\}$. The *expansion* of (ϕ, R_ϕ) is defined as $|V'|/|V|$. The *2-dimensional grid* of height h and width w, denoted by $h \times w$, is the graph (V, E) with node set $V = \{(a, b) \mid 0 \le a < h, 0 \le b < w\}$ and edge set $E = \{\{(a, b), (a', b')\} \mid (a, b), (a', b') \in V, |a - a'| + |b - b'| = 1\}$. The *aspect ratio* of the $h \times w$ grid G is defined as the quantity $\min\{h, w\}/\max\{h, w\}$.

Among others, embedding a guest graph into a host graph is used to model area-efficient graph layouts for VLSI [13] or to model the problem of processor allocation in a distributed system [15]. In the latter case the most important cost-measures to rate the quality of embeddings are the edge-congestion, i.e. the amount of possible contention in the system for the same link, and the dilation, i.e. the distance between communicating processes in the system. Several other applications, especially for embedding grids into grids, are listed in [2, 7, 14].

* This work was supported by the DFG-Sonderforschungsbereich 376 "Massive Parallelität: Algorithmen, Entwurfsmethoden, Anwendungen" and the EC ESPRIT Long Term Research Project 20244 "ALCOM-IT".

In this paper we consider one-to-one embeddings of $h \times w$ grids $G = (V, E)$ into $h' \times w'$ grids $H = (V', E')$ where we assume without loss of generality (and for the rest of the paper) that $h \leq w$ and $h' \leq w'$. We call H an ideal grid for G if $h'(w' - 1) < hw \leq h'w'$ holds. Note that there may exist different ideal grids H for the same grid G. Here, we will focus on embeddings where H is an *ideal grid* for G, since these are the hardest instances of this (one-to-one) embedding problem. Avoiding the trivial case (i.e., $h = h'$ and $w = w'$) we distinguish two cases: $h' < h \leq w < w'$ or $h < h' \leq w' < w$. In the first case the aspect ratio of H is smaller than the aspect ratio of G, in the latter case vice versa. In the following of this section we summarize previous results obtained in this area and compare them with our new results.

Embedding into Grids of Smaller Aspect Ratio

The special case where the $h \times w$ grid G is embedded into a linear array, i.e., H is a $1 \times (hw)$ grid, was studied successfully. Chvátalová [6] proved in 1975 that in this case h is a tight bound for the dilation. It can be derived from a result proved in 1995 by Ahlswede and Bezrukov [1] that the $h \times w$ grid, with $h \neq 2$ and $w \neq 2$, can be embedded with optimal edge-congestion $h + 1$ into the $1 \times (hw)$ grid. In 1988 Kosaraju and Atallah [12] showed that $\Theta(h/h')$ is a bound for the dilation of embedding $G = h \times w$ into $H = h' \times w'$, but they did not specify the constants involved. Römke et al. [16] showed in 1995 that G can be embedded into H with dilation $\lceil h/h' \rceil + 1$. In 1996 Huang et al. [10] improved this result. They constructed embeddings with dilation of at most $\lceil h/h' \rceil$. Shen et al. [20] proved in 1997 for the special case that the guest grid and host grid are of the same size, i.e., $hw = h'w'$, $\lceil h/h' \rceil$ as a lower bound for the dilation and the edge-congestion. Additionally, they proposed $\lceil h/h' \rceil + 3$ as an upper bound for the edge-congestion.

In the following we will show that $\lceil h/h' \rceil$ is the optimal value for the dilation for any embedding of grids into grids of smaller aspect ratio. Additionally, we prove $\lceil (h + 1)/h' \rceil$ as tight lower bound for the edge-congestion and establish $\lceil h/h' \rceil + 1$ as an upper bound.

Embedding into Grids of Larger Aspect Ratio

Most attempts to embed grids into grids of larger aspect ratio were restricted on embedding grids into square grids, i.e., it was assumed that $h' = w'$. Aleliunas and Rosenberg [2] showed in 1982 that, if $h \geq 25$, G can be embedded into H with dilation 15 and expansion 1.2. Furthermore, they conjectured that there may be an inherent expansion-dilation tradeoff. In 1991 Ellis [7] showed that for small values of the *compression ratio* w/w' there is no significant tradeoff between dilation and expansion. He proved that, if $w/w' \leq 3$, each $h \times w$ grid G can be embedded into each of its ideal $h' \times w'$ grids of larger aspect ratio with dilation three. Additionally, it was shown by Ellis, that each grid G can be embedded into its *nearly ideal* square grid $H = (\lceil \sqrt{hw} \rceil + 1) \times (\lceil \sqrt{hw} \rceil + 1)$ with dilation three. Moreover, Melhem and Hwang [14] constructed embeddings of each $h \times w$ grid into square grids with dilation two and expansion of at most 1.2.

In 1996, Ellis [8] introduced a technique called folding with compression which can be used to embed the $h \times w$ grid into each of its ideal $h' \times w'$ grids of larger aspect ratio with dilation two if $w/w' \geq h > 7$, i.e., if the compression ratio and h are sufficiently large. Huang et al. [11] showed in 1997 that each $h \times w$ grid can be embedded into each of its ideal $h' \times w'$ grids with dilation six. Recently, Shen et al. [20] constructed embeddings with edge-congestion four for some instances.

In the following we will prove that five is an upper bound for the dilation of embedding grids into grids of larger aspect-ratio. Moreover, we will show that each grid can be embedded into each of its ideal grids with edge-congestion four.

The paper is organized as follows. In Sect. 2 we present a general matrix-based embedding technique which is fundamental for our upper bound results. Then, we compute lower and upper bounds for the dilation and edge-congestion where we at first consider embedding grids into grids of smaller aspect ratio (Sect. 3), and then into grids of larger aspect ratio (Sect. 4).

2   Matrix-Based Embeddings

In the following we describe a general matrix-based technique to embed an $h \times w$ grid G into an $h' \times w'$ grid H with $hw \leq h'w'$, and w.l.o.g. $w > w'$. Aleliunas and Rosenberg [2] were the first to use matrices to describe embeddings of grids. More general matrix-based techniques were introduced in [5, 10, 11, 16, 18–20]. Now we will shortly repeat the most important facts, since we will use them in the following sections of the paper. Additionally, a new upper bound for the edge-congestion of such matrix-based embeddings is proposed.

To embed G into H we construct a so-called *embedding matrix* $M_{h \times w'}$ which entries m_{ij}, $0 \leq i < h, 0 \leq j < w'$, are nonnegative integers. Let us explain the correspondence of the matrix and the embedding by an example, the 5×31 grid as the guest and the 12×13 grid as the host. We align the guest in the host grid compressing 5 rows of length 31 to length 13. This is exemplified in Fig. 1.

Note that each row of the guest – called *chain* – can be described by a vector of length 13 where the jth entry of the vector, $0 \leq j < 13$, determines how many nodes of the chain are placed in the column j of the host, e.g. the vector

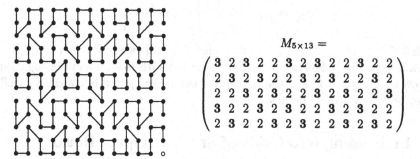

$$M_{5 \times 13} = \begin{pmatrix} 3 & 2 & 3 & 2 & 2 & 3 & 2 & 3 & 2 & 2 & 3 & 2 & 2 \\ 2 & 3 & 2 & 3 & 2 & 2 & 3 & 2 & 3 & 2 & 2 & 3 & 2 \\ 2 & 2 & 3 & 2 & 3 & 2 & 2 & 3 & 2 & 3 & 2 & 2 & 3 \\ 3 & 2 & 2 & 3 & 2 & 3 & 2 & 2 & 3 & 2 & 3 & 2 & 2 \\ 2 & 3 & 2 & 2 & 3 & 2 & 3 & 2 & 2 & 3 & 2 & 3 & 2 \end{pmatrix}$$

Fig. 1. Embedding the 5×31 grid into the 12×13 grid and the corresponding embedding matrix. The white node is no image of a node of the guest.

$(3, 2, 3, 2, 2, 3, 2, 3, 2, 2, 3, 2, 2)$ corresponds to the first chain. The definition of such a vector for each chain leads to the embedding matrix shown in Fig. 1.

Additionally, we must define how the elements of each chain are connected. We will choose the following interconnection scheme: Nodes of even columns $(0, 2, \dots)$ are connected from top to bottom, nodes of odd columns from bottom to top. The bottom node of each even column is connected to the bottom node of the next column. The top node of each odd column is connected to the top node of the next column (cf. Fig. 1).

Following this scheme for each node (a, b) of the grid G we define its image in H as (a', b') where

$$b' = \min\{l : \sum_{j=0}^{l} m_{aj} \geq b + 1\}, \tag{1}$$

$$a' = \sum_{i=0}^{a-1} m_{ib'} + \begin{cases} b - \sum_{j=0}^{b'-1} m_{aj} & \text{if } b' \text{ is even,} \\ \left(\sum_{j=0}^{b'} m_{aj}\right) - b - 1 & \text{if } b' \text{ is odd.} \end{cases} \tag{2}$$

Consider embedding the $h \times w$ grid G into the $h' \times w'$ grid H (where $w > w'$) by using a matrix $M_{h \times w'}$ that satisfies the following conditions:

(C1) $m_{ij} \in \{\lfloor w/w' \rfloor, \lceil w/w' \rceil\}$, for $0 \leq i < h$, $0 \leq j < w'$,

(C2) $m_{ij} = m_{i+1,j+1}$, for $0 \leq i < h - 1$, $0 \leq j < w' - 1$,

(C3) $\sum_{i=0}^{h-1} m_{ij} \leq h'$, for $0 \leq j < w'$,

(C4) $\sum_{j=0}^{w'-1} m_{ij} \geq w$, for $0 \leq i < h$.

Theorem 1. *Suppose $w > w'$. Any embedding matrix $M_{h \times w'}$ that satisfies (C1)–(C4) defines an embedding of an $h \times w$ grid into an $h' \times w'$ grid with dilation and edge-congestion of at most $\lceil w/w' \rceil + 1$.*

In the interest of brevity, we omit the very technical proof and refer to [17].

A simple set-up of an embedding matrix $M_{h \times w'}$ satisfying (C1)–(C4) was given by Römke *et al.* in [16], namely

$$m_{ij} = \left\lceil w \, \frac{j - i + 1}{w'} \right\rceil - \left\lceil w \, \frac{j - i}{w'} \right\rceil,$$

with $0 \leq i < h$, $0 \leq j < w'$. The embedding matrix shown in Fig. 1 is constructed according to this definition. Note that we get the ith row by a cyclic right shift of row $i - 1$, $1 \leq i < h$, and the embedding can be computed in constant parallel time (see [16] for more details).

3   Embedding into Grids of Smaller Aspect Ratio

Throughout this section we consider embedding $G = h \times w$ into $H = h' \times w'$ where H is an ideal grid for G and $h' < h \leq w < w'$.

3.1 Lower Bounds

Consider an arbitrary one-to-one embedding of $G = h \times w$ into $H = h' \times w'$, and let $\mathrm{dil}(G : H)$ denote its dilation and let $\mathrm{con}(G : H)$ denote its edge-congestion. To compute lower bounds for $\mathrm{dil}(G : H)$ and $\mathrm{con}(G : H)$ we consider some discrete isoperimetric problems on grids (see [3,9] for an introduction on isoperimetric problems). A numbering of the nodes of a grid $\bar{G} = (\bar{V}, \bar{E})$ is an injective function $\eta : \bar{V} \mapsto \{1, 2, \ldots, |\bar{V}|\}$. Given a grid $\bar{G}$ and a numbering η, then for each l, $0 \leq l \leq |\bar{V}|$, we define $S_l(\eta) = \eta^{-1}(\{1, \ldots, l\})$. $S_l(\eta)$ is the set of the first l nodes of $\bar{G}$ corresponding to the numbering η.

Lower Bound for the Dilation
Let $\bar{G} = (\bar{V}, \bar{E})$ be an arbitrary $\bar{h} \times \bar{w}$ grid with $\bar{h} \leq \bar{w}$. For $S \subseteq \bar{V}$ define

$$\Gamma_{\bar{G}}(S) = |\{u \in \bar{V} \setminus S \mid \exists v \in S : (u, v) \in \bar{E}\}|,$$
$$\Gamma_{\bar{G}}(l) = \min_{\eta} \Gamma_{\bar{G}}(S_l(\eta)).$$

$\Gamma_{\bar{G}}(S)$ is the number of nodes in the direct neighborhood of S and for a fixed l the problem of minimizing $\Gamma_{\bar{G}}(S_l(\eta))$ over all numberings η is known as the discrete vertex isoperimetric problem for grids.

For the above defined grid $\bar{G}$ we define an ordering $\mathcal{F}$ as follows: We say that $(a, b) \in \bar{V}$ is w.r.t. $\mathcal{F}$ greater than $(a', b') \in \bar{V}$, iff $(a + b > a' + b')$ or $(a + b = a' + b' \wedge a < a')$.

It can be derived from results presented by Chvátalová [6], that for any l with $0 \leq l \leq |\bar{V}|$ the set of nodes given by the first l nodes w.r.t. $\mathcal{F}$ has a minimal number of nodes in distance one in the grid outside of the set itself (among all subsets of $\bar{V}$ of the same cardinality). In other words, the node set defined in such a way solves the vertex isoperimetric problem on the $\bar{h} \times \bar{w}$ grid.

Theorem 2. *The dilation* $\mathrm{dil}(G : H)$ *of any embedding of an* $h \times w$ *grid* $G = (V, E)$ *into each of its ideal* $h' \times w'$ *grids* $H = (V', E')$, *with* $2 \leq h' < h \leq w < w'$, *is at least* $\lceil h/h' \rceil$.

Proof. Let l be an integer with $0 \leq l \leq |V'|$ and let $D \subseteq V'$ be the node set of H given by the first l nodes w.r.t. $\mathcal{F}$. Furthermore, let $A \subseteq V$ be the node set of G assigned to the nodes of D by an embedding (ϕ, R_ϕ) (see Fig. 2), i.e. $A = \{u \in V \mid \exists v \in D : \phi(u) = v\} = \phi^{-1}(D)$. Note that since the considered embedding is not necessarily surjective and we still assume that H is an ideal grid for G, at most $h' - 1$ nodes of H may be unused. Thus, it follows with $|A| = m$ that $l - h' < m \leq l$ holds. If the set A minimizes $\Gamma_G(m)$, we get a lower bound for $\mathrm{dil}(G : H)$ by estimating for all $l, 0 \leq l \leq |V'|$, the width of the smallest stripe $W(l)$ in H that is large enough that at least $\Gamma_G(A)$ nodes can be assigned to it (see Fig. 2). Therefore, we get

$$\mathrm{dil}(G : H) \geq \max_{0 \leq l \leq |V'|} W(l).$$

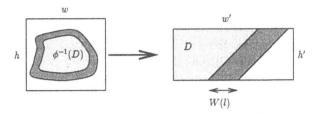

Fig. 2. Illustrating the proof of the lower bound for the dilation.

Since in a stripe of width 1 it is possible to place at most h' nodes, we get

$$\max_{0 \leq l \leq |V'|} W(l) = \left\lceil \frac{\max_{0 \leq m \leq |V|} \Gamma_G(m)}{h'} \right\rceil.$$

Obviously, w.r.t. the above defined ordering $\mathcal{F}$ it holds $\Gamma_G(m) \leq h$, with $0 \leq m \leq |V|$. Furthermore, $\Gamma_G(m)$ equals h for all m with $\frac{1}{2}(h-1)(h-2) < m < |V| - \frac{1}{2}h(h-1)$. Let us choose $l = \lceil h'w'/2 \rceil$. One can show that for all m, with $l - h' < m \leq l$, it holds $\Gamma_G(m) = h$. Thus, we get

$$\left\lceil \frac{\max_{0 \leq m \leq |V|} \Gamma_G(m)}{h'} \right\rceil = \left\lceil \frac{h}{h'} \right\rceil.$$

$\square$

Lower Bound for the Congestion

Let $\bar{G} = (\bar{V}, \bar{E})$ be an arbitrary $\bar{h} \times \bar{w}$ grid with $\bar{h} \leq \bar{w}$. For $S \subseteq \bar{V}$, define

$$\theta_{\bar{G}}(S) = |\{(u, v) \in \bar{E} \mid u \in S, \, v \in \bar{V} \setminus S\}|,$$
$$\theta_{\bar{G}}(l) = \min_{\eta} \theta_{\bar{G}}(S_l(\eta)).$$

$\theta_{\bar{G}}(S)$ is the number of edges separating S from $\bar{V} \setminus S$. For a fixed l, the problem of minimizing $\theta_{\bar{G}}(S_l(\eta))$ over all numberings η is known as the discrete edge isoperimetric problem for grids.

For the above defined grid $\bar{G}$ we define an ordering $\mathcal{L}$ as follows: We say that $(a, b) \in \bar{V}$ is w.r.t. $\mathcal{L}$ greater than $(a', b') \in \bar{V}$ iff $(b > b')$ or $(b = b' \wedge a > a')$.

Ahlswede and Bezrukov [1] showed that for any l with $\lfloor (\bar{h}/2)^2 \rfloor < l \leq \bar{h}\bar{w} - \lfloor (\bar{h}/2)^2 \rfloor$, the set of nodes given by the first l nodes w.r.t. $\mathcal{L}$ has a minimal number of edges connecting these l nodes with the remaining nodes of $\bar{G}$ (among all subsets of $\bar{V}$ of cardinality l). In other words, the node set defined in such a way solves the edge isoperimetric problem on the $\bar{h} \times \bar{w}$ grid for the mentioned values l.

Theorem 3. *The edge-congestion* $\mathrm{con}(G : H)$ *of any embedding of an* $h \times w$ *grid* $G = (V, E)$ *into each of its ideal* $h' \times w'$ *grids* $H = (V', E')$, *with* $2 \leq h' < h \leq w < w'$, *is at least* $\lceil (h+1)/h' \rceil$.

Proof. Let (ϕ, R_ϕ) be an embedding of G into H. Furthermore, suppose that D is any subset of V'. Then,

$$\theta_H(D) \cdot \mathrm{con}(G : H) \geq \theta_G(\phi^{-1}(D)),$$

since each of the edges from $\phi^{-1}(D)$ to $\phi^{-1}(V' \setminus D)$ must be assigned to a path from D to $V' \setminus D$, which contains at least one edge counted by $\theta_H(D)$. Each such edge $e' \in E'$ can be contained in at most $\mathrm{con}(G : H)$ such paths. If the set D minimizes $\theta_H(l)$ with $|D| = l$, and $m = |\phi^{-1}(D)|$ with $l - h' < m \leq l$ (with the same argument as in the proof of Theorem 2), we get

$$\theta_H(l) \cdot \mathrm{con}(G : H) = \theta_H(D) \cdot \mathrm{con}(G : H) \geq \theta_G(\phi^{-1}(D)) \geq \theta_G(m).$$

Thus, we have

$$\mathrm{con}(G : H) \geq \max_{\substack{0 < l < |V'| \\ l-h' < m \leq l}} \frac{\theta_G(m)}{\theta_H(l)}.$$

On the one hand it is obvious w.r.t. the above defined ordering $\mathcal{L}$ that for $1 \leq h' < h$ the fraction $\theta_G(m)/\theta_H(l)$ is at most $(h+1)/h'$. On the other hand one can show that it is always possible to find an l and a corresponding m, with $0 < l < |V'|$ and $l - h' < m \leq l$, such that $\theta_H(l) = h'$ and $\theta_G(m) = h + 1$. See [17] for a detailed proof. □

3.2 Upper Bounds

Following Theorem 1 (in which we exchange h and w, h' and w' respectively, since we have here $h > h'$), each $h \times w$ grid G can be embedded into any ideal $h' \times w'$ grid H of smaller aspect ratio with dilation and edge-congestion of at most $\lceil h/h' \rceil + 1$. Note that the achieved edge-congestion is optimal or differs from the lower bound by one (see Theorem 3). Huang *et al.* [10] improved the dilation up to one, i.e., they showed that the dilation is at most $\lceil h/h' \rceil$, which is optimal (cf. Theorem 2).

4 Embedding into Grids of Larger Aspect Ratio

Throughout this section we consider embedding $G = h \times w$ into $H = h' \times w'$ where H is an ideal grid for G and $h < h' \leq w' < w$. Obviously, each embedding of G into H needs at least a dilation of two and an edge-congestion of one.

In the following we combine three methods to embed G into H with dilation of at most five. The first method is the well-known *folding technique*, introduced by Aleliunas and Rosenberg [2]. The second method is new and called *folding with skipping*. It is based on a technique called 90° *folding with compression*, developed by Ellis [8]. Finally, the matrix-based approach as described in Sect. 2 is used.

Before presenting how the three methods are combined we describe 90° folding with compression and folding with skipping. In fact we develop two different techniques called folding with skipping (i.e. part 1 and part 2). We choose one of them depending on the size of the grids.

Lemma 1. 90° *folding with compression (Ellis'96 [8])*
For all $h \geq 1$, there exists a dilation two embedding of the $h \times (h+1)$ grid G into the $h \times (h+1)$ grid H, such that the nodes in the left column of G are mapped onto the left h nodes in the bottom row of H and the nodes in the right column of G are mapped onto the right column of H.

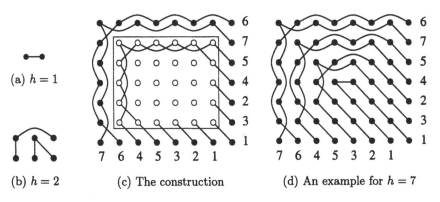

Fig. 3. Constructing 90° folds with compression.

Here we give a short sketch of the construction of 90° folds. Figure 3(a) and (b) illustrate solutions for the base cases, $h = 1$ and $h = 2$. Figure 3(c) illustrates a technique which constructs a solution for $h + 2$ from a solution for h. The box containing the white nodes represents an $h \times (h + 1)$ solution. The addition of black nodes around the entire periphery of the box represents a solution for $(h+2) \times (h+3)$. Consequently, a solution can be built for any h. Let the nodes from the leftmost column of G be named 1, 2, $\cdots$ from the bottom to top. If h is odd, then their order in the bottom host row of H, right to left is 1, 2, 3, 5, 4, 6, 7, 9, 8, $\cdots$, i.e., starting with (4,5), every second pair is reversed. If h is even, then their order in the bottom host row of H, right to left is 1, 2, 3, 4, 6, 5, 7, 8, 10, 9, $\cdots$, i.e., starting with (5,6), every second pair is reversed. Figure 3(d) illustrates the 90° fold for $h = 7$. For a detailed proof see [8].

Lemma 2. *Folding with skipping (part 1)*
For all $h, k \geq 1$ with $\lceil h/2 \rceil \leq k < h$, there exists an embedding with dilation four of the $h \times (2h + k)$ grid G into the $h \times (2h + k)$ grid H, such that

(a) *the nodes in the leftmost column of G (called start nodes) are mapped onto the leftmost h nodes of the bottom row of H where the order of the images of the start nodes is as specified in the remark to Lemma 1,*

(b) *the nodes in the rightmost column of G (called end nodes) are mapped onto the rightmost $h + k$ nodes of the bottom row of H where the order of the images of the end nodes is the same as the order of the images of the start nodes,*

(c) some end node is mapped onto the $(h+1)$st node from left of the bottom row of H and no end node is mapped onto the rightmost node of the bottom row of H,

(d) in the bottom row of H at most two end nodes are mapped onto any three adjacent nodes of the rightmost $h+k$ nodes, and

(e) the distance in H between the images of any two adjacent end nodes is at most four.

Proof. Take a 90° fold with compression with parameter h as it is described in Lemma 1 omitting the rightmost column of this fold. This construct corresponds to the leftmost h columns of H. In Fig. 4, we have illustrated this by an example, where the gray nodes and edges depict the 90° fold. Obviously, by this construct (a) is fulfilled and the grid induced by the first h columns of G is embedded into H providing dilation of at most two (see Lemma 1). Now create another 90° fold with compression with parameter h. The leftmost column of this construct is mapped onto the $(h+1)$st column of H and the rightmost column of this construct is mapped onto the rightmost column of H. The remaining $h-1$ columns of this construct are mapped onto certain columns of the remaining $h+k-2$ columns of H minding the properties (b), (d), and (e). This means that we "stretch" the second construct in a certain way. This is exemplified in Fig. 4, where the black nodes and edges correspond to this second 90° fold with compression. Up to now there are $k-1$ unused columns in H. We use these columns to connect the two constructs defined so far. Figure 4 illustrates these $k-1$ unused columns with white nodes. In the following we call these columns white columns and the columns of the second 90° fold with compression black columns. Since $\lceil h/2 \rceil \leq k$ holds, at least 1/3 columns of the rightmost $h+k$ columns are white columns. Thus, property (d) can be fulfilled, since it is always possible to place at most two black columns between two white columns avoiding the case that three adjacent columns of the rightmost $h+k$ columns are black. As a special case, the rightmost white column must be placed at a distance of at most three from the rightmost column of H (which is black) and the leftmost white column must be placed at a distance of at most three from the first fold. It follows that the connection between the two folds can be accomplished with dilation of at most four. An example is given by the dotted lines in Fig. 4. Additionally,

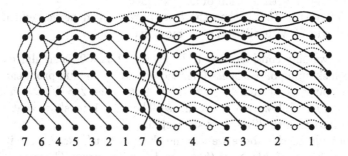

7 6 4 5 3 2 1 7 6   4   5 3   2   1

Fig. 4. Folding with skipping, exemplified by embedding $G = 7 \times 19$ into $H = 7 \times 19$.

since $k < h$ holds, at most $1/2$ columns of the rightmost $h+k$ columns are white columns. Thus, property (e) can be fulfilled, since it is always possible to place at most one white column between two black columns. According to Lemma 1 the grid induced by the rightmost $h + 1$ columns of G can be embedded into the $h \times (h + 1)$ grid with dilation of at most two by using the 90° fold with compression. By our construction we embed this 90° fold into the $h \times (2h + k)$ grid. This increases the dilation by a factor of two, since each edge of the fold is stretched by at most two. Thus, we have constructed a dilation four embedding of G into H that satisfies the properties (a)–(e).   □

Lemma 3. *Folding with skipping (part 2)*
For all $h, k \geq 1$ with $0 < k < \lceil h/2 \rceil$, there exists an embedding with dilation five of the $h \times (3h + k)$ grid G into the $h \times (3h + k)$ grid H, such that

(a) *the nodes in the leftmost column of G (called start nodes) are mapped onto the leftmost h nodes of the bottom row of H where the order of the images of the start nodes is as specified in the remark to Lemma 1,*

(b) *the nodes in the rightmost column of G (called end nodes) are mapped onto the rightmost $2h + k$ nodes of the bottom row of H where the order of the images of the end nodes is the same as the order of the images of the start nodes,*

(c) *some end node is mapped onto the $(h+1)$st node from left of the bottom row of H and no end node is mapped onto the rightmost node of the bottom row of H,*

(d) *in the bottom row of H at most one end node is mapped onto any two adjacent nodes of the rightmost $2h + k$ nodes, and*

(e) *the distance in H between the images of any two adjacent end nodes is at most five.*

Proof. The proof is similar to the one for Lemma 2. It can be found in [17].

Theorem 4. *Suppose $h < h' \leq w' < w$ and $hw \leq h'w'$. Any $h \times w$ grid G can be embedded into the $h' \times w'$ grid H with dilation of at most five.*

Proof. We will show the following facts:

1. If $(w' \bmod h) \geq \lceil h/2 \rceil$, then the $h \times w$ grid G can be embedded into the $h' \times w'$ grid H with dilation of at most four.
2. If $(w' \bmod h) < \lceil h/2 \rceil$, then the $h \times w$ grid G can be embedded into the $h' \times w'$ grid H with dilation of at most five.

For brevity, we only consider case 1. The proof of case 2. is similar except that folding with skipping part 2 is used instead of part 1. The full proof can be found in [17]. We can assume that $2h \leq h'$. If $2h > h'$, then $2 > h'/h = h'w'/hw' \geq hw/hw' = w/w'$ follows and we can use Theorem 1 to show that G can be embedded into H with dilation of at most three.

Additionally to the above we assume that $3h \leq w'$. Note, that if $3h > w'$, then $3 > w'/h = w'h'/hh' \geq wh/hh' = w/h'$ and by using Theorem 1 it follows that G can be embedded into H with dilation of at most four.

A   B

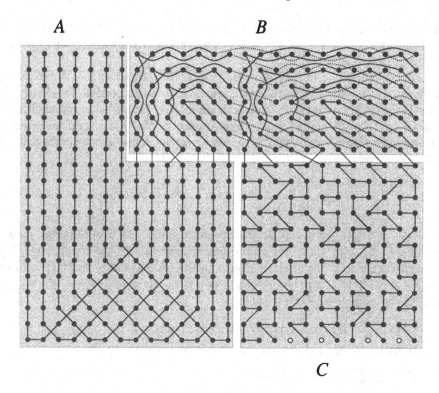

C

Fig. 5. An embedding of the 7×70 into the 19×26 grid with dilation four. The four white nodes correspond to the nodes that are no images of nodes of the guest.

To embed G into H we use a three step approach. We divide G into three parts: Part A consists of the first $h'(\lfloor w'/h \rfloor - 1) - h$ columns of G, part B of the next $2h + k$ columns of G, where $k = (w' \bmod h)$. The third part C consists of the remaining $r = \frac{1}{h}((h+k)(h'-h) - (w'h' - wh))$ columns of G. The nodes of A are mapped onto H by using the folding technique (see [2]), the nodes of B by using folding with skipping (see Lemma 2), and the nodes of C by using the matrix-based approach (see Theorem 1). This three step approach is illustrated in Fig. 5.

Note that the folding with skipping according to Lemma 2 can be applied since $3h \leq w'$ holds. The images of the nodes of A can be connected to images of the nodes of B with dilation of at most two, since under folding with skipping only adjacent rows are skewed, and since $2h \leq h'$ holds, the folding process mapping part A is finished before the images of nodes of B are reached. It remains to show that the nodes of part C can be mapped onto the remaining nodes of H with dilation of at most four and that the connection between the images of the nodes of B and the images of the nodes of C can be accomplished with dilation four.

Let us now consider part C. The nodes that must be mapped onto the remaining rectangle of H are isomorphic to the $h \times r$ grid. Note, that we can assume without loss of generality that $r > h' - h$ holds. Otherwise we get an embedding with dilation of at most four by vertically continuing where the end nodes of part B are mapped.

Let us define $l = \min\{s \mid s(h + k) \geq hr\}$. Note that the $(h + k) \times l$ grid is an ideal grid for the $h \times r$ grid, which remains to be embedded. Obviously, the $(h + k) \times l$ grid is a subgraph of the remaining $(h + k) \times (h' - h)$ rectangle of H, since $l \leq h' - h$. To embed the $h \times r$ grid into the $(h + k) \times l$ grid we compress all rows from length r to length l. In doing so, we define an embedding matrix $M_{h \times l}$ as follows:

$$m_{ij} = \left\lceil \frac{h + k}{h}(i - j + 1) \right\rceil - \left\lceil \frac{h + k}{h}(i - j) \right\rceil, \tag{3}$$

with $0 \leq i < h, 0 \leq j < l$. Note that we get column j by a cyclic down shift of column $j - 1, 1 \leq j < l$. If the embedding matrix satisfies the conditions (C1)–(C4) of Theorem 1, the dilation of such an embedding is at most $\lceil r/l \rceil + 1$. Note that $r/l \leq \frac{l(h+k)}{lh} = \frac{h+k}{h} < 2$. Thus, the dilation is at most three. Let us now consider the conditions (C1)–(C4):

(C1) $m_{ij} \in \{\lfloor r/l \rfloor, \lceil r/l \rceil\}$, for $0 \leq i < h, 0 \leq j < l$,

(C2) $m_{ij} = m_{i+1,j+1}$, for $0 \leq i < h - 1, 0 \leq j < l - 1$,

(C3) $\sum_{i=0}^{h-1} m_{ij} \leq h + k$, for $0 \leq j < l$,

(C4) $\sum_{j=0}^{l-1} m_{ij} \geq r$, for $0 \leq i < h$.

Ad (C1). Following the definition of l, $r \leq \frac{l(h+k)}{h}$ holds. Thus, $\frac{r}{l} \leq \frac{l(h+k)}{lh} = \frac{h+k}{h} < 2$. Additionally, $\frac{r}{l} \geq \frac{r}{h'-h} > 1$. It follows that $\{\lfloor \frac{r}{l} \rfloor, \lceil \frac{r}{l} \rceil\} = \{1, 2\}$. Using $\lceil \frac{a \pm b}{c} \rceil - \lceil \frac{b}{c} \rceil \in \{\lfloor \frac{a}{c} \rfloor, \lceil \frac{a}{c} \rceil\}$ for any integers a, b, and c, it follows that $m_{ij} = \lceil \frac{h+k}{h}(i - j + 1) \rceil - \lceil \frac{h+k}{h}(i - j) \rceil \in \{\lfloor \frac{h+k}{h} \rfloor, \lceil \frac{h+k}{h} \rceil\} = \{1, 2\}$, for $0 \leq i < h, 0 \leq j < l$.
Ad (C2). $m_{i+1,j+1} = \lceil \frac{h+k}{h}((i + 1) - (j + 1) + 1) \rceil - \lceil \frac{h+k}{h}((i + 1) - (j + 1)) \rceil = \lceil \frac{h+k}{h}(i - j + 1) \rceil - \lceil \frac{h+k}{h}(i - j) \rceil = m_{ij}$, for $0 \leq i < h - 1, 0 \leq j < l - 1$.
Ad (C3). $\sum_{i=0}^{h-1} m_{ij} = \lceil \frac{h+k}{h}(h - j) \rceil - \lceil \frac{-j(h+k)}{h} \rceil = \lceil \frac{-j(h+k)}{h} \rceil - \lceil \frac{-j(h+k)}{h} \rceil + h + k = h + k$, for $0 \leq j < l$.
Ad (C4). $\sum_{j=0}^{l-1} m_{ij} = \lceil \frac{h+k}{h}(i + 1) \rceil - \lceil \frac{h+k}{h}(i - l + 1) \rceil = \lceil \frac{(h+k)l+(h+k)(i-l+1)}{h} \rceil - \lceil \frac{(h+k)(i-l+1)}{h} \rceil \in \{\lfloor \frac{l(h+k)}{h} \rfloor, \lceil \frac{l(h+k)}{h} \rceil\}$, for $0 \leq i < h$. Thus, $\sum_{j=0}^{l-1} m_{ij} \geq \lfloor \frac{l(h+k)}{h} \rfloor \geq r$, for $0 \leq i < h$.

It remains to show that the connections between the images of the nodes of B and the images of the nodes of C can be accomplished with dilation four. An example is shown in Fig. 5, where we have connected the construct shown in Fig. 4 (mapping of part B) with the construct shown in Fig. 6 (mapping of part C).

We will show the following: If we choose the distribution of the white and black columns (see the proof of Lemma 2) in accordance with the distribution of

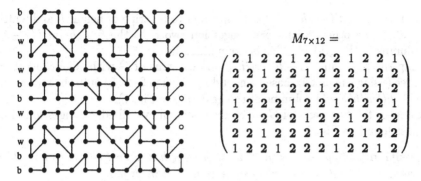

$$M_{7\times12} =$$

$$\begin{pmatrix}
2 & 1 & 2 & 2 & 1 & 2 & 2 & 2 & 1 & 2 & 2 & 1 \\
2 & 2 & 1 & 2 & 2 & 1 & 2 & 2 & 2 & 1 & 2 & 2 \\
2 & 2 & 2 & 1 & 2 & 2 & 1 & 2 & 2 & 2 & 1 & 2 \\
1 & 2 & 2 & 2 & 1 & 2 & 2 & 1 & 2 & 2 & 2 & 1 \\
2 & 1 & 2 & 2 & 2 & 1 & 2 & 2 & 1 & 2 & 2 & 2 \\
2 & 2 & 1 & 2 & 2 & 2 & 1 & 2 & 2 & 1 & 2 & 2 \\
1 & 2 & 2 & 1 & 2 & 2 & 2 & 1 & 2 & 2 & 1 & 2
\end{pmatrix}$$

Fig. 6. An embedding of the 7×20 into the 12×12 grid and the corresponding matrix $M_{7\times12}$. The four white nodes depict the nodes that are no images of nodes of the guest. The letters "w" and "b" show the distribution of white and black columns in correspondence with Fig. 4.

the 1's and 2's in the first column of the embedding matrix $M_{h\times l}$, then the nodes of part B can be mapped with dilation four, satisfying the properties (a)–(e) of Lemma 2. Additionally, the connections to the images of the nodes of C can be accomplished with dilation four. For each 1 in the first column of $M_{h\times l}$ we put a black column in the corresponding position of the folding with skipping construct. For each 2 we put a white column followed by a black column in the corresponding construct. A special case is the matrix entry $m_{00} = \lceil \frac{h+k}{h} \rceil = 2$, that corresponds to two black columns. An example is shown in Fig. 6, where the distribution of white and black columns is given by the letters "b" for black column and "w" for white column. The corresponding folding with skipping is shown in Fig. 4. In Fig. 5 the connection of these two constructs is shown.

Since there are no consecutive 1's in the first column of the embedding matrix, at least each third column of the rightmost $h + k$ columns of the folding with skipping is a white column. To see this, note that $m_{i0} + m_{i+1,0} = \lceil \frac{h+k}{h}(i+1) \rceil - \lceil \frac{h+k}{h}i \rceil + \lceil \frac{h+k}{h}(i+2) \rceil - \lceil \frac{h+k}{h}(i+1) \rceil = \lceil \frac{2(h+k)+i(h+k)}{h} \rceil - \lceil \frac{i(h+k)}{h} \rceil \geq \lfloor \frac{2(h+k)}{h} \rfloor \geq 3$, for $0 \leq i < h-1$. Additionally, since at most all but one entry of the first column of $M_{h\times l}$ are 2's, at most each second column of the rightmost $h + k$ columns of the folding with skipping is a white column. Thus, the properties (a)–(e) of Lemma 2 can be satisfied, i.e., a suitable distribution of the white and black columns can be determined by the first column of the embedding matrix. Since only adjacent rows of part B may be skewed (see Lemma 1), we can connect the nodes of B with the nodes of C with dilation of at most four. □

Unfortunately, using a shortest paths routing scheme, the edge-congestion achieved by the above mentioned embeddings is five and cannot be reduced in general. Therefore, we will shortly describe a much simpler embedding providing edge-congestion of at most four. Note that, if w/w' is at most 2, we can use the technique described in Sect. 2 (cf. Theorem 1) to embed G into H with edge-congestion of at most three. If $w/w' > 2$, we use the following approach:

1. Map the first $h'(\lfloor w/h' \rfloor - 1)$ columns of G (one stripe of height h and width h' after the other) with edge-congestion two onto the first $h(\lfloor w/h' \rfloor - 1)$ columns of the grid H. This is exemplified in Fig. 7.
2. Map the remaining $w - h'(\lfloor w/h' \rfloor - 1)$ columns of G with edge-congestion three (using the matrix-based embedding given in Sect. 2) onto the remaining $w' - h(\lfloor w/h' \rfloor - 1)$ columns of the grid H (see Fig. 7).
3. Connect the two parts of G which results in an overall edge-congestion of at most four (see Fig. 7).

Proposition 1. *Suppose $h < h' \leq w' < w$. Any $h \times w$ grid G can be embedded into the $h' \times w'$ grid H with edge-congestion of at most four.*

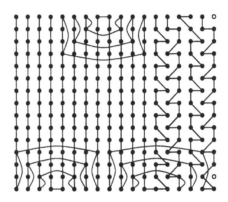

Fig. 7. Embedding $G = 4 \times 67$ into $H = 15 \times 18$ with edge-congestion of at most four. The dashed lines represent the routing scheme for the edges connecting the stripes.

Note that the above mentioned technique is applicable, if $2h \leq h'$ holds. If $2h > h'$, $w/w' < 2$ follows and G can be embedded into H with edge-congestion three (see Theorem 1). If $h|h'$ or $h|w'$, we only need the first step of the above algorithm. Otherwise, we additionally execute the second step using the matrix-based approach shown in Sect. 2 where the edge-congestion is bounded by $\lceil (w - h'(\lfloor w/h' \rfloor - 1))/h' \rceil + 1 \leq 3$. Executing the third step increases the overall edge-congestion to at most four.

5   Conclusion

The table 1 gives a complete overview about the results achieved in this paper (except the result marked with the reference that is due to Huang *et al.* [10]).

The lower bound techniques proposed in this paper can be used to achieve corresponding results for other embedding problems. For instance, in [4] we give an exact solution for the congestion problem of embedding a hypercube into d-dimensional grids of the same size.

Acknowledgments: The authors would like to thank Burkhard Monien and Sergej L. Bezrukov for many insightful comments and suggestions.

Table 1. Bounds for the dilation (dil.) and edge-congestion (cong.) of embedding $h \times w$ into any of its ideal $h' \times w'$ grids.

cases	low. bound dil.	upp. bound dil.	low. bound cong.	upp. bound cong.
$h = h'$, $w = w'$	1	1	1	1
$h' < h \leq w < w'$	$\lceil h/h' \rceil$	$\lceil h/h' \rceil$ see [10]	$\lceil (h+1)/h' \rceil$	$\lceil h/h' \rceil + 1$
$h < h' \leq w' < w$	2	5	1	4

References

1. R. Ahlswede, S.L. Bezrukov: Edge Isoperimetric Theorems for Integer Point Arrays. *Appl. Math. Lett.*, Vol. 8, No. 2, pp. 75–80, 1995.
2. R. Aleliunas, A. Rosenberg: On Embedding Rectangular Grids in Square Grids. *IEEE Transact. on Computers.* Vol. C-31, No. 9, pp. 907–913, 1982.
3. S.L. Bezrukov: Isoperimetric Problems in Discrete Spaces. *Extremal Problems for Finite Sets, Bolyai Soc. Math. Stud.* 3, pp. 59–91, 1994.
4. S.L. Bezrukov, J.D. Chavez, L.H. Harper, M. Röttger, U.-P. Schroeder: The Congestion of n-Cube Layout on a Rectangular Grid. *Discrete Math.*, to appear.
5. M.Y. Chan: Embedding of Grids into Optimal Hypercubes. *SIAM J. Computing*, Vol. 20, No. 5, pp. 834–864, 1991.
6. J. Chvátalová: Optimal Labelling of a Product of two Paths. *Discrete Math.* 11, pp. 249–253, 1975.
7. J.A. Ellis: Embedding Rectangular Grids into Square Grids. *IEEE Transact. on Computers*, Vol. 40, No. 1, pp. 46–52, 1991.
8. J.A. Ellis: Embedding Grids into Grids: Techniques for Large Compression Ratios. *Networks*, Vol. 27, pp. 1–17, 1996.
9. L.H. Harper: Optimal Numberings and Isoperimetric Problems on Graphs. *J. Comb. Theory* 1, No. 3, pp. 385–393, 1966.
10. S.-H. Huang, H.L. Liu, R.M. Verma: A New Combinatorial Approach to Optimal Embeddings of Rectangles. *Algorithmica* 16, pp. 161–180, 1996.
11. S.-H. Huang, H.L. Liu, R.M. Verma: On Embedding Rectangular Meshes into Rectangular Meshes of Smaller Aspect Ratio. *Inf. Proc. Let.* 63, pp. 123–129, 1997.
12. S.R. Kosaraju, M.J. Atallah: Optimal Simulations Between Mesh-Connected Arrays of Processors. *J. Assoc. Comput. Mach.*, Vol. 35, No. 3, pp. 635–650, 1988.
13. C. L. Leiserson: Area-Efficient Graph Layouts (for VLSI). *Proc. 21st IEEE Symp. on Foundations on Computer Science*, pp. 270–281, 1980.
14. R.G. Melhem, G.-Y. Hwang: Embedding Rectangular Grids into Square Grids with Dilation Two. *IEEE Transact. on Computers*, Vol. 39, No. 12, pp. 1446–1455, 1990.
15. B. Monien, I.H. Sudborough: Embedding one Interconnection Network in Another. *Computing Suppl.* 7, pp. 257–282, 1990.
16. T. Römke, M. Röttger, U.-P. Schroeder, J. Simon: On Efficient Embeddings of Grids into Grids in PARIX. *Proc. EURO-PAR'95*, Lecture Notes in Computer Science 966, Springer-Verlag, pp. 181–204, 1995.
17. M. Röttger, U.-P. Schroeder: Efficient Embeddings of Grids into Grids. Tech. Rep. tr-rsfb-97-054, University of Paderborn, 1997.
18. M. Röttger, U.-P. Schroeder: Embedding 2-Dimensional Grids into Optimal Hypercubes with Edge-Congestion 1 or 2. *Parallel Processing Letters*, to appear.
19. F.C. Sang, I.H. Sudborough: Embedding Large Meshes into Small Ones. *Proc. of the IEEE Symposium on Circuits and Systems*, Vol. 1, pp. 323–326, 1990.
20. X. Shen, W. Liang, Q. Hu: On Embedding Between 2D Meshes of the Same Size. *IEEE Transact. on Computers*, Vol. 46, No. 8, pp. 880–889, 1997.

Integral Uniform Flows in Symmetric Networks [*]
(Extended Abstract)

Farhad Shahrokhi[1] and László A. Székely[2]

[1] Department of Computer Science, University of North Texas
P.O.Box 311366, Denton, TX 76203-1366, USA
farhad@cs.unt.edu
[2] Department of Mathematics, University of South Carolina
Columbia, SC 29208, USA
laszlo@math.sc.edu

Abstract. We study the integral uniform (multicommodity) flow problem in a graph G and construct a fractional solution whose properties are invariant under the action of the automorphism group $Aut(G)$ of G. The fractional solution is shown to be close to an integral solution (depending on properties of $Aut(G)$), and in particular becomes an integral solution for a class of graphs containing Cayley graphs. As an application we estimate asymptotically (up to error terms) the edge congestion of an optimal integral uniform flow (edge forwarding index) in the cube connected cycles and the butterfly.

1 Introduction

The uniform concurrent multicommodity flow (uniform flow) problem [12, 15] is the problem of supplying one unit of (fractional) flow between all ordered pairs of vertices in a graph; the objective is to minimize the largest flow through any edge which is called the *congestion*. The integral version of this problem has been studied under the name *edge forwarding index* [2, 7, 14], and calls for the assignment of one path per each ordered pair of vertices to minimize the congestion. The (fractional) uniform flow problem is known to be solvable in polynomial time [15], and starting from the work of Shahrokhi and Matula [15], there have been a series of papers on how to approximately solve this problem faster [1], [9], [5], and [8]. The integral version is known to be NP-hard [3, 14].

There is a need for estimating the value of the congestion, since many important graph theoretical parameters are related to the congestion. For instance the congestion of a uniform flow provides for lower bounds for the bisection width [11, 19], and expansion (isoperimetric) rates [18, 19], whereas, the congestion of an integral uniform flow, or the forwarding index, provides for lower bounds for the the crossing number [11, 19–22]. More over, the close connection between

[*] The research of the first author was supported by NSF grant CCR-9528228. Research of the second author was supported in part by the Hungarian NSF contracts T 016 358 and T 019 367, and by the NSF contract DMS 970 1211.

the integral multicommodity flow problems and packet routing was discovered by Leighton, Maggs, and Rao [10], who showed the existence of a near optimal offline schedule for routing the packets on a set of paths involving a near optimal solution to the integral multicommodity flow problem.

In this paper (Section 3), we present an algorithm for constructing uniform flows which exhibit invariance under the action of the automorphism group of the graph (Theorem 1). The uniform flows constructed here are shown to be "near integral", in the sense that the number of paths hosting flow are bounded as a function of the order of the stabilizer of a two-tuple of vertices in the automorphism group. In particular, for a class of graphs containing Cayley graphs, the constructed flow is integral. Previously, we have been able to construct invariant uniform flows [18, 19]. Our previous methods, however, would construct uniform flows which use too many paths, and hence were very far from being an integral uniform flow. In [17] we were able to construct integral flows only in Cayley graphs. Our simple construction in the present paper (Theorem 1) implies all those previous ad-hoc results in Theorems 3.2, and 3.3 of [17] for off line computation of packet routes. For the class of orbit proportional graphs, defined in Section 3, the algorithm is shown to construct an optimal (fractional) flow (Theorem 2). This approach also provides formulas for the congestion of the optimal uniform flow.

Section 4 is a technical section and is dedicated to the structure of butterflies (with wrap around) and cube connected cycles, which are among the popular architectures in parallel computing. It is shown that the cube connected cycles is orbit proportional, whereas the butterfly is not. Section 5 is the most important part of the paper. In this section we construct optimal integral flows in a cube connected cycle and a butterfly and use probabilistic methods to derive asymptotic formulas (including the constants and error terms) for the optimal congestion in these graphs. To be more precise we show that the edge forwarding indices of n-dimensional cube connected cycles, and n-dimensional butterfly, are $\frac{5}{4}n^2 2^n(1 - o(1))$, and $\frac{5}{4}n^2 2^{n-1}(1 + o(1))$, respectively.

We close the introduction with an application of our formula for the edge forwarding index of the butterfly. We derive the currently best lower bound for the crossing number of the butterfly B_n, $(\frac{16}{125} + o(1))4^n$. One easily can apply the randomized rounding technique from our paper [19] to obtain an embedding of K_{n2^n} into B_n (see definitions there!) with congestion μ which is $(1/2 + o(1))$ times the edge forwarding index. Using our other theorem from [19], $cr(B_n) \geq \frac{cr(K_{n2^n})}{\mu^2}$ minus negligible error terms. Finally, use from [24] that $cr(K_N) \geq (\frac{1}{80} - o(1))N^4$. Merging all these results we obtain $cr(B_n) \geq (\frac{16}{125} + o(1))4^n$.

This paper is based on the technical report [16], proofs not spelled out here can be found there in details.

2   Definitions

Let $G = <V, E>$ be a connected simple finite graph. Let $L(p)$ denote the number of edges in the path p and let $L(i, j)$ denote the length of the *shortest* ij path. (We preserve the word *distance* for something else.) We set $L = \sum_{(i,j) \in V \times V} L(i,j)$. Let p be a path with end vertices a and b in the graph $G =< V, E >$. Then p will give rise to an oriented path from a to b; and to another from b to a. For any ordered pair of vertices $(i,j) \in V \times V$, we denote by P_{ij} the set of all oriented paths from i to j; any $p \in P_{ij}$ is termed an ij path. Let $P = \cup_{(i,j) \in V \times V} P_{ij}$ be the set of all oriented paths. Throughout this paper the term path means oriented path, unless stated otherwise. For $e \in E$, let P_e denote the collection of all paths containing e. Finally, let R^+ denote the set of non-negative real numbers.

A *uniform concurrent multicommodity flow* (shortly *uniform flow*) f is a function $f : P \to R^+$, such that $\sum_{p \in P_{ij}} f(p) = 1$ for any $(i,j) \in V \times V, i \neq j$. We call $f(p)$ the flow on path p; if $f(p) > 0$, then p is called an *active path*. We set $f(e) = \sum_{p \in P_e} f(p)$ for any edge $e = xy$, and call $f(e)$ the flow on the edge e. For a uniform flow f we denote $\max_{e \in E} f(e)$ by μ_f and call μ_f the *congestion* of f. A uniform flow f is called *integral*, if $f(p) = 1$ for any active path p. Let μ_G denote the smallest congestion achieved by a uniform flow in G. A uniform flow f in $G =< V, E >$ is *edge optimal*, if $\mu_f = \mu_G$. An edge optimal uniform flow can be computed using a node-arc form linear program [4] in polynomial time. Computing the integral versions of the multicommodity flows and uniform flows have been known to be NP-hard [3].

A *distance function* [13] on $G =< V, E >$ is a function $d : E \to R^+$ such that $d(e) > 0$ for at least one edge e. For any path $p \in P$, we define $d(p) = \sum_{e \in p} d(e)$, moreover for any $(i,j) \in V \times V$, we define $d(i,j) = d(j,i) = \min\{d(p) : p \in P_{ij}\}$ for all $(i,j) \in V \times V$. We further assume that $d(i,i) = 0$, for any $i \in V$. We term $d(p)$ the *distance* of path p. We define the *distance congestion* μ_d for the distance function d by

$$\mu_d = \frac{\sum_{(i,j) \in V \times V} d(i,j)}{\sum_{e \in E} d(e)}.$$

3   Uniform flows and graph symmetries

Our reference to algebraic graph theory is [23]. It is well known that the set of all permutations on V constitute a group on V which is called the *automorphism group* of G. Let $Aut(G)$ denote the automorphism group of G, and let Γ be a subgroup of $Aut(G)$, then we write $\Gamma < Aut(G)$. Note that the action of any $\Gamma < Aut(G)$ on E partitions E into equivalent classes. We call each class a Γ-*edge orbit*.

A uniform flow f is called a Γ-*invariant* [18,19] if for any $g \in \Gamma$ and any $p \in P$, we have $f(p) = f(g(p))$. Next, we show how to construct an invariant uniform flow in which the number of active paths depends on the structure of Γ, thus in certain desirable cases which includes Cayley graphs we will have integral uniform flows. Let $(a,b) \in V \times V$, we define the ab stabilizer of Γ, denoted by Γ_{ab} to be the set of all automorphisms which map a to a and b to b. Formally, $\Gamma_{ab} = \{\gamma \in \Gamma | \gamma(a) = a, \gamma(b) = b\}$. For $p \in P$, and $(a,b) \in V \times V$, let $\Gamma(p)$ and $\Gamma(a,b)$, denote $\{\gamma(p) | \gamma \in \Gamma\}$, and $\{(\gamma(a), \gamma(b)) | \gamma \in \Gamma\}$.

Theorem 1. *A Γ-invariant uniform flow f^* in G can be computed in a polynomial time of $|V|$ and $|\Gamma|$ so that the number of active paths for any pair $(a, b) \in V \times V$ is at most $|\Gamma_{ab}|$. Moreover, any active ab path p has $L(p) = L(a, b)$, and $\mu_{f^*} \leq \frac{L}{|E_1|}$, where E_1 is the smallest Γ-edge orbit.*

Proof. The action of Γ partitions $V \times V$ into equivalence classes $R^1, R^2, ..., R^k$; thus (a, b) and (c, d) are in the same equivalent class, if $c = \gamma(a)$, and $d = \gamma(b)$, for some $\gamma \in \Gamma$. Moreover, for any $(a, b) \in V \times V$, and any two shortest ab paths, p_1, p_2 define

$$p_1 \ R_{ab} \ p_2, \quad \text{iff} \quad p_2 = \gamma(p_1) \ \text{for some } \gamma \in \Gamma.$$

It is easily seen that, for any $(a, b) \in V \times V$, R_{ab} is an equivalence relation on the set of shortest ab paths; let R_{ab}^p denote the equivalence class containing the shortest ab path p and note that $|R_{ab}^p| \leq |\Gamma_{ab}|$.

We now describe the construction of f^* in each R^i. For $i = 1, 2, ..., k$, select a vertex pair (a_i, b_i) in R^i, and also select one shortest path p_i from a_i to b_i. Define for $i = 1, 2, ..., k$,

$$f^*(p) = \begin{cases} \frac{1}{|R_{ab}^{p_i}|}, & p \in \Gamma(p_i), \\ 0, & otherwise. \end{cases}$$

The claim regarding the time complexity is easy to verify. Moreover, the invariance of f^*, and the claim regarding the number of active paths are direct consequence of the construction. Now let $(a, b) \in R^i, i = 1, 2, ..., k$ and note that $\sum_{p \in P_{ab}} f^*(p) = \sum_{p \in P_{a_i b_i}} f^*(p) = |R_{a_i b_i}^{p_i}| \frac{1}{|R_{a_i b_i}^{p_i}|} = 1$. Finally, the upper bound on the congestion follows by observing that any two edges in E_1 must host the same amount of flow, and applying a simple averaging argument same as in [19]. $\square$

Note that in any edge transitive graph G, $E = E_1$, and indeed in this case f^* is edge optimal with $\mu_{f^*} = \frac{L}{|E|}$, since by the duality theory of linear programming [13] $\frac{L}{|E|}$ is a lower bound on the congestion of any uniform flow. Moreover, when G is a Cayley graph, $|\Gamma_{ab}| = 1$, hence f^* is an integral uniform flow (it has exactly one active path per vertex pair) and can be used for packet routing. Indeed in this case the general construction in Theorem 1 implies our previous ad-hoc results in Theorems 3.2, and 3.3 of [17] for off line computation of packet routes.

For $G = \langle V, E \rangle$ and $\Gamma < Aut(G)$, let $\{E_1, E_2, ..., E_k\}$ be the set of Γ-orbits of E. We say that G is Γ-orbit proportional (or orbit proportional when the context is clear) if for all $(i, j) \in V \times V$, any ij path p with $L(p) = L(i, j)$ and any ij path q, we have

$$|q \cap E_m| \geq |p \cap E_m|, \quad m = 1, 2, ..., k.$$

To see examples, note that any edge transitive graph is orbit proportional with respect to its automorphism group and any tree is orbit proportional with respect to the trivial group. We previously proved the following Theorem [18].

Theorem 2. *Assume that $G = \langle V, E \rangle$ is Γ-orbit proportional. Let $\hat{f}$ be a Γ-invariant uniform flow on G such that every ij active path p has $L(p) = L(i, j)$. Then we have:*
(i) $\hat{f}$ is edge optimal in G.
(ii) Assume that $\{E_1, E_2, ..., E_t\}$ is the set of Γ-orbits of E and for any $i = 1, 2, ..., t$, let d_i be a distance function with $d_i(e) = 1$, if $e \in E_i$, and $d_i(e) = 0$, otherwise. Then $\mu_G = \mu_{\hat{f}} = \max_i \mu_{d_i}$.

Observe that the flow f^* constructed in Theorem 1, satisfies the condition of Theorem 2, and hence when G is orbit proportional f^* is edge optimal. Indeed, the Theorems allow to estimate the optimal congestion of f^* in an orbit proportional graph. For instance for Q_k (k-dimension cube), which is edge transitive and hence orbit proportional, Theorems 1, 2 give $\mu_{f^*} = 2^k$. Moreover, since Q_k is a Cayley graph [23], for any vertex pair a, b, Γ_{ab} is the identity, and thus f^* is an optimal integral uniform flow. Finally, as we have shown in [18] the class of vertex transitive orbit proportional graphs is closed under Cartesian product. Hence, the class of orbit proportional graphs for which f^* is edge optimal is fairly large.

4   The structure of cube connected cycles and butterfly

It is well known that the cube connected cycles and the butterfly (with wrap-around) are Cayley graphs with the same underlying group Γ which is the *wreath product* of Z_n and Z_2 but with different generating sets. We exploit this structure in the following. Let $N = \{0, 1, 2, ..., n-1\}$ and $\Theta_n = \{g_{W,i} : W \subseteq N, i \in N\}$. For $i, j \in N$, let $i \oplus j$ denote $i + j$ modulo n. For $U \subseteq N$ and $i \in N$, let $U \oplus i = \{j \oplus i : j \in U\}$. Set $V \triangle U = (V \cup U) \setminus (V \cap U)$. Now Θ_n is a group with identity $g_{\emptyset, 0}$ and operations

$$g_{W,t} g_{U,i} = g_{W \triangle (U \oplus t), i \oplus t} \quad \text{and} \quad g_{W,t}^{-1} = g_{W \oplus (n-t), n-t}.$$

The n-dimensional cube-connected cycles CC_n is a Cayley graph over Θ_n with the generating set
$$H = \{g_{\emptyset, 1}, g_{\{0\}, 0}\}.$$
We term the edges produced by the first generator *cyclic edges* and the edges produced by the second generator *cubic edges*. It is easy to see that $CC_n = \langle V, E \rangle$, where
$$V = \{(W, i) : W \subseteq N, i \in N\}$$
and $(W, i)(U, j) \in E$ if $i = j = W \triangle U$ (cubic edges in dimension i) or if $|i - j| \equiv 1$ mod n and $U = V$ (cyclic edges).
Let BB_n denote the n-dimensional butterfly with wrap-around. It is well known that BB_n is a Cayley graph over Θ_n with the generating set $H = \{g_{\emptyset, 1}, g_{\{0\}, n-1}\}$. We term the edges arising from the first generator *cyclic edges* and the edges arising from the second generator *cubic edges*. It is easy to see that $BB_n = \langle V, E \rangle$, where
$$V = \{(X, i) : X \subseteq N, i \in N\},$$

and $(X,i)(Y,j) \in E$ if $X = Y$ and $|i - j| \equiv 1 \bmod n$ (cyclic edges) or $X \triangle Y = i$ and $j \equiv i - 1 \bmod n$ (cubic edges in dimension i).

Let C_n be the cycle on the vertex set $N = \{0, 1, \ldots, n - 1\}$ with the edge set $\{0, 1\}, \{1, 2\}, \ldots, \{n-1, 0\}$. Define C_n^+ be C_n with one loop added at every vertex. For any walk w in CC_n or BB_n, let $Cyclic(w)$ and $Cubic(w)$ denote the multiset of cyclic edges and the multiset of cubic edges, respectively, in w. Any $i, j \in N$ ($i \neq j$) split C_n (C_n^+) into two edge disjoint ij paths. We refer to these paths as *left side* and *right side*, where the vertices of the left side precede, and the vertices of the right side follow i in the cyclic order of N. For convenience, we assume that the right side is the short side and has length $L(i, j)$.

Let i and j be two vertices of C_n^+ (C_n) and $T \subseteq N$. A *gap* induced by T is any ab path p such that (i) $a, b \in T \cup \{i, j\}$, (ii) no intermediate vertex of p is in $T \cup \{i, j\}$. The *length* of any gap is the number of edges in this gap. A *gap* induced by $F \subseteq E(C_n)$ is the gap induced by the set of vertices of edges of F, such that the gap does not use edges of F. For $i \neq j$, it makes sense to speak about gaps on the left side and gaps on the right side.

We analyze the structure of shortest paths in CC_n first. Let $X = (W, i)$ be a vertex of CC_n; the projection of X on C_n^+ is the vertex i. This projection can be extended to the edges and therefore to the walks of CC_n in the following fashion: the cyclic edges of CC_n are projected to the edges of C_n^+, whereas the cubic edges of CC_n are projected to the loops of C_n^+. Given two vertices $X_1 = (U, i)$ and $X_2 = (W, j)$ in CC_n, it is convenient to analyze the structure of any $X_1 X_2$ walk p in CC_n by projecting it on C_n^+ to get an ij walk q in C_n^+. Notice that $L(p) = L(q)$, since each edge or loop contributes one to the length of the walk in C_n^+. Let $X_1 = (W, i)$ and $X_2 = (U, j)$ be two distinct vertices of CC_n. Any loop of an ij walk in C_n^+ at a vertex $a \in N$ is called an *essential loop* if $a \in W \triangle U$, otherwise the loop is *non-essential*. An ij walk w in C_n^+ is called an *essential walk*, if w has the following properties : (i) every essential loop of C_n^+ is traversed by w exactly once, and (ii) every non-essential loop of C_n^+ is traversed by w an even number of times.

Lemma 1. *Let $X_1 = (W, i)$ and $X_2 = (U, j)$ be two distinct vertices of CC_n and p be a shortest $X_1 X_2$ path in CC_n that is projecting to a walk q of C_n^+. The following hold:*

(i) Any $X_1 X_2$ walk in CC_n contains an odd number of edges from each dimension $i \in W \triangle U$, and an even number of edges from any dimension $j \notin W \triangle U$.

(ii) Any essential ij walk in C_n^+ is the projection of an $X_1 X_2$ walk in CC_n.

(iii) $|Cubic(p)| = |U \triangle W|$, with one cubic edge in each dimension $i \in U \triangle W$.

(iv) q contains any edge of C_n^+ at most twice.

(v) Assume that $i \neq j$ and let l_1 and l_2 be the lengths of the longest gaps induced by $U \triangle W$ on the right side and the left side of C_n^+, respectively, then, $|Cyclic(p)| = n + \min\{L(i, j) - 2l_1, n - L(i, j) - 2l_2\}$.

(vi) Assume that $i = j$ and let l be the length of the longest gap induced by $U \triangle W$ on C_n^+. Then, $|Cyclic(p)| = \min\{n, 2n - 2l\}$.

Proof. Proof omitted. $\square$

Now we continue with the structure of shortest paths in the butterfly. A walk w in C_n is called a *labeled walk*, if the edges in w are labeled cubic or cyclic. If an edge is contained more than once in w, we allow different labels at different occurrences of the edge. Let $X = (W, i)$ be a vertex of BB_n; the projection of X on C_n is defined to be the vertex i of C_n. Given two vertices $X_1 = (U, i)$ and $X_2 = (W, j)$ in BB_n, it is convenient to analyze the structure of any $X_1 X_2$ walk p in BB_n by projecting it on C_n to get a labeled ij walk q in C_n. Any edge of q which is the projection of a cyclic edge of p is labeled *cyclic*, any edge of q which projection of a cubic edge of p is labeled *cubic*. Let $X_1 = (W, i), X_2 = (U, j)$ be two distinct vertices in BB_n; an edge $(i, i \oplus 1)$ in C_n is called *essential*, if $i \oplus 1 \in W \triangle U$. A labeled ij walk w in C_n is called an *essential walk*, if it has the following properties: (i) every essential edge is assigned the cubic label exactly once in w, and (ii) the number of occurrences of any non-essential edge with cubic label in w is even. Note that an essential walk w can use an essential edge e several times with cyclic label, as long as e is labeled cubic in w only once.

Lemma 2. *Let $X_1 = (W, i)$ and $X_2 = (U, j)$ be two distinct vertices of BB_n. Assume p is a shortest $X_1 X_2$ path in BB_n projecting to a walk q of C_n. The following hold:*

(i) Let w be the projection of any $X_1 X_2$ walk in BB_n to C_n, then, the number of occurrences of any non-essential edge with cubic label in w is even.

(ii) Any ij essential walk w in C_n is the projection of an $X_1 X_2$ walk in BB_n.

(iii) q is a shortest ij essential walk in C_n.

(iv) q does not use any edge of C_n more than twice.

(v) Assume that $i \neq j$ and let l_1 (l_2) be the lengths of the longest gaps induced by the set of edges $\{(m, m \oplus 1) : m \oplus 1 \in U \triangle W\}$ on the right (left) side. Then, $L(p) = L(q) = n + \min(L(i, j) - 2l_1, n - L(i, j) - 2l_2)$.

(vi) Assume that $i = j$, and let l be the length of the longest gap induced by the set of edges $\{(m, m \oplus 1) : m \oplus 1 \in U \triangle W\}$ on C_n, then $L(p) = \min\{n, 2n - 2l\}$.

Proof. Proof omitted.   □

Lemma 3. *CC_n is Θ_n-orbit proportional, while BB_n is not.*

Proof. Note that the edge orbits of CC_n under Θ_n are the set of cyclic edges and the set of cubic edges. Let $X_1 = (W, i)$ and $X_2 = (U, j)$ be two vertices of CC_n. By Lemma 1(iii) any $X_1 X_2$ path p with $L(p) = L(X_1, X_2)$ must have $|Cubic(p)| = |U \triangle W|$. Now assume that p' is any $X_1 X_2$ path in CC_n; by Lemma 1(i) p' must have an odd number of cubic edges from each dimension $i, i \in U \triangle W$ and even number of cubic edges from any dimension $i, i \notin U \triangle W$. Thus, $|Cubic(p')| \geq |Cubic(p)|$. Assume to the contrary that, $|Cyclic(p')| < |Cyclic(p)|$, and consider q' the projection of p' on C_n^+. Then, q' can be converted to an essential ij walk $\hat{q}$ in C_n^+ by applying the method in Lemma 1(iii) to remove the unnecessary essential and non-essential loops. For the path $\hat{p}$ whose projection is $\hat{q}$ we have, $|Cubic(\hat{p})| = |U \triangle W| = |Cubic(p)|$, and $|Cyclic(\hat{p})| = |Cyclic(p')| < |Cyclic(p)|$. It follows that $L(\hat{p}) < L(p)$, a contradiction.

To show that BB_n is not orbit proportional, take BB_4, $X_1 = (\{2,4\},1)$, $X_2 = (\emptyset, 2)$. Consider two shortest $X_1 X_2$ paths p_1 and p_2,

$p_1 : (\{2,4\},1)\text{cubic}(\{4\},2)\text{cyclic}(\{4\},3)\text{cubic}(\emptyset,4)\text{cyclic}(\emptyset,3)\text{cyclic}(\emptyset,2)$

$p_2 : (\{2,4\},1)\text{cubic}(\{4\},2)\text{cubic}(\{3,4\},3)\text{cubic}(\{3\},4)\text{cyclic}(\{3\},3)\text{cubic}(\emptyset,2)$.

Observe that p_1 has 2 cubic and 3 cyclic edges, while p_2 has 1 cyclic and 4 cubic edges; thus BB_4 is not Θ_n-orbit proportional. $\qquad\square$

Lemma 4. *Assume $U \subseteq N$ is chosen randomly with uniform distribution. Then,*
(i) $Prob(\left||U| - \frac{n}{2}\right| \leq n^{2/3}) = 1 - o(1)$.
(ii) $Prob(\text{ length of the longest gap induced by } U \text{ on } C_n < \log^2 n) = 1 - o(1)$.
(iii) Let E' be a random subset of edges of C_n chosen with the probability 2^{-n}. Then

$$Prob(\text{ length of the longest gap induced by } E' \text{ on } C_n < \log^2 n) = 1 - o(1).$$

(iv) Assume that p is any ij path in C_n. Then

$$Prob(\left||p \cap U| - \frac{L(p)}{2}\right| \leq n^{2/3}) = 1 - o(1).$$

Proof. Proof omitted. $\qquad\square$

5   Optimal integral uniform flows in CC_n and BB_n

To estimate the congestion of an optimal integral flow in CC_n and BB_n, we will use probabilistic methods. It should be noted that although the tools involve usage of probability, the final outcome is completely deterministic and does not involve probability.

Theorem 3. *For $CC_n = < V, E >$, there exists an edge optimal integral uniform flow f, such that $\mu_f = \frac{5}{4}n^2 2^n(1 - o(1))$.*

Proof. Since CC_c is a Cayley graph, our construction in Theorem 1 gives an integral uniform flow. By Lemma 3 CC_n is orbit proportional, hence by Theorem 2(i) the flow f is edge optimal. To evaluate μ_f we use Theorem 2(ii). Define, $d_1(e) = 0$, if e is cyclic and $d_1(e) = 1$, if e is cubic. Similarly, define $d_2(e) = 0$, if e is cubic, and $d_2(e) = 1$, if e is cyclic.
By Theorem 2(ii) we have, $\mu_f = \max(\mu_{d_1}, \mu_{d_2})$. Note that, for $X_1 = (U,i) \in V$ and $X_2 = (W,j) \in V$, by of Lemma 1(iii) we have, $d_1(X_1, X_2) = U \triangle W$. It is easy to see that $\sum_{U \subseteq N} \sum_{W \subseteq N} |U \triangle W| = \frac{n}{2}4^n$ and hence that

$$\sum_{(X_1,X_2)\in V\times V} d_1(X_1, X_2) = \frac{n^3}{2}4^n, \qquad (1)$$

thus, $\mu_{d_1} = \frac{\sum_{(X_1,X_2)\in V\times V} d_1(X_1,X_2)}{\sum_{e\in E} d_1(e)} = \frac{n^2 4^n n/2}{n2^{n-1}} = n^2 2^n$. Let $X_0 = (\emptyset,0) \in V$ and $X = (U,i) \in V$, and assume that p is a shortest $X_0 X$ path. Orbit proportionality implies $d_2(X_0,X) = |Cyclic(p)|$. By Lemma 1(v)-(vi),

$$d_2(X_0,X) = |Cyclic(p)| \le n + L(0,i). \tag{2}$$

In order to study the distribution of distances $d_j(X_0,X)$, we think about X as a random variable and use facts from probability theory (the normal convergence) to estimate the distribution. Next consider any vertex $X = (U,i)$, such that U is selected randomly with the probability 2^{-n}; we refer to X as a random vertex. For any random vertex $X = (U,i)$, by Lemma 4(i), $||U| - n/2| = o(n)$ with probability $1 - o(1)$. It follows from Lemma 4(ii), that U does not induce any gaps longer than $\log^2 n$ on C_n^+ with probability $1 - o(1)$. Therefore by Lemma 1(v)-(vi) we have,

$$d_2(X_0,X) = |Cyclic(X_0,X)| = (n + L(0,i))(1 - o(1)) \tag{3}$$

with the probability $1 - o(1)$. It follows from (2) and (3) that,

$$\sum_{X=(W,i)\in V} d_2(X_0,X) = (1 - o(1)) \sum_{X=(W,i)\in V} (n + L(0,i)). \tag{4}$$

(The sums in (4) are taken over all vertices!) It is easy to verify that, $\sum_{X=(W,i)\in V}(n + L(0,i)) = \frac{5}{4}n^2 2^n(1 - o(1))$, therefore, $\sum_{X\in V} d_2(X_0,X) = \frac{5}{4}n^2 2^n(1 - o(1))$. It easily follows from the vertex transitivity of CC_n that

$$\sum_{(X_1,X_2)\in V\times V} d_2(X_1,X_2) = n2^n \sum_{X\in V} d_2(X_0,X) = \frac{5}{4}n^3 4^n((1 - o(1)). \tag{5}$$

However, $\mu_{d_2} = \sum_{(X_1,X_2)\in V\times V} d_2(X_1,X_2)/(n2^n) = \frac{5}{4}n^2 2^n(1 - o(1)) \ge \mu_{d_1}$, for large n. Therefore, $\mu_f = \frac{5}{4}n^2 2^n(1 - o(1))$.   □

Since BB_n is not Θ_n-orbit proportional, the construction of Theorem 1 only gives an integral approximate solution. (Our results in [17] can be used to show that the congestion is within a multiplicative factor of 2 from the optimal.) We will now present an algorithm which computes an integral flow with asymptotically optimal congestion. The key point behind our near-optimal uniform flow for BB_n is a collection of shortest paths, which uses each cyclic edge and each cubic edge about the same times. We note that the complexity of the algorithm is $O(n^3 4^n)$

Butterfly Flow Algorithm
INPUT: $< V, E >= BB_n$
OUTPUT: An integral uniform flow f.
Let $X_0 \leftarrow (\emptyset,0)$ and compute a shortest $X_0 X$ path $q_{X_0 X}$ for every $X \ne X_0$.
For all $X = (A,i) \in V$, $X \ne X_0$ **Do**

Denote by W_{X_0X} the $0i$ walk in C_n which is the projection of q_{X_0X}. (Recall that $L(0,i)$ is the length of a shortest $0i$ path on C_n.)

Case

$L(0,i) < \frac{n}{\sqrt{8}}$:   Consider any non-essential edge e which appears in W_{X_0X} with cubic label. Notice that e must appear twice with cubic label in W_{X_0X} and change the label of both occurrences of e in W_{X_0X} to cyclic. Denote this new labeled walk in C_n by W'_{X_0X}. By Lemma 2(i), W'_{X_0X} is a $0i$ essential walk and let h_{X_0X} be an X_0X path in BB_n which projects to W'_{X_0X}.

$L(0,i) > \frac{n}{\sqrt{8}}$:   Consider any non-essential edge e which appears twice in W_{X_0X} with cyclic label; change the label of both occurrences of e in W_{X_0X} to cubic. Denote this new walk in C_n by W'_{X_0X}. By Lemma 2(i), W'_{X_0X} is a $0i$ essential walk and let h_{X_0X} be an X_0X path in BB_n which projects to W'_{X_0X}.

EndCase

EndFor

Extend the set of paths $S = \{h_{X_0X} : X \in V,\ X \neq X_0\}$ to a Θ_n-invariant integral flow using the action of Θ_n. That is, compute $\Theta_n(S)$.

End.

Theorem 4. *The Butterfly Flow Algorithm constructs an integral uniform flow f in $BB_n =< V, E >$ with $\mu_f = \frac{5}{4}n^2 2^{n-1}(1 + o(1))$, which is asymptotically optimal for large n.*

Proof. It is easy to verify that the last step of the algorithm produces a flow f which is integral Θ_n-invariant using shortest paths. Assume that $X = (U, i)$ is a random vertex of BB_n, that is, U is selected randomly with the probability 2^{-n}, when i is arbitrary. Let $X_0 = (\emptyset, 0)$. Consider the X_0X path q_{X_0X} computed at the initial step of the algorithm. Let W_{X_0X} be the projection of q_{X_0X} on C_n, then by Lemma 2(iv)-(v) the portion of C_n which is not traversed by W_{X_0X} must be a (longest) gap. Since U is selected randomly, by Lemma 4(iv) the length of this gap is at most $\log^2 n$ with probability $1 - o(1)$. Therefore this gap is located with probability $1 - o(1)$ on the shorter side of C_n. By Lemma 2, any edge e, which is located on the shorter side of C_n and is contained in W_{X_0X} will appear in W_{X_0X} exactly twice. Also, any edge e located on the long side will appear in W_{X_0X} exactly once. Next we estimate (with probability $1 - o(1)$) the number of essential edges labeled cubic, the number of essential edges labeled cyclic, and the total number of non-essential edges in W_{X_0X}. These values are easily estimated using Lemma 2, Lemma 4, and the topological properties of W_{X_0X} are recorded in the following Table.

	long side of C_n	short side of C_n
Number of essential edges labeled cubic in W_{X_0X}	$\frac{1}{2}(n - L(0,i)) - o(n)$	$\frac{1}{2}L(0,i) - o(n)$
Number of essential edges labeled cyclic in W_{X_0X}	0	$\frac{1}{2}L(0,i) - o(n)$
Total number of non-essential edges in W_{X_0X}	$\frac{1}{2}(n - L(0,i)) - o(n)$	$\frac{1}{2}L(0,i) - o(n)$

If $L(0, i) < n/\sqrt{8}$, the **Case** statement in the algorithm guarantees that any non-essential edge of W_{X_0X} which is located on the short side will be labeled cyclic in W'_{X_0X}. (Notice that any non-essential edge e of W_{X_0X} which is located in the left side appears exactly once in W_{X_0X}. Thus, by Lemma 2(i), e must have been labeled cyclic in W_{X_0X}. Consequently the label of e does not change.) Therefore, employing the last two rows of Table , we will get

$$|Cyclic(h_{X_0X})| = |Cyclic(W_{X_0X})| = ((n/2 + L(0, i))(1 - o(1)), \qquad (6)$$

with probability $1 - o(1)$. Similarly, using the first row of Table , we have

$$|Cubic(h_{X_0X})| = |Cubic(W_{X_0X})| = n/2 - o(n) \qquad (7)$$

with probability $1 - o(1)$. Now assume that $L(0, i) > n/\sqrt{8}$, then, the **Case** statement in the algorithm guarantees that any non-essential edge of W_{X_0X} which is located on the short side of C_n will have a cubic label in W'_{X_0X}. Using rows one and three in Table , it is easily shown that with probability $1 - o(1)$

$$|Cubic(h_{X_0X})| = |Cubic(W'_{X_0X})| = n/2 + L(0, i) - o(n),$$

whereas using rows two and three,

$$|Cyclic(h_{X_0X})| = |Cyclic(W'_{X_0X})| = n/2 - o(n), \qquad (8)$$

with probability $1 - o(1)$. Next, we claim that

$$\sum_{X \in V} |Cyclic(h_{X_0X})| = (2 - o(1))2^n \{ \sum_{l=0}^{\lfloor \frac{n}{\sqrt{8}} \rfloor} (\frac{n}{2} + l) + \sum_{l=\lceil \frac{n}{\sqrt{8}} \rceil}^{\lfloor \frac{n}{2} \rfloor} \frac{n}{2} \} = \frac{5}{4} n^2 2^{n-1}(1 + o(1)),$$
$$(9)$$

$$\sum_{X \in V} |Cubic(h_{X_0X})| = (2 - o(1))2^n \{ \sum_{l=0}^{\lfloor \frac{n}{\sqrt{8}} \rfloor} \frac{n}{2} + \sum_{l=\lceil \frac{n}{\sqrt{8}} \rceil}^{\lfloor \frac{n}{2} \rfloor} (\frac{n}{2} + l) \} = \frac{5}{4} n^2 2^{n-1}(1 + o(1)).$$
$$(10)$$

We now justify (9) and leave (10) to the reader. Consider a random vertex $X = (U, i)$ and let $l = L(0, i)$. If $l < n/\sqrt{8}$, we can count with high accuracy $|Cyclic(h_{X_0X})|$ using (6); likewise, if $l > n/\sqrt{8}$, we can count with high accuracy $|Cyclic(h_{X_0X})|$ using (8). Now observe that there are 2^n choices for the random U, and typically 2 choices for a vertex at distance l from the vertex 0 on C_n. This justifies the existence of two sums and in (9). The evaluation of the sums is just algebra. Our estimates in (9) went through for random vertices. However, the number of cyclic and cubic edges for atypical vertices is not too large either, since the diameter of the butterfly is $O(n)$. The contribution of the neglected case $i = 0$ is negligible. Denote p_{XY} the unique active XY path in f, use the fact that the orbits of Θ_n are the set of cyclic edges and the set of cubic edges to obtain

$$CU = \sum_{(X,Y)} |Cubic(p_{XY})| = n2^n \sum_{X \in V} |Cubic(h_{X_0X})| = \frac{5}{2} n^3 4^{n-1}(1 + o(1)),$$

$$CY = \sum_{(X,Y)} |Cyclic(p_{XY})| = n2^n \sum_{X \in V} |Cyclic(h_{X_0 X})| = \frac{5}{2} n^3 4^{n-1}(1 + o(1)).$$

Since f is Θ_n-invariant by construction, the value of f on any cyclic or cubic edge is

$$\frac{CY}{n2^n} = \frac{5}{4} n^2 2^{n-1}(1 + o(1)), \quad \text{and} \quad \frac{CU}{n2^n} = \frac{5}{4} n^2 2^{n-1}(1 + o(1)),$$

and $\mu_f = \frac{5}{4} n^2 2^{n-1}(1 + o(1))$. The identically one distance function d yields

$$\sum_{(X_1, X_2) \in V \times V} d(X_1, X_2) = \sum_{(X_1, X_2) \in V \times V} L(X_1, X_2) = CY + CU. \tag{11}$$

Consequently, $\mu_d = \frac{\sum_{(X_1, X_2) \in V \times V} d(X_1, X_2)}{\sum_{e \in E} d(e)} = \frac{CU + CY}{2n2^n} = \frac{5}{4} n^2 2^{n-1}(1 + o(1))$. This verifies the asymptotic edge optimality of f, since by duality theory of linear programming [13] μ_d is a lower bound on the congestion of an optimal flow. □

References

1. Awerbuch, B., Leighton, T., A simple local control approximation algorithm for multicommodity flow, Proc. 34th Annual Symp. on Foundations of Computer Science, 1993, 459–468.
2. Chung, E., Coffman, E., Reiman M., Simon, B., The forwarding index of communication networks, IEEE Trans. on Information Theory, **33**(1987), 224–232.
3. Even, S., Itai, A., and Shamir, A., On the complexity of timeable and multicomodity flow problems, SIAM Journal on comp., **4:5**(1976), 691–703.
4. Ford, L. R, and Fulkerson, D. R., Flows in network, Princeton University Press, 1962.
5. Grigoriadis, M.D., Khachiyan, L., Fast approximation schemes for convex programs, SIAM Journal on Optimization 4(1994), 86-107.
6. Kainen, P. C., White, A. T., On stable crossing numbers, J. Graph Theory **2**(1978) 181–187.
7. Heydemann, M. C., Meyer, J. C., Opatrny J., Sotteau, D., On forwarding indices of k-connected graphs, Discrete Appl. Math. **37/38**(1992), 287–296.
8. Klein, P., Plotkin, S., Stein, C., and Tardos, E., Faster approximation algorithms for unit capacity concurrent flow problems with applications to routing and sparsest cuts, SIAM Journal on Computing **3:23** (1994), 466-488.
9. Leighton, T., Makedon, F., Plotkin S., Stein, C., Tardos, E., Tragoudas, S., Fast approximation algorithms for multicommodity flow problems, Proc. 23rd ACM Symp. on Theory of Computing, 1991, 101–111.
10. Leighton, T., Maggs, B., Rao, S., Universal packet routing algorithms, Proc. 29th IEEE Symp. on Foundations of Computer Science, 1988, 256-269.
11. Leighton, F. T., Complexity Issues in VLSI, MIT Press, Cambridge, Massachusetts, 1983.
12. Matula, D. W., Concurrent flow and concurrent connectivity in graphs, in Graph Theory and its Applications to Algorithms and Computer Science, ed. Alavi, Y., et al., Wiley, New York, 1985, 543–559.

13. Onaga, K., A multicommodity flow theorem, Trans. IECE Japan, 53-A, **7**(1970), 350–356.
14. Saad, R., Complexity of the forwarding index problem, SIAM J. Discrete Math. **6**(1993), 413–427.
15. Shahrokhi, F., and Matula, D. W., The maximum concurrent flow problem, J. Assoc. for Computing Machinery **2:37**(1990), 318–334.
16. Shahrokhi, F., and Székely, L. A., An algebraic approach to the uniform concurrent multicommodity flow problem: theory and applications, Technical Report CRPDC-91-4, Dept. Computer Science, Univ. of North Texas, 1991.
17. Shahrokhi, F., and Székely, L. A., Concurrent flows and packet routing in Cayley graphs, in: Graph-Theoretic Concepts in Computer Science (Utrecht, 1993), ed. J. van Leeuwen, Lecture Notes in Computer Science Vol. 790, Springer Verlag, Berlin, 1994, 327–338.
18. Shahrokhi, F., and Székely, L. A., On group invariant flows and applications, in: *Graph Theory, Combinatorics, and Applications: Proceedings of the Seventh Quadrennial International Conference on the Theory and Applications of Graphs, Volume 2* eds. Y. Alavi and A. Schwenk, John Wiley and Sons, Inc., 1995, 1033–1042.
19. Shahrokhi, F., Székely, L. A., Canonical concurrent flows, crossing number and graph expansion, Combinatorics, Probability, and Computing **3** (1994), 523–543.
20. Shahrokhi, F., Sýkora, O., Székely, L. A., Vrťo, I., The crossing number of a graph on a compact 2-manifold, Adv. Math. **123**(1996), 105–119.
21. Sýkora, O., Vrťo, I., On the crossing number of hypercubes and cube connected cycles, in: Proc. 17th Intl. Workshop on Graph Theoretic Concepts in Computer Science WG'91, eds. G. Schmidt and R. Berghammer, LNCS 570, Spinger Verlag, Berlin, 1992, 214–218.
22. Sýkora, O., Vrťo, I., Edge separators for graphs of bounded genus with applications, in: Proc. 17th Intl. Workshop on Graph Theoretic Concepts in Computer Science WG'91, eds. G. Schmidt and R. Berghammer, LNCS 570, Springer Verlag, Berlin, 1992, 159–168, also in Theoretical Computer Science **112**(1993), 419–429.
23. White, A. T., Graphs, Groups and Surfaces, North-Holland, Amsterdam, 1984.
24. White, A. T., Beineke, L. W., Topological graph theory, in: Selected Topics in Graph Theory, eds. L. W. Beineke and R. J. Wilson, Academic Press, 1978, 15–50.

Splitting Number is NP-complete*

L. Faria[1], C. M. H. de Figueiredo[2], and C. F. X. Mendonça[3]

[1] Faculdade de Formação de Professores/UERJ and COPPE/UFRJ, Brazil
luerbio@cos.ufrj.br
[2] Instituto de Matemática and COPPE, Universidade Federal do Rio de Janeiro,
Brazil celina@cos.ufrj.br
[3] Instituto de Computação, Universidade Estadual de Campinas, Brazil
xavier@dcc.unicamp.br

Abstract. We consider two graph invariants that are used as a measure of nonplanarity: the splitting number of a graph and the size of a maximum planar subgraph. The splitting number of a graph G is the smallest integer $k \geq 0$, such that a planar graph can be obtained from G by k splitting operations. Such operation replaces a vertex v by two nonadjacent vertices v_1 and v_2, and attaches the neighbors of v either to v_1 or to v_2. We prove that the SPLITTING NUMBER decision problem is NP-complete, even when restricted to cubic graphs. We obtain as a consequence that PLANAR SUBGRAPH remains NP-complete when restricted to cubic graphs. Note that NP-completeness for cubic graphs also implies NP-completeness for graphs not containing a subdivision of K_5 as a subgraph.

1 Introduction

Applications in Computer Science are frequently modeled with nonplanar graphs. Graph visualization and VLSI projects many times require strategies of layout techniques. Layout algorithms are limited to special classes of graphs. For instance, there is a wealth of layout algorithms for planar graphs; however, these algorithms are useless for nonplanar graphs. One approach to handling nonplanarity in layout algorithms is to consider another topological invariant of the graph, the splitting number. The splitting number is a graph invariant that is used as a measure of nonplanarity in many applications such as graph drawing.

The *splitting number* $\sigma(G)$ of a graph G is the smallest integer $k \geq 0$ such that a planar graph can be obtained from G by k vertex splitting operations. A *vertex splitting operation*, or simply *splitting*, of a vertex $v \in V(G)$ partitions the set of neighbors of v into two nonempty sets P_1 and P_2 and adds to $G \setminus v$ two new and nonadjacent vertices v_1 and v_2, such that P_1 is the set of neighbors of v_1 and P_2 is the set of neighbors of v_2. If a graph H is obtained from G by a set of k splittings, we say that H is the *resulting graph* of this set of k splittings in G.

---

* This work was partially supported by CNPq, CAPES, FAEP, FAPESP and FAPERJ, Brazilian research agencies.

Note that the resulting graph H can be obtained either by splitting only vertices of G, or by splitting vertices of G and vertices created by former splittings.

Two aspects of the study of splitting numbers have been considered recently by Eades and Mendonça [2, 3, 12]: they established the NP-completeness of a related problem — ELIGIBLE SPLIT SET —, and they successfully used splitting numbers in layout algorithm design. The splitting number has been computed for the class of complete graphs [6] and for the class of complete bipartite graphs [8]. For a recent survey on splitting numbers, see [10].

The knowledge of the value of nonplanarity invariants for the smallest non-planar member in a class of graphs can help to find the values or bounds for this invariant for every member in the class. For instance, we have recently established [4] that the splitting number of the 4-cube is 4. We also showed that this result implies that the splitting number of the n-cube is in fact $\Theta(2^n)$.

Liu and Geldmacher [11] proved that the PLANAR SUBGRAPH decision problem is NP-complete. Note that, for a fixed value of k, PLANAR SUBGRAPH is easily seen to be polynomial whereas the number of all possible splittings for a vertex in a graph G being of order $\Omega(2^{|V(G)|})$ may indicate that SPLITTING NUMBER, even for a fixed value of k, is not a polynomial problem.

In this paper we prove that SPLITTING NUMBER is NP-complete. We obtain as a consequence that SPLITTING NUMBER remains NP-complete when restricted to graphs with maximum degree 3 or to graphs with no subdivision of K_5. We also prove that SPLITTING NUMBER remains NP-complete when restricted to cubic graphs. This result is used in turn to prove that PLANAR SUBGRAPH remains NP-complete when restricted to cubic graphs. We obtain as a consequence that PLANAR SUBGRAPH remains NP-complete when restricted to graphs with no subdivision of K_5. These variants of PLANAR SUBGRAPH had been open since 1979 [11].

2   Preliminaries

A *graph* G is an ordered triple $G = (V(G), E(G), \psi(G))$ consisting of a nonempty set $V(G)$ of *vertices*, a set of *edges* $E(G)$ disjoint from $V(G)$ and an *incidence function* $\psi(G)$ that associates to each edge of $E(G)$ an unordered pair of distinct vertices of $V(G)$. A graph with *multiple edges* is a graph that admits two edges associated to the same pair of vertices. We shall omit the incidence function of a graph by writing only $G = (V(G), E(G))$. We say that a graph $G' = (V(G'), E(G'))$ is a subgraph of G if $V(G') \subseteq V(G)$ and $E(G') \subseteq E(G)$.

A graph is *planar* when it admits a plane drawing, that is, a drawing in the plane such that no edges cross. There are efficient, linear-time algorithms for testing whether a graph G has $\sigma(G) = 0$, that is, for testing whether a graph is planar [7]. We use strongly the characterization of Kuratowski [9]: a graph is planar if and only if it does not contain a subdivision of K_5 or of $K_{3,3}$ as a subgraph, in particular the nonplanarity of the subdivisions of $K_{3,3}$.

In this way, for a better understanding of our proof it is important that the reader familiarizes himself with the special drawing of $K_{3,3}$ defined in Fig. 1a.

We say that a graph G is a $K_{3,3} \setminus \{e\}$ *linked to* the vertices v and w if G is defined by the drawing in Fig. 1b. This graph is an important *tool* in our proof, and the main property used is that a graph is not planar, if it contains as subgraph a $K_{3,3} \setminus \{e\}$ linked to vertices v and w and a path P joining v to w, where P is disjoint in vertices of this $K_{3,3} \setminus \{e\}$, because this is a subdivision of $K_{3,3}$.

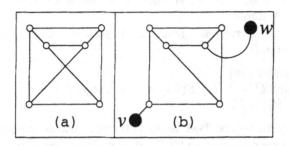

(a) v (b)

Fig. 1. $K_{3,3}$ and $K_{3,3} \setminus \{e\}$ linked to the vertices v and w.

A *simple drawing* $D(G)$ is a drawing of the graph G on the plane such that no edge crosses itself, adjacent edges do not cross, crossing edges do so only once, edges do not cross vertices, and no more than two edges cross at a common point. In what follows all drawings are assumed to be simple. Let $D(G)$ be a simple drawing of G and v be a vertex in $V(G)$, with degree $d(v)$. Because $D(G)$ is simple and in a simple drawing edges incident to the same vertex cannot share crossings, $D(G)$ defines for each vertex v an *ordered adjacency list* in the clockwise direction $\vec{Adj}(v) = (v_1, v_2, \ldots, v_{d(v)})$, where $\{v_1, v_2, \ldots, v_{d(v)}\}$ is the neighborhood of v. Thus, each such ordered adjacency list $\vec{Adj}(v)$ is a circular permutation of the set of edges incident to v. An example of the set of ordered adjacency lists with respect to a drawing $D(G)$ is shown in Fig. 2. Note that we may have multiple edges.

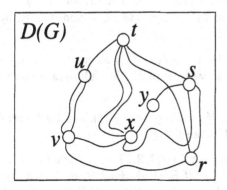

$$\vec{Adj}(r) = (v, t, s)$$
$$\vec{Adj}(s) = (y, t, r, x)$$
$$\vec{Adj}(t) = (u, s, r, x, x)$$
$$\vec{Adj}(u) = (t, v, v)$$
$$\vec{Adj}(v) = (u, u, x, r)$$
$$\vec{Adj}(x) = (t, y, s, v, t)$$
$$\vec{Adj}(y) = (s, x)$$

Fig. 2. The ordered adjacency lists of G with respect to $D(G)$.

3   The NP-completeness of SPLITTING NUMBER

In this section we prove that SPLITTING NUMBER is NP-complete, by reducing the NP-complete problem 3–SATISFIABILITY [1] to SPLITTING NUMBER. These problems are defined as follows:

3–SATISFIABILITY (3SAT)
INSTANCE: Set U of variables, collection C of clauses over U such that each clause $c \in C$ has $|c| = 3$ literals.
QUESTION: Is there a truth assignment for U such that each clause in C has at least one true literal?

SPLITTING NUMBER (SN)
INSTANCE: Graph $G = (V(G), E(G))$ and an integer $k \geq 0$.
QUESTION: Is $\sigma(G) \leq k$?

The strategy to reduce 3SAT to SN is to construct an integer $k \geq 0$ and a graph G from a generic instance (U, C) of 3SAT, such that C is satisfiable if and only if $\sigma(G) \leq k$. The graph G is composed of two types of subgraphs: Truth-Setting subgraphs corresponding to the variables of U and Satisfaction–Testing subgraphs corresponding to the clauses of C. The definition of the Satisfaction-Testing subgraphs requires some topological properties of a certain class $\mathcal{A}$ of graphs that we are about to define and study.

A graph G is a member of the class $\mathcal{A}$ if G has two subgraphs P_G and Q_G, such that $V(P_G) \cup V(Q_G) = V(G)$ and $V(P_G) \cap V(Q_G) = \{f_1, f_2, \ldots, f_6, g_1, g_2, \cdots, g_6\}$ with P_G and Q_G satisfying:

- The subgraph P_G is defined by the drawing in Fig. 3. In this figure the subset $\Sigma = \{a_1, a_2, \ldots, a_6, d_1, d_2, \ldots, d_6\}$ is depicted with black vertices. There are exactly $q \geq 2$ edges linking two adjacent vertices of $P_G \setminus \Sigma$, and a single edge linking a white vertex of $P_G \setminus \Sigma$ to a vertex of Σ. Note that we draw between two adjacent vertices in $P_G \setminus \Sigma$ only two edges: one drawn with a continuous line, and one drawn with a dashed line, the other $(q - 2)$ edges are omitted, but considered drawn in the region without vertices bounded by those two edges.
- The subgraph Q_G is a connected planar graph not drawn in Fig. 3, such that Q_G admits a plane drawing within the exterior region defined by the drawing of P_G depicted in Fig. 3.

The following four lemmas consider how a planar graph can be obtained from $G \in \mathcal{A}$ by a set Z of splittings only in vertices of Σ. The full details and proofs are in the technical report [5].

Lemma 1. *Let G be a graph in $\mathcal{A}$. If H is a planar graph obtained from G by a set Z of splittings in vertices of Σ, then $|Z| \geq 6$.*   □

Lemma 2. *Let G be a graph in $\mathcal{A}$. Let $i \in \{1, 2, 3\}$ be a fixed index and let $M_i = \{a_1, a_2, a_3, a_4, a_5, a_6, d_{2i-1}, d_{2i}\}$. If H is obtained from G by a set Z of splittings, with $|Z| = 8$, such that there is one splitting of Z in each vertex in the set M_i, then H is nonplanar.*   □

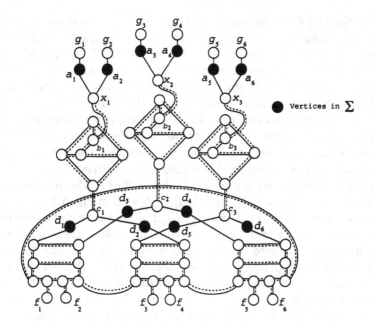

Fig. 3. A drawing for the subgraph P_G of G in class $\mathcal{A}$.

Lemma 3. *Let G be a graph in $\mathcal{A}$. Let Z be a nonempty set of splittings in the set $\{a_1, a_2, \ldots, a_6\}$, such that at most one splitting is done in each set $\{a_1, a_2\}$, $\{a_3, a_4\}$ and $\{a_5, a_6\}$ yielding a resulting graph G' from G. If H is a planar graph obtained in turn from G' by a set Z' of splittings in vertices of Σ, then $|Z'| \geq 5$ and there is such a set Z' satisfying $|Z'| = 5$. In addition, a drawing $D(H)$ can be constructed such that, the nonsplit vertices of G in H have the same ordered adjacency lists with respect to $D(H)$ and with respect to the drawing in Fig. 3.* □

Lemma 4. *Let G be a graph in $\mathcal{A}$. Let $l \in \{1, 2, 3\}$ be a fixed index. Let G' be the graph obtained from G by a set Z of splittings in vertices of $\{a_1, a_2, \ldots, a_6\}$, with $|Z| \geq 2l$, such that $2l$ splittings are in vertices of l of the pairs: $\{a_1, a_2\}$, $\{a_3, a_4\}$, $\{a_5, a_6\}$, and such that $3 - l$ of the pairs: $\{a_1, a_2\}$, $\{a_3, a_4\}$, $\{a_5, a_6\}$ have each one at most one splitting of Z. If H is a planar graph obtained in turn from G' by a set Z' of splittings in vertices of Σ, then $|Z'| \geq 5 - l$.* □

Theorem 1. SN *is NP-Complete.*

Proof. It is easy to see that SN is in NP, because once a non–deterministic algorithm guesses a set of splittings, we need only to check in linear time [7] whether the resulting graph is planar. We reduce 3SAT to SN as follows. Let $U = \{u_1, u_2, \ldots, u_n\}$ and $C = \{c_1, c_2, \ldots, c_m\}$ be an instance of 3SAT. We construct in polynomial time a graph G and an integer $k \geq 0$, such that $\sigma(G) \leq k$ if and only if C is satisfiable. In order to define G we construct first an auxiliary graph G^*.

Construction of G^*. The graph G^* is made up of two types of subgraphs: *Truth–Setting* subgraphs and *Satisfaction–Testing* subgraphs, and of a set of edges used to connect these subgraphs. We shall define G^* by ordered adjacency lists. We need to give drawings corresponding to each one of the two types of subgraphs in order to define their corresponding ordered adjacency lists. The two drawings we are about to describe, have the following common strategy: First, we use the drawings in Fig. 4c and 5 respectively to define each one of the subgraphs. Second, in Fig. 4c and 5 we partition the vertices of the subgraphs of G^* into white, black and striped vertices, such that every black vertex has degree 2 and every white vertex has degree 3. The striped vertices are linking vertices between subgraphs and may have degree 2 or 3. The vertices $e_i[1], e_i[2]$ in Fig. 4c and the vertices $f_j[1], f_j[6], b_j[1], b_j[2], b_j[3]$ in Fig. 5, have an incident edge with a missing endpoint. These edges will be used later and indicate striped vertices that necessarily have degree 3 in G^*. Third, the edges of G^* in each one of the subgraphs are defined by continuous lines. Observe that in Fig. 4c and 5, for each continuous edge linking two vertices of degree 3 there is also a dashed edge. This dashed edge is not used in the construction of G^*, it should be ignored in the construction of G^*, because it will be used later only in the construction of G. Now we describe the two types of subgraphs used to construct G^*.

- **Truth–Setting Subgraph.** For each variable $u_i \in U$, there is a Truth–Setting subgraph T_i defined by the drawing in Fig. 4c. The subgraph T_i is obtained from a $K_{3,3}$ (Fig. 4a) by replacing two edges and subdividing a third one as shown in Fig. 4b. Note that we have two additional vertices $e_i[1]$ and $e_i[2]$ (Fig. 4b and 4c). The two replaced edges give place to two graphs called R_{u_i} and $R_{\bar{u}_i}$. Let p be the positive integer that satisfies $2^p > 5m > 2^{p-1}$. Graphs R_{u_i} and $R_{\bar{u}_i}$ are complete binary trees, respectively with roots $r_{u_i[1]}, r_{u_i[2]}$, and $r_{\bar{u}_i[1]}, r_{\bar{u}_i[2]}$, linked by their leaves through vertices $\bar{u}_i[1], \bar{u}_i[2], \ldots, \bar{u}_i[2^p], u_i[1], u_i[2], \ldots, u_i[2^p]$ as shown in Fig. 4c. Note that the greatest level in each one of these trees has $O(m)$ vertices:
- **Satisfaction–Testing Subgraph.** For each clause $c_j \in C$, there is a Satisfaction–Testing subgraph S_j consisting of the graph defined by Fig. 5.

There is a set of edges connecting Truth–Setting subgraphs to Satisfaction–Testing subgraphs:

$$E' = \bigcup_{i=1}^{n-1} \{(e_i[2], e_{i+1}[1])\} \cup \{(e_n[2], f_1[1])\} \cup \bigcup_{j=1}^{m-1} \{(f_j[6], f_{j+1}[1])\} \cup \{(f_m[6], e_1[1])\}.$$

The only part in the construction of G^* that depends on which literals occur in which clauses is the following collection of edges produced sequentially when j grows from 1 until m. Let x_{i_1}, x_{i_2} and x_{i_3}, with $i_1, i_2, i_3 \in \{1, 2, \ldots, n\}$ be the three literals in clause c_j. We have the following sets of edges emanating of the subgraphs T_i and S_j: $E''_j = \{(b_j[1], x_{i_1}[l_1]), (b_j[2], x_{i_2}[l_2]), (b_j[3], x_{i_3}[l_3])\}$, where $l_s(s = 1, 2, 3)$ is the minimum number in the set $\{1, 2, \ldots, 2^p\}$ such that there is no vertex $b_{j'}[h], h \in \{1, 2, 3\}$ linked to $x_{i_s}[l_s]$ with $j' \leq j$.

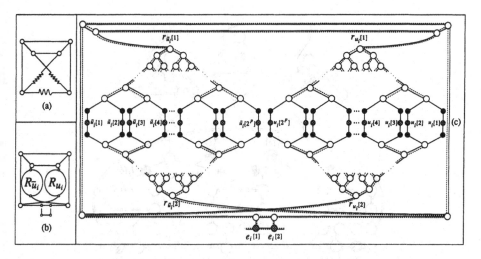

Fig. 4. The Truth–Setting Subgraph T_i.

The construction of G^* is completed by setting: $G^* = (V(G^*), E(G^*))$, where:

$$V(G^*) = (\bigcup_{i=1}^{n} V(T_i)) \cup (\bigcup_{j=1}^{m} V(S_j)),$$

$$E(G^*) = (\bigcup_{i=1}^{n} E(T_i)) \cup (\bigcup_{j=1}^{m} E(S_j)) \cup E' \cup (\bigcup_{j=1}^{m} E''_j).$$

Complexity of the construction of G^*. The size of T_i is in $\Theta(m)$ and therefore the size of G^* is in $\Theta(mn)$. The construction of the ordered adjacency lists for each subgraph S_j depends only on its drawing given in Fig. 5 and it is not dependent on the size of the instance of 3SAT. On the other hand, we can construct the ordered adjacency lists for each T_i in time $O(m)$ as follows. We obtain a total order of the vertices in a complete binary tree by using a Breadth First Search (BFS) from the root to the leaves and from the left side to the right side. The ordered adjacency lists are constructed in linear time by considering this total order restricted to the neighborhood of each vertex. Thus we can construct the ordered adjacency lists for a complete binary tree with 2^p vertices in the greatest level in time $O(m)$. Because of the tests for connecting the subgraphs S_j's and T_i's, we have total time of order $O(m^2 n)$. Hence, it is possible to construct G^* in polynomial time in the size of the instance of 3SAT.

We shall see next that the graph G^* satisfies that C is satisfiable if and only if there exists a set of splittings in the black vertices of G^* of size $2^p n + 5m$ that obtains from G^* a planar graph. In order to conclude our reduction from 3SAT to SN, we want a graph G, such that C is satisfiable if and only if $\sigma(G) \le 2^p n + 5m$. We show how to modify G^* in order to obtain a graph G, quite close to G^*, where the fact that a planar graph is obtained from G with at most

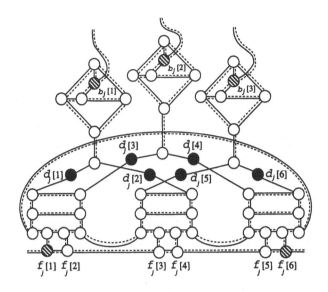

Fig. 5. The Satisfaction–Testing Subgraph S_j.

$2^p n + 5m$ splittings forces in G that these splittings occur close to the black vertices of G which in turn defines a truth assignment that satisfies C. Our strategy to define G is to modify the subgraph of G^* induced by vertices of degree 3 by replacing these vertices and corresponding edges by supervertices as we define next.

Construction of G. Let B be the subgraph of G^* induced by the vertices of degree 3. We shall exhibit a partition $V_1(B)$, $V_2(B)$ for $V(B)$, showing that B is a bipartite graph and we shall use this partition to define G from G^*. To prove that B is in fact a bipartite graph it is enough to prove that each connected component of B is a bipartite graph.

Observe first that there are exactly $3m + 1$ connected components in B, $3m$ of them are each isomorphic to $K_{3,3} \setminus \{e\}$ linked to two vertices, and the other one contains all edges of E'.

We define the partition of B into $V_1(B), V_2(B)$ in three steps:

- For each $h \in \{1, 2, 3\}$ and $j \in \{1, 2, \ldots, m\}$ take a Breadth First Search (BFS) from $b_j[h]$ in the connected component of B containing $b_j[h]$.
- For each $i \in \{1, 2, \ldots, n\}$, take a BFS from $e_i[1]$ in the subgraph of G^* induced by the set $V(T_i) \cap V(B)$. And, for each $j \in \{1, 2, \ldots, m\}$ take a BFS from $f_j[1]$ in the connected component of the subgraph of G^* containing $f_j[1]$ induced by the set $V(S_j) \cap V(B)$.
- For each one of the $n + 4m$ produced BFS–trees, add to $V_1(B)$ the vertices in the even level and add to $V_2(B)$ the vertices in the odd level.

The $3m$ components of B isomorphic each to $K_{3,3} \setminus \{e\}$ linked to two vertices are trivially bipartite graphs. To show that there is no conflict in the definition of the bipartition of B, it remains to analize the connected component of B containing the edges in E'. For note that $e_i[2] \in V_2(B)$ and $e_{i+1}[1] \in V_1(B)$, for $i \in \{1, 2, \ldots, (n-1)\}$; $e_n[2] \in V_2(B)$ and $f_1[1] \in V_1(B)$. And note that $f_j[6] \in V_2(B)$ and $f_{j+1}[1] \in V_1(B)$ for $j \in \{1, 2, \ldots, (m-1)\}$; $f_m[6] \in V_2(B)$ and $e_1[1] \in V_1(B)$.

Now we are ready to define the supervertices of G corresponding to the degree 3 vertices in G^*. For each such vertex v of G^* with ordered adjacency list $\vec{Adj}(v) = (w, t, s)$, we add to G one *Clockwise supervertex* N_v (Fig. 6a), if v is a vertex in partition $V_1(B)$ and one *Counterclockwise supervertex* N_v (Fig. 6b), if v is a vertex in partition $V_2(B)$. The $3(2^p n + 5m + 1)$ labeled vertices in each supervertex will be used later as endpoints of edges linking adjacent supervertices.

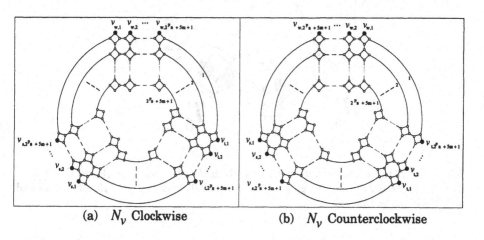

(a) N_v Clockwise   (b) N_v Counterclockwise

Fig. 6. Non-Splitting subgraph N_v, with $\vec{Adj}(v) = (w, t, s)$.

Finally, we end the construction of the simple graph G as follows. For each edge $(v, w) \in E(B)$, let $E_{vw} = \{(v_{w,s}, w_{v,s}) | s \in \{1, 2, \ldots, 2^p n + 5m + 1\}\}$. For each edge $(v, w) \in E(G^*)$ where v has degree 2 in G^*, let $E_{vw} = \{(v, w_{v,1})\}$, if w has degree 3; and $E_{vw} = \{(v, w)\}$, if w has degree 2.

The construction of the instance (G, k) of SN is completed by setting $k = 2^p n + 5m$ and $V(G) = (\bigcup_{v \in V(B)} V(N_v)) \cup (V(G^*) \backslash V(B))$ and $E(G) = (\bigcup_{v \in V(B)} E(N_v)) \cup (\bigcup_{(u,v) \in E(G^*)} E_{uv})$. Note that the simple graph G has a large number of vertices and edges. For N_v has $3 \times 4(k+1)^2 = 12(k+1)^2$ vertices and $3[(k+1)^2 + 4(k+1)^2 + k(k+1)] = 3(k+1)(6k+5)$ edges. In the set of T_i's there are $[(12(k+1)^2(10+4(2^p-1))) + (6 \times 2^p)]n - 3m$ vertices and $[((k+1)(16+4(2^p-2))) + (3(k+1)(6k+5)(10+4(2^p-1))) + (8 \times 2^p)]n$ edges. In the set of S_j's there are $[(12(k+1)^2 54) + (6)]m$ vertices and $[((k+1)71) + 3(k+1)(6k+5)54 + (12)]m$ edges. And in the set of communication edges there are $(k+1)(m+n)$ edges. Figure 7 shows an example of the construction of an instance (G, k) of SN.

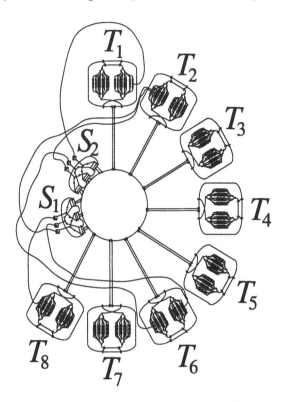

Fig. 7. SN instance obtained from 3SAT in which $U = \{u_1, u_2, \ldots, u_8\}$ and $C = \{(\bar{u}_2 \vee \bar{u}_6 \vee u_8), (\bar{u}_2 \vee u_6 \vee u_1)\}$. Here, graph G has 154.877.910 vertices, 232.139.388 edges and $k = 2^p n + 5m = 138$.

Complexity of the construction of G. As G^* can be constructed in polynomial time in the size of the instance of 3SAT and the size of G^* is $O((m^2 n)^2)$, we have that G can be constructed in time $O((2^p n)^2.(m^2 n)^2) = O((mn)^2.(m^2 n)^2)$, which is polynomial in the size of the instance of 3SAT.

It remains to prove that C is satisfiable if and only if $\sigma(G) \leq k$. See the technical report [5] for the proofs of Claims 1, 2 and 3.

Claim 1 *There is a drawing $D(G)$ for G such that:*

   (i) For every $v \in V(B)$, every edge of the corresponding N_v is in no crossing;

   (ii) For every $(u, v) \in E(B)$, there are no crossings between two edges linking vertices of N_u to vertices of N_v.   □

Consider the drawing for a subgraph N_v of G depicted in Fig. 6. We define the 1–*meridian* of N_v to be the cycle contained in the exterior face of this drawing of N_v. Recursively, for $i = 1, 2, \ldots, k$ we remove the vertices of the exterior face (vertices of the i–meridian plus pendant vertices) obtaining a new drawing and define the $(i+1)$–*meridian* to be the current cycle contained in the exterior face of this drawing. Observe that by construction, if $i < j$, with $i, j \in \{1, 2, \ldots, k+1\}$, then the i–meridian and the j–meridian are disjoint in vertices.

Claim 2 *If G' is obtained from G by a set Z of splittings, where $|Z| \leq k$, then there is a subgraph B_c of G' contractible to B, such that B_c contains a meridian of N_v as subgraph, for all $v \in B$.* □

From now on we refer to the subgraphs of G corresponding to the subgraphs $T_i, S_j, R_{\bar{u}_i}$ and R_{u_i} of G^* by saying, respectively, $T_i, S_j, R_{\bar{u}_i}$ and R_{u_i}.

- If $\sigma(G) \leq k$, then C is satisfiable.
  Suppose there is a planar graph H obtained from G by a set Z, with $|Z| \leq k$ splittings. By Claim 2, H has a subgraph contractible to B. In order to make each T_i planar Z must admit a subset with 2^p splittings in the black or striped vertices of T_i, or in the supervertices N_v with vertices adjacent to some black vertex of T_i. Note that there can be no T_i in which Z has simultaneously 2^p splittings in R_{u_i} and 2^p splittings in $R_{\bar{u}_i}$ because: there are n disjoint subgraphs T_i's in G; each one of the T_i's requires at least 2^p splittings in Z; and H is obtained from G by the set Z of splittings with $|Z| \leq 2^p n + 5m$, where $5m < 2^p$. Let $i \in \{1, 2, \ldots, n\}$ be an index. Let L_i be the subgraph in the pair $R_{u_i}, R_{\bar{u}_i}$ of subgraphs of T_i, that contains at least 2^p splittings of Z. We denote by Z_i the subset of Z consisting of all splittings of Z in L_i. A truth assignment for U can be obtained by setting the variable $u_i = T$, if $L_i = R_{u_i}$. On the other hand we set the variable $u_i = F$, if $L_i = R_{\bar{u}_i}$. Note that this truth assignment can be obtained in polynomial time in the size of G, that is, in the size of the instance of 3SAT.

Claim 3 *The following truth assignment satisfies C: set $u_i = T$, if $L_i = R_{u_i}$; set $u_i = F$, if $L_i = R_{\bar{u}_i}$.* □

- If C is satisfiable, then $\sigma(G) \leq k$.
  We shall define a set of size k of splittings that obtains a planar graph from G. Consider a truth assignment for U that satisfies C. If the literal u_i has value T, then split in G, the 2^p leaves of one of the two trees of R_{u_i}. If the literal u_i has value F, then split in G, the 2^p leaves of one of the two trees of $R_{\bar{u}_i}$. Let G' be the resulting graph obtained from G by this set of $2^p n$ splittings. Now consider $D(G)$, the drawing for G defined in Claim 1. Consider $D(G')$ a drawing for G' obtained from $D(G)$, such that all corresponding drawings for Truth–Setting subgraphs are plane. Thus in $D(G')$ the remaining crossings occur in edges linking vertices of some N_v in S_j to vertices not in this same N_v. As there is at least one literal with the value T for each clause in C, by applying Lemma 3 we define in polynomial time, for each graph S_j, a corresponding set of five splittings, such that we have no crossings in the edges of the resulting graph from S_j. Let G'' be the resulting graph obtained from G' by this set of $5m$ splittings. A plane drawing $D(G'')$ for G'' is obtained from $D(G')$, where each one of the three $K_{3,3} \setminus \{e\}$'s of each S_j, is located either inside a region of the plane drawing corresponding to S_j, or inside a region of the plane drawing corresponding to some T_i. Therefore, we obtain a planar graph from G with exactly $2^p n + 5m$ splittings. □

Corollary 1. SN *is NP-complete when restricted to cubic graphs.*

Proof. We use the strategy of Theorem 1 by modifying locally the graph G in Theorem 1 as follows. Consider the auxiliary graph C_v depicted in Fig. 8(a). For each vertex v of degree 2 in G, we add to G a copy of C_v, such that w_v is the vertex of C_v adjacent to v, as show in Fig. 8(b).   □

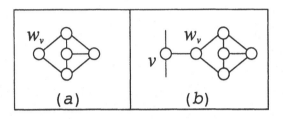

Fig. 8. Auxiliary graph for the proof of Corollary 1.

Corollary 2. SN *is NP-complete when restricted to graphs not containing a subdivision of K_5 as a subgraph.*

Proof. It follows from Corollary 1 because a cubic graph does not have vertices of degree 4.   □

As an application of the NP-completeness of SPLITTING NUMBER, consider another nonplanarity measure of a graph: the size of its maximum planar subgraph, and the decision problem:

PLANAR SUBGRAPH (PS)
INSTANCE: Graph $G = (V(G), E(G))$ and an integer $0 \leq p \leq |E(G)|$.
QUESTION: Is there a subset $E' \subseteq E(G)$ with $|E'| \geq p$ such that $G' = (V(G), E')$ is planar?

Liu and Geldmacher [11] proved that PLANAR SUBGRAPH is NP-complete, but it was not known until now whether this problem remains NP-complete when restricted to cubic graphs. Our final result shows how to use that SPLITTING NUMBER FOR CUBIC GRAPHS (SN$\Delta3$) is NP-complete to prove that PLANAR SUBGRAPH FOR CUBIC GRAPHS (PS$\Delta3$) is NP-complete.

Corollary 3. PS$\Delta3$ *is NP-complete.*

Proof. PS$\Delta3$ is in NP because PS is in NP. Let G, k be an instance for SN$\Delta3$. We may assume $k \leq |E(G)|$. Consider the instance of PS$\Delta3$ consisting of G and the integer $(|E(G)| - k)$. Note that any splitting in a graph of maximum degree 3 yields one or two leaves. In addition, a crossing in an edge incident to a leaf can always be removed by considering a different drawing in the plane. Thus, if L is the set of the leaves of G, then G has the same splitting number as $G \setminus L$.

Assume there exists a set Z of splittings, $|Z| \leq k$, obtaining a planar graph H from G. Define a subset L of $V(H)$, $|L| = |Z|$, such that L is obtained from Z by adding to L one leaf obtained in each splitting of Z. By construction, the graph $H \setminus L$ is isomorphic to a subgraph of G with $|E(H \setminus L)| \geq |E(G)| - k$, i.e., we have the answer yes to PSΔ3. Now suppose that G has a planar subgraph $G' = (V(G), E')$, with $|E'| \geq |E(G)| - k$. Consider the subset $Z = (E(G) \setminus E')$ of $E(G)$. A planar graph H is obtained from G by splitting, for each edge (u, v) of Z one of its endpoints, say v, with degree greater than 1, into v_1 and v_2, such that $\{u\}$ is the neighborhood of v_1. Thus, a set of splittings of size k or less is produced obtaining a planar graph H from G, as required. $\square$

References

1. S. A. Cook (1971). "The complexity of theorem–proving procedures", *Proceedings of the 3rd. ACM Symposium on Theory of Computing*, Association for Computing Machinery, New York, 151–158.
2. P. Eades and C. F. X. Mendonça (1993). "Heuristics for Planarization by Vertex Splitting". *Proc. ALCOM Int. Workshop on Graph Drawing, GD'93*, 83–85.
3. P. Eades and C. F. X. Mendonça (1996). "Vertex Splitting and Tension–Free Layout". *Proc. GD'95, Lecture Notes in Computer Science* **1027**, 202–211.
4. L. Faria, C. M. H. Figueiredo and C. F. X. Mendonça (1998). "The splitting number of the 4–cube", *Proc. LATIN'98, Lecture Notes in Computer Science* **1380**, 141–150.
5. L. Faria, C. M. H. Figueiredo and C. F. X. Mendonça (1997). "Splitting number is NP–Complete", Technical Report **ES-443/97**, COPPE/UFRJ, Brazil. Available at ftp://ftp.cos.ufrj.br/pub/tech_reps/es44397.ps.gz.
6. N. Hartfield, B. Jackson and G. Ringel (1985). "The splitting number of the complete graph", *Graphs and Combinatorics* **1**, 311–329.
7. J. E. Hopcroft and R. E. Tarjan (1974). "Efficient Planarity Testing", *J. ACM* **21**, 549–568.
8. B. Jackson and G. Ringel (1984). "The splitting number of complete bipartite graphs", *Arch. Math.* **42**, 178–184.
9. K. Kuratowski (1930). "Sur le problème des courbes gauches en topologie", *Fund. Math.* **15**, 271–283.
10. A. Liebers (1996). "Methods for Planarizing Graphs – A Survey and Annotated Bibliography". Available at ftp://ftp.informatik.uni-konstanz.de/pub/preprints/ /1996/preprint-012.ps.Z.
11. P. C. Liu and R. C. Geldmacher (1979). "On the deletion of nonplanar edges of a graph", *Cong. Num.* **24**, 727–738.
12. C. F. X. Mendonça (1994). "A Layout System for Information System Diagrams", PhD thesis, Univ. of Queensland, Australia.

Tree Spanners in Planar Graphs
(Extended Abstract)

Sándor P. Fekete[1] and Jana Kremer[2]

[1] Center for Parallel Computing, Universität zu Köln
D–50923 Köln, GERMANY
sandor@zpr.uni-koeln.de

[2] Lehrstuhl für Volkswirtschaftslehre
Otto-Friedrich Universität Bamberg
D–96045 Bamberg
GERMANY
jana.kremer@sowi.uni-bamberg.de

Abstract. A tree t-spanner of a graph G is a spanning subtree T of G in which the distance between every pair of vertices is at most t times their distance in G. Spanner problems have received some attention, mostly in the context of communication networks. It is known that for general unweighted graphs, the problem of deciding the existence of a tree t-spanner can be solved in polynomial time for $t = 2$, while it is NP-hard for any $t \geq 4$; the case $t = 3$ is open, but has been conjectured to be hard.

In this paper, we consider tree spanners in planar graphs. We show that even for planar unweighted graphs, it is NP-hard to determine the minimum t for which a tree t-spanner exists. On the other hand, we give a polynomial algorithm for any fixed t that decides for planar unweighted graphs with bounded face length whether there is a tree t-spanner. Furthermore, we prove that it can be decided in polynomial time whether a planar unweighted graph has a tree t-spanner for $t = 3$.

1 Introduction

A t-spanner of a graph G is a spanning subgraph H of G in which the distance between every pair of vertices is at most t times their distance in G. We can think of the "stretch factor" t as the relative price increase that may incur for individual connections after replacing the network G by a cheaper subnetwork H. Spanners were first considered in the context of practical motivations from communication networks (see Peleg and Ullman [20], who introduced spanners to synchronize asynchronous networks). They have also been used for simplifying geometric data structures – see Chew [11], Dobkin, Friedman, and Supowit [12], and Arikati et al. [2]. Surveys of results on the existence and efficient constructibility can be found in [19] and [23].

Depending on the objective for choosing a subnetwork, various kinds of spanners have been considered – see the list of references for a selection of variants.

Since the main motivation is to obtain a network of small total weight, particular attention has focused on *tree spanners*, where the subnetwork H is minimal with respect to edge removal. As Cai [8], and Cai and Corneil [10] showed, the problem of deciding the existence of a tree t-spanner in an unweighted graph G can be solved in polynomial time for $t = 2$; on the other hand, the problem is NP-complete for any $t \geq 4$. The case $t = 3$ is still open, but it was conjectured in [10] to be NP-complete.

As noted above, spanners have been considered in the context of geometric distance queries – see [11, 12, 2]. Since planar graphs form a particularly well-understood class of sparse graphs with a number of structural and algorithmic properties that make them interesting as spanners, the focus of those works has been on *planar spanners*, where the spanning graph H is required to be planar. Also, see Brandes and Handke [7] for a proof that it is NP-hard to determine a minimum weight planar t-spanner in a graph. They also showed that determining a minimum weight t-spanner in a planar graph is an NP-hard problem.

Between considering tree spanners in general graphs and planar spanners in general graphs, it is natural to consider tree spanners in planar graphs. Not only does this allow a better understanding of the properties of graph spanners, but results on the stretch factors of tree spanners in planar graphs combine with bounds on the stretch factors of planar spanners in general graphs to yield estimates on tree spanners in general graphs.

In this paper, we show that deciding the existence of a tree t-spanner in a graph G is NP-complete, if t is part of the input, even when restricted to the situation where G is planar and unweighted. On the other hand, we prove that this problem can be solved in polynomial time for planar unweighted graphs with bounded face length and fixed t.

For some purposes, not all pairs of connections have the same importance. This motivates the concept of s, t-spanners: For a partition of $E(G)$ into two given sets of edges E_1 and E_2, a tree s, t-spanner consists of edges in E_1, and it replaces any edge $(v_1, v_2) \in E_1$ by a path of at most s times its length, and any edge $(v_1, v_2) \in E_2$ by a path of at most t times its length. We show that for fixed s and t, the existence of a tree s, t-spanner in planar unweighted graphs with bounded face length can be checked in polynomial time. By a detailed analysis of the neighborhood structures of planar graphs with tree 3-spanners, we are able to show that a planar graph has a tree 3-spanner, iff it is a subgraph of a planar graph with bounded face length that has a tree 3,12-spanner. This implies a polynomial algorithm for deciding whether a planar graph G has a tree 3-spanner.

The rest of this paper is organized as follows: In Section 2, we introduce some basic concepts. Section 3 sketches the NP-completeness of deciding the existence of a tree t-spanner in a planar graph. In Section 4, we describe the polynomial algorithm for deciding whether a planar graph with bounded face length has a tree s, t-spanner. Section 5 gives an overview of the polynomial algorithm for deciding whether a planar graph has a tree 3-spanner. In Section 6 we conclude with some open problems.

2   Preliminaries

Throughout this paper, we use the terminology of Bondy and Murty [5]. A graph G has edge set $E(G)$ and vertex set $V(G)$; we may simply write E and V when the meaning is clear. If H is a subgraph of G, then $G - H$ denotes the graph obtained by deleting from G all edges of H. For a pair of vertices v_1 and v_2 in a connected graph G, we denote the length of a shortest path from v_1 to v_2 by $d_G(v_1, v_2)$. We will concentrate on the case of unweighted graphs without loops, so for any edge $(v_1, v_2) \in E(G)$, we have $d_G(v_1, v_2) = 1$. For a planar graph G, we write G^* for the dual graph. For $S \subset V$, the number of the edges leaving S in the graph G is denoted by $\delta_G(S)$. For $S \subset V$, we denote by $N(S)$ the set of neighbors of S, i.e., the set of vertices $v \in V \setminus S$ with a $w \in S$, such that $(v, w) \in E$. For a set of vertices $S \subseteq V$, the subgraph induced by S is denoted by $G[S]$.

For a real number $t \geq 1$, a subgraph H of a connected graph G is a *t-spanner* if $d_H(v_1, v_2) \leq t \cdot d_G(v_1, v_2)$ for all $v_1, v_2 \in E(G)$. A *tree t-spanner* is a t-spanner that is a tree. The parameter t is called the *stretch factor*; the smallest value t for which a graph G has a tree t-spanner is called the *tree stretch index* of G, denoted by $\sigma_T(G)$. It was shown in [10] that the following holds:

Lemma 1 *A subgraph H of a connected graph G is a t-spanner, iff for all edges $(v_1, v_2) \in E(G) - E(H)$, we have $d_H(v_1, v_2) \leq t$*

This allows us to consider only integer stretch factors for unweighted graphs. If the condition $d_H(v_1, v_2) \leq t$ is satisfied for a particular edge $e = (v_1, v_2) \in E(G) - E(H)$, we say that e has a *short detour* in H; for the case of tree spanners T, there is a unique corresponding shortest path, denoted by $p_T(e)$.

3   An NP-completeness result

It was shown in [10] that it is NP-complete to decide whether $\sigma_T(G) \leq t$ for a general unweighted graph, as long as $t \geq 4$. In this section, we sketch our proof that it is NP-complete to decide $\sigma_T(G) \leq t$ for a planar unweighted graph, where t is part of the input. Our reduction is from a special subclass of 3-SAT instances, called PLANAR 3SAT, which was shown to be NP-complete by Lichtenstein [16].

A 3SAT instance I is said to be an instance of PLANAR 3SAT, if the following bipartite graph G_I is planar: Every variable and every clause in I is represented by a vertex in G_I; two vertices are connected, if and only if one of them represents a variable that appears in the clause that is represented by the other vertex. See Figure 1 (a) for an example.

In the following, we sketch the necessary gadgets for our hardness proof. Details are contained in the full version of the paper, see also [15].

3.1   The basic setup

In a first step, the graph G_I is transformed into a graph G'_I. As shown in Figure 1, each set of three edges adjacent to the same clause vertex is replaced by three

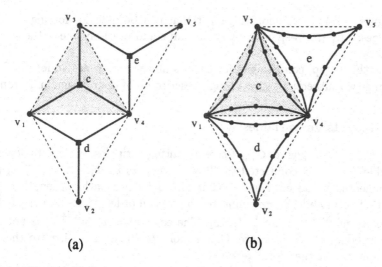

Fig. 1. (a) The graph G_I representing the PLANAR 3SAT instance $(x_1 \vee x_3 \vee x_4) \wedge (\bar{x}_1 \vee \bar{x}_2 \vee x_4) \wedge (\bar{x}_3 \vee x_4 \vee \bar{x}_5)$; (b) The transformed graph G'_I

paths of length 4. From this graph G'_I, *any* spanning tree T' is chosen. This spanning tree has a certain stretch factor t', which is polynomially bounded by the size of I.

For the second step, we add edges and vertices to G'_I to get a graph G''_I. In particular, we use the gadgets shown in Figure 2 to make sure that for $t = t' + 1$, all the edges of T' must be contained in a potential tree t-spanner of G''_I, if there is one.

The gadget shown in (a) has been used extensively in the proofs of [10] and [7]. It is easy to see that any tree 5-spanner of the graph G shown in the figure must contain the edge e. In the following, edges *forced* in this way are indicated by bold drawing.

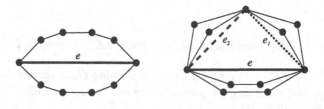

Fig. 2. (a) A forced edge; (b) a forced pair

Figure 2 (b) shows another gadget that can be used for forcing one out of two edges: Any tree 3-spanner must contain e and precisely one of the two edges e_1 and e_2.

In a third step, components for clauses and variables are added to G''_I. The following two subsections give a rough description of their design and properties.

3.2 Gadgets for variables

Figure 3 shows the gadget G_{var} for representing variables. It consists of a central "variable" vertex v, connected to "literal" vertices $v_1, \overline{v_1}, \ldots, v_s, \overline{v_s}$. v_i and $\overline{v_i}$ are connected by an edge $w^{(i)}$ that is forced by two paths of length t. $\overline{v_i}$ and v_{i+1} (indices modulo s) are connected by a path of length $t-2$, containing the vertices $\overline{v_i}, w_1^{(i,i+1)}, \ldots, w_{t-3}^{(i,i+1)}, v_{i+1}$. The edge $f_i = (\overline{v_i}, w_1^{(i,i+1)})$ is not forced, all other edges of the path are. Connections to the outside, i.e., to the rest of the graph, are at the literal vertices.

Furthermore, no two literal vertices are adjacent and there is no outside vertex that is connected for all $1 \leq i \leq s$ to at least one af the vertices $v_i, \overline{v_i}$.

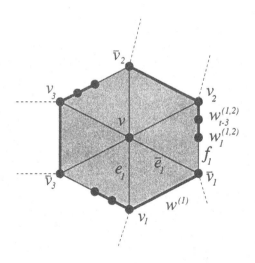

Fig. 3. Variable component G_{var}

Using straightforward induction, it is not hard to prove the following:

Lemma 2 *A tree t-spanner of a graph containing G_{var} cannot contain any of the edges f_i and must contain precisely one of the edges $e_i, \overline{e_i}$. Furthermore, e_1 is contained iff all e_i are contained.*

Containment of e_i or $\overline{e_i}$ corresponds to a truth assignment of the represented variable.

3.3   Gadgets for clauses

Due to space limitations, we cannot give full technical details of the clause gadgets, but the basic idea is shown in Figure 4 (a). Figure 4 (b) shows the general layout for combining clauses and variables.

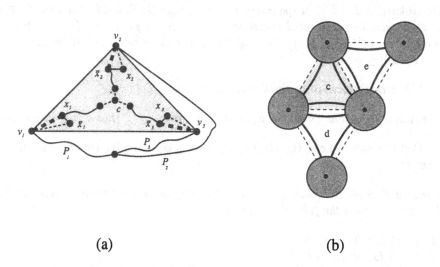

(a)   (b)

Fig. 4. (a) Idea of the clause component, shown for $(x_1 \vee \overline{x_2} \vee \overline{x_3})$; (b) combination of clause gadgets (triangles) with variable gadgets (circles)

Around a central node c, we group three forced paths of appropriate length, starting with edges (c, u_1), (c, u_2), (c, u_3). These paths connect to literal nodes of the corresponding variables. The choice of path lengths, forced edges, forced pairs and connections to variable components is done in a way that forces c to be a leaf of a tree t-spanner, if there is one. Furthermore, the existence of a tree t-spanner hinges on the existence of short detours $p_T(c, u_i)$, $p_T(c, u_j)$ for the two edges (c, u_i), (c, u_j) adjacent to c that are not contained in a spanner T.

Each non-true literal forces an extra edge into a potential short detour $p_T(c, u_i)$. The path lengths are set up in a way that allows one extra edge, but not two of them. This forces at least one satisfying literal to be in each clause. Conversely, if there is a truth assignment, we can keep c connected to the path that leads directly to the satisfying literal, making sure that there can be at most one extra edge for the detours $p_T(c, u_i)$, $p_T(c, u_j)$.

We summarize:

Theorem 3. *It is NP-complete to decide $\sigma_T(G) \leq t$ for planar unweighted graphs G and integers t.*

4   Planar graphs with bounded face length

In this section, we show that deciding the existence of a tree t-spanner in a planar graph with all faces of bounded length can be performed in polynomial time.

For this purpose, we introduce the notion of a *c-cut tree* in a graph:

Definition 4 *Let T be a spanning tree in a graph G. Removing any edge $e \in T$ splits T into two connected components, inducing a partition of the vertex set into $P_T(e) = (V_T(e), V_T'(e))$. We say that T is a c-cut tree in G, if for all $e \in T$, $|\delta_G(V_T(e))| \le c$.*

It is straightforward to show that the following holds:

Lemma 5 *A planar graph G has a tree t-spanner, iff G^* has a $(t+1)$-cut tree.*

Furthermore, we can establish the following constructive characterization of c-cut trees:

Lemma 6 *A planar graph G has a c-cut tree, iff there is a "rooted nested family" $F \subseteq 2^V \times V$ with the following properties:*

1. *$(V, r) \in F$ for an $r \in V$*
2. *$r \in S$ for all $(S, r) \in F$,*
3. *$|\delta_G(S)| \le c$ for all $(S, r) \in F$,*
4. *for all $(S_1, r_1), (S_2, r_2) \in F$ we have $S_1 \subseteq S_2$ or $S_1 \subseteq V \setminus S_2$,*
5. *for all $(S, r) \in F$ there are $l \ge 1$ and $(S_i, r_i) \in F$, $1 \le i \le l$, with $S \setminus \{r\} = \dot\bigcup S_i$ and $(r, r_i) \in E$.*

The vertex sets S correspond to the subsets of a partition induced by the removal of an edge $e \in T$ from T, while $r \in S$ is the vertex adjacent to e. The proof is straightforward and omitted.

Using the characterization from Lemma 6, we get the following result:

Theorem 7. *For fixed t, it can be decided in polynomial time for planar unweighted graphs G with bounded face length whether $\sigma_T(G) \le t$.*

Sketch: Consider the existence of a rooted nested family F of G^* as described in Lemma 6. Since t is fixed, there are only polynomially many possible cuts in G^* of size not larger than $t+1$, implying we only have to consider polynomially many sets (S, r) that can be used for F. Since all faces in G have bounded length, the dual graph G^* has bounded degree, so there is a polynomial number of possible partitions for any (S, r). Using dynamic programming and proceeding by increasing size of S, we can decide the existence of a rooted nested family as described in Lemma 6 in polynomial time.

$\square$

As described in the introduction, the concept of tree t-spanners can be generalized:

Definition 8 *Let G be a graph with $E(G) = E_1 \dot\cup E_2$. Then a spanning tree T of G is a tree s,t-spanner for $G = (V, E_1 \dot\cup E_2)$, iff T is a subgraph of (V, E_1), and for all edges $(v_1, v_2) \in E_1 - T$, we have $d_T(v_1, v_2) \leq s$, and for all edges $(v_1, v_2) \in E_2 - T$, we have $d_T(v_1, v_2) \leq t$.*

With an analogous approach to the one for tree t-spanners, we can establish the following result for tree s,t-spanners:

Theorem 9. *For fixed s and t, it can be decided in polynomial time for planar unweighted graphs $G = (V, E_1 \dot\cup E_2)$ of bounded face length, possibly with multi-edges, whether there is a tree s,t-spanner.*

5   Deciding the existence of tree 3-spanners in planar graphs

In this section, we sketch the polynomial algorithm for deciding whether a planar unweighted graph G has a 3-spanner. The key idea is to add a set of edges E' to obtain a graph $G_{\leq 4}$ with face length bounded by 4, such that $G_{\leq 4} = (V, E \dot\cup E')$ has a tree $3,12$-spanner, iff G has a tree 3-spanner. The existence of a $3, 12$-spanner in $G_{\leq 4}$ can be decided in polynomial time by the algorithm from the previous section.

If there is no face of length more than 4, we are done, so consider a face bounded by the chordless cycle $C = v_1, \ldots, v_s$ with $|C| \geq 5$. Now assume there is a tree 3-spanner in G. Since $|C| \geq 5$, for any edge $e = (v_i, v_{i+1})$ in $C - T$, there must be a path in T that is not longer than 3 and not fully contained in C. The different possibilities for such a path are shown in Figure 5.

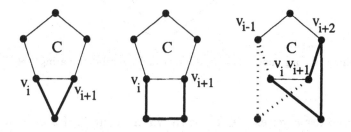

Fig. 5. Different possibilities for a short detour $p_T(v_i, v_{i+1})$ of an edge (v_i, v_{i+1}) in $C - T$

Now we can analyze the structure of T in the neighborhood of C: consider the edges in $p_T(v_i, v_{i+1}) - C$. It is not hard to see that each of these edges must be contained in $p_T(v_j, v_{j+1})$ for an edge $(v_j, v_{j+1}) \in ((C - T) - (v_i, v_{i+1}))$, since T cannot contain a cycle. Because $p_T(v_j, v_{j+1})$ contains at most three edges, (v_j, v_{j+1}) is adjacent to (v_i, v_{i+1}) or both are adjacent to the same edge in C.

From this, we can derive the following lemma:

Lemma 10 *Let G be a planar graph with a tree 3-spanner T. If C is a chordless cycle in G, $|C| \geq 5$, then there is a "semi-dominating" tree T_C in T, such that*

1. *C is "weakly dominated" by $V(T_C)$, i. e., for any vertex $v_i \in C$, v_i is adjacent to T_C, or both its neighbors v_{i-1} and v_{i+1} are.*
2. *If a vertex $v \in C$ is not adjacent to T_C, then both of its neighbors are adjacent to the same vertex of T_C.*

Furthermore, for a given cycle C that bounds a face F of G, the semi-dominating tree is uniquely determined.

An example is shown in Figure 6. Bold lines show the semi-dominating tree.

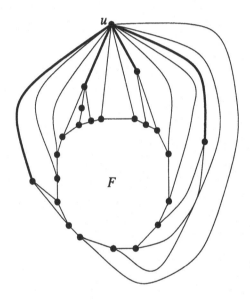

Fig. 6. A long chordless cycle C in G and a semi-dominating tree T_C

Now consider a vertex $u \in T_C$ as shown in Figure 7. If u does not weakly dominate C (which would imply $u = T$), then it induces a subdivision (called the *u-subdivision*) of C as follows. Let D_i be a maximal path weakly dominated by u. The first vertex of D_i is denoted by d_i^h, the last by d_i^t. Between any two D_i and D_{i+1}, there is a path P_i, consisting of vertices that are non-adjacent to u. Clearly, any P_i must contain at least two vertices. For any i, let P_i^1 be the path (d_i^t, P_i, d_{i+1}^h), while P_i^2 is the path $(D_{i+1}, P_{i+1}, \ldots, P_{i-1}, D_i)$.

Now we insert a set $E'(u)$ of new edges as follows. For any i, insert the edge (d_i^t, d_{i+1}^h) – shown by broken lines in Figure 7. This yields a face $F(u)$ that is dominated by u. This face is then triangulated by further new edges. Using the structure of the semi-dominating tree, we can show that the face bounded by C is subdivided into faces of length at most 4. Furthermore, the end vertices of the

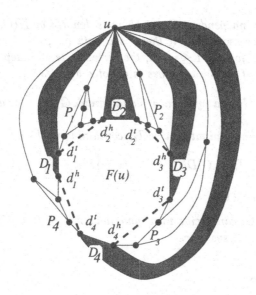

Fig. 7. A vertex u in a semi-dominating tree T_C, its u-partition, and the subface $F(u)$

new edges are connected by paths with at most 12 edges in a tree 3-spanner T of G. Note that in the process of introducing new edges, we may create multi-edges.

After inserting a new chord for any chord of a face, this subdivision is carried out for every face that is bounded by a chordless cycle C with more than four edges and for all vertices of the semi-dominating tree T_C of C. Eventually we get the planar supergraph $G_{\leq 4}$ with the desired properties.

Conversely, any tree $3, 12$-spanner in the expanded graph $(V, E(G) \dot\cup E')$ induces a tree 3-spanner in G. (Full details can be found in [15].)

The following definition and Lemma 12 show how to find a semi-dominating tree of a cycle in polynomial time. Once the semi-dominating trees are found, the procedure of inserting the edges, and testing for the existence of a tree $3, 12$-spanner yields a polynomial algorithm – recall Theorem 9.

Definition 11 *Let $u \in N(C)$ be a vertex that does not weakly dominate the cycle C. Let $D_1, P_1, \ldots, D_r, P_r$ be the u-subdivision of C.*

A vertex $w \in N(C)$ is an independent C-neighbor of u, if it is adjacent to u in G and if there is an index $1 \leq i \leq r$ such that the following conditions hold:

1. *There is a path of at most two edges in G that connects w with a vertex of P_i, and*
2. *there are vertices w_i^h, w_i^t from P_i^1 that are adjacent to w in G and vertices u_i^h, u_i^t from P_i^2 that are adjacent to u in G, such that $w_i^h, w_i^t, u_i^h,$ and u_i^t are pairwise disjoint and $u_i^h w_i^t, w_i^h u_i^t \in E(C)$ holds.*

(Note that the path does not contain vertex u, since u is not adjacent to any vertex in P_i.)

The set of all independent C-neighbors is denoted by $N(C, u)$. A vertex $w \in N(C)$ is a C-successor of u, if there is a path $(w_0, w_1, \ldots, w_k$. with $w_0 = u$, $w_k = w$, such that for any $1 \leq i \leq k$, the vertex w_i is an independent C-neighbor of w_{i-1}. The set of all C-successors is denoted by $D(C, u)$.

Lemma 12 *Let C be a cycle in a planar graph G, and let u be adjacent to a vertex in C.*

If C has a semi-dominating tree T_C containing u, then

$$T_C = G[D(C, u)] - \{(v, w) : w \notin N(C, v)\}.$$

Summarizing, we get

Theorem 13. *We can decide in polynomial time whether a planar unweighted graph G has a tree 3-spanner.*

6 Conclusion

In this paper, we have shown that for planar graphs, it is possible to decide the existence of a tree 3-spanner in polynomial time. Our method makes strong use of planarity, yet the resulting algorithm is rather complicated. It has been conjectured that deciding the existence of a tree 3-spanner is an NP-complete problem, and our impression from the experience with planar graphs seems to support this belief.

On the other hand, we could prove that deciding the existence of a tree t-spanner is NP-complete, as long as t is part of the input. The complexity for fixed t is unclear, but there may be a polynomial method of deciding the question, possibly using a combination of dynamic programming and an analysis of neighborhood structures, as we did for the case $t = 3$. Unfortunately, this analysis appears to become rather tedious even for $t = 4$.

Acknowledgment

We would like to thank Dorothea Wagner, Ulrik Brandes, and Dagmar Handke for helpful discussions.

References

1. I. Althöfer, G. Das, D. Dobkin, D. Joseph, and J. Soares. On sparse spanners of weighted graphs. *Discrete and Computational Geometry*, **9** (1993), pp. 81–100.
2. S. Arikati, D. Z. Chen, L. P. Chew, G. Das, M. Smid, and D. Zaroliagis. Planar spanners and approximate shortest path queries among obstacles in the plane.. In: J. Diaz, ed., *Algorithms - ESA '96*, Springer Lecture Notes in Computer Science #1136, 1996, pp. 514–528.
3. B. Awerbuch, A. Baratz, and D. Peleg. Efficient broadcast and light-weight spanners. Manuscript, 1992.

4. S. Bhatt, F. Chung, F. Leighton, and A. Rosenberg. Optimal simulations of tree machines. In *Proceedings of the 27th IEEE Symposium on Foundations of Computer Science* (FOCS 1986), pp. 274-282.

5. J. A. Bondy and U. S. R. Murty. *Graph Theory with Applications.* North-Holland, New York, 1976.

6. K. S. Booth and G. S. Lueker. Testing for the consecutive ones property, interval graphs, and planarity using PQ-tree algorithms. *Journal of Computer and System Sciences,* **13** (1976), pp. 335-379.

7. U. Brandes and D. Handke. NP-completeness results for minimum planar spanners. In *Proceedings of the 23th Workshop on Graph-Theoretic Concepts in Computer Science* (WG '97), Springer Lecture Notes in Computer Science # 1335, 1997, pp. 85-99.

8. L. Cai. Tree spanners: spanning trees that approximate distances. Ph.D. thesis, University of Toronto, Toronto, Canada, 1992. Available as *Technical Report 260/92,* Department of Computer Science, University of Toronto.

9. L. Cai. NP-completeness of minimum spanner problems, *Discrete Applied Mathematics,* **48** (1994), pp. 187-194.

10. L. Cai and D. G. Corneil. Tree spanners. *SIAM Journal of Discrete Mathematics,* **8** (1995), pp. 359-387.

11. L. P. Chew. There is a planar graph almost as good as the complete graph. In *Proceedings of the 2nd ACM Symposium on Computational Geometry,* pp. 169-177, 1986.

12. D. P. Dobkin, S. J. Friedman, and K. J. Supowit. Delaunay graphs are almost as good as complete graphs. *Discrete and Computational Geometry,* **5** (1990), pp. 399-407.

13. J. E. Hopcroft and R. E. Tarjan. Efficient planarity testing. *Journal of the ACM,* **21,** (1974), 549-568.

14. G. Kortsarz and D. Peleg. Generating sparse 2-spanners. In: *Proceedings of the Third Scandinavian Workshop on Algorithm Theory* (SWAT 1992).

15. J. Kremer. Baumspanner in planaren Graphen. *Diploma thesis,* Mathematisches Institut, Universität zu Köln, 1997.

16. D. Lichtenstein. Planar formulae and their uses. *SIAM Journal of Computing,* **11** (1982), pp. 329-343.

17. A. L. Liestman and T. Shermer. Grid spanners. *Networks,* **23** (1993), pp. 123-133.

18. A. Mansfield. Determining the thickness of graphs is NP-hard. *Mathematical Proceedings of the Cambridge Philosophical Society,* **93** (1983), pp. 9-23.

19. D. Peleg and A. A. Schäffer. Graph spanners. *Journal of Graph Theory,* **13** (1989), pp. 99-116.

20. D. Peleg and J. D. Ullman. An optimal synchronizer for the hypercube. In *Proceedings of the 6th ACM Symposium on Principles of Distributed Computing,* 1987, pp. 77-85.

21. D. Peleg and E. Upfal. A tradeoff between space and efficiency for routing tables. In *Proceedings of the 20th ACM Symposium on the Theory of Computing,* (STOC 1988), pp. 43-52.

22. D. Richards and A. L. Liestman. Degree-constrained pyramid spanners. *Parallel and Distributed Computing,* **25** (1995), pp. 1-6.

23. J. Soares. Graph spanners: a survey. *Congressus Numerantium,* **89** (1992), pp. 225-238.

A Linear-Time Algorithm
to Find Four Independent Spanning
Trees
in Four-Connected Planar Graphs

Kazuyuki Miura, Daishiro Takahashi,
Shin-ichi Nakano, and Takao Nishizeki

Graduate School of Information Sciences
Tohoku University, Sendai 980-8579, Japan
e-mail:{miura,daishiro,nakano,nishi}@nishizeki.ecei.tohoku.ac.jp

Abstract. Given a graph G, a designated vertex r and a natural number k, we wish to find k "independent" spanning trees of G rooted at r, that is, k spanning trees such that, for any vertex v, the k paths connecting r and v in the k trees are internally disjoint in G. In this paper we give a linear-time algorithm to find four independent spanning trees in a 4-connected planar graph rooted at any vertex.

1 Introduction

Given a graph $G = (V, E)$, a designated vertex $r \in V$ and a natural number k, we wish to find k spanning trees $T_1, T_2, \cdots, T_k$ of G such that, for any vertex v, the k paths connecting r and v in $T_1, T_2, \cdots, T_k$ are internally disjoint in G, that is, any two of them have no common intermediate vertices. Such k trees are called k *independent spanning trees of G rooted at r*. Four independent spanning trees are drawn in Fig. 1 by thick lines. Independent spanning trees have applications to fault-tolerant protocols in networks [BI96,DHSS84,IR88,OIBI96].

Given a graph $G = (V, E)$ of n vertices and m edges, and a designated vertex $r \in V$, one can find two independent spanning trees of G rooted at any vertex in linear time if G is biconnected [BTV96,IR88], and find three independent spanning trees of G rooted at any vertex in $O(mn)$ and $O(n^2)$ time if G is triconnected [BTV96,CM88]. It is

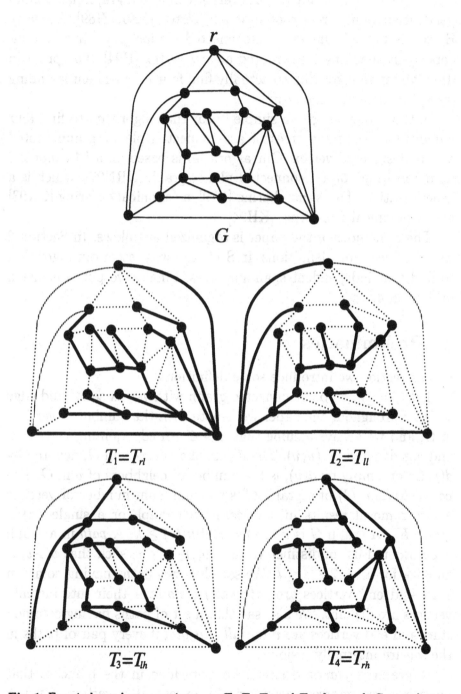

Fig. 1. Four independent spanning trees T_1, T_2, T_3 and T_4 of a graph G rooted at r.

conjectured that, for any $k \geq 1$, every k-connected graph has k independent spanning trees rooted at any vertex [KS92,ZI89]. Recently Huck has proved that every 4-connected planar graph has four independent spanning trees rooted at any vertex [H94]. The proof in [H94] yields an algorithm to actually find four independent spanning trees, but it takes time $O(n^3)$.

In this paper we give a simple linear-time algorithm to find four independent spanning trees of a 4-connected planar graph rooted at any designated vertex. Our algorithm is based on a "4-canonical decomposition" of a 4-connected planar graph [NRN97], which is a generalization of an st-numbering [E79], a canonical ordering [CK93] and a canonical 4-ordering [KH94].

The remainder of the paper is organized as follows. In Section 2 we introduce some definitions. In Section 3 we present our algorithm to find four independent spanning trees. Finally we put conclusion in Section 4.

2   Preliminaries

In this section we introduce some definitions.

Let $G = (V, E)$ be a connected graph with vertex set V and edge set E. Throughout the paper we denote by n the number of vertices in G, and we always assume that $n > 4$. An edge joining vertices u and v is denoted by (u, v). The *degree* of a vertex v in G, denoted by $d(v, G)$ or simply by $d(v)$, is the number of neighbors of v in G. The *connectivity* $\kappa(G)$ of a graph G is the minimum number of vertices whose removal results in a disconnected graph or a single-vertex graph K_1. A graph G is k-*connected* if $\kappa(G) \geq k$. A *path* in a graph is an ordered list of distinct vertices $v_1, v_2, \cdots, v_l$ such that $v_{i-1}v_i$ is an edge for all i, $2 \leq i \leq l$. We say that two paths having common start and end vertices are *internally disjoint* if their intermediate vertices are disjoint. We also say that a set of paths having common start and end vertices are *internally disjoint* if every pair of paths in the set are internally disjoint.

A graph is *planar* if it can be embedded in the plane so that no two edges intersect geometrically except at a vertex to which they are both incident. A *plane graph* is a planar graph with a fixed embedding. The *contour* $C_o(G)$ of a biconnected plane graph G is the clockwise (simple) cycle on the outer face. We write $C_o(G) = (w_1, w_2, \cdots, w_h)$ if the vertices $w_1, w_2, \cdots, w_h$ on $C_o(G)$ appear in this order.

3   Algorithm

In this section we give our algorithm to find four independent spanning trees of a 4-connected planar graph rooted at any designated vertex.

Given a 4-connected planar graph $G = (V, E)$ and a designated vertex $r \in V$, we first find a planar embedding of G in which r is located on $C_o(G)$. Let $G' = G - \{r\}$ be the subgraph of the plane graph G induced by $V - \{r\}$. In Fig. 2 (a) G is drawn by solid and dotted lines, and G' by solid lines. Since G is 4-connected, $d(r) \geq 4$. We may assume that all the neighbors $r_1, r_2, \cdots, r_{d(r)}$ of r in G appear on $C_o(G')$ clockwise in this order. Let $C_o(G') = (w_1, w_2, \cdots, w_h)$, $r_1 = w_1, r_2 = w_a, r_3 = w_b$ and $r_4 = w_c$, where $1 < a < b < c \leq d(r)$. We add to G' two new vertices r_b and r_t, join r_b with r_1 and r_2, and join r_t with $r_3, r_4, \cdots, r_{d(r)}$. Let G'' be the resulting plane graph, where vertices r_1, r_b, r_2, r_3, r_t and $r_{d(r)}$ appear on $C_o(G'')$ clockwise in this order. Fig. 2 (b) illustrates G''.

Let $\Pi = (W_1, W_2, \cdots, W_m)$ be a partition of the vertex set $V - \{r\}$ of G'. We denote by G_k, $1 \leq k \leq m$, the plane subgraph of G'' induced by $\{r_b\} \cup W_1 \cup W_2 \cup \cdots \cup W_k$. We denote by $\overline{G_k}$, $0 \leq k \leq m - 1$, the plane subgraph of G'' induced by $W_{k+1} \cup W_{k+2} \cup \cdots \cup W_m \cup \{r_t\}$. We assume that if $1 \leq k \leq m$ and $W_k = \{u_1, u_2, \cdots, u_l\}$ then vertices $u_1, u_2, \cdots, u_l$ consecutively appear on $C_o(G_k)$ clockwise in this order. A partition $\Pi = (W_1, W_2, \cdots, W_m)$ of $V - \{r\}$ is called a *4-canonical decomposition* of G' if the following three conditions (co1)–(co3) are satisfied.

(co1) $W_1 = \{w_a, w_{a-1}, \cdots, w_1\}$ and $W_m = \{w_b, w_{b+1}, \cdots, w_c\}$;

(co2) For each k, $1 \leq k \leq m - 1$, both G_k and $\overline{G_{k-1}}$ are biconnected (See Fig. 3.); and

(co3) For each k, $1 < k < m$, one of the following three conditions holds (See Fig. 3.):

    (a) $|W_k| \geq 2$, and each vertex $u \in W_k$ satisfies $d(u, G_k) = 2$ and $d(u, \overline{G_{k-1}}) \geq 3$;

(b) $|W_k| = 1$, and the vertex $u \in W_k$ satisfies $d(u, G_k) \geq 2$ and $d(u, \overline{G_{k-1}}) \geq 2$; and

(c) $|W_k| \geq 2$, and each vertex $u \in W_k$ satisfies $d(u, G_k) \geq 3$ and $d(u, \overline{G_{k-1}}) = 2$.

Fig. 2 (b) illustrates a 4-canonical decomposition of $G' = G - \{r\}$, where G' are drawn in solid lines and each set W_i is indicated by an oval drawn in a dotted line. A 4-canonical decomposition is a generalization of an "st-numbering" [E79], a "canonical decomposition" [CK93] and a "canonical 4-ordering" [KH94]. Although the definition of a 4-canonical decomposition above is slightly different from one in [NRN97], they are effectively equivalent each other.

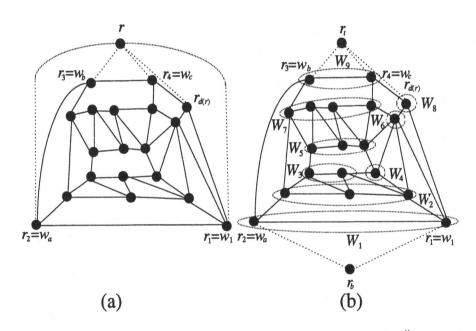

(a)   (b)

Fig. 2. (a) Four-connected plane graph G and (b) plane graph G''.

We have the following lemma.

Lemma 1. Let $G = (V, E)$ be a 4-connected plane graph, and let r be a designated vertex on $C_o(G)$. Then $G' = G - \{r\}$ has a 4-canonical decomposition Π. Furthermore Π can be found in linear time.

Proof. Similar to the proof of Lemma 3 in [NRN97]. $Q.\mathcal{E}.\mathcal{D}.$

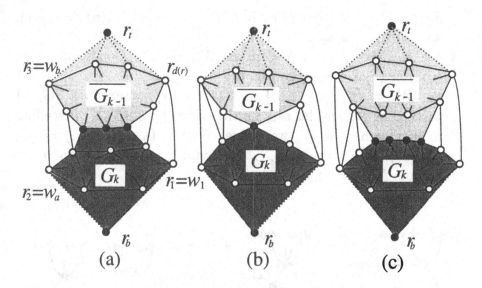

Fig. 3. Three conditions for (co3).

We need a few more definitions to describe our algorithm. For a vertex $v \in V - \{r\}$ we write $N(v) = \{v_1, v_2, \cdots, v_{d(v)}\}$ if $v_1, v_2, \cdots, v_{d(v)}$ are the neighbors of vertex v in G'' and appear around v clockwise in this order. To each vertex $v \in V - \{r\}$ we assign four edges incident to v in G'' as *the left hand* $lh(v)$, *the right hand* $rh(v)$, *the left leg* $ll(v)$ and *the right leg* $rl(v)$ as follows. We will show later that such an assignment immediately yields four independent spanning trees of G. Let $v \in W_k$ for some k, $1 \le k \le m$, then there are the following three cases to consider.

Case 1: either (i) $1 < k < m$ and W_k satisfies Condition (a) of (co3) or (ii) $k = 1$. (See Fig. 4.)
Let $W_k = \{u_1, u_2, \cdots, u_l\}$. Let u_0 be the vertex on $C_o(G_k)$ preceding u_1, and let u_{l+1} be the vertex on $C_o(G_k)$ succeeding u_l. For each $u_i \in W_k$ we define $rl(u_i) = (u_i, u_{i+1})$, $ll(u_i) = (u_i, u_{i-1})$, $lh(u_i) = (u_i, v_1)$, and $rh(u_i) = (u_i, v_{d(u_i)-2})$ where we assume $N(u_i) = \{u_{i-1}, v_1, v_2, \cdots, v_{d(u_i)-2}, u_{i+1}\}$.

Case 2: W_k satisfies Condition (b) of (co3). (See Fig. 5.)
Let $W_k = \{u\}$, let u' be the vertex on $C_o(G_k)$ preceding u, and let u'' be the vertex on $C_o(G_k)$ succeeding u. Let $N(u) = \{u', v_1, v_2, \cdots, v_{d(u)-1}\}$, and let $u'' = v_x$ for some x, $3 \le x \le$

$d(u) - 1$. Then $rl(u) = (u, u'')$, $ll(u) = (u, u')$, $lh(u) = (u, v_1)$, and $rh(u) = (u, v_{x-1})$.

Case 3: either (i) $1 < k < m$ and W_k satisfies Condition (c) of (co3) or (ii) $k = m$. (See Fig. 6.)

Let $W_k = \{u_1, u_2, \cdots, u_l\}$. Let u_0 be the vertex on $\overline{C_o(G_{k-1})}$ succeeding u_1, and let u_{l+1} be the vertex on $\overline{C_o(G_{k-1})}$ preceding u_l. For each $u_i \in W_k$ we define $rl(u_i) = (u_i, v_1)$, $ll(u_i) = (u_i, v_{d(u_i)-2})$, $lh(u_i) = (u_i, u_{i-1})$, and $rh(u_i) = (u_i, u_{i+1})$ where we assume $N(u_i) = \{u_{i+1}, v_1, v_2, \cdots, v_{d(u_i)-2}, u_{i-1}\}$.

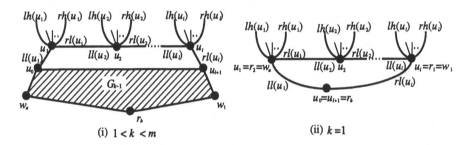

(i) $1 < k < m$ (ii) $k = 1$

Fig. 4. Assignment for Case 1.

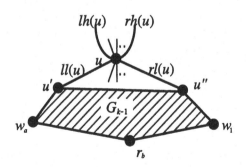

Fig. 5. Assignment for Case 2.

We are now ready to give our algorithm.

Procedure FourTrees(G, r)
begin
1   Find a planar embedding of G such that $r \in C_o(G)$;

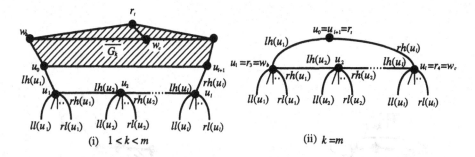

Fig. 6. Assignment for Case 3.

2 Find a 4-canonical decomposition $\Pi = (W_1, W_2, \cdots, W_m)$ of $G - \{r\}$;

3 For each vertex $v \in V - \{r\}$ find $rl(v), ll(v), rh(v)$ and $lh(v)$;

4 Let T_{rl} be a graph induced by the right legs of all vertices in $V - \{r\}$;

5 Let T_{ll} be a graph induced by the left legs of all vertices in $V - \{r\}$;

6 Let T_{lh} be a graph induced by the left hands of all vertices in $V - \{r\}$;

7 Let T_{rh} be a graph induced by the right hands of all vertices in $V - \{r\}$;

8 Regard vertex r_b in trees T_{rl} and T_{ll} as vertex r;

9 Regard vertex r_t in trees T_{lh} and T_{rh} as vertex r;

10 **return** T_{rl}, T_{ll}, T_{lh} and T_{rh} as four independent spanning trees of G.

end

We then verify the correctness of our algorithm. Assume that $G = (V, E)$ is a 4-connected planar graph with a designated vertex $r \in V$, and that Algorithm FourTrees finds a 4-canonical decomposition $\Pi = (W_1, W_2, \cdots, W_m)$ of $G - \{r\}$ and outputs T_{rl}, T_{ll}, T_{lh} and T_{rh}. We first have the following lemma.

Lemma 2. Let $1 \leq k \leq m$, and let T_{rl}^k be a graph induced by the right legs of all vertices in $G_k - \{r_b\}$. Then T_{rl}^k is a spanning tree of G_k.

Proof. We prove the claim by induction on k.

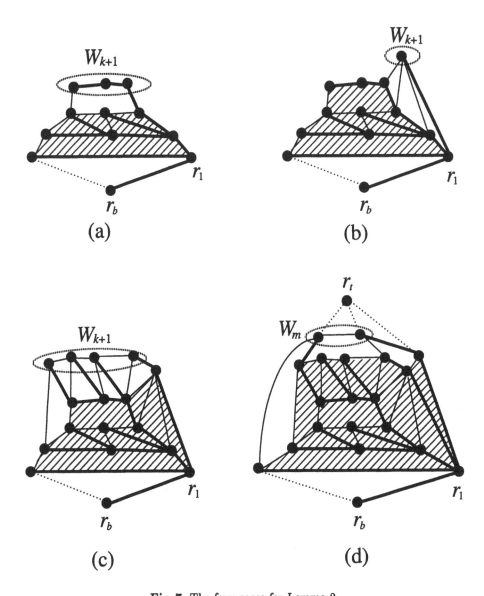

Fig. 7. The four cases for Lemma 2.

Clearly the claim holds for $k = 1$.

We assume that $1 \leq k \leq m - 1$ and T_{rl}^k is a spanning tree of G_k, and we shall prove that T_{rl}^{k+1} is a spanning tree of G_{k+1}. There are the following four cases to consider.

Case 1: $k \leq m - 2$ and W_{k+1} satisfies Condition (a) of (co3).
Case 2: $k \leq m - 2$ and W_{k+1} satisfies Condition (b) of (co3).
Case 3: $k \leq m - 2$ and W_{k+1} satisfies Condition (c) of (co3).
Case 4: $k = m - 1$.

For each case T_{rl}^{k+1} is a spanning tree of G_{k+1} as shown in Fig. 7; (a) for Case 1; (b) for Case 2; (c) for Case 3; and (d) for Case 4. $Q.\mathcal{E}.\mathcal{D}.$

We then have the following lemma.

Lemma 3. T_{rl}, T_{ll}, T_{lh} and T_{rh} are spanning trees of G.

Proof. By Lemma 3.2 T_{rl}^m is a spanning tree of G_m, and hence T_{rl} in which r_b is regarded as r is a spanning tree of G.

Similarly T_{ll}, T_{lh} and T_{rh} are spanning trees of G. $Q.\mathcal{E}.\mathcal{D}.$

Let v be any vertex in $V - \{r\}$, and let P_{rl}, P_{ll}, P_{lh} and P_{rh} be the paths connecting r and v in T_{rl}, T_{ll}, T_{lh} and T_{rh}, respectively. For any vertex u in $V - \{r\}$ we write $rank(u) = k$ if $u \in W_k$; $rank(r)$ is undefined. If an edge (v, u) of G' is a leg of vertex v, and (v, w) of G' is a hand of v, then $rank(u) \leq rank(v) \leq rank(w)$ and $rank(u) < rank(w)$.

Lemma 4. Each of the four pairs of paths, P_{rl} and P_{lh}, P_{rl} and P_{rh}, P_{ll} and P_{lh}, P_{ll} and P_{rh}, are internally disjoint.

Proof. We prove only that P_{rl} and P_{lh} are internally disjoint. Proofs for the other pairs are similar. If $v = r_1$ then $P_{rl} = (v, r)$. If $v = r_3$ then $P_{lh} = (v, r)$. Therefor P_{rl} and P_{lh} are internally disjoint if v is r_1 or r_3. Thus we may assume that $v \neq r_1, r_3$. Let $P_{rl} = (v, v_1, v_2, \cdots, v_l, r)$, then $v_l = r_1$. Let $P_{lh} = (v, u_1, u_2, \cdots, u_{l'}, r)$, then $u_{l'} = r_3$. The definition of a right leg implies that $rank(v) \geq rank(v_1) \geq rank(v_2) \geq \cdots \geq rank(v_l)$, and the definition of a left hand implies that $rank(v) \leq rank(u_1) \leq rank(u_2) \leq \cdots \leq rank(u_{l'})$. Thus $rank(v_l) \leq \cdots \leq rank(v_2) \leq rank(v_1) \leq rank(v) \leq rank(u_1) \leq rank(u_2) \leq \cdots \leq rank(u_{l'})$. We furthermore have $rank(v_1) < rank(u_1)$. Therefore P_{rl} and P_{lh} are internally disjoint. $Q.\mathcal{E}.\mathcal{D}.$

We next have the following lemma.

Lemma 5. Let $u \in V - \{r\}$, $ll(u) = (u, u')$, $rl(u) = (u, u'')$, and $N(u) = \{v_1, v_2, \cdots, v_{d(u)}\}$. One may assume that $u' = v_1$ and $u'' = v_s$ for some s, $1 < s \leq d(u)$. Then there exists t, $1 \leq t \leq s$, such that $rl(v_i) = (v_i, u)$ for each i, $2 \leq i \leq t - 1$, and $ll(v_j) = (v_j, u)$ for each j, $t + 1 \leq j \leq s - 1$. (Thus either (i) $rl(v_t) = (v_t, u) \neq ll(v_t)$, (ii) $rl(v_t) \neq (v_t, u) = ll(v_t)$, or (iii) $rl(v_t) \neq (v_t, u) \neq ll(v_t)$. See Fig. 8.)

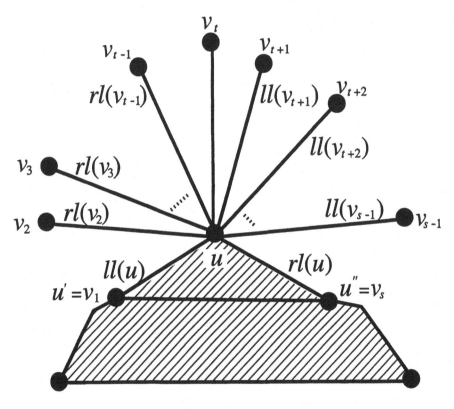

Fig. 8. Illustration for Lemma 5.

Proof. From the definitions of a 4-canonical decomposition and a right leg, one can observe that if $2 \leq i \leq s - 1$ and $rl(v_i) = (v_i, u)$ then $rank(v_{i-1}) < rank(v_i)$. Similarly, if $2 \leq i \leq s - 1$ and $ll(v_j) = (v_j, u)$ then $rank(v_j) > rank(v_{j+1})$.

Assume for a contradiction that the claim does not hold. Then $rl(v_i) = (v_i, u)$ and $ll(v_j) = (v_j, u)$ for some i and j, $1 \leq j < i \leq s$. Let $v_i \in W_{i'}$ and $v_j \in W_{j'}$ for some i' and j', $1 \leq i', j' \leq m$. Thus $rank(v_i) = i'$, $rank(v_j) = j'$, and both $G_{i'}$ and $G_{j'}$ are biconnected. There are the following three cases.

Case 1: $i' = j'$. In this case, $G_{i'}$ has edges (u, v_j) and (v_i, u), and all vertices in $W_{i'}$ appear on $C_o(G_{i'})$. Therefore, vertex u and the vertices in $W_{i'}$ from v_j to v_i form a cycle in $G_{i'}$, and $G_{i'}$ has at least one vertex in the proper inside of the cycle. None of the edges of G in the outside of the cycle is incident to any vertex on the cycle other than u, v_j and v_i. Hence the removal of three vertices u, v_j and v_i from G results in a disconnected graph, contrary to the 4-connectivity of G.

Case 2: $i' < j'$. Since $rl(v_i) = (v_i, u)$, v_i precedes u on $C_o(G_{i'})$. Since $ll(v_j) = (v_j, u)$, v_j succeeds u on $C_o(G_{j'})$. Since $G_{i'}$ is a subgraph of $G_{j'}$, v_i must precede v_j in $N(u)$, contrary to the assumption $j < i$.

Case 3: $i' > j'$. Similar to Case 2 above. $Q.\mathcal{E}.\mathcal{D}.$

Lemma 5 immediately implies the following lemma.

Lemma 6. P_{rl} and P_{ll} may cross at a vertex u, but do not share a vertex u without crossing at u.

From the definitions of a left leg and a right leg one can immediately have the following lemma.

Lemma 7. Let $1 \leq k \leq m$ and $u \in W_k$. Then u is on $C_o(G_k)$. Let u' be the succeeding vertex of u on $C_o(G_k)$. Assume that the ordered set $N(u)$ starts with u'. Let $ll(u) = (u, v')$ and $rl(u) = (u, v'')$. Then v'' precedes v' in $N(u)$.

We then have the following lemma.

Lemma 8. Each of the two pairs of paths, P_{rl} and P_{ll}, P_{lh} and P_{rh}, are internally disjoint.

Proof. We prove only that P_{rl} and P_{ll} are internally disjoint. Proof for the other case is similar. Suppose for a contradiction that P_{rl} and

P_{ll} share an intermediate vertex. Let w be the intermediate vertex that is shared by P_{rl} and P_{ll} and appear last on the path P_{rl} going from r to v. Then P_{rl} and P_{ll} cross at w by Lemma 6. However, the claim in Lemma 7 holds both for $k = rank(v)$ and $u = v$ and for $k = rank(w)$ and $u = w$, and hence P_{rl} and P_{ll} do not cross at w, a contradiction. $Q.E.D.$

By Lemmas 3, 4 and 8 we have the following lemma.

Lemma 9. T_{rl}, T_{ll}, T_{lh} and T_{rh} are four independent spanning trees of G rooted at r.

Clearly the running time of Algorithm FourTrees is $O(n)$. Thus we have the following theorem.

Theorem 1. Four independent spanning trees of any 4-connected plane graph rooted at any designated vertex can be found in linear time.

4   Conclusion

In this paper we give a linear-time algorithm to find four independent spanning trees of a 4-connected planar graph rooted at any designated vertex. Using four independent spanning trees, one can efficiently solve the 4-path query problem for 4-connected planar graphs.

It is remained as future work to find a linear-time algorithm for a larger class of graphs, say 4-connected graphs which are not always planar.

References

[BI96]   F. Bao and Y. Igarashi, *Reliable broadcasting in product networks with Byzantine faults*, Proc. 26th Annual International Symposium on Fault-Tolelant Computing (FTCS'96) (1996) 262-271.

[BTV96] G. Di Battista, R. Tamassia and L.Vismara, *Output-sensitive reporting of disjoint paths*, Technical Report CS-96-25, Department of Computer Science, Brown University (1996).

[CK93]   M. Chrobak and G. Kant, *Convex grid drawings of 3-connected planar graphs*, Technical Report RUU-CS-93-45, Department of Computer Science, Utrecht University (1993).

[CM88]   J. Cheriyan and S. N. Maheshwari, *Finding nonseparating induced cycles and independent spanning trees in 3-connected graphs*, J. Algorithms, 9 (1988) 507-537.

[DHSS84] D. Dolev, J. Y. Halpern, B. Simons, and R. Strong, *A new look at fault tolerant network routing*, Proc. 16th Annual ACM Symposium on Theory of Computing (1984) 526-535.

[E79]   S. Even, *Graph Algorithms*, Computer Science Press, Potomac (1979).

[H94]   A. Huck, *Independent trees in Graphs*, Graphs and Combinatorics, 10 (1994) 29-45.

[IR88]   A. Itai and M. Rodeh, *The multi-tree approach to reliability in distributed networks*, Information and Computation, 79 (1988) 43-59.

[KH94]   G. Kant and X. He, *Two algorithms for finding rectangular duals of planar graphs*, Proc. 19th Workshop on Graph-Theoretic Concepts in Computer Science (WG'93), LNCS 790 (1994) 396-410.

[KS92]   S. Khuller and B. Schieber, *On independent spanning trees*, Information Processing Letters, 42 (1992) 321-323.

[NRN97]   S. Nakano, M. S. Rahman and T. Nishizeki, *A linear time algorithm for four-partitioning four-connected planar graphs*, Information Processing Letters, 62 (1997) 315-322.

[OIBI96] K. Obokata, Y. Iwasaki, F. Bao and Y. Igarashi, *Independent spanning trees of product graphs and their construction*, Proc. 22nd Workshop on Graph-Theoretic Concepts in Computer Science (WG'96), LNCS 1197 (1996) 338-351.

[ZI89]   A. Zehavi and A. Itai, *Three tree-paths*, J. Graph Theory, 13 (1989) 175-188.

Linear Algorithms for a k-partition Problem of Planar Graphs without Specifying Bases

Koichi Wada[1] and Wei Chen[1]

Nagoya Institute of Technology
Gokiso-cho, Syowa-ku, Nagoya 466, JAPAN
e-mail:(wada|chen)@elcom.nitech.ac.jp

Abstract. This paper describes linear algorithms for partitioning a planar graph into k edge-disjoint connected subgraphs, each of which has a specified number of vertices and edges. If $\ell(\leq k)$ subgraphs contain the specified elements(called bases), we call this problem the k-partition problem with ℓ-base (denoted by k-PART-B(ℓ)). In this paper, we obtain the following results: (1)for any $k \geq 2$, k-PART-B(1) can be solved in $O(|E|)$ time for every 4-edge-connected planar graph $G = (V, E)$, (2)3-PART-B(1) can be solved in $O(|E|)$ time for every 2-edge-connected planar graph $G = (V, E)$ and (3)5-PART-B(1) can be solved in $O(|E|)$ time for every 3-edge-connected planar graph $G = (V, E)$.

1 Introduction

In this paper, we consider the following k-partition problem.
Input:
(1) an undirected graph $G = (V, E)$ with $n = |V|$ vertices and $m = |E|$ edges;
(2) $S \subseteq (V \cup E)(|S| \geq k)$;
(3) ℓ distinct vertices and/or edges $a_i(1 \leq i \leq \ell \leq k) \in S$; and
(4) k natural numbers $n_1, n_2, \ldots, n_k$ such that $\sum_{i=1}^{k} n_i = |S|$.
Output:
a partition $S_1 \cup S_2 \cup \ldots \cup S_k$ of the specified set S such that for each $i(1 \leq i \leq k)$
(a) $a_i \in S_i(1 \leq i \leq \ell)$;
(b) $|S_i| = n_i$; and
(c) there is a connected subgraph $G_i = (V_i, E_i)$ of G such that $S_i \subseteq (V_i \cup E_i)$ and $G_1, G_2, \ldots,$ and G_k are mutually edge-disjoint.
The problem is called the k-partition problem with respect to edge-disjointness and it is simply called the k-partition problem unless confusion arises. Each a_i is called a base of the subgraph G_i. If $\ell(\leq k)$ bases are specified for the k-partition problem, that is, $a_i(1 \leq i \leq \ell)$ must be in G_i, the problem is called the k-partition problem with ℓ-base.

It has been shown that the k-partition problem with k-base has a solution for every k-edge-connected graph and the edge-connectivity k is necessary to solve the k-partition problem [3, 7, 10]. It is an interesting graph-theoretic question to reveal relation between the number of partitions for these problems and the edge-connectivity of input graphs for the cases that some bases are not specified. For the k-partition problem with one-base, the following results have been obtained [11].

1. For any $k \geq 2$, the k-partition problem with one-base can be solved in $O(|V|\sqrt{|V|\log_2|V|} + |E|)$ time for every 4-edge-connected graph $G = (V, E)$.
2. The tripartition problem with one-base can be solved in $O(|V|^2)$ time for every 2-edge-connected graph $G = (V, E)$.
3. The 4-partition problem with one-base can be solved in $O(|E|^2)$ time for every 3-edge-connected graph $G = (V, E)$.

In this paper, we show that if the input graph is planar, all the k-partition problems stated above can be solved in linear time. In order to get these results, we use the similar methods in [11]. We can reduce the time complexity to linear time by several properties of planar graphs. Furthermore, we derive a new relation between the number of partitions and the edge connectivity of the input graph. We show that we can solve the 5-partition problem with one-base for every 3-edge-connected planar graph in linear time. Our algorithm uses "canonical ordering" known in the area of graph drawings [6].

2   Preliminaries

We deal with a connected undirected graph $G = (V, E)$ with a vertex set V and an edge set E. For a graph G, the vertex set is denoted by $V(G)$ and the edge set is denoted by $E(G)$. For a graph $G = (V, E)$ and a vertex subset V', the *induced subgraph* is denoted by $G[V']$. For two graphs $G = (V, E)$ and $G' = (V', E')$, the graph $(V \cup V', E \cup E')$ is denoted by $G \cup G'$. For a graph $G = (V, E)$ and a set E' of edges, the graph $(V, E - E')$ is denoted by $G - E'$, and if $E' = \{e\}$ then it is denoted by $G - e$.

A *cut-vertex* of G is a vertex whose removal disconnects G. A *bridge* of G is an edge whose removal disconnects G. A biconnected component of G is a maximal set of edges such that any two edges in the set lie on a common cycle. A *block* of G is a bridge or a biconnected component of G. An *Eulerian cycle* of a connected graph G is a cycle that traverses each edge of G exactly once, although it may visit a vertex more than once. A *Hamiltonian cycle* of a graph G is a cycle which contains all the vertices of G. A graph G is *k-connected(k-edge-connected)* if there exist k internally vertex-disjoint(edge-disjoint) paths between every pair of distinct nodes in G. Usually 2-connected graphs are called *biconnected graphs* and 3-connected graphs are called *triconnected graphs*.

A graph is *planar* if it can be embedded in the plane so that no two edges intersect except at a endvertex in common. A *plane* graph is a planar graph with a fixed embedding.

In order to reduce the partition problem for k-edge-connected graphs into the one for k-connected graphs, we used some transformation from a k-edge-connected graph G to a k-connected graph $\psi_k(G)$ [11]. If $k = 2, 3$ then the transformation $\psi_k(G)$ preserves the planarity. Thus, using the result in [11], we can reduce the partition problem for k-edge-connected planar graphs to the same problem for k-connected planar graphs. However, the transformation $\psi_4(G)$ does not preserve the planarity. In Section 3, we will introduce another transformation from 4-edge-connected graphs to 4-connected graphs preserving planarity.

3   The k-partition for 4-edge-connected graphs

Theorem 1. [11] *Let $k \geq 2$. If $G = (V, E)$ has an Eulerian cycle as a spanning subgraph, the k-partition problem with one-base can be solved in $O(T_{ec}(G) + |E|)$ time, where $T_{ec}(G)$ is a computation time to find a spanning Eulerian cycle in G.*

It is known that if G is 4-edge-connected, then G has a spanning Eulerian cycle. This spanning Eulerian cycle can be computed in $O(|V|\sqrt{|V|}\log_2|V| + |E|)$ time [2]. Therefore, the k-partition problem with one-base can be solved for every 4-edge-connected graph in $O(|V|\sqrt{|V|}\log_2|V| + |E|)$ time from Theorem 1. In the following, we show that if the input graph is 4-edge-connected and planar, the k-partition problem with one-base can be solved in linear time.

Given a 4-edge-connected plane graph $G = (V, E)$, we transform it to a 4-connected graph $\pi(G) = (V_\pi, E_\pi)$ preserving the planarity. Intuitively, $\pi(G)$ is obtained from G so that we add every vertex v of G with two concentric circles and we introduce new vertices corresponding to the crosspoints between edges and the concentric circles and new edges around the circles and between the inner and the outer circle. Figure 1 shows an example of $\pi(G)$. Clearly, $\pi(G)$ is

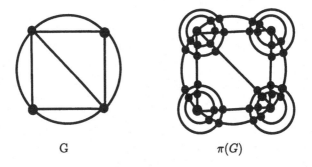

G $\pi(G)$

Fig. 1. An example of $\pi(G)$.

a plane graph with $|V_\pi| = |V| + 4|E|$ and $|E_\pi| = 9|E|$. From the construction of $\pi(G)$, we can obtain the following lemma. The proof can be done to show that there are 4 vertex-disjoint paths between any pair of vertices in $\pi(G)$ by using 4 edge-disjoint paths between the corresponding pair of vertices in G [9].

Lemma 1. *If G is 4-edge-connected planar , then $\pi(G)$ is 4-connected planar.*

Since every 4-connected planar graph has a Hamiltonian cycle [1], we can derive a spanning Eulerian cycle in G using the Hamiltonian cycle in $\pi(G)$. Since the Hamiltonian cycle is computed in linear time [1] and the derivation can be obtained in linear time, we have the following.

Theorem 2. *Let $k \geq 2$. If $G = (V, E)$ is 4-edge-connected planar, the k-partition problem with one-base can be solved in $O(|E|)$ time.*

4   The Tripartition for 2-edge-connected graphs

In this section, we prove that we can solve the tripartition problem with one-base for every 2-edge-connected planar graph in linear time. Since the transformation $\psi_2(G)$ [11] preserves the planarity, it is sufficient to solve the tripartition problem with one-base for every biconnected planar graph.

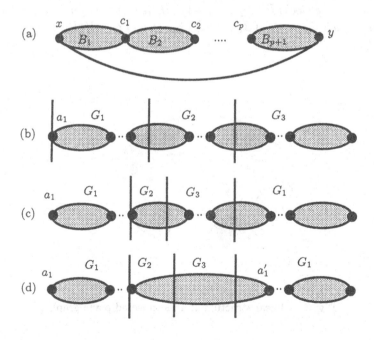

Fig. 2. The idea of the tripartition algorithm.

The tripartition algorithm for a biconnected graph is based on the following idea [11]. The algorithm makes use of a linear structure of blocks shown in

Figure 2(a). The algorithm consists of three cases illustrated in Figure 2(b)-(d). Since each block B_i can be bipartitioned into two connected graphs B_{i1} and B_{i2} such that $c_i \in B_{i1}$ and $c_{i+1} \in B_{i2}$ [10], if there are no cases that some block is partitioned into three pieces shown in Figure 2(b) and (c), the desired tripartition can be done. For the case shown in Figure 2(d), the algorithm is called for the block recursively. In this algorithm, one vertex a_1 can be specified to be contained in one subgraph G_1. This enables us to call it recursively.

Therefore, it is sufficient to compute a linear structure in each block and the size of these blocks(the number of specified elements in the blocks). For the general case, such a linear structure can be computed in $O(n)$ time. However, since such linear structures have to be recomputed at every recursive call in the algorithm and the recursive calls have to be done $\Omega(n)$ times in the worst case, it takes $O(n^2)$ time. On the other hand, we show that we can compute a linear structure in block and the size of each block for biconnected planar graphs in linear time using the next lemma. Therefore, the tripartition problem for biconnected planar graphs can be solved in linear time.

Lemma 2. *Let G be a biconnected plane graph with the outerface F. Let (x,y) be an arbitrary edge on F and let $G' = G-(x,y)$. Then G' is biconnected, or G' can be divided into the blocks $B_i(1 \leq i \leq p+1)$, where $c_i(1 \leq i \leq p)$ is the cutvertices of G' on F such that $x \in V(B_1)$, $y \in V(B_{p+1})$ and $c_i = V(B_i) \cap V(B_{i+1})$ (See Figure 3).*

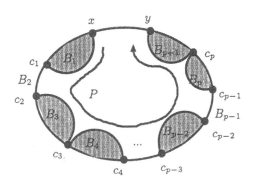

Fig. 3. A linear structure of a biconnected plane graph.

We can determine the linear structure of blocks shown in Lemma 2 and also the linear structures of blocks for all B_i's recursively, and count the size of each block in linear time as follows.

Let P be the path from x to y in the outerface F' of G' which is newly appears on F'. Traversing P, we can find (a) the cutvertices $c_i(1 \leq i \leq p)$, which appear on both F and F', (b) the outerface F_i of B_i and (c) the size of F_i.

We can do this traversal for each block B_i recursively. Thus, we can find the linear structures and count the size of each block. Since every edge appears on a boundary of an outerface at most twice, the number of edges traversed during the computation is at most $2|E|$ in total.

We can implement this algorithm to attain the required time using doubly linked adjacency list as a data structure for a plane graph [8].

Theorem 3. *Let $G = (V, E)$ be a biconnected planer graph. The tripartition problem with one-base can be solved in $O(|E|)$ time.*

We have the following theorem from Theorem 3 and the transformation $\psi_2(G)$.

Theorem 4. *The tripartition problem with one-base for 2-edge-connected planar graph $G = (V, E)$ can be solved in $O(|E|)$ time.*

Remark: We can not 4-partition 2-edge-connected planar graphs in general. There is a 2-edge-connected planar graph(in fact a biconnected planar graph) which cannot be 4-partitioned with one-base. Figure 4 shows an example of an instance which cannot be 4-partitioned under the assumption. The next section will show some sufficient condition for some class of 2-edge-connected planar graph which can be 4-partitioned with one-base.

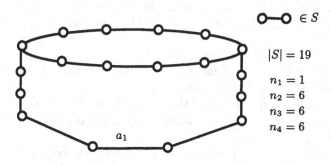

Fig. 4. An example of a biconnected plane graph which cannot be 4-partitioned.

5   The 5-partition for 3-edge-connected graphs

This section shows that the 5-partition problem with one-base has a solution for every 3-edge-connected planar graph and it can be obtained in linear time. Same as the 3-partition problem in the preceding section, it is sufficient to solve the 5-partition problem for triconnected planar graphs by considering the transformation.

The 4-partition problem with one-base for triconnected graphs can be solved with the nonseparating ear decomposition for triconnected graphs and the tripartition with one-base for biconnected graphs [11]. The algorithm first bipartition the input graph G into a biconnected graph G' and a connected graph with the base a_1, which are connected with a path P. This can be done by using the nonseparating ear decomposition. We can adjust the number of elements in the graph with a_1 using P. Then it can tripartition G' with the path by using the tripartition algorithm with one-base for biconnected graphs.

Since the nonseparating ear decomposition of triconnected planar graphs can be computed in linear time [5], the 4-partition problem with one-base for triconnected planar graphs can be solved in linear time. In this section, we show that we can solve the 5-partition problem with one-base for triconnected planar graphs in linear time by using *canonical ordering* which is a special case of nonseparating ear decomposition and is useful for some graph drawing [6].

5.1   Canonical Ordering

Let $G = (V, E)$ be a triconnected plane graph with a vertex v_1 on the outerface. Let $\pi = (V_1, \ldots, V_K)$ be an ordered partition of V, that is, $V_1 \cup \cdots \cup V_K = V$ and $V_i \cap V_j = \phi$ for $i \neq j$. Define G_k to be the subgraph of G induced by $V_1 \cup \cdots \cup V_k$ and denote by C_k the outerface of G_k. π is said to be a *canonical ordering* of G if:

1. V_1 consists of $\{v_1, v_2\}$, where v_2 lies on the outerface and $(v_1, v_2) \in E$.
2. V_K is a singleton $\{v_n\}$, where v_n lies on the outerface, $(v_1, v_2) \in E$ and $v_n \neq v_2$.
3. Each $C_k(k > 1)$ is a cycle containing (v_1, v_2).
4. Each G_k is biconnected and internally triconnected, that is, removing two interior vertices of G_k does not disconnect it.
5. For each k in $2, \ldots, K - 1$, one of the following condition holds:
   V_k is a singleton, $\{z_1\}$, where z_1 belongs to C_k and has at least one neighbor in $G - G_k$.
   V_k is a chain, $\{z_1, \ldots, z_\ell\}$, where each z_i has one neighbor in $G - G_k$, and where z_1 and z_ℓ each have one neighbor on C_{k-1}, and these are the only two neighbors of V_k in G_{k-1}.

Proposition 1. [6] *Every triconnected planar graph G with predefined vertex v_1 on the outerface has a canonical ordering. And it can be computed in linear time and space.*

For $k(1 \leq k < K)$, define P_k as the path $(z_s, z_1, \ldots, z_\ell, z_t)$, where $z_i(1 \leq i \leq \ell) \in V_k$ and z_s and z_t are the only two neighbors of V_k in G_{k-1}. The canonical ordering has the following properties. The proof is omitted here and is shown in the final paper [9].

Lemma 3. *The canonical ordering for every triconnected planar graph has the following properties:*

1. $G - G_k$ is connected for any $k(1 \leq k \leq K - 1)$.
2. P_k contains at most one edge on the outerface of G and it must be (z_s, z_1) or (z_ℓ, z_t) if it exists.

Since canonical ordering is a special case of nonseparating ear decomposition, we can use the similar idea in [11]. The properties in Lemma 3 enable us to 5-partition triconnected planar graphs.

5.2   The 4-partition with one-base and the tripartition with two-base

In this subsection, we solve the 4-partition problem with one-base and the tripartition problem with two-base for *almost triconnected* planar graphs. Not only this algorithm will be used as subroutines but also the idea of this algorithm will be used repeatedly in the next subsections

A graph G' is said to be a *subdivision* of a graph G if G' is obtained from G by replacing some of edges by paths having at most their endvertices in common. In the subdivision, if an edge is replaced by a path, this edge is said to *be subdivided*. First we give an algorithm for the 4-partition problem with one-base.

Let n_1, n_2, n_3 and n_4 be the numbers of elements specified in the 4-partition problem for G'. Let S be the set of partitioned elements in G'. Let a_1 be the base contained in the subgraph with n_1 elements and assume that $a_1 \in V$ and it is located on the outerface of G. [1]

G' is 4-partitioned by using a canonical ordering of G as follows:

Let v_1 be a neighbor of a_1 and also belong to the outerface. Let $\pi = (V_1, \ldots, V_K)$ be a canonical ordering of G with the vertex v_1, where $V_K = \{a_1\}$.

By using the canonical ordering of G, the similar structure for G' can be constructed in which for each path $P_k = (z_s, z_1, \ldots, z_\ell, z_t)$ of V_k in G either (z_s, z_1) or (z_ℓ, z_t) may be subdivided in G' from Lemma 3(2). If the path contains the subdivision of an edge, without loss of generality, this edge is assumed to be (z_ℓ, z_t). This structure for G' is also called the canonical ordering of G'.

There are two cases: Let $P(a_1, v_1)$ be the path between a_1 and v_1 on the outerface. The first case is that the two edges adjacent to a_1 and on the outerface are subdivided and the number of elements of $P(a_1, v_1)$ in S is at least n_1. Otherwise is the second one.

For the first case, $P(a_1, v_1)$ can contain the first subgraph, say S_1, of the solution containing the base a_1. Since $G' - E(S_1)$ consists of a biconnected plane graph and the path $P(a_1, v_1) - E(S_1)$ with the endvertex v_1, $G' - E(S_1)$ can be tripartitioned. Because $G' - E(P(a_1, v_1))$ can be tripartitioned with the base v_1.

The second case utilizes the canonical ordering of G'. Let k be the largest number i such that the number of elements of $G' - G_{i-1}$ in S is at least n_1. That is, the first subgraph S_1 of the solution containing the base a_1 consists of $G - G_k$ and the subpath of P_k (See Figure 5). Thus, G' is partitioned as illustrated in Figure 5 . Since G_{k-1} is biconnected, it can be tripartitioned with the base z_t.

---

[1] Although the base is assumed to be a vertex, it may be an edge.

Therefore the desired partition can be obtained for G_{k-1} and the path from z_p to z_t.

Since the canonical ordering of G' can be computed in linear time, we can get the following theorem.

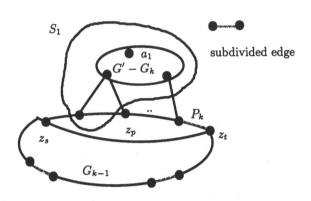

Fig. 5. The partition of G'.

Theorem 5. *Let* $G' = (V', E')$ *be a subdivision of a triconnected plane graph* $G = (V, E)$ *in which only the edges on the outerface of* G *are subdivided. If the base is in* G *and is located on the outerface of* G*, the 4-partition problem with one-base for* G' *can be solved in* $O(|E'|)$.

We can similarly treat the tripartition problem with two-base (See [9]).

Theorem 6. *Let* $G' = (V', E')$ *be a subdivision of a triconnected plane graph* $G = (V, E)$ *in which only the edges on the outerface of* G *are subdivided. If the bases are in* G *and are located on the outerface of* G*, the tripartition problem with two-base for* G' *can be solved in* $O(|E'|)$.

5.3   The 4-partition for internally triconnected graphs

In this subsection, using the result in the preceding subsection and the triconnected decomposition of biconnected graphs, we solve the 4-partition problem with one-base for internally triconnected and biconnected planar graphs which appear in canonical orderings of triconnected plane graphs.

The *triconnected components* of a biconnected graph G are defined as follows: If G is triconnected itself it is the unique triconnected component. Otherwise let (x, y) be a separation pair of G. G is partitioned into two subgraphs G_1 and G_2, which have only vertices x and y in common. The decomposition process is continued recursively on $G'_1 = G_1 \cup (x, y)$ and $G'_2 = G_2 \cup (x, y)$ until no decomposition is possible. The added edges are called *virtual edges*. The resulting graphs are each either a triconnected simple graph, or a set of three multiple edges(called triple bond) or a cycle of length 3 (triangle). The triconnected components of G are obtained from such graphs by merging the triple bonds into maximal sets of multiple edges, and the triangles into maximal simple cycles. The triconnected components are uniquely determined and they are obtained in linear time [4].

Here we describe the outline of the algorithm and the detail will be shown in the final paper[9].

Let $\pi = (V_1, \ldots, V_K)$ be a canonical ordering of a triconnected plane graph. Let $G = (V, E)$ be a graph induced by $V_1 \cup \cdots \cup V_k$ for some $k (1 \leq k \leq K - 1)$. Since G is an internally triconnected and biconnected plane graph, all vertices of the separation pairs belong to the outerface of G. Moreover, there are no vertices with degree 2 inside G because G is obtained from a triconnected graph. Therefore, the triconnected components of G is generally shown as Figure 6.

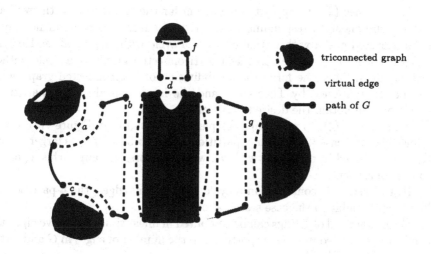

Fig. 6. A triconnected decomposition of an internally triconnected and biconnected plane graph.

In Figure 6, the shaded areas represent triconnected graphs and the dotted lines represent virtual edges. Here virtual edges corresponding to paths in G are replaced by the paths. Note that these paths appear only on the outerface of G. We can observe that in each triconnected component every virtual edge is located on the outerface of the component because G is internally triconnected.

In order to 4-partition G with one-base, it is sufficient to 4-partition triconnected graphs with virtual edges located on the outerface and cycles with virtual edges.

Let n_1, n_2, n_3 and n_4 be the numbers of elements specified in the 4-partition problem for G. Let S be the set of partitioned elements in G. Let a_1 be the base contained in the subgraph with n_1 elements and assume that a_1 is located on the outerface of G.

If the base a_1 is contained in a triconnected graph with virtual edges, say G', we 4-partition G' as follows:

Since the virtual edges are located only on the outerface, we can use the same method mentioned in the preceding subsection by considering virtual edges as subdivisions of edges. Assume that the partition proceeds as shown in Figure 7(a) according to a canonical ordering of G'. From Lemma 3(2) the virtual edge must be (z_ℓ, z_t) if it exists.

Unless $P_k = (z_s, \ldots, z_t)$ contains the virtual edge, we have done the desired 4-partition. Otherwise, there are three cases:(1) the virtual edge is bipartitioned, (2) the virtual edge is tripartitioned and (3) the virtual edge is 4-partitioned (Figure 7(b)).

We can treat the case (1) easily because we can bipartition each component corresponding to the virtual edge with two bases.

For the case (2), we apply the algorithm for the tripartition with two bases to the component corresponding to the virtual edge. This component may be a triconnected graph with virtual edges or a cycle with virtual edges. However, repeating this process, we can reach a triconnected graph or a cycle without virtual edges. Since the former is a subdivision of a triconnected graph which can be tripartitioned by Theorem 6 and the latter can also be tripartitioned easily, we can obtain the desired partition.

For the case (3), as we change the partition as shown in Figure 7(c), we can reduce it to the case (2) and the bipartition with two-base. If n_1 is larger, that is, G_{k-1} is included in the graph with a_1, we can recursively apply this algorithm to the component.

If the base a_1 is contained in a cycle with virtual edges, the 4-partition can be treated similar to the case stated above [9].

Since canonical orderings can be computed in linear time, the above algorithm can be implemented in time proportional to the number of edges in G and virtual edges, which is $O(|E|)$.

Theorem 7. *Let $\pi = (V_1, \ldots, V_K)$ be a canonical ordering of a triconnected plane graph. Let $G = (V, E)$ be a graph induced by $V_1 \cup \cdots \cup V_k$ for some $k(1 \leq k \leq K - 1)$. If the base is located on the outerface of G, the 4-partition problem with one-base for G can be solved in $O(|E|)$ time.*

5.4   The 5-partition for triconnected graphs

We are ready to solve the 5-partition problem with one-base for triconnected planar graphs.

Let $G = (V, E)$ be a triconnected plane graph. Let n_1, n_2, n_3, n_4 and n_5 be the numbers of elements specified in the 5-partition problem with one-base for G. Let a_1 be the base contained in the subgraph with n_1 elements. Without loss

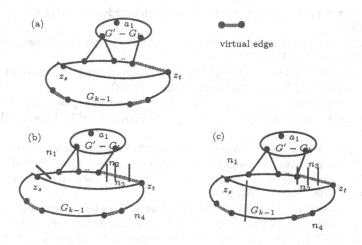

Fig. 7. The partition of the triconnected graph with virtual edges

of generality, G is assumed to be embedded in the plane such that a_1 is located on the outerface. Using the method described in Section 5.2 for the canonical ordering of G and Theorem 7, we have the following theorems.

Theorem 8. *Let $G = (V, E)$ be a triconnected planar graph. The 5-partition problem with one-base for G can be solved in $O(|E|)$ time.*

Theorem 9. *Let $G = (V, E)$ be a 3-edge-connected planar graph. The 5-partition problem with one-base for G can be solved in $O(|E|)$ time.*

Acknowledgement This research is supported in part by the Grant-in-Aid of Scientific Research (C)(2) of the Ministry of Education, Science and Culture of Japan under Grant: 10680352.

References

1. N. Chiba and T. Nishizeki: "The Hamiltonian cycle problem is linear-time solvable for 4-connected planar graphs," *Journal of Algorithms*, 10, 187–211 (1989).
2. H.N.Gabow and H.H.Westermann: "Forests frames and games: algorithms for matroid sums and applications," Algorithmica, 7, 5/6, 465–497 (1992).
3. E. Győri: "On division of connected subgraphs," in: Combinatorics(Proc. 5th Hungarian Combinational Coll., 1976, Keszthely) North-Holland, Amsterdam, 485–494 (1978).
4. J.E.Hopcroft and R.E.Tarjan: "Dividing a graph into triconnected components," SIAM J. Computing, 2, 3, 135-158 (1973).
5. L.Jou, H.Suzuki and T.Nishizeki: "A linear algorithm for finding a nonseparating ear decomposition of triconnected planar graphs," Tech. Rep. of Information Processing Society of Japan, AL40-3 (1994).

6. G. Kant: "Drawing planar graphs using the canonical ordering," *Algorithmica*, 16, 1, 4-32 (1996).
7. L. Lovász: "A homology theory for spanning trees of a graph," *Acta math. Acad. Sci. Hunger*, 30, 241–251 (1977).
8. T. Nishizeki and N. Chiba: Planar graphs: Theory and algorithms, Annals of Discrete Mathematics 32 Monograph, North-Holland (1988).
9. K.Wada and W.Chen: "Linear algorithms for a k-partition problem of planar graphs without specifying bases," Tech. Rep. of Wada–Lab. of ECE, NIT, TR98-01(1998).
10. K.Wada and K.Kawaguchi: "Efficient algorithms for tripartitioning triconnected graphs and 3-edge-connected graphs, Proc. of the 19th International Workshop on Graph-Theoretic Concepts in Computer Science(WG'93), Lecture Notes in Computer Science, 790, 132-143(1994).
11. K.Wada A.Takaki and K.Kawaguchi: "Efficient algorithms for a mixed k-partition problem of graphs without specifying bases," Proc. of the 20th International Workshop on Graph-Theoretic Concepts in Computer Science(WG'94), Lecture Notes in Computer Science, 903, 319–330(1994), also to appear in Theoretical Computer Science.

Domination and Steiner Tree Problems on Graphs with Few P_4s

Luitpold Babel[1] and Stephan Olariu[2]*

[1] Zentrum Mathematik
Technische Universität München
D-80290 München, Germany
babel@mathematik.tu-muenchen.de
[2] Department of Computer Science
Old Dominion University
Norfolk, VA 23529, U.S.A.
olariu@cs.odu.edu

Abstract. The contribution of this work is to show that the recently-proposed primeval and homogeneous decompositions of graphs can be used to solve efficiently various types of weighted domination and Steiner tree problems. Furthermore, we point out that these results imply linear-time algorithms for large classes of graphs which, in some local sense, contain only a small number of induced P_4s.

1 Introduction

A set D of vertices of a graph $G = (V, E)$ is a *dominating set* if every vertex in $V \setminus D$ has a neighbor in D. The *minimum dominating set* problem asks to determine a dominating set of smallest cardinality. In the weighted version, each vertex v of the graph is assigned a nonnegative weight $c(v)$ and the problem is to find a dominating set of smallest total weight.

In many applications (see e.g. [8, 9]) dominating sets are subject to additional constraints. In particular, one is frequently interested in dominating sets which are either *independent* or *cliques* or induce *connected* subgraphs. In our paper, the parameters $\alpha(G)$, $\beta(G)$, and $\gamma(G)$ denote, respectively, the *independent domination number*, the *dominating clique number*, and the *connected domination number*, that is, the smallest weight of an independent, complete, and connected dominating set in G.

The *Steiner tree problem* bears some similarity to the minimum connected dominating set problem. Given a set T of *target vertices* in a graph $G = (V, E)$, we are interested in finding a smallest set $S \subseteq V \setminus T$ of *Steiner vertices* such that $S \cup T$ induces a connected subgraph of G. Naturally, the weighted version of the problem asks for a set S of smallest total weight. Such a set S is usually called a *Steiner set*.

---

* This author was supported in part by NSF grant CCR-9522093 and by ONR grant
N00014-97-1-0526.

In this work we demonstrate that the above problems can be solved in a very efficient manner given the *homogeneous decomposition* of graphs. This new type of graph decomposition extends the well known *modular decomposition* [18]. As a direct application, we present linear-time algorithms for $(q, q-4)$-graphs, that is, graphs where no set of at most q vertices induces more than $q-4$ P_4s. These graphs properly contain, among others, all cographs, P_4-reducible and P_4-sparse graphs.

2 p-connected graphs and the homogeneous decomposition

Let $G = (V, E)$ be a finite graph with no loops nor multiple edges. In order to simplify our exposition, we shall often blur the distinction between sets of vertices and the subgraphs they induce, using the same notation for both. For a set $U \subseteq V$ we let $N(U)$ denote the set of vertices outside U that are adjacent to vertices in U. The complement of G is denoted by $\overline{G}$. A *clique* is a set of pairwise adjacent vertices, an *independent set* is a set of pairwise nonadjacent vertices. G is termed a *split* graph if its vertices can be partitioned into a clique and an independent set.

The P_4 is the chordless path with four vertices and three edges. In a P_4 consisting of vertices u, v, w, x and edges uv, vw and wx we refer to v and w as *midpoints* whereas u and x are called *endpoints*.

In [13] B. Jamison and S. Olariu introduced the notion of p-connectedness. Specifically, a graph is said to be *p-connected* if for every partition of its vertex set into two nonempty, disjoint, sets some P_4 in the graph has vertices from both sets in the partition. Such a P_4 is also called a *crossing P_4*. The *p-connected components* of a graph are the maximal induced subgraphs which are p-connected. It is easy to confirm that the p-connected components of a graph G are closed under complementation and are connected subgraphs of G and $\overline{G}$. Moreover, each graph has a unique partition into p-connected components.

A p-connected graph G is *separable* if there is a partition of its vertices into nonempty, disjoint, sets V_1 and V_2 such that every crossing P_4 between V_1 and V_2 has its midpoints in V_1 and its endpoints in V_2. We shall say that (V_1, V_2) is a *separation* of G. In order to make the paper self-contained we now review some important properties of p-connected graphs.

Theorem 1 ([13]). *Every separable p-connected graph has a unique separation. Furthermore, every vertex belongs to a crossing P_4 with respect to the separation.*

$\square$

A proper subset H of at least two vertices of a graph G is termed *homogeneous* if every vertex outside H is either adjacent to all the vertices in H or to none of them. A homogeneous set H is *maximal* if no homogeneous set properly contains H. The graph obtained from a p-connected graph G by shrinking every maximal homogeneous set to a single vertex is called the *characteristic graph* of G.

The structure of separable p-connected graphs is revealed very clearly in the following statement.

Theorem 2 ([13]). *Let G be separable p-connected with separation (V_1, V_2). The subgraph of G (respectively $\overline{G}$) induced by V_2 (respectively V_1) is disconnected. Furthermore, every component of the subgraph of G (respectively $\overline{G}$) induced by V_2 (respectively V_1) with at least two vertices is a homogeneous set in G.* □

This result immediately implies the following useful characterization of separable p-connected graphs.

Corollary 1 ([13]). *A p-connected graph is separable if and only if its characteristic graph is a split graph.* □

The introduction and study of p-connected and separable p-connected graphs is justified by the following general structure theorem.

Theorem 3 ((Structure Theorem) [13]). *For an arbitrary graph G, exactly one of the following conditions is satisfied:*

(1) G is disconnected;
(2) $\overline{G}$ is disconnected;
(3) There is a unique proper separable p-connected component H of G with separation (V_1, V_2) such that every vertex outside H is adjacent to all vertices in V_1 and to no vertex in V_2;
(4) G is p-connected. □

This theorem suggests, in a quite natural way, a tree representation for arbitrary graphs which is unique up to isomorphism. The tree associated with a graph G is called the *primeval tree* of G. The internal nodes of the primeval tree are labeled by integers $i \in \{0, 1, 2\}$, where an i-node indicates that the graph associated with the subtree rooted at this node is obtained from the graphs corresponding to its children by an ⓘ operation. The leaves of the tree are the p-connected components of G.

The operations ⓪ , ① and ② reflect the first three cases of the Structure Theorem. A graph G arises from disjoint graphs G_i, $(1 \leq i \leq k)$, by means of a ⓪ operation if G is the disjoint union of the graphs G_i (i.e. there are no edges between two different graphs G_i). If G arises by a ① operation then G is the disjoint sum of the graphs G_i (i.e. all edges between different graphs G_i are present). Finally, let G_1 be a separable p-connected graph with separation (V_1, V_2) and let G_2 be an arbitrary graph disjoint from G_1. The graph G arises from G_1 and G_2 by means of a ② operation if every vertex of G_2 is adjacent to all the vertices in V_1 and to no vertex in V_2. In the primeval tree, the subtree associated with G_1 becomes the left child, while the subtree associated with G_2 becomes the right child of a 2-node.

The *homogeneous decomposition* [13] involves, additionally, the homogeneous sets of the graph. Given the primeval tree, it constructs a new tree representation by – loosely speaking – replacing homogeneous sets by single vertices.

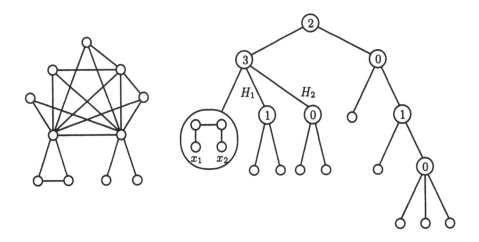

Fig. 1. *A graph and the associated homogeneous decomposition tree*

This substitution is reflected by a ③ operation. Let $G_0, H_1, \ldots, H_k$ be disjoint graphs and let $\{x_1, \ldots, x_k\}$ be a set of vertices of G_0. The graph G arises from $G_0, H_1, \ldots, H_k$ by means of a ③ operation if every vertex x_i in G_0 is replaced by the graph H_i in the obvious way. In the resulting decomposition tree, the leftmost child of a 3-node represents the characteristic graph, the other children are the subtrees which represent the maximal homogeneous sets as illustrated in Fig. 1.

Recently, linear-time algorithms have been proposed which construct the primeval and the homogeneous decomposition tree for arbitrary graphs (see [2]). Both these algorithms rely, in a crucial way, on known linear-time algorithms for the modular decomposition [6, 7, 17].

3   Independent dominating sets

Assume that we are given an arbitrary graph G along with its homogeneous decomposition tree $T(G)$. We first consider the case where G arises by a ⓪ operation, that is, G is the disjoint union of graphs $G_1, \ldots, G_k$. Clearly, any independent dominating set in G is the union of independent dominating sets in the subgraphs G_i. Hence, we write

$$\alpha(G) = \alpha(G_1) + \ldots + \alpha(G_k).$$

If G arises by a ① operation, that is, if G is the disjoint sum of graphs $G_1, \ldots, G_k$ then we obtain immediately

$$\alpha(G) = \min\{\alpha(G_1), \ldots, \alpha(G_k)\},$$

since an independent dominating set in G contains vertices from exactly one of the subgraphs G_i.

More generally, consider a homogeneous set H in G. If an independent dominating set contains no vertices from $N(H)$ then it must consist of an independent dominating set in H and an independent dominating set in $G \setminus H \setminus N(H)$. On the other hand, if it contains a vertex from $N(H)$ then it does not contain a vertex from H. This observation implies that

$$\alpha(G) = \alpha(G_H),$$

where G_H denotes the graph obtained from G by replacing H by a single vertex of weight $\alpha(H)$.

As a consequence, if G arises by a ③ operation involving homogeneous sets $H_1, \ldots, H_k$, then it suffices to solve the problem on the characteristic graph of G. Naturally, the weight of the vertex representing the homogeneous set H_i is $\alpha(H_i)$.

Assume, finally, that G arises by a ② operation. Then G consists of a separable p-connected component G_0 and a subgraph H_0 outside G_0 which is adjacent to G_0 as stipulated in the Structure Theorem. By Theorem 2, it follows that the characteristic graph of G is a split graph. If the weights of all vertices representing homogeneous sets are already known, then we can easily solve the problem for G. For that purpose, denote the vertices of the clique by $y_1, \ldots, y_r$ and the vertices of the independent set by $z_0, z_1, \ldots, z_s$. The vertex z_0 represents the set H_0 (which is homogeneous whenever it contains two or more vertices) and, by convention, belongs to the independent set. If an independent dominating set in the split graph contains no vertex from the clique then it must contain all vertices from the independent set. If an independent dominating set contains a vertex from the clique, say y_i, then it must contain all vertices from the independent set that are nonadjacent to y_i. This shows that

$$\alpha(G) = \min\{\sum_{j=0}^{s} c(z_j);\ \min_{1 \le i \le r}\{c(y_i) + \sum_{z_j \notin N(y_i)} c(z_j)\}\}.$$

These results suggest the following algorithm for computing $\alpha(G)$ for an arbitrary graph G.

Algorithm
    Input: an arbitrary graph G;
    Output: the independent domination number $\alpha(G)$ of G;
begin
    construct the homogeneous decomposition tree $T(G)$ of G;
    let r denote the root of $T(G)$;
    call INDEPENDENT-DOMINATION(T, r);
end.

Procedure INDEPENDENT-DOMINATION traverses the homogeneous decomposition tree $T(G)$ in a Depth-First manner, starting at the root. The details are spelled out as follows.

Procedure INDEPENDENT-DOMINATION(T, v);
begin
    Let G denote the graph represented by v;
    If v is a leaf of T **then**
        Compute $\alpha(G)$;
    If v is a 0-node **then**
        Let G_i denote the subgraphs of G represented
        by the children w_i, $(1 \le i \le k)$ of v;
        For every child w_i of v **do**
            INDEPENDENT-DOMINATION(T, w_i);
        Set $\alpha(G) = \sum_{i=1}^{k} \alpha(G_i)$;
    If v is a 1-node **then**
        Let G_i denote the subgraphs of G represented
        by the children w_i, $(1 \le i \le k)$ of v;
        For every child w_i of v **do**
            INDEPENDENT-DOMINATION(T, w_i);
        Set $\alpha(G) = \min_{1 \le i \le k} \alpha(G_i)$;
    If v is a 2-node **then**
        Let G_0 and H_0 denote the subgraphs of G represented
        by the children u and w_0 of v;
        Let further S and H_i denote the subgraphs of G_0 represented
        by the children w and w_i, $(1 \le i \le k)$ of u;
        For every child w_i of u and for w_0 **do**
            INDEPENDENT-DOMINATION(T, w_i);
        Let x_i denote the representing vertices from H_i in S;
        Define vertex-weights $c(x_i) = \alpha(H_i)$ $(i = 0, \ldots, k)$;
        Let $y_1, \ldots, y_r$ resp. $z_1, \ldots, z_s$ denote the vertices from
        the clique resp. the independent set from S and let $z_0 = x_0$;
        Set $\alpha(G) = \min\{\sum_{j=0}^{s} c(z_j); \ \min_{1 \le i \le r}\{c(y_i) + \sum_{z_j \notin N(y_i)} c(z_j)\}\}$;
    If v is a 3-node **then**
        Let G_0 and H_i denote the subgraphs of G represented
        by the children w and w_i, $(1 \le i \le k)$ of v;
        For every child w_i of v **do**
            INDEPENDENT-DOMINATION(T, w_i);
        Let x_i denote the representing vertices from H_i in G_0;
        Define vertex-weights $c(x_i) = \alpha(H_i)$ $(i = 1, \ldots, k)$;
        Compute $\alpha(G_0)$;
end.

The correctness of the algorithm follows immediately from our previous discussion. It is straightforward to modify the algorithm such that it returns not only the independent domination number but also the corresponding independent dominating set. Clearly, given the solutions for all subgraphs corresponding to the (nonseparable) leaves of the tree, the solution for the graph can be computed in linear time.

4   Connected dominating sets and dominating cliques

We now consider the problem of finding a smallest weight connected dominating set in an arbitrary graph G. If G arises by a ⓪ operation then no connected dominating set exists. This is indicated by writing

$$\gamma(G) = \infty.$$

Assume that G arises by a ① operation. A smallest weight connected dominating set in G consists either of a smallest weight connected dominating set in one of the subgraphs G_i or it consists of precisely two vertices from different subgraphs G_i and G_j. Therefore, we have

$$\gamma(G) = \min_{1 \le i < j \le k} \{\gamma(G_i); c_{min}(G_i) + c_{min}(G_j)\},$$

where $c_{min}(G_i)$ denotes the smallest weight of a vertex from G_i.

Now, assume that G arises by a ③ operation involving the homogeneous sets $H_1, \ldots, H_k$. Consider an arbitrary set H_i. Clearly, any connected dominating set in G must contain at least one vertex from $N(H_i)$. Therefore, a smallest weight connected dominating set contains at most one vertex from H_i and, if this is the case, this vertex has smallest weight in H_i. Hence, we can restrict the problem to the characteristic graph of G. Naturally, the vertex which represents H_i has weight $c_{min}(H_i)$.

The same idea applies if G arises by a ② operation. We can restrict our attention to the characteristic graph which is now a split graph.

It is easy to verify that the dominating clique problem can be treated in a quite analogous manner. In case of a ⓪ operation no dominating clique exists. In case of a ① operation we either choose a dominating clique in one of the subgraphs or two vertices from two different subgraphs. In case of a ② or a ③ operation we have to find a dominating clique in the characteristic graph. Hence, for the computation of $\beta(G)$ we can adopt the ideas used to determine $\gamma(G)$.

As seen in the previous section, the weighted independent dominating set problem can be solved very easily when restricted to split graphs. However, it is known that the connected dominating set problem is NP-complete for split graphs [19]. The latter result implies that the dominating clique problem is also NP-complete for split graphs: a smallest connected dominating set in a split graph is always a clique.

Hence, in the case of a 2-node, we do not obtain $\gamma(G)$ and $\beta(G)$ as easily as this was possible for $\alpha(G)$. Here, it is still an open problem to compute these parameters for the characteristic graph of G. Except for this case, the procedures CONNECTED-DOMINATION and DOMINATING-CLIQUE can be formulated quite similarly to procedure INDEPENDENT-DOMINATION. We only need to ensure that the weight of the vertex x_i representing the homogeneous set H_i is now $c_{min}(H_i)$. Furthermore, in every node of the tree we have to determine the smallest weight $c_{min}(G)$ of all vertices belonging to G.

5   Steiner sets

Let $G = (V, E)$ be an arbitrary graph and let $T \subset V$ be a set of target vertices. We assume again that the homogeneous decomposition tree $T(G)$ of G is available.

Consider first the case where G is the disjoint union of graphs $G_1, \ldots, G_k$. If there are two different subgraphs G_i and G_j which both contain vertices from T then, obviously, no Steiner set S exists. On the other hand, if all vertices of T belong to one of the subgraphs, say G_i, then S is completely contained in G_i. Hence, we can restrict the problem to G_i: a Steiner set for G_i is also a Steiner set for G.

If G is the disjoint sum of graphs $G_1, \ldots, G_k$ then we need to distinguish between two cases: If different subgraphs G_i and G_j contain vertices of T, then T induces a connected graph, and so S is empty. If all vertices from T belong to some subgraph G_i then either S is completely contained in G_i or S contains precisely one vertex, namely a vertex of smallest weight, outside of G_i. Hence, we solve the problem restricted to G_i and determine a vertex with smallest weight outside G_i. The Steiner set S is the one of the two resulting sets which has the smallest weight.

Now, let H be a homogeneous set in G. We consider three cases. First, assume that $T \subseteq H$ holds. In this case, S is completely contained in H or consists of one vertex of smallest weight in $N(H)$.

Next, assume that $T \cap H = \emptyset$. Now, S contains at most one vertex from H and, if this is the case, this vertex has smallest weight. Hence, we can restrict the problem to the graph obtained from G by replacing H by some vertex of smallest weight in H. Clearly, a Steiner set in the new graph is also a Steiner set in the original graph.

Finally, assume that $T \cap H \neq \emptyset$ and $T \nsubseteq H$. It is an easy observation that S contains no vertices from H (if no vertex from T belongs to $N(H)$ then at least one vertex from $N(H)$ must belong to S). This shows that, as before, it suffices to study the graph where H is replaced by a single vertex which represents the set $T \cap H$. A Steiner set in the latter graph is again a Steiner set in the original graph.

Hence, if G arises by a ② or a ③ operation with homogeneous sets $H_1, \ldots, H_k$ then we have to check whether $T \subseteq H_i$ holds for some $i \in \{1, \ldots, k\}$. In this case, we compute a Steiner set in the subgraph H_i and determine a set containing a single vertex of smallest weight from $N(H_i)$. By the previous arguments, S is the set with smaller weight. Otherwise, we have to compute a Steiner set in the characteristic graph of G where each homogeneous set H_i with $T \cap H_i = \emptyset$ is represented by one of its vertices of smallest weight. The vertices representing sets H_j with $T \cap H_j \neq \emptyset$ belong to the new set T' of target vertices, together with the vertices from T which belong to none of the homogeneous sets.

6 Applications to graphs with few P_4s

A large number of computational problems (see e.g. [3, 5, 11, 16]) suggest the study of graphs featuring certain "local density" properties. These properties are traditionally equated with the absence of induced paths of length three. In this context, Corneil and his co-workers introduced and investigated the class of *cographs* characterized by the absence of induced P_4s (see [3–5] for a discussion of structural and algorithmic results). The cographs have been extended by Jamison and Olariu to different graph classes which contain a restricted number of induced P_4s. The most familiar of these classes are the P_4-reducible, P_4-sparse and P_4-extendible graphs. In particular, P_4-*reducible graphs* [10] are graphs where no vertex belongs to more than one P_4. A graph is termed P_4-*sparse* [12] if no set of five vertices induces more than one P_4. Obviously, P_4-sparse graphs generalize both cographs and P_4-reducible graphs. P_4-*extendible graphs* [11] are graphs where each p-connected component consists of at most five vertices.

A *spider* is a split graph G consisting of a clique and an independent set of equal size at least two, such that either in G or in $\overline{G}$ each vertex of the independent set has precisely one neighbor in the clique and each vertex of the clique has precisely one neighbor in the independent set. In the first case G is called a *thin* spider, otherwise a *thick* spider. Clearly, the complement of a thin spider is a thick spider and vice versa.

In [1] we proposed to call a graph a (q, t)-*graph* if no set of at most q vertices induces more than t distinct P_4s. In this terminology, the cographs are precisely the $(4, 0)$-graphs and the P_4-sparse graphs coincide with the $(5, 1)$-graphs. Furthermore, it turns out that the C_5-free P_4-extendible graphs are exactly the $(6, 2)$-graphs. It has been shown in [1] that $(q, q-4)$-graphs can be recognized in linear time for every fixed value of q. The following theorem of [1] characterizes the p-connected components of $(q, q-4)$-graphs.

Theorem 4 ([1]). *A p-connected component of a $(q, q-4)$-graph is either isomorphic to a spider or it contains less than q vertices.*   □

It is an immediate consequence of this result that for a $(q, q-4)$-graph G the leaves of the primeval decomposition tree, and hence also for the homogeneous decomposition tree, represent spiders or graphs of restricted size. We shall now demonstrate how to solve domination problems for these simple graphs.

If G contains fewer than q vertices, for some fixed q, then the parameters $\alpha(G)$, $\beta(G)$ and $\gamma(G)$ can be determined in constant time. Assume that G is a spider. Denote the vertices of the clique by $y_1, \ldots, y_r$ and the vertices of the independent set by $z_1, \ldots, z_r$, such that y_i is adjacent (resp. nonadjacent) to z_i if G is a thin (resp. thick) spider. If G is a thin spider then an independent dominating set consists either of all vertices from the independent set or of one vertex from the clique together with all nonadjacent vertices from the independent set. Hence,

$$\alpha(G) = \min_{1 \le j \le r} \{c(y_j) + \sum_{k \ne j} c(z_k); \sum_{i=1}^{r} c(z_i)\}.$$

If G is a thick spider then an independent dominating set either contains all vertices from the independent set or one vertex from the clique together with the corresponding nonadjacent vertex from the independent set. This implies that

$$\alpha(G) = \min_{1 \le j \le r} \{c(y_j) + c(z_j); \sum_{i=1}^{r} c(z_i)\}.$$

If G is a thin spider then the clique is the only minimal connected dominating set. Hence

$$\beta(G) = \gamma(G) = \sum_{i=1}^{r} c(y_i).$$

If G is a thick spider then a minimal connected dominating set must consist of two vertices from the clique. Therefore, in this case,

$$\beta(G) = \gamma(G) = \min_{1 \le i < j \le r} \{c(y_i) + c(y_j)\}.$$

It is easy to see that all computations can be performed in time proportional to the number of vertices of the spider.

Since the values of the parameters belonging to the leaves of the primeval decomposition tree can be computed efficiently, we can efficiently compute the independent domination number of a $(q, q - 4)$-graph using the algorithm from Section 3. For the other two problems it remains to investigate the ② operation.

With Theorem 4 it is easy to see that, if G is the result of a ② operation, then the characteristic graph of G either has at most q vertices or is isomorphic to a spider with one additional vertex which is adjacent precisely to the clique. In the first case we can solve the problem in constant time, in the second case the parameters are determined in the same way as this has been done for spiders.

Altogether we obtain the following result.

Theorem 5. *Once the primeval decomposition tree $T(G)$ is available, the independent domination number, the connected domination number and the dominating clique number of a $(q, q - 4)$-graph $G = (V, E)$, for some fixed q, can be computed in $O(|V|)$ time.* □

We now consider the Steiner tree problem with target set T. First we have to show that the problem can be solved efficiently when restricted to the graphs corresponding to the leaves of the primeval decomposition tree.

Let G be such a graph. If G has less than q vertices then the problem can be solved in constant time. Assume that G is a thin spider. If $|T| = 1$ then, clearly, $S = \emptyset$. Therefore, assume that $|T| \geq 2$. If a vertex z_i from the independent set of the spider belongs to T and if y_i is not in T then y_i must belong to S. This fact suffices to construct the Steiner set S.

Now, let G be a thick spider and $|T| \geq 2$. If at least two vertices y_i and y_j of the clique belong to T then T induces a connected graph and $S = \emptyset$. If only one vertex y_i of the clique belongs to T then we have to consider two cases. If z_i is not in T then T is connected and $S = \emptyset$. If z_i belongs to T then S must contain one vertex from the clique with smallest weight. Finally, if no vertex from the clique belongs to T then again we have two cases. If all vertices from the independent set are in T then S consists of two vertices from the clique having smallest weights. Otherwise, write $T = \{z_1, \ldots, z_k\}$. Then S either contains only one vertex, namely a vertex of smallest weight from $\{y_{k+1}, \ldots, y_r\}$, or it contains two vertices of smallest weight from $\{y_1, \ldots, y_k\}$. These observations imply that the Steiner set can be computed in time linear in the number of vertices of the spider.

If G is the result of a ⓪ or a ① operation then we proceed as shown in Section 5. If G is the result of a ② operation then the characteristic graph of G either has at most q vertices or is isomorphic to a spider with one additional vertex which is adjacent precisely to the clique. In the first case we can solve the problem in constant time, in the second case we proceed analogously as above.

To summarize our findings we state the following result.

Theorem 6. *Once the primeval decomposition tree $T(G)$ is available, the Steiner tree problem for a $(q, q-4)$-graph $G = (V, E)$, for some fixed q, can be solved in $O(|V|)$ time.* □

7 Dominating and total dominating sets

A further problem which has attracted considerable attention in recent years is the *minimum total dominating set problem* (see e.g. [14, 15]). The task involves finding a dominating set which contains no isolated vertices. We denote by $\delta(G)$ and $\theta(G)$, respectively, the *domination number* and the *total domination number*, that is, the smallest weight of a dominating set and of a total dominating set in G.

The minimum dominating set and the minimum total dominating set problem resist an analogous treatment which, given the homogeneous decomposition tree, computes the solution in a simple bottom-up manner from the values of the leaves. Nevertheless, as we demonstrate next, we can also solve these problems in linear time for $(q, q-4)$-graphs (with fixed q) by applying only slightly more sophisticated techniques.

We assume that a $(q, q-4)$-graph G is given along with its primeval decomposition tree. If G corresponds to a leaf and if G has fewer than q vertices then both $\delta(G)$ and $\theta(G)$ can be determined in constant time (we write $\theta(G) = \infty$ in case G contains no total dominating set). If G is a thin spider then it is easy to verify that

$$\delta(G) = \sum_{i=1}^{r} \min\{c(y_i); c(z_i)\},$$

and

$$\theta(G) = \sum_{i=1}^{r} c(y_i).$$

On the other hand, if G is a thick spider then

$$\delta(G) = \min_{1 \leq j < k \leq r} \{\sum_{i=1}^{r} c(z_i); c(y_j) + c(y_k); c(y_j) + c(z_j)\},$$

and

$$\theta(G) = \min_{1 \leq j < k \leq r} \{c(y_j) + c(y_k)\}.$$

If G arises from graphs $G_1, \ldots, G_k$ by means of a ⓪ operation then we obtain

$$\delta(G) = \delta(G_1) + \ldots + \delta(G_k).$$

If G arises by a ① operation then

$$\delta(G) = \min_{1 \leq i < j \leq k} \{\delta(G_i); c_{min}(G_i) + c_{min}(G_j)\},$$

with $c_{min}(G_i)$ denoting the smallest weight of a vertex in G_i. For $\theta(G)$ we obtain analogous statements. Note that the last equality follows from the fact that a minimum dominating set (and also a minimum total dominating set) in the disjoint sum of graphs consists either of such a set in one of the graphs or of two vertices from two different graphs.

Let now H be a homogeneous set in G. If a minimum dominating set D in G contains no vertices from $N(H)$ then it consists of dominating sets in H and in $G \setminus H \setminus N(H)$. If D contains a vertex from $N(H)$ then at most one vertex from H belongs to D and this vertex has smallest weight. Hence, we have

$$\delta(G) = \min\{\delta(H) + \delta(G \setminus H \setminus N(H)); \delta(G_H)\},$$

where G_H is the graph which arises from G by replacing H by an independent set of size two with one vertex having smallest weight in H and the other vertex having weight ∞ (the second vertex guarantees that a vertex from $N(H)$ must belong to a minimum dominating set). An analogous equality holds for $\theta(G)$, however, in G_H it suffices to replace H by one vertex of smallest weight. Since a total dominating set D contains no isolated vertices, it is clear that at least one vertex from $N(H)$ must belong to D.

Let G arise by a ② operation with a separable p-connected component G^0 with separation (G_1^0, G_2^0) and a subgraph H outside G^0 adjacent to all vertices of G_1^0 and to no vertex of G_2^0. By the previous arguments, we obtain

$$\delta(G) = \min\{\delta(H) + \delta(G_2^0); \delta(G_H)\}.$$

Since G is a $(q, q - 4)$-graph it follows that G^0 has fewer than q vertices or G is a spider. In the first case, both $\delta(G_2^0)$ and $\delta(G_H)$ can be computed in constant time since G_2^0 and G_H have at most $q + 1$ vertices. If G^0 is a spider then

$$\delta(G_2^0) = \sum_{i=1}^{r} c(z_i).$$

If G^0 is a thin spider then

$$\delta(G_H) = \min_{1 \le i \le r} \{c(y_i) + \sum_{j \ne i} \min\{c(y_j); c(z_j)\}\}.$$

If G^0 is a thick spider then

$$\delta(G_H) = \min_{1 \le i < j \le r} \{c(y_i) + c(y_j); c(y_i) + c(z_i)\}.$$

Similarly, if G^0 is a thin spider then we obtain

$$\theta(G) = \sum_{i=1}^{r} c(y_i),$$

and, if G^0 is a thick spider

$$\theta(G) = \min_{1 \le i < j \le r} \{c(y_i) + c(y_j)\}.$$

The previous considerations immediately imply the following statement.

Theorem 7. *Once the primeval decomposition tree $T(G)$ is available, the domination number and the total domination number of a $(q, q - 4)$-graph $G = (V, E)$ with fixed q can be computed in $O(|V|)$ time.* $\square$

8   Conclusions

We studied several types of domination problems and the Steiner tree problem in weighted graphs. In particular, we showed that the homogeneous and the primeval decomposition are valuable tools for the resolution of these problems. We demonstrated that the problems are solvable in polynomial time whenever they can be solved efficiently for the subgraphs associated to the leaves of the primeval or homogeneous decomposition trees. In particular, we obtained linear-time algorithms for $(q, q - 4)$-graphs which contain such familiar classes as cographs, P_4-reducible graphs and P_4-sparse graphs, among others.

References

1. L. Babel, S. Olariu: On the structure of graphs with few P_4s. Discrete Applied Mathematics **84** (1998) 1–13

2. S. Baumann: A linear algorithm for the homogeneous decomposition of graphs. Report No. M-9615. Zentrum Mathematik, Technische Universität München (1996)
3. D.G. Corneil, H. Lerchs, L. Stewart Burlingham: Complement reducible graphs. Discrete Applied Mathematics **3** (1981) 163–174
4. D.G. Corneil, Y. Perl: Clustering and domination in perfect graphs. Discrete Applied Mathematics **9** (1984) 27–39
5. D.G. Corneil, Y. Perl, L.K. Stewart: A linear recognition algorithm for cographs. SIAM Journal on Computing **14** (1985) 926–934
6. A. Cournier, M. Habib: A new linear time algorithm for modular decomposition. Trees in Algebra and Programming, Lecture Notes in Computer Science Vol. 787, Springer (1994) 68–84
7. E. Dahlhaus, J. Gustedt, R. McConnell: Efficient and practical modular decomposition. In: Eight Annual ACM-SIAM Symposium on Discrete Algorithms, New Orleans, Louisiana (1997) 26–35
8. T. Haynes, S. Hedetniemi, P. Slater (eds.): Domination in Graphs: Advanced Topics. Marcel Dekker (1998)
9. S. Hedetniemi, R. Laskar: Topics on Domination. Annals of Discrete Mathematics **48**, North-Holland, Amsterdam (1991)
10. B. Jamison, S. Olariu: P_4-reducible graphs, a class of uniquely tree representable graphs. Studies in Applied Mathematics **81** (1989) 79–87
11. B. Jamison, S. Olariu: On a unique tree representation for P_4-extendible graphs. Discrete Applied Mathematics **34** (1991) 151–164
12. B. Jamison, S. Olariu: A unique tree representation for P_4-sparse graphs. Discrete Applied Mathematics **35** (1992) 115–129
13. B. Jamison, S. Olariu: p-components and the homogeneous decomposition of graphs. SIAM J. Discrete Mathematics **8** (1995) 448–463
14. D. Kratsch, L. Stewart: Domination on cocomparability graphs. SIAM J. Discrete Mathematics **6** (1993) 400–417
15. R. Laskar, J. Pfaff, S.M. Hedetniemi, S.T. Hedetniemi: On the algorithmic complexity of total domination. SIAM J. Algebraic Discrete Methods **5** (1984) 420–425
16. R. Lin, S. Olariu: A fast parallel algorithm to recognize P_4-sparse graphs. Discrete Applied Mathematics **81** (1998) 191–215
17. R. McConnell, J. Spinrad: Linear-time modular decomposition and efficient transitive orientation of comparability graphs. Fifth Annual ACM-SIAM Symposium on Discrete Algorithms, Arlington, VA (1994) 536–545
18. R.H. Möhring: Algorithmic aspects of comparability graphs and interval graphs. In: I. Rival (ed.), Graphs and Orders. Dordrecht, Holland (1985)
19. K. White, M. Farber, W. Pulleyblank: Steiner trees, connected domination and strongly chordal graphs. Networks **15** (1985) 109–124

Minimum Fill-in and Treewidth for Graphs Modularly Decomposable into Chordal Graphs

Elias Dahlhaus[1]

Department of Mathematics and Department of Computer Science,
University of Cologne
and
Department of Computer Science
University of Bonn,
Germany
e-mail: dahlhaus@cs.uni-bonn.de and dahlhaus@informatik.uni-koeln.de

Abstract. We show that a minimum fill-in ordering of a graph can be determined in linear time if it can be modularly decomposed into chordal graphs. These graphs are also called *weak bipolarizable* [12]. This generalizes results of [2]. We show that the treewidth of these graphs can be determined in $O((n + m) \log n)$ time.

1 Introduction

One of the major problems in computational linear algebra is that of sparse Gauss elimination. The problem is to find a pivoting, such that the number of zero entries of the original matrix that become non zero entries in the elimination process is minimized. In case of symmetric matrices, we would like to restrict pivoting along the diagonal. The problem translates to the following graph theory problem [13].

Minimum Elimination Ordering: For an ordering $<$ on the vertices, we consider the fill-in graph $G'_< = (V, E')$ of $G = (V, E)$. $G'_<$ contains first the edges in E and secondly two vertices x and y form an edge in $G'_<$ if they have a common smaller neighbor in $G'_<$. *The problem of Minimum Elimination ordering is, given a graph $G = (V, E)$, find an ordering $<$, such that $G'_<$ has a minimum number of fill-in edges.*

Note that this problem is NP-complete [15] in general. There is a polynomial time solution for this problem for so called HHD-free graphs [4]. Moreover, for distance hereditary graphs and for certain graph classes with few P4-s, there are linear time solutions [3, 2]. Here we generalize the result of [2] and show that a minimum fill-in ordering can be determined in linear time if the graph can be modularly decomposed into chordal graphs. Graphs that are modularly decomposable into chordal graphs coincide with the class of weak bipolarizable graphs [12]

Another problem is to find an elimination scheme, such that the size of "dense matrices" is as small as possible. This is related to the problem of treewidth.

Treewidth: Find an ordering $<$, such that the maximum clique size of $G'_<$ is minimized.

Also the problem of minimum treewidth is NP-complete [1]. The problem has a polynomial time solution for HHD-free graphs [4].

We will show that we can compute a minimum elimination ordering and the treewidth for graphs that can be modularly decomposed into chordal graphs in linear time or almost in linear time.

In section 2, we introduce the notation of the paper. Section 3 presents a linear time algorithm for minimum fill-in for graphs that can be modularly decomposed into chordal graphs. Section 4 discusses the treewidth of graphs that can be modularly decomposed into chordal graphs.

2   Notation

A *graph* $G = (V, E)$ consists of a *vertex set* V and an *edge set* E. Multiple edges and loops are not allowed. The edge joining x and y is denoted by xy.

We say that x is a *neighbor* of y iff $xy \in E$. The set of neighbors of x is denoted by $N(x)$ and is called the *neighborhood*. Analogously, for a set X of vertices, $N(X)$ is the set of neighbors of some vertex in X that are not in X and $N[X]$ is the set of neighbors of vertices in X together with the vertices in X.

Trees are always directed to the root. The notion of the *parent, child, ancestor,* and *descendent* are defined as usual.

A *subgraph* of (V, E) is a graph (V', E') such that $V' \subseteq V$, $E' \subseteq E$. The graph $G[X]$ is the subgraph *induced* by X consisting of all vertices in X and all edges $xy \in E$ with $x, y \in E$.

We denote by n the number of vertices and by m the number of edges of G.

A graph is called *chordal* iff each cycle of length greater than three has a chord, i.e. an edge that joins two nonconsecutive vertices of the cycle. Note that chordal graphs are exactly those graphs having a *perfect elimination ordering* $<$, i.e. for each vertex v the neighbors $w > v$ induce a complete subgraph, i.e. they are pairwise joined by an edge [8].

Note that in any chordal graph, the number of maximal cliques is bounded by n and the number of pairs (x, c) such that x is in the clique c is bounded by $n + m$.

The *fill-in* of a graph $G = (V, E)$ and an ordering $<$ is the smallest edge set E', such that $E \subseteq E'$ and $<$ is a perfect elimination ordering of $G_< := (V, E')$. Note that $G_<$ is chordal. The problem to get a minimum fill-in or a minimum elimination ordering is to get an ordering $<$, such that the fill-in is minimum.

The *treewidth* of $G = (V, E)$ is the minimum maximum clique size of a chordal graph $G' = (V, E')$ with $E \subseteq E'$.

By a *module* of a graph $G = (V, E)$, we define a subset V' of the vertex set V such that all vertices in V' have the same neighbors outside V'. We do not compute all modules but those modules X which do not *overlap* with other modules, i.e. there is no module Y that has a nonempty intersection with X and neither $X \subset Y$ nor $Y \subset X$. We call such modules also *overlap free*. Note that the overlap-free modules of any graph form a tree with respect to set inclusion, i.e. two overlap free modules are disjoint or comparable with respect to set inclusion. The notions of parent and child modules can be defined in an obvious way. The parent $P(X)$ of an overlap-free module X is the unique smallest overlap free module that is a proper superset of X. Y is a child module of X if and only if Y is an inclusion maximal proper overlap free submodule of X. The system of overlap-free modules is also called the *modular decomposition* of the graph G.

Note that a modular decomposition can be determined in linear time [11, 6, 7].

Let G be a graph with modules $V_1, \ldots, V_k$. Then $G/(V_1, \ldots, V_k)$ is the graph we obtain from G by shrinking each V_i to one vertex. Let X be a module with child modules $Y_1, \ldots, Y_k$. $G_X = G[X]/(Y_1, \ldots, Y_k)$ is the graph we get from $G[X]$ by shrinking each child module of Y_j of X to one vertex.

We call a graph *modularly decomposable into chordal graphs* or shortly *modulated chordal* if all graphs G_X are chordal. Note that this class is identical with the class of *weak bipolarizable graphs* [12].

3 Minimum Fill-in of Graphs with Modular Decomposition into Chordal Graphs

We follow the ideas of [2]. The key lemma is the following.

Lemma 1. *Let V' be a module of G and let $N(V')$ be the set of neighbors of V' that do not belong to V'. Then in any fill-in E' of G, V' is complete or $N(V')$ is complete.*

Proof. Assume V' and $N(V')$ are both not complete in E', i.e. there are $u, v \in V'$ and $u', v' \in N(V')$ that are not joint by an edge in E'. Then they form a cycle of length four in G and also in E'.
Q.E.D.

Corollary 1. *Suppose V_1 and V_2 are disjoint modules of G and all vertices in V_1 are adjacent with all vertices in V_2. Then in any fill-in E' of G, V_1 or V_2 is a complete set.*

We immediately get the following.

Corollary 2. *Let $V_1, \ldots, V_k$ be the child modules of G. Then for any fill-in E', the set of V_i that are not complete in E' form an independent set in*

$$G/(V_1, \ldots, V_k).$$

Now we assume that $V_1, \ldots, V_k$ are the child modules of G and

$$G' = G/(V_1, \ldots, V_k)$$

is a chordal graph.

To get a minimum fill-in, we first recursively compute a minimum fill-in E'_i, for all V_i. Then we select an independent set V' of vertices of G' as those modules V_i that are not made complete. The neighborhoods of the modules in V' are made complete. The modules corresponding to vertices of G' not belonging to V' are made complete. The number of resulting fill-in edges has to be minimized.

The following result proves that the resulting graph is a chordal graph and is easy to check.

Lemma 2. *1. Let G be a chordal graph and v be a vertex of G. Then the graph that comes up by replacing v by a complete set V_1 with the same neighbors outside V_1 as v is a chordal graph.*

2. Let G be a chordal graph and v be a vertex of G. Then the graph G' that comes up by making the neighborhood of v complete is a chordal graph.

3. Let G be a chordal graph and v be a simplicial vertex (i.e. the neighborhood is complete) of G. Then the graph that comes up by replacing v by a module that is a chordal graph is chordal.

Proof. We always can assume that G has a perfect elimination ordering $<$. When we replcace v by consecutive pairwise adjacent vertices $v_1, \ldots, v_k$, i.e. $v_1 < v_2 < \ldots < v_k$, $w < v$ iff $w < v_1$, and $v < w$ iff $v_k < w$, for each vertex $w \neq v$ of G, $<$ remains a perfect elimination ordering. This proves the first statement of the lemma.

The second part is proved as follows. We show that $<$ remains a perfect elimination ordering. Let $w < w_1, w_2$ and $ww_\nu \in E$ or w and w_μ are both neighbors of v. It has to be shown that $w_1 w_2 \in E$ or that w_1 and w_2 are both neighbors of v. If w is not a neighbor of v then $w_1 w_2 \in E$. If w is a neighbor of v and $v \leq w$ then $ww_1 \in E$ and $ww_2 \in E$, because they are neighbors of w or greater neighbors of v (and therefore adjacent to the greater neighbor w of v). Therefore also $w_1 w_2 \in E$. Finally let $w < v$ be a neighbor of v. If w_1 or w_2 is a neighbor of w then it is also a neighbor of v (v and w_1 or w_2 arte greater neighbors of w). In any case, w_1 and w_2 belong to the neighborhood of v.

The third part is proved as follows. We always have a perfect elimination ordering $<$ of G that starts with the simplicial vertex v. Let M be another chordal graph with a perfect elimination ordering $<'$. When we replace v by M as a module then the concatenation of $<'$ and $<$ restricted to $G - v$ is a perfect elimination ordering.

Q.E.D.

To get the right independent set V', we proceed as follows. For each module V_i, let f_i be the number of fill-in edges one gets if V_i is made complete, i.e. the number of non edges in V_i, and let g_i be the number of fill-in edges one gets if V_i is not made complete, i.e. the number of fill-in edges of a minimum fill-in of $G[V_i]$ plus the number of non edges in the neighborhood of V_i that join vertices that appear in different V_j.

Lemma 3. *The number of fill-in edges that are created by making exactly the modules in V' not complete (and making the remaining modules complete) is*

$$\Sigma_{V_i \in V'} g_i + \Sigma_{V_i \notin V'} f_i.$$

Proof. Note that V' is an independent set in $G' = G/(V_1, \ldots, V_k)$ and that fill-in edges might be created by two modules V_i and V_j only in case that they are common non edges of the neighborhoods of V_i and V_j. This can only be the case if V_i and V_j belong to V'. Now V_i and V_j are not joint by an edge in G'. Since G' is a chordal graph, all vertices in the joint neighborhood of V_i and V_j are pairwise joint by an edge (otherwise G' had a chordless cycle of length four. Therefore no fill-in edge is created by two V_i's. This proves the lemma.
Q.E.D.

To get the size of a minimum fill-in of G, one has to compute a maximum weighted independent set of $G' = G/(V_1, \ldots, V_k)$, where the weight of V_i is $f_i - g_i$.

A. Frank [9] stated an algorithm to determine a maximum weighted independent set in a chordal graph. He proved that the algorithm has a polynomial time bound. The algorithm has in fact a linear time bound.

Lemma 4. *[9] We can determine a maximum weighted independent set in a chordal graph G' in linear time.*

Since the number of vertices of V_i and non edges of V_i is known, one gets f_i immediately.

To get g_i, one has to compute the number of non edges in the neighborhood of V_i that are not in the same V_j. We have the number of non edges of the neighborhood of V_i if we have the number of edges in the neighborhood of V_i that are not in the same V_j. We consider the chordal graph $G/(V_1, \ldots, V_k)$ and weight each edge V_iV_j by $|V_i||V_j|$.

Lemma 5. *For a chordal graph G' with vertex weights $w(v)$, for each vertex v and edge weights $w(e) = w(vw) = w(v)w(w)$, for each edge $e = vw$, we can compute, for all vertices v of G' simultaneously, the sum of edge weights in the neighborhood of v in linear time.*

Proof. We assume that a perfect elimination ordering of G' is known. Let $h(v)$ be the sum of edge weights of edges that join neighbors x and y of v that are greater than v. Since greater neighbors of v are pairwise adjacent,

$$h(v) = \Sigma_{x,y>v, xv, yv \in E, x \neq y} w(x)w(y).$$

This can be replaced by

$$h(v) = ((\Sigma_{xv \in E, x>v} w(x))^2 - \Sigma_{xv \in E} w(x)^2)/2.$$

Therefore all $h(v)$ can be determined in linear time.

Next we have to consider neighbors x and y, such that at least one of x or y is smaller than v. Without loss of generality, $x < y$ and $x < v$. Note that if y is a neighbor of x then y is a neighbor of v.

Let $w'(x)$ be the sum of edge weights $w(xy)$ with $xy \in E$ and $x < y$. Note that all $w'(x)$ can be determined simultaneously in linear time.

The sum of all edge weights in the neighborhood of v is determined by

$$h(v) + \Sigma_{x<v, xv \in E}(w'(x) - w(xv)).$$

Q.E.D.

Theorem 1. *The size of a minimum fill-in of a modulated chordal graph can be determined in linear time.*

It remains to get a minimum fill-in ordering i.e. a perfect elimination ordering of the fill-in. Note that we did not compute the edges of the fill-in graph explicitly. What we still can get in linear time is the set of cliques of the fill-in graph. We create the set C of cliques of the fill-in of G recursively as follows.

1. We assume that we know the set C' cliques of G' and the sets C_i of the fill-ins of $G[V_i]$.
2. Suppose V_i belongs to V'. We create cliques $c_i \cup \bigcup_{c, V_i \in c \in C'} c$.
3. Suppose V_i does not belong to V'. Then in each clique that contains V_i, we replace V_i by the vertices in V_i.

To get a minimum fill-in ordering of G, we only have to know that a perfect elimination ordering can be determined in linear time if the cliques of a chordal graph are known (perfect elimination ordering of an α-acyclic hypergraph) [14].

4   Treewidth of Modulated Chordal Graphs

We will show the following.

Theorem 2. *The treewidth of a modulated chordal graph can be determined in $O(n + m) \log n$ time.*

Proof. We proceed in a similar way as in the case of minimum fill-in. We recursively determine the treewidths of the maximal modules $V_1, \ldots, V_k$ and select an appropriate independent set I of $G/(V_1, \ldots, V_k)$, such that the $V_i \in I$ are exactly those modules that are not made complete. We may assume that a perfect elimination ordering of $G/(V_1, \ldots, V_k)$ is known. In any fill-in G' of G, there are two kinds of cliques.

1. Cliques c that are unions of V_i, i.e. there is a V_i, such that $c = V_i \cup \bigcup_{V_j > V_i, V_i V_j \in E} V_j$. Note that in this case, all V_j in c are made complete.
2. Cliques c that are not unions of V_i, i.e. there is a clique c_1 of $G'[V_i]$, such that $c = c_1 \cup N(V_i)$.

We assume that we know the treewidth t_i of $G[V_i]$. Let $s_i^1 := t_i + |N(V_i)|$ and $s_i^2 := |V_i| + \Sigma_{V_j > V_i, V_i V_j \in E}|V_j|$. s_i^1 is the maximum size of a clique that intersects V_i if V_i is not made complete. s_i^2 is the size of V_i together with its greater neighborhood. For an independent set I of $G/(V_1, \ldots, V_k)$, let J be the set of $V_i \notin I$, such that all modules V_j in the greater neighborhood of V_i are not in I. The each clique that is a union of V_j is of the size s_i^2, for some $i \in J$ and each clique that is not the union of some V_j is of the size s_i^1, for some $i \in I$. The maximum clique size of the fill-in G_I associated with I is denoted by S_I and can be determined as follows.

$$S_I = \max(s_i^1 : V_i \in I, s_i^2 : V_i \in J).$$

The treewidth is therefore determined by

$$S := \min_I S_I.$$

We again consider the elimination tree T with parent function Par where $Par(V_i)$ is the next greater neighbor of V_i in $G/(V_1, \ldots, V_k)$. Let D_i be the set of descendents of V_i including V_i in T. For an independent set I of D_i, let J be the set of $V_j \notin I$ in D_i, such that no greater neighbor of V_j is in I. Then $S_I^i := \max(s_j^1 | V_j \in D_i \cap I, s_j^2 | V_j \in J)$ and $S^i = \min_I S_I^i$.

To get S, we determine the S^i recursively. Let $S_i^1 := \min_{V_i \in I} S_I^i$ and $S_i^2 := \min_{V_i \notin I} S_I^i$. Then

$$S_i^1 = \max(s_i^1, (S^j : V_j \in D_i, V_j V_i \notin E, Par(V_j)V_i \in E))$$

and

$$S_i^2 = \max(s_i^2, (S^j : Par(V_j) = V_i)).$$

Note that $S^i = \min(S_i^1, S_i^2)$.

Obviously, the time bound is $O(n^2)$. But we also can get $O(n + m) \log n$ as follows. We sort the children V_j of each V_i by the numbers S^j as soon as we know the S^j, for all children V_j of V_i. This takes $O(n + m) + O(n \log n)$ time. For each descendent V_j of V_i (i.e. $V_j < V_i$) that is a neighbor of V_i, we now can determine a child V_l of V_j that is not a neighbor of V_i of maximum S^l in the order of the number of children of V_j that are neighbors of V_i. This gives a time bound in the order of the number of children of V_j that are neighbors of V_i. This gives a time bound in the order of the number of neighbors of V_i times a logarithmic factor. The overall complexity of the algorithm is therefore $O(n + m) \log n$.

Q.E.D.

Remark 1. The additional logarithmic factor comes from the fact that we have to sort the children of any node of the elimination tree. It might be interesting to improve the algorithm in such a way that we can circumvent sorting.

5   Conclusions

It might also be possible to extend the ideas also to HHD-free graphs. One should mention that in HHD-free graphs, there is always a vertex, such that the in the neighborhood, the connected components of the complement form modules [10]. A polynomial time algorithm to get a minimum fill-in for HHD-free graphs is due to [4]. The same methods as in this paper also apply if all prime modules modules with the exception of "small" leaf modules are chordal (compare [2]).

6   Acknowledgements

I am very grateful for fruitful discussions with Ton Kloks.

References

1. Arnborg, S., D. G. Corneil and A. Proskurowski, Complexity of finding embeddings in a k-tree, *SIAM J. Alg. Disc. Meth.* **8**, (1987), pp. 277–284.
2. L. Babel, *Triangulating Graphs with Few P_4s*, submitted.
3. H.J. Broersma, E. Dahlhaus, T. Kloks, *A Linear Time Algorithm for Minimum Fill-in and Treewidth for Distance Hereditary Graphs* , submitted.
4. H.J. Broersma, E. Dahlhaus, T. Kloks, *Algorithms for the Treewidth and Minimum Fill-in of HHD-Free Graphs*, WG 97, LNCS 1335, pp. 109-117.
5. P. Bunemann, *A Characterization of Rigid Circuit Graphs*, Discrete Mathematics 9 (1974), pp. 205-212.
6. A. Cournier and M. Habib, *A new linear algorithm for modular decomposition*, in CAAP '94: 19th International Colloquium, Lecture Notes in Computer Science, Sophie Tison, ed., 1994, pp. 68-82.
7. Elias Dahlhaus, Jens Gustedt, Ross McConnell, *Efficient and Practical Modular Decomposition*, Eighth Annual ACM-SIAM Symposium on Discrete Algorithms (1997), pp. 26-35.
8. M. Farber, *Characterizations of Strongly Chordal Graphs*, Discrete Mathematics 43 (1983), pp. 173-189.
9. A. Frank, *Some Polynomial Time Algorithms for Certain Graphs and Hypergraphs*, Proceedings of the 5th British Combinatorial Conference, Congressus Numerantium XV, Untilitas Mathematicae, Winnipeg (1976), pp. 211-226.
10. C.T. Hoáng and N. Khouzam, *On Brittle Graphs*, Journal of Graph Theory 12 (1988), pp. 391-404.
11. R. M. McConnell and J. P. Spinrad, *Linear-time modular decomposition and efficient transitive orientation of undirected graphs*, in Proceedings of the fifth annual ACM-SIAM Symposium on Discrete Algorithms, D. D. Sleator et al., eds., Society of Industrial and Applied Mathematics (SIAM), 1994, pp. 536–545.
12. S. Olariu, *Weak Bipolarizable Graphs*, Discrete Mathematics 74 (1989), pp. 159-171.
13. D. Rose, *Triangulated Graphs and the Elimination Process*, Journal of Mathematical Analysis and Applications 32 (1970), pp. 597-609.
14. R. Tarjan, M. Yannakakis, *Simple Linear Time Algorithms to Test Chordality of Graphs, Test Acyclicity of Hypergraphs, and Selectively Reduce Acyclic Hypergraphs*, SIAM Journal on Computing 13 (1984), pp. 566-579.
    Addendum: SIAM Journal on Computing 14 (1985), pp. 254-255.
15. M. Yannakakis, Computing the Minimum Fill-in is NP-complete, *SIAM Journal on Algebraic and Discrete Methods* 2 (1981), pp. 77-79.

Interval Completion with the Smallest Max-degree

(extended abstract)

Fedor V. Fomin[1]* and Petr A. Golovach[2]**

[1] Department of Operations Research, Faculty of Mathematics and Mechanics, St.Petersburg State University, Bibliotechnaya sq.2, St.Petersburg, 198904, Russia, fomin@gamma.math.spbu.ru.
[2] Department of Applied Mathematics, Faculty of Mathematics, Syktyvkar State University, Oktyabrsky pr., 55, Syktyvkar, 167001, Russia, golov@ssu.komitex.ru.

Abstract. The interval degree of a graph is defined to be the smallest max-degree of any of its interval supergraphs. We find various bounds for this parameter. We prove that for any graph G the interval degree of G is at least the bandwidth of G, the pathwidth of G^2 and at most twice the bandwidth of G. Also we show that if G is an AT-free claw-free graph, then the interval degree of G is equal to the clique number of G^2 minus one. Finally, we show that there is a polynomial time algorithm for computing the interval degree of AT-free claw-free graphs.

1 Introduction and statement of the problem

There are two interval completion problems which were studied intensively because of the large number of practical applications (see [4, 13] for a survey). The first (the profile problem) is for a given graph G to find an interval supergraph of G with the minimum number of edges. The second (the pathwidth problem) is to find an interval supergraph of G with the smallest clique number. Here we introduce the related problem of finding an interval supergraph with the smallest max-degree. The problem formulated arouse our interest because of its close association with the bandwidth, the pathwidth and the treewidth minimization problems and owing to its natural statement.

We use the standard graph-theoretic terminology compatible with [3], to which we refer the reader for basic definitions. G is an undirected, simple (without loops and multiple edges) and finite graph with the vertex set $V(G)$ and the edge set $E(G)$. Unless otherwise specified, n denotes the number of vertices of G. Let $\omega(G)$ denotes the clique number (maximum clique-size) of G. The degree of a vertex v in a graph G is denoted by $deg_G(v)$ and the maximum degree of the vertices of a graph G by $\Delta(G)$. The closed neighbourhood of a vertex v in a graph G (the set of neighbours of v in G with v) is denoted by $N_G[v]$. The

* The research of this author was partially supported by RFBR grant 98-01-00934.
** The research of this author was partially supported by RFBR grant 96-01-00285.

distance $d_G(u, v)$ between two vertices u and v of G is the length of the shortest path between u and v in the graph G. Let G^k be the graph with vertex set $V(G)$ and two vertices u and v are adjacent in G^k if and only if $d_G(u, v) \leq k$.

A graph G is called an *interval graph* provided we can assign to each $v \in V(G)$ an interval I_v so that $(u, v) \in E(G)$ if and only if $I_v \cap I_u \neq \emptyset$. Some interesting graph parameters can be defined in terms of interval supergraph. For example, the *pathwidth* $pw(G)$ of a graph which was introduced by Robertson and Seymour [18] can be defined (see, e.g. [13]) as

$$pw(G) \triangleq \min\{\omega(G') - 1 : G' \text{ is an interval supergraph of } G\}.$$

The problem of finding the profile of a graph has applications in sparse matrix computations (see [4]). In terms of interval supergraphs the profile of a graph G can be defined (see, e.g. [1]) as

$$p(G) \triangleq \min\{|E(G')| : G' \text{ is an interval supergraph of } G\}.$$

In the same manner we define a new graph parameter, namely the interval degree of a graph. The *interval degree* of a graph G is

$$id(G) \triangleq \min\{\Delta(G') : G' \text{ is an interval supergraph of } G\}.$$

The problem of *interval completion with the smallest max-degree* is for a given graph G to find an interval supergraph I of G such that $\Delta(I) = id(G)$.

2   Linear layouts

Since an interval representation of a graph naturally induces an ordering of its vertices then it is not surprising that sometimes interval completion problems can be 'rewritten' in terms of vertex orderings or linear layouts.

A *linear layout* of a graph G is a one-to-one mapping $f : V(G) \to \{1, \ldots, n\}$. Letting f be a linear layout of a graph G and $i \in \{1, \ldots, n\}$, we define

$$S_i(G, f) = |\{v \in V(G) : f(v) \leq i \text{ and } \exists (u, v) \in E(G), \text{ such that } f(u) > i\}|$$

and

$$vs(G, f) = \max\{S_i(G, f) : i \in \{1, \ldots, n\}\}.$$

The *vertex separation number* of G [5] is

$$vs(G) \triangleq \min\{vs(G, f) : f \text{ is a linear layout of } G\}.$$

As noted by a number of researchers for any graph G, $vs(G) = pw(G)$ (see, e.g. [9] or [8]).

Profile also may be 'redefined' [1] as a graph invariant $p(G)$ by finding a layout f of G which minimises the sum

$$\sum_{u \in V(G)} (f(u) - \min\{f(v) : v \in V(G), v \in N_G[u]\}).$$

Another very important graph parameter, namely bandwidth, is defined by making use of linear layouts. For a linear layout f of G setting

$$W_i(G, f) = \max\{j - i : j \in \{i, \ldots, n\}, f^{-1}(j) \in N_G[f^{-1}(i)]\}$$

and

$$bw(G, f) = \max\{W_i(G, f) : i \in \{1, \ldots, n\}\}$$

one can define *bandwidth* of G as

$$bw(G) \triangleq \min\{bw(G, f) : f \text{ is a linear layout of } G\}.$$

The bandwidth problem has a long history and a number of practical applications (see [4] for a survey).

In order to define the interval degree in terms of linear layouts, we introduce a 'hybrid' of the bandwidth and the vertex separation number. Let f be a linear layout of a graph G. For $i \in \{1, \ldots, n\}$ we define

$$sw(G, f) = \max\{S_{i-1}(G, f) + W_i(G, f) : i \in \{1, \ldots, n\}\}$$

(putting $S_0(G, f) = 0$) and

$$sw(G) \triangleq \min\{sw(G) : f \text{ is a linear layout of } G\}.$$

Notice that S_i is from the definition of the vertex separation number and W_i is from the definition of the bandwidth.

Theorem 1. *For any graph G, $id(G) = sw(G)$.*

Proof. (Sketch). First we prove that $id(G) \geq sw(G)$. Let G_I be an interval supergraph of G such that $\Delta(G_I) = id(G)$. Without loss of generality we can assume that the left endpoints of the intervals that represent G_I are distinct integers $1, 2, \ldots, n$. Such a representation leads to a linear layout f of G: for $v \in V(G)$ $f(v) = i$ if and only if i is the left point of the corresponding interval of v. For a vertex v with $f(v) = i$ let us define

$$d_1(v) = |\{u \in V(G) : f(u) < i \text{ and } (u, v) \in E(G_I)\}|$$

and

$$d_2(v) = |\{u \in V(G) : f(u) > i \text{ and } (u, v) \in E(G_I)\}|.$$

Obviously, $S_{i-1}(G, f) \leq d_1(v)$, $W_i(G, f) \leq d_2(v)$ and $deg_{G_I}(v) = d_1(v) + d_2(v)$. Choose a vertex v with a number i such that

$$S_{i-1}(G, f) + W_i(G, f) \text{ is maximum.}$$

Then $\Delta(G_I) \geq deg_{G_I}(v) = d_1(v) + d_2(v) \geq S_{i-1}(G, f) + W_i(G, f) = sw(G, f) \geq sw(G)$.

We now turn to $id(G) \leq sw(G)$. Let f be a linear layout of G such that $sw(G) = sw(G, f)$. We assign to each vertex $v \in V(G)$ the interval $(f(v), r(v))$, where $r(v) = \max\{i \in \{f(v), \ldots, n\} \colon f^{-1}(i) \in N_G[v]\} + \frac{1}{2}$. Denote the corresponding interval graph by G_I. If $(u, v) \in E(G)$ and $f(u) < f(v)$ then $f(v) < r(u)$; hence $(f(v), r(v)) \cap (f(u), r(u)) \neq \emptyset$. Consequently, $(u, v) \in E(G_I)$ and G_I is the interval supergraph of G. The further proof is straightforward once the following observations are made: for a vertex v with a number i, $S_{i-1}(G, f) \geq d_1(v)$ and $W_i(G, f) \geq d_2(v)$.   □

The following Corollary is the main reason of our interest to the interval completion problem with the smallest max-degree.

Corollary 1. *For any graph* G, $bw(G) \leq id(G) \leq 2\, bw(G)$.

Proof. It easy to check that for any graph G and linear layout f, $vs(G, f) \leq bw(G, f)$. Then Corollary follows immediately from Theorem 1.   □

The lower bound of Corollary is sharp

Lemma 1. *If* G *is connected, then* $bw(G) = id(G)$ *if and only if* G *is complete.*

Proof. (We omit the proof here).   □

The upper bound for the interval degree in terms of the bandwidth is tight as well. For example, for any star $K_{1,n}$, where $n = 2k$, $id(K_{1,n}) = 2bw(K_{1,n}) = n$ and for any path P_n with $n > 2$ vertices, $id(P_n) = 2bw(P_n) = 2$.

Corollary 2. *For any graph* G, $pw(G^2) \leq id(G)$.

Proof. (Sketch). Since $pw(G^2) = vs(G^2)$ we show that for any linear layout f, $vs(G^2, f) \leq sw(G, f)$. Let f be a linear layout and let v be a vertex of G, $i = f(v)$. Let j be the smallest integer for which $f^{-1}(j)$ has a neighbour w in G with $f(w) \geq i$. Define $D^{j\leq}$ $(D^{<j})$ to be the set of all vertices x of G such that x is adjacent in G^2 to a vertex y having $f(y) > i$ and $j \leq f(x) \leq i$ $(f(x) < j)$. It should be noted that $S_i(G^2, f) = |D^{j\leq}| + |D^{<j}|$. Inequalities $|D^{j\leq}| \leq W_j(G, f)$, $|D^{<j}| \leq S_{j-1}(G, f)$ are almost obvious and we arrive at $S_i(G^2, f) \leq W_j(G, f) + S_{j-1}(G, f)$.   □

The bound in Corollary 2 is sharp. Let us give (without proofs) some simple examples.

A graph G is a *cograph* if it does not contain P_4 (a path with four vertices) as an induced subgraph.

Example 1. Let G be a connected cograph. Then $vs(G^2) = id(G) = n - 1$.

A graph G is said to be *cobipartite* if it is the complement of a bipartite graph. Let a cobipartite graph G be the complement of a bipartite graph with bipartition (X, Y). We define $n_1 = |X|$ and $n_2 = |Y|$. The number of vertices of X (Y) that are adjacent in G to some vertices of Y (X) is denoted by m_1 (m_2).

Example 2. Let G be a cobipartite graph. Then $vs(G^2) = id(G) = \max\{n_1 + m_2, n_2 + m_1\} - 1$.

(The proof is not difficult and we omit it here). We find Example 2 to be interesting since, as shown by Parra and Scheffler in [17] the bandwidth problem is NP-hard even for cobipartite graphs. We generalise this Example in the next section.

3 Minimal Triangulations and AT-free claw-free graphs

A *chord* of a cycle C is an edge not in C that has endpoints in C. A *chordless cycle* in G is a cycle of length more than three in G that has no chord. A graph G is *chordal* if it does not contain a chordless cycle. A set of three vertices x, y, z of a graph G is called an *asteroidal triple* (abbr. AT) if for any two of these vertices there exists a path joining them that avoids the (closed) neighbourhood of the third. A graph G is called an *asteroidal triple-free* (abbr. AT-free) graph if G does not contain an asteroidal triple. This notion was introduced by Lekkerkerker an Boland for the following characterisations of interval graphs.

Theorem 2. [12] *G is an interval graph if and only if it is chordal and AT-free.*

A graph isomorphic to $K_{1,3}$ is referred to as a *claw*, and a graph that does not contain an induced claw is said to be *claw-free*. Notice that cobipartite graphs form a subclass of AT-free claw-free graphs. Another subclass of AT-free claw-free graphs form proper interval graphs. An interval graph G is a *proper interval graph* if it is claw-free. Thus G is a proper interval graph if and only if it is chordal and AT-free claw-free.

Because of Example 2, one can conjecture that $vs(G^2) = id(G)$ for any AT-free claw-free graph G. In order to prove this conjecture we restate the interval completion problem in terms of minimal triangulations. A *triangulation* of a graph G is a graph H on the same vertex set as G that contains all edges of G and is chordal. A *minimal* triangulation of G is a triangulation H such that no proper subgraph of H is a triangulation of G.

Möhring generalised Theorem 2 in the following way.

Theorem 3. [14] *Every minimal triangulation of an AT-free graph is an interval graph.*

Möhring's theorem implies

Corollary 3. *For any AT-free graph G, $id(G)$ is equal to the smallest max-degree over all minimal triangulations of G.*

Lemma 5 provides us with some information on the structure of minimal triangulations. This information is strongly used in the proof of Theorem 4. In order to obtain Lemma 5 we need additional 'tools'. A subset S of vertices of a connected graph G is called an *a, b-separator* for non adjacent vertices a and b in $V(G) \setminus S$ if a and b are in different connected component of the subgraph of G induced by $V(G) \setminus S$. If no proper subset of an a, b-separator S separates a and b in this way, then S is called a *minimal a, b-separator*. A subset S is referred to as a *minimal separator*, if there exist non adjacent vertices a and b for which S is a minimal a, b-separator.

The following characterisation of minimal separators is well-known (see e.g. [6]).

Lemma 2. *Let S be an a, b-separator of G and let G_a, G_b be two components of $G \setminus S$ containing a and b, respectively. Then S is a minimal a, b-separator if and only if every vertex $s \in S$ is adjacent to a vertex in each of these components.*

The following lemma can be found in [10], see also [11].

Lemma 3. *Let H be a minimal triangulation of G and S be a minimal a, b-separator of H. Then S is a minimal a, b-separator of G.*

The following lemma is an immediate consequence of characterisations of a minimal triangulation by 'completing' minimal separators, see [11].

Lemma 4. *Let H be a minimal triangulation of G. If an edge $e = (x, y) \in E(H) \setminus E(G)$ then there is a minimal separator of G containing x and y.*

The next lemma is related to a well-known theorem of Rose, Tarjan and Lueker [19] on minimal triangulations.

Lemma 5. *Let H be a minimal triangulation of G. If an edge $e = (x, y) \in E(H) \setminus E(G)$, then there is an induced cycle of length ≥ 4 in G such that $x, y \in V(C)$.*

Proof. By lemma 4 there exists a minimal a, b-separator S in G containing x and y. Let G_a, G_b be components of $G \setminus S$ containing a and b, respectively. By Lemma 2 vertices x and y have neighbours in G_a and G_b. Hence there exist inclusion-minimal paths $(x, a_1, \ldots, a_k, y)$, $a_i \in V(G_a)$, and $(x, b_1, \ldots, b_l, y)$, $b_i \in V(G_a)$. Since for no pair of vertices a_i and b_j $(a_i, b_j) \in E(G)$, vertices $(x, a_1, \ldots, a_k, y, b_1, \ldots, b_l)$ induce a cycle of length ≥ 4 in G. $\square$

Lemma 5 implies some interesting corollaries.

Corollary 4. *Let G be an AT-free graph. Then $id(G) \leq \omega(G^4) - 1$. In particular, $pw(G^2) \leq id(G) \leq pw(G^4)$.*

Proof. Let G_I be an interval supergraph of G such that $\Delta(G_I) = id(G)$. By Corollary 3 G_I is a minimal triangulation of G. Since G is AT-free then it does not contain a chordless cycle of length at least 6. Let O be a vertex of the maximal degree in G_I. By Lemma 5 for any $u, w \in N_I[v]$ $d_G(u, w) \leq 4$. Hence $deg_{G_I}(O) \leq \omega(G^4) - 1 \leq pw(G^4)$. $\square$

Corollary 5. *Let G be an AT-free graph. Then $\frac{\Delta(G^2)}{4} \leq bw(G) \leq \Delta(G^2)$.*

Proof. From Theorem 1 and Lemma 5 it follows that $bw(G) \leq id(G) \leq \Delta(G^2)$. It is known (see [4]) that for any graph G $\Delta(G) \leq 2bw(G)$ and $bw(G^2) \leq 2bw(G)$. Hence $\Delta(G^2) \leq 4bw(G)$. $\square$

Now we are ready to prove the main result of this section.

Theorem 4. *Let G be an AT-free claw-free graph. Then $\omega(G^2) - 1 \geq id(G)$.*

Proof. Let G_I be an interval supergraph of G such that $\Delta(G_I) = id(G)$. Notice that owing to Corollary 3 G_I may be treated as a minimal triangulation of G. Let O be a vertex of the maximal degree in G_I. Then $|N_{G_I}[O]| - 1 = \Delta(G_I) = id(G)$. Let a, b be vertices of $N_{G_I}[O]$. We show that $d_G(a, b) \leq 2$ which implies the existence of the clique in G^2 containing all vertices of $N_{G_I}[O]$.

Case 1. If $a, b \in N_G[O]$ then obviously $d_G(a, b) \leq 2$.

Case 2. Assume that $a \in N_G[O]$ and $b \notin N_G[O]$. Then $(O, b) \in E(G_I) \setminus E(G)$ and by Lemma 5 there is an induced cycle C_b in G of length at least four such that $O, b \in V(C_b)$. Because G is AT-free, the length of C_b is at most five. Hence if $a \in V(C_b)$ then $d_G(a, b) \leq 2$. Suppose that $a \notin V(C_b)$. C_b is a chordless cycle and G is claw-free; hence a is adjacent in G to at least one neighbour of O in C_b. Let b_1 be such a neighbour. If $d_G(b_1, b) = 1$ (see the left graph in Fig. 1) then $d_G(a, b) = 2$. Let $d_G(b_1, b) = 2$. Denote the second neighbour of O in C_b by b_2 and the vertex that is placed between b_1 and b in C_b by b_3 (see the right graph in Fig. 1). Then a is adjacent to at least one of the vertices b_2, b_3 and b because a, b_2, b_3 cannot form an AT in G. Thus $d_G(a, b) \leq 2$.

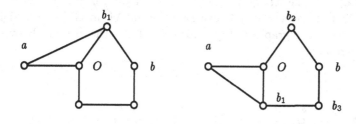

Fig. 1. Case 2

Case 3. Suppose that $a, b \notin N_G[O]$. Let C_a and C_b be chordless cycles of length at least four in G that contain vertices O, a and O, b, respectively.

The proof is obvious if $|E(C_a \cap C_b)| \geq 3$ (the diameter of the graph $C_a \cup C_b$ is at most two). Supposing $|E(C_a \cap C_b)| < 3$, we arrive at three possible cases.

Case 3.1. $|E(C_a \cap C_b)| = 2$. It is easy to check that if C_a and C_b have a common edge that is not incident to O then the diameter of $C_b \cup C_a$ is at most two. Because of this, we can assume that O has the same neighbours in C_a and C_b, i.e. $V(C_a) \cap N_G[O] = V(C_b) \cap N_G[O]$. We denote these neighbours by x and y (see the left graph in Fig. 2). Since G is claw-free and C_a, C_b are chordless cycles, the neighbour of x in $C_a \setminus C_b$ is adjacent to the neighbour of x in $C_b \setminus C_a$ and the neighbour of y in $C_a \setminus C_b$ is adjacent to the neighbour of y in $C_b \setminus C_a$ (see the right graph in Fig. 2). Then the distance in G between any two vertices from $V(C_b) \cup V(C_a)$ is at most two.

Case 3.2. C_a and C_b have only one common edge. If this edge is not incident to O then it is easy to check that the diameter of $C_a \cup C_b$ is at most two. Suppose that $V(C_b) = \{O, b_1, b_2, b_3, b_4\}$ and $V(C_a) = \{O, a_1, a_2, a_3, a_4\}$ (see the left graph on Fig. 3) (if C_a or C_b is the cycle of length four then the proof is the same). Since G is claw-free, then $(a_1, b_1) \in E(G)$ and $(a_3, b_3) \in E(G)$ (see the right graph on Fig. 3). If $(a_3, b_1) \in E(G)$ then a_2 is adjacent to b_1 in G (a_2, a_3, b_1, b_4 induce

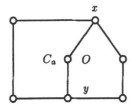 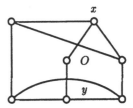

Fig. 2. Case 3.1

a claw otherwise) (see the left graph in Fig. 4). If $(a_3, b_1) \notin E(G)$ then a_3 is adjacent to b_2 in G because a_3, b_2, O cannot form an AT in G (see the right graph in Fig. 4). For both graphs in Fig. 4 for $i, j = \{2, 3\}$ the distance between a_i and b_j is at most two.

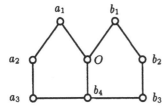 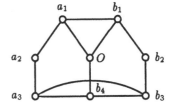

Fig. 3. Case 3.2

Case 3.3. $E(C_a \cap C_b) = \emptyset$. It is easy to see that if $|V(C_a \cap C_b)| \geq 2$ then the distance between any two vertices of the set $V(C_a \cup C_b) \setminus N_G[O]$ is at most two.

Suppose that $V(C_a \cap C_b) = O$ and $V(C_b) = \{O, b_1, b_2, b_3, b_4\}$, $V(C_a) = \{O, a_1, a_2, a_3, a_4\}$ (if C_a or C_b is the cycle of length four the proof is similar). Because G is claw-free and C_a, C_b are chordless in G then $(a_1, b_1), (a_4, b_4) \in E(G)$ (or $(a_1, b_4), (a_4, b_1) \in E(G)$ but this is the 'symmetric' case). (See the left graph in Fig. 5.)

Vertices a_1, b_2, b_4 and a_4, b_1, b_3 do not form ATs in G, so at least one pair from each triple $\{(a_1, b_2), (a_1, b_3), (a_1, b_4)\}$ and $\{(a_4, b_1), (a_4, b_2), (a_4, b_3)\}$ is an edge of G. If $(a_1, b_3) \in E(G)$ (or $(a_4, b_2) \in E(G)$ for the second triple) then (a_1, b_2) or (a_1, b_4) is in $E(G)$ $((a_4, b_1)$ or (a_4, b_3) is in $E(G))$ because vertices b_2, b_3, b_4 and a_1 $(b_1, b_2, b_3$ and $a_4)$ otherwise induce a claw in G. Thus at least one edge of each pair $\{(a_1, b_2), (a_1, b_4)\}$, $\{(a_4, b_1), (a_4, b_3)\}$ is in $E(G)$.

There is a need to examine the following cases:

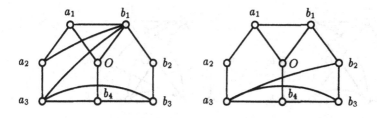

Fig. 4. Case 3.2: $(a_3, b_1) \in E(G)$ and $(a_3, b_1) \notin E(G)$

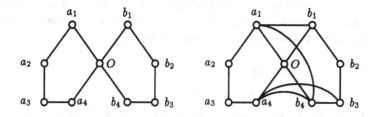

Fig. 5. Case 3.3: $V(C_a \cap C_b) = O$ and Case 3.3.1

Case 3.3.1. $(a_1, b_4), (a_4, b_3) \in E(G)$ (see the right graph in Fig. 5.). $(a_4, b_3) \in E(G)$ implies (vertices a_3, a_4, O, b_3 cannot induce a claw) $(a_3, b_3) \in E(G)$. From $(a_1, b_4) \in E(G)$ it follows (a_2, a_1, b_4, b_1 do not induce a claw) that $(a_2, b_2) \in E(G)$ or $(a_2, b_4) \in E(G)$). If $(a_2, b_2) \in E(G)$ then the diameter of $C_a \cup C_b$ is at most two. If $(a_2, b_4) \in E(G)$ then (a_2, b_4, O, b_3 do not induce a claw) $(a_2, b_3) \in E(G)$. Thus the distance between any two vertices $a_i, b_j \in V(C_a \cup C_b) \setminus N_G[O]$ in G is at most two.

Case 3.3.2. $(a_1, b_2), (a_4, b_1) \in E(G)$. This case is 'symmetric' about the previous case.

Case 3.3.3. $(a_1, b_4), (a_4, b_1) \in E(G)$ (see the left graph in Fig. 6). If $(a_1, b_2) \in E(G)$ then we arrive at Case 3.3.2. If $(a_4, b_3) \in E(G)$ then this is Case 3.3.1. Supposing that $(a_1, b_2), (a_4, b_3) \notin E(G)$ we obtain $(a_1, b_3) \in E(G)$ (a_4, b_4, b_3, a_1 induce a claw otherwise) and $d_G(a_2, b_3) \leq 2$. Furthermore, vertices a_2, a_1, b_3, b_1 do not induce a claw in G; hence $(a_2, b_3) \in E(G)$ or $(a_2, b_2) \in E(G)$. Therefore, $d_G(a_2, b_2) \leq 2$. Vertices $a_4 b_2$ are adjacent in G because a_1, b_1, b_2, a_4 otherwise induce a claw (see the right graph in Fig. 6). Then $d_G(a_3, b_2) \leq 2$. The graph induced by a_3, a_4, b_2, b_4 is not a claw; hence $d_G(a_3, b_3) \leq 2$. Thus for $i, j = \{2, 3\}$ $d_G(a_i, b_j) \leq 2$.

Case 3.3.4. $(a_1, b_2), (a_4, b_3) \in E(G)$. The claw-free condition for a_1, a_2, O, b_2 implies $(a_2, b_2) \in E(G)$ and the claw-free condition for a_3, a_4, O, b_3 implies $(a_3, b_3) \in E(G)$. □

Lemma 6. *If G is AT-free then G^2 is AT-free.*

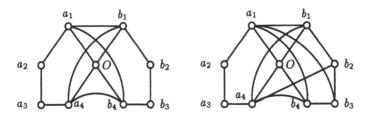

Fig. 6. Case 3.3.3

Proof. Let x, y, z be distinct vertices of G^2. Suppose that there is a path $S = (y = s_1, s_2, \ldots, s_k = z)$ from y to z in G^2 avoiding the closed neighbourhood of x (in G^2). Obviously, the subgraph of G induced by $\cup_{i=1}^{k} N_G[s_i]$ is connected and

$$\cup_{i=1}^{k} N_G[s_i] \cap N_G[x] = \emptyset.$$

Therefore, every AT in G^2 is also an AT in G. □

Lemma 7. *Let G be an AT-free claw-free graph. Then G^2 is AT-free claw-free.*

Proof. By Lemma 6 G^2 is AT-free. Suppose that there exist vertices b, c, d and a inducing a claw K in G^2, where a is the vertex of degree three. Note that at least two edges of K are from $E(G^2) \setminus E(G)$. If all edges of K are from $E(G^2) \setminus E(G)$ then vertices b, c, d form an AT in G.

Assume that only two edges, say (b, a) and (c, a), are in $E(G^2) \setminus E(G)$. Then $d_G(b, a) = d_G(c, a) = 2$. Let x be a vertex adjacent to vertices a, b and y be a vertex adjacent to a and c in G. Vertices b and c are not adjacent to d in G^2; hence x and y are not adjacent to d in G. Because G is claw-free, (x, y) is an edge in G; whence it follows that b, c, d form an AT in G. This is a contradiction and concludes the proof of Lemma 7. □

The following statement is owed to Parra and Scheffler.

Theorem 5. [17] *Let G be an AT-free claw-free graph. Then $bw(G) = pw(G)$.*

There are different ways to define the treewidth of a graph (see, e.g. [10]). For more information on this parameter the reader is referred to the recent survey paper of Bodlaender [2]. The following definition is more convenient for our purposes. The *treewidth* $tw(G)$ of a graph G is the smallest clique number over all triangulations of G decreased by one.

The next Theorem summarise the results of this section.

Theorem 6. *For any AT-free claw-free graph G,*

$$id(G) = \omega(G^2) - 1 = pw(G^2) = tw(G^2) = bw(G^2).$$

Proof. Let G be an AT-free claw-free graph. By Lemma 7 G^2 is also AT-free claw-free. By Theorem 3 $pw(G^2) = tw(G^2)$ and by Theorem 5 $pw(G^2) = bw(G^2)$. Since for any graph G, $pw(G) \geq \omega(G) - 1$ then Corollary 2 and Theorem 4 imply $pw(G^2) = \omega(G^2) - 1 = id(G)$. □

4   Concluding remarks

The following result is due to Müller.

Lemma 8. [16] *Let G be an AT-free claw-free graph. Then G^2 is a chordal graph.*

Lemma 6, Lemma 7 and Lemma 8 imply

Corollary 6. *Let G be an AT-free claw-free graph. Then G^2 is a proper interval graph.*

It easy to check that G^2 can be constructed in $O(n|E(G)|)$ time. Since the the clique number of an interval graph can be calculated in a linear time (see, e.g. [6]) then Theorem 6 and Lemma 8 implies

Corollary 7. *For any AT-free claw-free graph G $id(G)$ can be calculated in $O(n|E(G)| + |E(G^2)|)$ time. Therefore, there is an $O(n|E(G)| + |E(G^2)|)$ algorithm to approximate the bandwidth of AT-free claw-free graphs with worst case performance ratio 2.*

Karpinski and Wirtgen [7] showed that for the bandwidth problem on cobipartite graphs (and hence on AT-free claw-free graphs) there is no polynomial time approximation algorithm with an absolute error guarantee of $n^{1-\varepsilon}$ for any $\varepsilon > 0$ (unless $P = NP$).

We leave many questions unanswered, a few of them are:

1. What about the computational complexity of the interval completion problem with the smallest max-degree? Notice that the bandwidth problem remains to be NP-hard even for a special class of trees called caterpillars with hairs of length at most three [15].
2. Figure 7 shows that the claw-free condition in Theorem 6 is necessary. Is it true that there exists $k \geq 0$ such that for any AT-free graph G $id(G) \leq pw(G^2) + k$?
3. Kloks, Kratsch and Müller [11] obtained an $O(|E(G)| + n \log n)$ algorithm to approximate the bandwidth of an AT-free graph within a factor 4. Because of Corollary 5 it is interesting to know whether calculation of the max-degree of an AT-free graph squared can be done faster.
4. Is it possible to improve the time bound in Corollary 7? Probably the construction of the square is not necessary for calculating $\omega(G^2)$.

Acknowledgements. We are grateful to Dieter Kratsch and Haiko Müller for fruitful discussions and suggestions. Also we thank anonymous referees for their useful comments.

References

1. A. BILLIONNET, *On interval graphs and matrice profiles*, RAIRO Rech. Opér., 20 (1986), pp. 245–256.

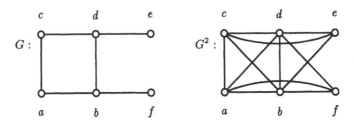

Fig. 7. Example of $id(G) = pw(G^2) + 1$. $id(G) = 4$; every triangulation of G contains (a,d) or (b,c). $pw(G^2) = 3$; G^2 is an interval graph and $\omega(G^2) = 4$.

2. H. L. BODLAENDER, *A partial k-arboretum of graphs with bounded treewidth*, Technical Report UU-CS-1996-02, Department of Computer Science, Utrecht University, Utrecht, the Netherlands, 1996.

3. J. A. BONDY, *Basic graph theory: Paths and circuits*, in Handbook of Combinatorics, Vol. 1, R. L. Graham, M. Grötschel, and L. Lovász, eds., Elsevier Science B.V., 1995, pp. 3–110.

4. P. Z. CHINN, J. CHVÁTALOVÁ, A. K. DEWDNEY, AND N. E. GIBBS, *The bandwidth problem for graphs and matrices — a survey*, J. Graph Theory, 6 (1982), pp. 223–254.

5. J. A. ELLIS, I. H. SUDBOROUGH, AND J. TURNER, *The vertex separation and search number of a graph*, Information and Computation, 113 (1994), pp. 50–79.

6. M. C. GOLUMBIC, *Algorithmic Graph Theory and Perfect Graphs*, Academic Press, New York, 1980.

7. M. KARPINSKI AND J. WIRTGEN, *On approximation hardness of the bandwidth problem*, Technical Report TR-97-041, ECCC, 1997.

8. N. G. KINNERSLEY, *The vertex separation number of a graph equals its path width*, Inform. Proc. Letters, 42 (1992), pp. 345–350.

9. L. M. KIROUSIS AND C. H. PAPADIMITRIOU, *Interval graphs and searching*, Disc. Math., 55 (1985), pp. 181–184.

10. T. KLOKS, *Treewidth. Computations and Approximations*, Lecture Notes in Computer Science, Vol. 842, Springer-Verlag, Berlin, 1994.

11. T. KLOKS, D. KRATSCH, AND H. MÜLLER, *Approximating the bandwidth for asteroidal triple-free graphs*, in Proceedings 3rd Annual European Symposium on Algorithms ESA'95, P. Spirakis, ed., Springer Verlag, Lecture Notes in Computer Science, vol. 979, 1995, pp. 434–447.

12. C. G. LEKKERKERKER AND J. C. BOLAND, *Representation of a finite graph by a set of intervals on the real line*, Fund. Math, 51 (1962), pp. 45–64.

13. R. H. MÖHRING, *Graph problems related to gate matrix layout and PLA folding*, in Computational Graph Theory, Comuting Suppl. 7, E. Mayr, H. Noltemeier, and M. Sysło, eds., Springer Verlag, 1990, pp. 17–51.

14. ———, *Triangulating graphs without asteroidal triples*, Disc. Appl. Math., 64 (1996), pp. 281–287.

15. B. MONIEN, *The bandwidth minimization problem for caterpillars with hair length 3 is NP-complete*, SIAM J. Alg. Disc. Meth., 7 (1986), pp. 505–512.

16. H. MÜLLER, 1998. Personal communication.

17. A. PARRA AND P. SCHEFFLER, *Treewidth equals bandwidth for AT-free claw-free graphs*, Technical Report 436/1995, Technische Universität Berlin, Fachbereich Mathematik, Berlin, Germany, 1995.

18. N. ROBERTSON AND P. D. SEYMOUR, *Graph minors. I. Excluding a forest*, J. Comb. Theory Series B, 35 (1983), pp. 39–61.

19. D. J. ROSE, R. E. TARJAN, AND G. S. LUEKER, *Algorithmic aspects of vertex elimination on graphs*, SIAM J. Comput., 5 (1976), pp. 266–283.

An Estimate of the Tree-Width of a Planar Graph Which Has Not a Given Planar Grid as a Minor.[*]

K.Yu. Gorbunov

Institute of New Technologies,Moscow,109004, Russia, Nizhnaya Radishevskaya 10, fax 7095 915-69-63; gorbunov@mccme.ru

Abstract. We give a more simple than in [8] proof of the fact that if a finite graph has no minors isomorphic to the planar grid of the size of $r \times r$, then the tree-width of this graph is less than $\exp(\mathrm{poly}(r))$. In the case of planar graphs we prove a linear upper bound which improves the quadratic estimate from [5].

1. **Introduction.** Neil Robertson and P.D. Seymour in [6] proved that for any r there exists $m = f(r)$ such that every graph has tree-width $\leq m$ provided it has no planar grid of size $r \times r$ as its minor. A nonelementary upper bound of $f(r)$ follows from their proof. In [3] we presented a proof giving an elementary upper bound. The method from [3] allows to obtain the bound $m \leq \exp(\mathrm{poly}(r))$, where $\exp(x)$ is function 2^x. N. Robertson, P.D. Seymour and R. Thomas [8] obtain a bound of less than 2^{9r^5}. When considering the case of planar graphs, N. Robertson and P.D. Seymour gave in [5] a proof with a quadratic upper bound of corresponding function $f(r)$. In Theorem 3 of the present paper we prove a linear upper bound for planar graphs. Incidentally (Theorem 2) we state in detail a shorter proof than in [8] for the bound $\exp(\mathrm{poly}(r))$ in general case. But let us remark that for this case a much simpler proof still, and with a better bound, can be found in [2].

The author does not know whether a polynomial upper bound is possible for the problem. If the answer to this question is affirmative, we will have the complete characterization of the graphs for which typical NP-problems (such as the problem of the existence of the Hamiltonian cycle) can be solved in polynomial time. This follows from the fact that such problems are solvable in polynomial time for any family of graphs with bounded tree-width, whereas for a family of graphs containing any plane grid they are NP-complete.

It is more convenient for us to use as in [3] the notion of n-divisibility instead of the notion of the tree-width. We prove in Theorem 1 that tree-width of a graph is related linearly (in both directions) with the minimal n for which the graph is n-divisible.

2. **Definitions and Theorems.** We will consider procedures of dividing of a finite graph into subgraphs: each subgraph arising in the process of dividing and

---

[*] This paper was supported in part by the INTAS project No. 93-0893.

having more than one vertex, at the next step is divided into two subgraphs, until all subgraphs have only one vertex (at the beginning of the process we have only one subgraph — the graph itself; here we mean by *subgraph* a subset of vertices of the graph together with all edges between them, but when we say that a subgraph is divided into two subgraphs we mean that the set of its vertices is partitioned into two parts). We'll say that a subgraph B of a graph G *is separable from its complement by no more than n vertices* iff there are no more than n vertices in the graph G such that any boundary for B (i.e. having only one end in B) edge is incident with at least one of these vertices. We'll call these "separating" vertices *marked for B*.

Definition. A graph is called *m-divisible*, if there exists a procedure of its dividing where each arising subgraph is separable from its complement by no more than m vertices. We say that a graph is *m-nondivisible* if it is not *m*-divisible.

We'll call *degree of nondivisibility* of a graph the minimal n such that the graph is n-divisible. Theorem 1 shows that the notions of tree-width and degree of nondivisibility are in fact equivalent. Let us recall the definition of tree-width from [4]. Let $V(G)$ denote the set of vertices of a graph G.

Definition. A *tree-decomposition* of a graph G is a family $(X_i | i \in I)$ of subsets of $V(G)$, together with a tree T with $V(T) = I$, with the following properties.

1. $\bigcup_{i \in I} X_i = V(G)$.
2. Every edge of G has both its ends in some X_i ($i \in I$).
3. For any $i, j, k \in I$, if j lies on the path of T from i to k, then $(X_i \cap X_k) \subseteq X_j$.

The *width of the tree-decomposition* is $\max_{i \in I}(|X_i| - 1)$. The *tree-width* of G is the minimum $m \geq 0$ such that G has a tree-decomposition of width $\leq m$.

Theorem 1. a) *Any graph having tree-width n is $(n+1)$-divisible.*
b) *Any n-divisible graph has tree-width no more than $3n$.*

Proof. Let us prove the item a). Consider the tree T of a tree-decomposition of a graph G. Let us describe a process of dividing of G. First, we separate from G the subgraph X_t corresponding to the root t of T. Remaining part is devided into parts that equal to $\bigcup_{v \in T'} X_v \setminus X_t$ for each subtree T' with a root in a son of t (we separate these parts one by one; they are pairwise disjoint because by Property 3 their intersection would lie in X_t; by Property 2 there are not edges in G connecting these parts). In each part we again sepapate the subgraph situated in the root of corresponding subtree, then again divide the rest into parts and so on.

Let $e = (a, b)$ be an edge boundary for some subgraph G' arising in the process and corresponding to a subtree T' with root t'. It is easy to see that e has its external end (say, b) in X_f, where f is farther of t'. Indeed, by Property 2, there is r such that $a, b \in X_r$. We have $r \in T'$ since by construction (and by Property 3) for any $j \notin T'$ $G' \cap X_j = \emptyset$ and $a \in G'$. As $b \notin G'$, there exists an ancestor s of r such that $b \in X_s$. Then, by Property 3, $b \in X_f$. Thus, G' is separable from its complement by vertices of X_f, the number of which is $\leq n+1$.

At the end of the process G will be devided into subgraphs having $\leq n+1$ vertices. We split from them vertices one by one untill all subgraphs consist of only one vertex. The item a) is proved.

Let us prove the item b). First, we prove the following lemma.

Lemma 1. *Let a graph G be n-divisible and a corresponding process of its dividing be given. Then for each arising non-one-vertex subgraph P we can mark $\leq n$ vertices separating P from its complement, such that at the next partitioning P into P_1 and P_2 the following conditions hold:*
1. *If a vertex a does not belong to P and a is marked for at least one of P_1, P_2 then a is marked for P.*
2. *If $a \in P_i$ and a is marked for P then a is marked for P_i $(i = 1, 2)$.*
3. *If $a \in P_i$ and a is marked for P_j where $j \neq i$ then a is marked for P_i $(i = 1, 2)$.*

Proof. For each subgraph P arising in the process let $n_P \leq n$ be the minimal number such that there exists a set consisting of n_P vertices separating P from $G \setminus P$. Among all such separating sets of cardinality n_P we select (and mark for P) a set M_P with minimal number of external (i.e. not belonging to P) vertices.

Let us prove the item 1. Consider the set M of vertices from $G \setminus P$ marked for P_1, say, but not for P. Let, contrary to the statement, $M \neq \emptyset$. Let C be the set of vertices in P_1 joint by an edge to M and unmarked for P_1.

Let $|C| > |M|$. Evidently, all vertices in C belong to M_P. Any boundary for P edge, incident with a vertex in C, has another end in M_{P_1} hence it leads either to M or to M_P. Therefore if we replace in M_P the subset C with the set M, we obtain a set of vertices separating P from $G \setminus P$ and having less elements than M_P. This contradicts to minimality of M_P.

Now, let $|C| \leq |M|$. Every vertex in P_1 which is adjacent with a vertex in M either lies in C or is marked for P_1. Therefore if we replace in M_{P_1} the subset M with the set C, we obtain a new set separating P_1 from $G \setminus P_1$. It has no more vertices than the initial set, but the number of external for P_1 vertices is reduced. This contradiction proves the item 1.

Let us prove the item 2. Let, say, $i = 1$. Consider the set M of vertices in P_1 marked for P but not for P_1. Let $M \neq \emptyset$. Let C be the set of vertices in $G \setminus P$ joint by an edge to M and not belonging to M_P. Let $|M| \leq |C|$. Evidently, all C is marked for P_1. Let us replace in the set M_{P_1} the subset C with the set M. We obtain a new set of vertices separating P_1 from $G \setminus P_1$. It has no more vertices than the old set but the number of external vertices is reduced. This contradicts to the choice of M_{P_1}. Now, let $|M| > |C|$. Replace in M_P the subset M with the set C. We obtain a new set of vertices separating P from $G \setminus P$ and having less vertices than M_P. This contradiction proves the item 2.

Let us prove the item 3. Let, say, $i = 1$, $j = 2$. Consider the set M of vertices in P_1 marked for P_2 but not for P_1. Let $M \neq \emptyset$. Let C be the set of vertices in P_2 joint by an edge to M and unmarked for P_2. Let $|M| \leq |C|$. Evidently, $C \subseteq M_{P_1}$. Replace in the set M_{P_1} the subset C with M. We obtain a new set of vertices separating P_1 from $G \setminus P_1$ with smaller number of external vertices. This contradicts to the choice of M_{P_1}. Now, let $|M| > |C|$. Replace in M_{P_2} M with

C. We obtain a new set of vertices separating P_2 from $G \setminus P_2$ and having less vertices than M_{P_2}. This contradiction proves the item 3. Lemma 1 is proved. $\square$

So, let a graph G be n-divisible. Let us take the process of its dividing and mark for each arising subgraph P the set M_P separating P from $G \setminus P$ as in the proof of Lemma 1. We'll represent this process in the form of a binary dividing tree D with subgraphs placing in its vertices (in a natural manner). A tree-decomposition tree T is obtained from D by ascribing to each vertex p the set X_p equal to union of all marked vertices for three subgraphs: the subgraph P and two its sons into which it is partitioned (if they exist).

It remaines to prove three properties from the definition of tree-decomposition. Property 1 is obvious, since by definition of M_P any vertex is marked for corresponding one-vertex subgraph. Let us prove Property 2. Let e be an edge of graph G. Consider a moment of dividing process when the ends of e turn out to be in different subgraphs and let A be that of them for which marked end of e is external (if there is no such subgraphs, Property 2 for e, clearly, holds). Consider the sequence of such descendants of subgraph A that have e as a boundary edge. Since for one-vertex subgraphs only one internal vertex is marked, there are two neighboring subgraphs in this sequence — a farther and a son such that for the farther an external end of e is marked and for the son — an internal end. This implies satisfaction of Property 2 for e.

Let us prove Property 3. Let a vertex a is marked for two subgraphs G_1 and G_2. It is sufficient to show that a is marked for all subgraphs on the path in dividing tree connecting G_1 and G_2 exept, maybe, their common ancestor. Let A be a subgraph on the path between G_1 and G_2. Consider two cases.

Case 1. One of G_1, G_2 is ancestor of another, say, G_1 is ancestor of G_2. Let the vertex a not belong to G_1. Then it follows from item 1 of Lemma 1 that if a is unmarked for A then a is unmarked for all descendants of A. Hence, as a is marked for G_2 then a is marked for A. Now, let $a \in G_2$. Then by item 2 of Lemma 1 as a is marked for G_1 then a is marked for any descendant of G_1 which contains a, including A. Now, let $a \in G_1$, $a \notin G_2$. Let B be the nearest to G_1 descendant of G_1 on the path to G_2 such that $a \notin B$. Then if A lies between B and G_2 (or if $A = B$), it is easy to see that from item 1 and the fact that a is marked for G_2 it follows that a is marked for A. And if A lies between G_1 and B ($A \neq B$) then from item 2 and the fact that a is marked for G_1 it follows that a is marked for A.

Case 2. None of subgraphs G_1, G_2 is ancestor of another. Let P be the nearest to them their common ancestor. It is easy to see that for both sons of P (as well as for all subgraphs between them and G_1, G_2) a is marked: if a son does not contain a this follows from item 1, if a son contains a then from item 3 of Lemma 1 and from the fact that another son does not contain a. Further, all considerations, evidently, are reduced to the case 1. Theorem 1 is proved. $\square$

N. Robertson and P.D. Seymour in [4] nonconstructively proved the existense of a polynomial algorithm to test if a graph has tree-width $\leq m$ for fixed m. We briefly describe a polynomial algorithm which for any fixed n decides if an input graph is n-divisible and if so, constructs a process of its n-dividing.

We'll mean by *n-divisibility of a subgraph* its n-divisibility as a graph but we take its boundary edges into consideration (in particular, the subgraph itself must be separable from its complement by no more than n vertices). We call a vertex g belonging to a subgraph B *saturated for* B if there are more than n boundary edges incident with g (multiple edges are considered only once). An external vertex which is incident with a boundary edge will be called *external boundary vertex*.

Lemma 2. *If a graph G is n-divisible, there exists a process of its n-dividing such that for every arising subgraph B we at first separate from it one by one no more than n vertices so that remaining subgraph B' either becomes one-vertex or has no saturated vertices and has no more than n^2 external boundary vertices. Then we divide B' into connected components and only after this we divide this components.*

Proof. Let a non-one-vertex subgraph B be n-divisible and let no more than n vertices separating B from its complement be marked. Let us separate out of B a saturated (and, hence, marked) vertex b. It is easy to see that the rest B_1 is n-divisible because a process of n-dividing of B induces a process of n-dividing of B_1. (Indeed, a subgraph C arising in the induced process and corresponding to the subgraph $C' = C + \{b\}$ in the main process is separable from its complement by $\leq n$ vertices — these vertices are the same as for C' including b.) We show that any saturated for B_1 vertex b_1 is marked for B. Eash incident with b_1 and boundary for B_1 edge either is boundary for B or leads to b. If b_1 is not marked for B then at least n adjacent with b_1 vertices out of B must be marked for B. Besides, b is marked for B, and we have a contradiction.

Separating b_1 out of B_1, we obtain B_2 and so on until $B_i = B'$ has no saturated vertices. Evidently, we have to separate $\leq n$ vertices. The fact that B' has $\leq n^2$ external boundary vertices is obvious enough. Lemma 2 is proved. $\square$

We'll call the process of dividing described in Lemma 2 *canonical process*. Now, we describe an algorithm. We consider the following totalities: either a one-vertex subgraph K or a pair $\langle K, P \rangle$ where P is a set of $\leq n^2$ vertices of an input graph G and K is a connected component of the subgraph $G \setminus P$. We will form step by step a list of all the totalities where K is n-divisible. Before the first step we put all one-vertex subgraphs down on the list. After the m-th step there will be all such pairs in our list that K is n-divisible by $\leq m$ partitioning (and, maybe, some other pairs with n-divisible K).

At the $(m+1)$-th step we look over all pairs $\langle K, P \rangle$ and for every pair which is not contained in our list we do the following. First, we verify that K is separable from its complement by $\leq n$ vertices. Let it be so. Then we suppose that K can be partitioned into two (unknown) parts K_1 and K_2 being n-divisible by $\leq m$ dividing. Look over all quadruples of sets of vertices $\langle O_1, O_2, P_1, P_2 \rangle$ where $|O_1| \leq n$, $|O_2| \leq n$, $|P_1| \leq n^2$, $|P_2| \leq n^2$. The meaning is: O_i — the set of those marked for K_i vertices which by Lemma 2 can be separated so that the subgraph $K_i \setminus O_i$ has the properties stated in Lemma 2; P_i — the set of all external boundary for $K_i \setminus O_i$ vertices. For a quadruple corresponding to a canonical process, the subgraphs $K_1 \setminus O_1$ and $K_2 \setminus O_2$ which we try to find must be a union of some connected components of the subgraphs $G \setminus P_1$ and $G \setminus P_2$ respectively.

Let $K' = K \setminus (O_1 \bigcup O_2)$. We call a path *clear* if all its vertices except, maybe, ends lie in $K' \setminus (P_1 \bigcup P_2)$. For each vertex a in K' consider two the following conditions.
1. Either $a \in P_1$ or there exists a clear path leading from a to some vertex $b \in P_1 \bigcap K'$.
2. $a \notin P_1 \bigcup P_2$ and there exists a clear path leading from a to $P_2 \setminus K'$.
We put a in K_2 if at least one of the conditions holds, otherwise we put a in K_1. (Note, that if both conditions are not satisfied then the conponent of $G \setminus P_1$ containing a either does not belong to K' or coincides with a component of $G \setminus P_2$.)

After this partitioning of K we verify that P_1 and P_2 really are the sets of all external boundary vertices for $K_1 \setminus O_1$ and $K_2 \setminus O_2$ respectivelly. It is easy to see that if it is not the case then the chosen quadruples of sets does not correspond to a canonical process of dividing. Finally, we verify that K_1 and K_2 are separable from their complements by a sets of $\leq n$ vertices including respectively O_1 and O_2.

We put $\langle K, P \rangle$ down on our list if and only if all the connected components of $G \setminus P_1$ and $G \setminus P_2$ contained in K' already present in the list. It is not difficult to see that the described algorithm is required.

Remark. There is also another notion being studied in literature — the *branchwidth* of a graph G. It is equal to the minimal t for which there exists a process of dividing of edges of G (like our process for vertices) such that for any arising set of edges E' it holds $|\mathrm{coup}(E')| \leq t$ where $\mathrm{coup}(E')$ is the set of vertices incident both with an edge in E' and with an edge not in E'. N. Robertson and P.D. Seymour in [7] proved linear equivalence of branchwidth and tree-width. Hans L. Bodlaender and Dimitrios M. Thilikos in [1] constructed a linear algorithm for recognition of the relation branchwidth $< T$ (for arbitrary fixed T).

Let us turn to our main result. Recall that a graph A is a *minor* of a graph B if we can map every vertex of the graph A to a nonempty connected subgraph of the graph B (moreover, different vertices correspond to disjoint subgraphs) and map every edge of the graph A to an edge of the graph B joining those two subgraphs which correspond to the ends of the edge in A.

Theorem 2. *For any natural $r \geq 2$ there exists $m \leq r^2 \exp(r^{20})$ such that if a finite graph G has no minors isomorphic to the planar grid of the size of $r \times r$, then this graph is m-divisible.*

Proof. We say that two subgraphs P_1 and P_2 of a graph G are *n-separable through a subgraph C* of the graph G if we can select $\leq n$ vertices in C with the following property: any path between P_1 and P_2 which has all interior vertices in C and contains at least two edges, passes through at least one of the selected vertices.

Lemma 3. *For any n, k in any (nk)-nondivisible graph there exist a connected subgraph C and k connected subgraphs, pairwise disjoint and disjoint from C, such that any two of these k subgraphs are n-nonseparable through C.*

Proof. Let $m = nk$. Let a graph G be m-nondivisible. We will carry out some procedure on G described below. Before the beginning of every stage of this procedure the conditions described in the following paragraph will be satisfied.

Some pairwise disjoint connected subgraphs are selected in the graph G. One of them is m-nondivisible. We'll call this subgraph "*central subgraph*" and denote it by C. The selected subgraphs joined by an edge to C will be called "*boundary subgraphs*". There are not more than k boundary subgraphs. Any edge boundary for C has external end in one of boundary subgraphs. For each boundary subgraph P we can select $\leq n$ vertices in $C \bigcup P$ such that any edge which joins P to C is incident with at least one of the selected vertices.

It follows from the conditions above that C is separable from its complement by $\leq m$ vertices. So, since C is m-nondivisible, for any partition of C into two subgraphs, at least one of them is m-nondivisible. Before the beginning of our procedure the subgraph C is a connected m-nondivisible component of the graph G. The boundary subgraphs are absent.

Before the beginning of every stage, the number of the boundary subgraphs is either strictly less than k or equal to k. In the first case let c be an arbitrary vertex in C. Then the subgraph $C_1 = C \setminus \{c\}$ is m-nondivisible and is separable from $G \setminus C_1$ by $\leq m$ vertices. Let C_0 be a m-nondivisible component of the subgraph C_1. C_0 becomes the new central subgraph, and $\{c\}$ becomes the new boundary subgraph. Clearly, the inductive conditions are satisfied.

In the second case if there is no pair of boundary subgraphs being n-separable through C then our procedure is completed, and we have found the required subgraphs. Otherwise let P_1 and P_2 be such a pair. Consider the set M of vertices in C which are joined by an edge to $P_1 \bigcup P_2$. If M consists of only one vertex c then we separate c in the same way as in the first case. In this case we exclude P_1 and P_2 from the set of the selected subgraphs. Clearly, inductive conditions are satisfied. If $|M| > 1$ then we mark $\leq n$ vertices in C separating P_1 and P_2. Let us prove the following fact:

there exists a partitioning of C into nonempty parts C_1 and C_2 such that the graphs $P_1 \bigcup C_1$ and $P_2 \bigcup C_2$ are connected and each edge connecting C_1 with $P_2 \bigcup C_2$ or C_2 with $P_1 \bigcup C_1$ is incident with one of the marked vertices.

Choose in C two different vertices c_1 and c_2 such that c_i is joined to P_i by an edge. Ascribe c_i to C_i. Ascribe to C_i the remaining vertices in C which can be joined to P_i by a path with all interior vertices unmarked and lying in C. (If both 1 and 2 can serve as i, we act arbitrary). Consider the subgraph C' in C consisting of vertices which were not ascribed neither to C_1 nor to C_2. Since C is connected, for each connected component K of the graph C' there is a vertex in $C \setminus C'$ which is joined to K by an edge. Fix such a vertex a. Ascribe K to C_i which contains a. Now, the stated fact became obvious enough.

One of C_i is m-nondivisible, let it be C_1. From the proven fact it follows that C_1 is separable from $G \setminus C_1$ by $\leq m$ vertices. Let C_0 be an m-nondivisible component of the subgraph C_1. It will be the new central subgraph. The subgraph $P_2 \bigcup C_2$ will be the new boundary subgraph replacing P_2. It is easy to verify that the inductive conditions are satisfied.

Our procedure will end in a construction of the required subgraphs. This completes the proof of Lemma 3.   □

Let us take $n = \exp(r^{20})$, $k = r^2$ in Lemma 3. We will use the following theorem of Menger.

Menger's Theorem. *Two given nonadjacent vertices a and b of a graph cannot be separated by deleting n vertices (different from a, b) if and only if there exist $n + 1$ pairwise vertex-disjoint paths between a and b.*

It follows from this theorem that for each pair of boundary subgraphs in G there exist $n + 1$ pairwise vertex-disjoint (except ends) paths between these subgraphs having all interior vertices in C. For all these pairs we fix n corresponding paths. Let us order the formed families of paths and denote them by $S_1, S_2, \ldots S_{\frac{k(k-1)}{2}}$. We will reconstruct these families as follows.

At the next stage we take the next family S_i in this ordering. By S_i we mean the family which was formed from the original S_i by the reconstruction made up to the current moment. We assume as an inductive condition that for each $j < i$ the family S_j consists of only one path and this path does not cross any path of any other family. For each $j > i$ we take for the new S_j some subfamily of the old S_j of cardinality $l = |S_i|/\exp(r^{10})$. Consider the graph $S_i \bigcup S_j \subseteq C$ which consists of all vertices and edges belonging to S_i or to S_j except for the end vertices and edges. Let us draw in $S_i \bigcup S_j$ a new family S_j of the cardinality l so that it joins the same boundary subgraphs as the old S_j and the number of edges in $S_i \bigcup S_j$ belonging to S_j but not to S_i is minimal. One of the two following cases holds.

Case 1. There is a path q in S_i which does not cross $\geq |S_j|/\exp(r^{10})$ paths in each S_j when $j > i$. In this case we take $\{q\}$ for the new S_i, and for each $j > i$ we take for the new S_j the subfamily of the old S_j which consists of all paths not crossed by q. Evidently, the inductive condition is satisfied.

Case 2. There is no path described in the case 1. In this case we stop our procedure.

If we have the case 1 at every stage than at the end of the procedure we will have the complete graph with k vertices (and, hence, the $r \times r$ grid) as a minor of our graph.

Assume that we have the case 2 at i-th stage. Then there exists $j > i$ such that not less than $|S_i|/k^2$ paths in S_i cross $\geq |S_j| - |S_j|/\exp(r^{10})$ paths in S_j. Fix such j and denote the set of $|S_i|/k^2$ described paths in S_i by S_i^1. We will find the $r \times r$ grid in $S_i^1 \bigcup S_j$.

We order paths in S_j in the order of the decrease the number of paths in S_i^1 crossed by the paths in S_j. Let $V = \{q_1, q_2, \ldots, q_k\}$ be the set of the initial k paths in this ordering.

Lemma 4. *There exist at least $|S_j| \exp(r^9)$ paths in S_i^1 crossing each path in V.*

Proof. Denote $b = |S_j|$. Let us show that the path q_k (and, hence, each path in V) crosses at least $N = |S_i^1| - b - \exp(r^{10})$ paths in S_i^1. Indeed, in all there exist at least $P = \left(1 - \frac{1}{\exp(r^{10})}\right) |S_i^1| b$ pairs of crossing paths. Even if the paths $q_1, \ldots, q_{k-1}$ cross all paths in S_i^1, it remains $E = P - (k-1)|S_i^1|$ such pairs for the other paths in S_j. Evidently, q_k must cross at least

$$\frac{E}{b} = |S_i^1| - \frac{|S_i^1|}{\exp(r^{10})} - (k-1)\frac{|S_i^1|}{b} \geq |S_i^1| - b - (k-1)\exp(r^{10}) \geq N$$

path in S_i^1 as we wanted. Hence, there exist

$$|S_i^1| - kN' = \frac{b\exp(r^{10})}{k^2} - kb - k\exp(r^{10}) \geq$$

$$\geq b\left(\frac{\exp(r^{10})}{r^4} - r^2 - \frac{r^2\exp(r^{10})}{\exp(r^{19})}\right) \geq b\exp(r^9)$$

paths in S_i^1, crossing each path in V. The set of such paths we denote by U. Lemma 4 is proved.   □

We'll call paths in U *vertical* and in V — *horizontal*. Consider a horizontal path q. Clearly, there is an edge $e \notin U$ on q such that the path q crosses equal (to within 1) number of different vertical paths on each side from e. It follows from the minimality of the number of edges in S_j that after removal of the edge e there will be no b (recall, $b = |S_j|$) pairwise vertex-disjoint paths between the boundary subgraphs joined by S_j. By Menger's theorem they are $(b-1)$-separable. Fix $(b-1)$ vertices separating these subgraphs. Clearly, on each path in S_j except q there is just one fixed vertex. There are no more than $(b-1)$ vertical paths passing through the fixed vertices. It is easy to see that any other path in U does not cross q on both sides from e, otherwise we could go from "the left" boundary subgraph to "the right" one not passing both through e and through the fixed vertices. Since on each side from e the path q crosses half of vertical paths, there are two large subfamilies U_l and U_r in U such that U_l crosses q only on "the left" side from e and U_r only on "the right" side. It is easy to see that on any horizontal path q' there is an edge e' such that U_l crosses q' only on "the left" side from e' and U_r — only on "the right" side. (Indeed, it is sufficient to show that for any $q_1 \in U_l$, $q_2 \in U_r$ there are not vertices a_l, a_r on q' such that $a_l \in q_2$, $a_r \in q_1$ and a_l lies on the left of a_r on q'. But if it is not the case we could easily bypass both e and all fixed vertices going from the left to the right.)

Similarly, we divide each of two "halves" of the path q (before e and after e) in two equal parts with respect to the corresponding part of vertical paths. We continue this procedure until the path q (and, hence, all horizontal paths) is divided into $r^2 \exp(r^4)$ segments. At the end of the procedure we have subfamily $U_1 \subseteq U, |U_1| = r^2 \exp(r^4)$ and the partition of each horizontal path into segments such that each path in U_1 crosses any horizontal path on only one segment, and different paths on different segments. All horizontal paths cross paths in U_1 in the same order. (Of course, at each step of dividing of a subset of vertical paths into two parts, we throw out $\leq b$ "bad" vertical paths. But b is small in comparison with $|U|$ which ensures realizability of the procedure.)

We'll say that a path q crosses a path p *only once* if their common vertices and edges constitute exactly one (maybe, one-vertex) path (thus, this path is a subpath of both p and q). We will use the following trivial fact. *Let S_1 and S_2 be* *families of n pairwise vertex-disjoint paths such that any two paths in different families cross only once and all paths in S_1 cross paths in S_2 in the same order and all paths in S_2 cross paths in S_1 in the same order. Then the graph $S_1 \bigcup S_2$ has as a minor the grid of size of $n \times n$.*

For each $\alpha \in U_1$ we consider the following graph. Its vertices are horizontal paths. Vertices x and y are joined by an edge if there is a segment of the path α such that its end vertices are on the paths x and y and all its interior vertices are not in V. Clearly, the constructed graph is connected. Consider the subfamily $U_2 \subseteq U_1$ of r^2 paths such that all paths in U_2 correspond to the same graph. Let us take a frame tree in this graph. Evidently, a tree with r^2 vertices has either the height $\geq r$ or the number of leaves $\geq r$. In the first case, clearly, we have the $r \times r$ grid as a minor of our graph. In the second case consider the linear ordering of U_2 in which paths in U_2 are crossed by horizontal paths. Let us divide U_2 into r groups of neighboring paths with respect to this ordering. We use every group for the passing a path which in some fixed order crosses only once horizontal paths corresponding to leaves of the tree. We use non-leaf vertices of the tree for a moving from a leaf to another leaf vertex. Each such moving takes place in individual tree. Thus we have the $r \times r$ grid as a minor. This completes the proof of Theorem 2.   □

Remark. It is shown in [2] that the degree 20 in the bound $r^2 \exp(r^{20})$ can be improved substantially while making the proof even simpler.

As we can see from the following theorem, for planar graphs there is a linear upper bound of the value of m.

Theorem 3. *For any r there exists $m \leq cr$ where $c = 2^{16}$ such that if a finite planar graph has no minors isomorphic to the planar $r \times r$ grid, then this graph is m-divisible.*

Proof. Let us take $k = 5$, $n = cr$ in Lemma 3, where $c = 2^{16}$. We construct the families $S_1, S_2, \ldots, S_{10}$ in the same way as in the proof of Theorem 2. We will carry out the same procedure with families of paths as in the proof of Theorem 2. At i-th stage we take the family S_i being a subfamily of the original S_i. There are two possible cases. In the first case there exists a path $q \in S_i$ which crosses less than half of paths in each S_j when $j > i$. Then for each $j > i$ we take for the new S_j the subfamily consisting of the paths of the old S_j which are not crossed by q. After that we proceed to the next stage.

Since the complete graph with five vertices can not be a minor of a planar graph, we will have at some i-th stage ($i \leq 9$) the second case, that is, there is no path described in the first case. Then there exists $j > i$ such that $\geq \frac{cr}{10 \cdot 2^{10}} \geq 4r$ paths in S_i are crossed by $\geq |S_j|/2$ paths in S_j. Let us fix such j and denote the set of $\geq 4r$ described paths corresponding to j by S_i^1.

Clearly, we can consider the connected graphs A and B joined by S_i to be trees. Then it is easy to see that paths in S_j together with A, B divides the plane into $|S_j|$ parts called *faces* and every face has exactly two paths on its boundary. Let us number paths in S_j by numbers $1, \ldots, |S_j|$ so that the pairs of paths $(i, i+1)$ where $i < |S_j|$ and $(|S_j|, 1)$ are neighboring i.e. some face has in its boundary both paths. This numbering gives a cycle order on S_j.

It is easy to see that to pass from some path in S_j to another path in S_j we must cross all paths of one of two sets between them. Therefore, each path in S_i^1 crosses $|S_j|/2$ paths in S_j which form a segment in the cycle order. Let us divide S_j into four equal segments in the order. Evidently, there exists a quarter such that $\geq |S_i^1|/4$ paths in S_i^1 contain a subpath crossing all paths of this quarter and having its ends on the two exterior paths q_1, q_2 of the quarter and having all its interior vertices out of q_1, q_2. Denote the set of such subpaths on paths in S_i^1 by S_i^2 and denote the considered quarter of paths in S_j by S_j^1. Clearly, $|S_i^2| \geq r$, $|S_j^1| \geq r$.

Let us draw in the graph $S_i^2 \bigcup S_j^1$ a family U of $|S_i^2|$ pairwise vertex-disjoint paths between q_1 and q_2 and a family V of $|S_j^1|$ pairwise vertex-disjoint paths between A and B, such that the number of edges of the graph $U \bigcup V$ is minimal. We'll call paths in U *vertical* and in V — *horizontal*. Clearly, each vertical path crosses each horizontal path. It is evident also that vertical paths divide the part of the plane bounded by q_1, q_2, A, B into parts and the set U (as well as the parts of the plane) are ordered in a natural way so that to pass from some vertical path to another vertical path we must cross all the paths between them. The same is true for horizontal paths. Therefore, for the proof of the existence of the grid it is sufficient to show that each vertical path crosses each horizontal path only once. Suppose that it is not true. Let α be the nearest to q_1 horizontal path which crosses some vertical path β in vertices a_1 and a_2 not connected by a path in $\alpha \cap \beta$.

We will show that the subpath $[a_1, a_2]$ of the path β does not pass through the part of the plane lying between q_1 and α. Assume that it is not true. Then either this subpath crosses the path α' neighboring to α from the side of q_1 or there exists a subpath l of the path β with the ends lying on α and the interior vertices lying out of V. The first case contradicts the condition of the choice of the path α, since β crosses α' not only in $[a_1, a_2]$. In the second case we can pass α along l and reduce the number of edges in $U \bigcup V$. This contradicts the minimality of this number.

If there are no vertices of vertical paths on the segment $r = [a_1, a_2]$ of the path α except the vertices of β, then we can pass β along r, which contradicts the minimality of the number of edges. Otherwise, assume that there is a vertex b on r belonging some vertical path β'. The subpath of β' from b to q_2 can not lie entirely between α and q_2 because it does not cross β. But this subpath can not pass through the part of the plane between α and q_1, because by the same argument as for β we obtain from this assumption a contradiction either with the condition of the choice of α or with the minimality of the number of edges in $U \bigcup V$. This contradiction completes the proof of Theorem 3. $\square$

Acknowledgement. The author gratefully thanks An.A. Muchnik and N.K. Vereshchagin for fruitful discussions, useful stimulations and a lot of help.

References

1. H.L. Bodlaender and D.M. Thilikos. Constructive linear time algorithms for branchwidth. In P. Degano, R. Gorrieri and A. Marchetti-Spaccamela, editors, *Proceedings 24th International Colloquium on Automata, Languages, and Programming, ICALP'97*, pp. 627-637. Springer Verlag, Lecture Notes in Computer Science, 1997, vol. 1256.
2. R. Diestel, T.R. Jensen, K.Yu. Gorbunov and C. Thomassen. Highly connected sets and the excluded grid theorem. Submitted.
3. K.Yu. Gorbunov. Context-free decidability of graphs that do not realize a plane lattice, *Soviet Math. Dokl*, **43** (1) (1991), 53-57.
4. N. Robertson and P.D. Seymour. Graph minors. II. Algorithmic Aspects of Tree-Width, *Journal of Algorithms*, **7** (1986), 309-322.
5. N. Robertson and P.D. Seymour. Graph minors. III. Planar tree-width, *Journal of Combinatorial Theory*, Series B, **36** (1984), 49-64.
6. N. Robertson and P.D. Seymour. Graph minors. V. Excluding a planar graph, *Journal of Combinatorial Theory*, Series B, **41** (1986), 92-114.
7. N. Robertson and P.D. Seymour Graph minors. X. Obstructions to tree-decomposition, *Journal of Combinatorial Theory*, Series B, **52** (1991), 153-190.
8. N. Robertson, P.D. Seymour & R. Thomas. Quickly excluding a planar graph, *Journal of Combinatorial Theory*, Series B, **62** (1994), 323-348.

Author Index

Springer
and the
environment

At Springer we firmly believe that an international science publisher has a special obligation to the environment, and our corporate policies consistently reflect this conviction.

We also expect our business partners – paper mills, printers, packaging manufacturers, etc. – to commit themselves to using materials and production processes that do not harm the environment. The paper in this book is made from low- or no-chlorine pulp and is acid free, in conformance with international standards for paper permanency.

Springer

Lecture Notes in Computer Science

For information about Vols. 1–1436

please contact your bookseller or Springer-Verlag

Vol. 1475: W. Litwin, T. Morzy, G. Vossen (Eds.), Advances in Databases and Information Systems. Proceedings, 1998. XIV, 369 pages. 1998.

Vol. 1476: J. Calmet, J. Plaza (Eds.), Artificial Intelligence and Symbolic Computation. Proceedings, 1998. XI, 309 pages. 1998. (Subseries LNAI).

Vol. 1477: K. Rothermel, F. Hohl (Eds.), Mobile Agents. Proceedings, 1998. VIII, 285 pages. 1998.

Vol. 1478: M. Sipper, D. Mange, A. Pérez-Uribe (Eds.), Evolvable Systems: From Biology to Hardware. Proceedings, 1998. IX, 382 pages. 1998.

Vol. 1479: J. Grundy, M. Newey (Eds.), Theorem Proving in Higher Order Logics. Proceedings, 1998. VIII, 497 pages. 1998.

Vol. 1480: F. Giunchiglia (Ed.), Artificial Intelligence: Methodology, Systems, and Applications. Proceedings, 1998. IX, 502 pages. 1998. (Subseries LNAI).

Vol. 1481: E.V. Munson, C. Nicholas, D. Wood (Eds.), Principles of Digital Document Processing. Proceedings, 1998. VII, 152 pages. 1998.

Vol. 1482: R.W. Hartenstein, A. Keevallik (Eds.), Field-Programmable Logic and Applications. Proceedings, 1998. XI, 533 pages. 1998.

Vol. 1483: T. Plagemann, V. Goebel (Eds.), Interactive Distributed Multimedia Systems and Telecommunication Services. Proceedings, 1998. XV, 326 pages. 1998.

Vol. 1484: H. Coelho (Ed.), Progress in Artificial Intelligence – IBERAMIA 98. Proceedings, 1998. XIII, 421 pages. 1998. (Subseries LNAI).

Vol. 1485: J.-J. Quisquater, Y. Deswarte, C. Meadows, D. Gollmann (Eds.), Computer Security – ESORICS 98. Proceedings, 1998. X, 377 pages. 1998.

Vol. 1486: A.P. Ravn, H. Rischel (Eds.), Formal Techniques in Real-Time and Fault-Tolerant Systems. Proceedings, 1998. VIII, 339 pages. 1998.

Vol. 1487: V. Gruhn (Ed.), Software Process Technology. Proceedings, 1998. VIII, 157 pages. 1998.

Vol. 1488: B. Smyth, P. Cunningham (Eds.), Advances in Case-Based Reasoning. Proceedings, 1998. XI, 482 pages. 1998. (Subseries LNAI).

Vol. 1489: J. Dix, L. Fariñas del Cerro, U. Furbach (Eds.), Logics in Artificial Intelligence. Proceedings, 1998. X, 391 pages. 1998. (Subseries LNAI).

Vol. 1490: C. Palamidessi, H. Glaser, K. Meinke (Eds.), Principles of Declarative Programming. Proceedings, 1998. XI, 497 pages. 1998.

Vol. 1493: J.P. Bowen, A. Fett, M.G. Hinchey (Eds.), ZUM '98: The Z Formal Specification Notation. Proceedings, 1998. XV, 417 pages. 1998.

Vol. 1494: G. Rozenberg, F. Vaandrager (Eds.), Lectures on Embedded Systems. Proceedings, 1996. VIII, 423 pages. 1998.

Vol. 1495: T. Andreasen, H. Christiansen, H.L. Larsen (Eds.), Flexible Query Answering Systems. IX, 393 pages. 1998. (Subseries LNAI).

Vol. 1496: W.M. Wells, A. Colchester, S. Delp (Eds.), Medical Image Computing and Computer-Assisted Intervention – MICCAI'98. Proceedings, 1998. XXII, 1256 pages. 1998.

Vol. 1497: V. Alexandrov, J. Dongarra (Eds.), Recent Advances in Parallel Virtual Machine and Message Passing Interface. Proceedings, 1998. XII, 412 pages. 1998.

Vol. 1498: A.E. Eiben, T. Bäck, M. Schoenauer, H.-P. Schwefel (Eds.), Parallel Problem Solving from Nature – PPSN V. Proceedings, 1998. XXIII, 1041 pages. 1998.

Vol. 1499: S. Kutten (Ed.), Distributed Computing. Proceedings, 1998. XII, 419 pages. 1998.

Vol. 1501: M.M. Richter, C.H. Smith, R. Wiehagen, T. Zeugmann (Eds.), Algorithmic Learning Theory. Proceedings, 1998. XI, 439 pages. 1998. (Subseries LNAI).

Vol. 1502: G. Antoniou, J. Slaney (Eds.), Advanced Topics in Artificial Intelligence. Proceedings, 1998. XI, 333 pages. 1998. (Subseries LNAI).

Vol. 1503: G. Levi (Ed.), Static Analysis. Proceedings, 1998. IX, 383 pages. 1998.

Vol. 1504: O. Herzog, A. Günter (Eds.), KI-98: Advances in Artificial Intelligence. Proceedings, 1998. XI, 355 pages. 1998. (Subseries LNAI).

Vol. 1507: T.W. Ling, S. Ram, M.L. Lee (Eds.), Conceptual Modeling – ER '98. Proceedings, 1998. XVI, 482 pages. 1998.

Vol. 1508: S. Jajodia, M.T. Özsu, A. Dogac (Eds.), Advances in Multimedia Information Systems. Proceedings, 1998. VIII, 207 pages. 1998.

Vol. 1510: J.M. Zytkow, M. Quafafou (Eds.), Principles of Data Mining and Knowledge Discovery. Proceedings, 1998. XI, 482 pages. 1998. (Subseries LNAI).

Vol. 1511: D. O'Hallaron (Ed.), Languages, Compilers, and Run-Time Systems for Scalable Computers. Proceedings, 1998. IX, 412 pages. 1998.

Vol. 1512: E. Giménez, C. Paulin-Mohring (Eds.), Types for Proofs and Programs. Proceedings, 1996. VIII, 373 pages. 1998.

Vol. 1513: C. Nikolaou, C. Stephanidis (Eds.), Research and Advanced Technology for Digital Libraries. Proceedings, 1998. XV, 912 pages. 1998.

Vol. 1514: K. Ohta,, D. Pei (Eds.), Advances in Cryptology – ASIACRYPT'98. Proceedings, 1998. XII, 436 pages. 1998.

Vol. 1515: F. Moreira de Oliveira (Ed.), Advances in Artificial Intelligence. Proceedings, 1998. X, 259 pages. 1998. (Subseries LNAI).

Vol. 1516: W. Ehrenberger (Ed.), Computer Safety, Reliability and Security. Proceedings, 1998. XVI, 392 pages. 1998.

Vol. 1517: J. Hromkovič, O. Sýkora (Eds.), Graph-Theoretic Concepts in Computer Science. Proceedings, 1998. X, 385 pages. 1998.

Vol. 1518: M. Luby, J. Rolim, M. Serna (Eds.), Randomization and Approximation Techniques in Computer Science. Proceedings, 1998. IX, 385 pages. 1998.

Vol. 1520: M. Maher, J.-F. Puget (Eds.), Principles and Practice of Constraint Programming - CP98. Proceedings, 1998. XI, 482 pages. 1998.

Vol. 1522: G. Gopalakrishnan, P. Windley (Eds.), Formal Methods in Computer-Aided Design. Proceedings, 1998. IX, 529 pages. 1998.